HANDBOOK ON RADIATION PROBING, GAUGING, IMAGING AND ANALYSIS

Handbook on Radiation Probing, Gauging, Imaging and Analysis

Volume I:
Basics and Techniques

by

ESAM M.A. HUSSEIN
Department of Mechanical Engineering,
University of New Brunswick,
Fredericton, New Brunswick, Canada

KLUWER ACADEMIC PUBLISHERS
DORDRECHT / BOSTON / LONDON

A C.I.P. Catalogue record for this book is available from the Library of Congress.

ISBN 1-4020-1294-2 (Vol. I)
ISBN 1-4020-1295-0 (Vol. II)
ISBN 1-4020-1296-9 (Set)

Published by Kluwer Academic Publishers,
P.O. Box 17, 3300 AA Dordrecht, The Netherlands.

Sold and distributed in North, Central and South America
by Kluwer Academic Publishers,
101 Philip Drive, Norwell, MA 02061, U.S.A.

In all other countries, sold and distributed
by Kluwer Academic Publishers,
P.O. Box 322, 3300 AH Dordrecht, The Netherlands.

Printed on acid-free paper

Printed in the Netherlands.

Dedicated in memory of
my Father Mahmoud,
my Uncle Roshdi, and
my Aunt Alia.

Contents

Contents

Preface

The need for this book arose from my teaching, engineering, and research experience in the non-power aspects of nuclear technology. The lack of a comprehensive textbook in industrial applications of radiation frustrated my students, who had to resort to a multitude of textbooks and research publications to familiarize themselves with the fundamental and practical aspects of radiation technology. As an engineer, I had to acquire the design aspects of radiation devices by trial-and-error, and often by accidental reading of a precious publication. As a researcher and a supervisor of graduate students, I found that the needed literature was either hard to find, or too scattered and diverse. More than once, I discovered that what appeared to be an exciting new idea was an old concept that was tried a few decades earlier during the golden era of "Atom for Peace". I am hoping, therefore, that this book will serve as a single comprehensive reference source in a growing field that I expect will continue to expand.

This book is directed to both neophytes and experts, and is written to combine the old and the new, the basic and the advanced, the simple and the complex. It is anticipated that this book will be of help in reviving older concepts, improving and expanding existing techniques and promoting the development of new ones. Hopefully, the consolidation of this material in one book will incite wider use and application of this powerful and useful technology. Therefore, the book is intended to be a single handy source of information for students, instructors, current and potential users of radiation technology, and its design engineers and researchers.

The book is divided into four parts to accommodate a wide spectrum of readers. Part I deals with the fundamental aspects of radiation sources, physics and detection, and is particularly helpful for students who are not familiar with nuclear and atomic radiation. Part II is directed to industrial physicists and engineers, as it provides an exposition of the different ways (techniques) by which radiation can be used to meet industrial measurement needs. Part III presents a large number of specific industrial applications. In order to assist the reader, an application index is given at the end of the book, identifying the areas in which radiation techniques are utilized. Those who would like to design a new device, or improve or alter an existing system, can refer to Part IV.

This book can serve both as a reference book and as a textbook. To assist readers in searching and locating specific information, two separate indices are provided, paragraphs are given titles, and wide use is made of cross-referencing. Extensive literature is cited, and references are listed in detail. Three one-semester courses can be based on this book. The first part of the book, along with appendices C, D, E and G, can be used in an introductory course on radiation fundamentals to students with no or little background in atomic and nuclear physics. Part II, introduced by Chapter 1 and supplemented by appendix F, can serve as a course on nondestructive examination and imaging with radiation, assisted by example applications from Part III. A design course can be based on Part IV, with design projects assigned for systems described in Part III. For a set of problems and solutions, instructors can contact the author by e-mail at hussein@unb.ca. The author welcomes any comments and suggestions for inclusion in future revisions of this book.

ESAM M. A. HUSSEIN

Acknowledgments

Dr. John H. Hubbell has given me the great honor of writing the "Foreword" to the book, and provided viable comments and suggestions. Dr. Hubbell needs no introduction; researchers, practitioners and students of x- and gamma-rays are all familiar with his comprehensive work and tables on photon cross sections.

The concept of this work was formulated through discussions with esteemed colleagues: Dr. Ned Kondic (formerly with the US Nuclear Regulatory Commission) in the late 1980's, and Dr. Richard C. Lanza (Massachusetts Institute of Technology) and Prof. Nares Chankow (Chulalongkorn University, Thailand) in the nineties. Although writing this manuscript was a solo endeavor, it would not have been possible without the effort of all the authors whose work is cited in the book. Mr. P. Jacob Arsenault, Mr. Hassan A. Jama, Dr. Edward J. Waller and Dr. Ilan Yaar, of the Laboratory for Threat Material Detection at University of New Brunswick (UNB), thankfully read some chapters and provided useful suggestions.

The prompt and professional response of the staff of the UNB Libraries, particularly at the Engineering Library and the Document Delivery Department, made it possible for me to readily access the needed literature. This book was written using LaTeX, based on MiKTeX platform, with WinTeX 2000 as the interfacing editor.

Finally a personal note to my wonderful children Mahmoud and Amina: without your love, forbearance, patience and understanding, the completion of this work would not have been possible.

Acknowledgments

Foreword

In browsing through a draft copy of this two-volume "how-to" desk reference on virtually all aspects of the use of photon and corpuscular radiations in the interrogation of materials and structures, I found the presentation format to be unique and useful. Although the variety and comprehensiveness is akin to a topical encyclopedia, the presentation reminded me of a thesaurus, in which the subtopics are not sequenced alphabetically, but, similar to in a thesaurus, are sequenced in a logical progression. Then, going "Roget" one better, at the end of the book are found not one, but two alphabetized indexes, first an "application index" and finally a conventional index alphabetically listing key words and their page numbers from throughout the text.

In **Volume One (Basics and Techniques)**, following a brief Chapter 1 (Introduction) surveying the unique features of radiation interrogation, often the only available tool for some NDE (nondestructive evaluation) challenges, **Part I (Basics)** begins this logical progression, in Chapter 2 (Radiation Types and Sources), with the definitions and nature of the various available radiations and how they can be obtained, then progressing in Chapter 3 (Modifying Physics) to the basic underlying physical processes by which these different radiations interact with atoms and with bulk materials. Part I concludes with a virtual Baedeker to the many types of radiation detectors and their underlying principles in Chapter 4 (Detection Methods) and a briefer Chapter 5 (Radiation Safety) addressing both common-sense and mandatory regulatory concerns.

In **Part II (Techniques)** the logical topical progression continues in Chapter 6 (Transmission Methods) with a comprehensive array of topics ranging from time-honored film radiography to the mathematical intricacies of tomography to the esoteric application of Mössbauer spectrometry. Chapters 7 (Scattering Methods), 8 (Emission Methods) and

9 (Absorption Methods) similarly provide encyclopedic coverage of the available methods and techniques in each of these categories.

Volume Two (Applications and Design) brings the above wealth of "why?" and "how-to" information "into the real world" beginning in **Part III (Applications)** with an introductory Chapter 10 (Probing, Inspection and Monitoring) with topics ranging from the function of the alpha-particle sources in most smoke detectors (such as the ones in my home) to monitoring package-filling in opaque containers. This is followed by Chapter 11 (Gauging) listing and exhaustively discussing the otherwise-difficult-or-impossible interrogations of bulk density, thickness, porosity and voidage, moisture content, fluid flow and finally dating including with carbon-14 (geological time scales), tritium (short term, such as for ground waters) and the use of thermoluminescence for TL dating (archeological time scales).

Chapter 12 (Elemental and Content Analysis) continues the logical progression of subtopics including nucleus-based analysis (e.g., activation analysis), atom-based analysis (e.g., x-ray fluorescence spectroscopy, XRFS), hydrogen measurement (mostly by neutrons), and material content analysis (e.g., applications of the above-mentioned Mössbauer spectroscopy). Concluding Part III, Chapter 13 (Imaging) catalogs and provides detailed information on the many different kinds of photon radiography, on neutron and charged-particle radiography, on tomography and on imaging for material content such as in dual-energy imaging.

Finally, in **Part IV (Design)** the author shares the experiences and knowledge accumulated in his long and distinguished teaching and research career, much of it involved in synthesizing the above material into the invention, production and putting into practice a significant fraction of the above principles and devices for carrying out NDE tasks. Thus, Chapter 14 (Performance Parameters and Design Aspects) opens Part IV with material on performance parameters, statistical optimization, design objectives, and source, technique, and detection system selections, followed by Chapter 15 (Source Modulation) with collimation considerations, filtration, and for neutrons, both moderation and multiplication.

The book's logical topical progression continues in Chapter 16 (Design Calculations) with subtopics on Monte Carlo simulations, and shielding requirements in the differing cases of x-ray machines, isotopic gamma sources, and for neutron sources. Finally, Chapter 17 (Experiments) treats the important considerations of licensing, background reduction and dynamic analysis to verify that the device will indeed perform its intended function, and the brief Chapter 18 (Finalization) discusses prototyping and intellectual property protection (trade secrets, copyright, trademarks and patents).

Volume Two closes with several appendices providing valuable information including (A) Basic Units and Constants, (B) List (alphabetically) of Elements and Natural Isotopes, (C) Relativistic Mechanics, (D) Quantum Mechanics, including the Schrödinger equation and the concept of cross-section, (E) Nuclear/Atomic Parameters for Compounds and Mixtures, (F) Effective Energy, and finally (G) Radiation Counting Statistics, including Poisson statistics, mean and variance. Following the appendices is the listing of the 1373 references including full titles and inclusive page numbers, followed in turn by the two indexes mentioned, an Application Index and a conventional Index concluding this monumental and uniquely useful "encyclopedic/thesauric" guide and companion through the thickets of radiation nondestructive probing, gauging, imaging and analysis.

John H. Hubbell
Ionizing Radiation Division "Emeritus"
National Institute of Standards and Technology
Gaithersburg, Maryland USA
October 31, 2002

VOLUME ONE:

BASICS AND TECHNIQUES

Chapter 1

INTRODUCTION

1.1. Why Radiation

Radiation, in the form of x-rays, gamma rays, or neutrons, is attractive for use in industrial measurements for a number on reasons. It is a nonintrusive, noninvasive and nondestructive examination method. Radiation can provide intrinsic properties that are not directly given by other methods, such as density and elemental composition. It can be used in harsh environments (high temperature, high pressure, corrosive, caustic, explosive, viscous media, etc.), since radiation examination does not require direct contact with the inspected object. Radiation can be used with any type of medium regardless of its nature; i.e. the material can be in the form of a gas, liquid or solid, it can be conductive or insulating of heat or electrically, it can be ferromagnetic or nonmagnetic, ceramic or metal, porous or impermeable, sealed or open, etc. Therefore, nondestructive examination with radiation is found in many applications, from radiography of industrial welds to inspection of airport luggage.

Radiography is now a standard and widely used nondestructive examination method. However, there are many other applications in which radiation examination is useful, and often the only means of obtaining certain measurements. Although some industrial users may view radiation as the last-resort solution to measurement problems that cannot be solved by conventional methods, radiation-based methods have some unique features. Being non-destructive and non-intrusive, radiation devices can be installed without design alterations of existing systems, and without shutting-down an operation; a considerable cost-saving in many cases. Radiation devices can also accommodate dusty and noisy industrial environments. The increased competitiveness, need for efficiency, quality and safety assurances, and the emphasis on optimization of industrial operations and processes to enhance energy efficiency and reduce cost, will no doubt increase reliance on such techniques. Security considerations, for the detection of explosive and contraband materials, is also increasingly relying on radiation techniques. Although, radiological constraints and regulations have traditionally discouraged industrial use of radiation, its medical utilization have flourished over the years and it is inevitable that industry will follow suit. Radiation safety precautions are not impossible to deal with, and indeed are manageable since radiation can be easily monitored and shielded against.

Management Information Services Inc. (Washington, DC) estimated in March 1994 that radiation applications created an industry that generated an economy within the USA of $357 billion per year and 3.7 million jobs. In a 1990 article, reference [1] summarized the status of radioisotope applications in industrial plant and process investigation as follows:

> applied radioisotope technology is flourishing world-wide—cause for some satisfaction both to those who practice the art and those who enjoy the undoubted benefits which result from successful radioisotope applications.
>
> However, it is a reasonable contention that, in spite of this growth, radioisotope applications are not as commonplace as their almost universal applicability (Charlton, 1986) [2]. It may be argued that this is due to ignorance on the part of industrialists of the very substantial economic benefits which may be realized from the use of this technology.
>
> If this is so, then it is hard to escape the conclusion that the problem lies with the radioisotope practitioner and his failure to make the industrialist sufficiently aware of what radioisotope technology has to offer.
>
> How might this be changed?

It is hoped that this book will help in bridging the gap between the frustration of the practitioner and the lack of awareness of the industrialist.

1.2. Nondestructive Examination (NDE)

Nondestructive testing is the process used to examine the condition of an object without damaging it, so that the usefulness and integrity of the object are not affected by the interrogation process. The common terminology used in industry to describe techniques for performing noninvasive measurements is "nondestructive examination", or NDE. This term is used to refer to various evaluation, testing, monitoring, or inspection techniques that are performed to "examine" materials or components in ways that do not impair future usefulness and serviceability in order to detect, locate, measure and evaluate discontinuities, defects and other imperfections; to assess integrity, properties and composition; and to measure geometric characteristic [3]. In this book the term non-destructive examination (NDE), is used to describe the processes of *probing*, *inspecting*, *monitoring*, *gauging*, performing elemental/composition analysis, or *imaging*; with these processes defined as follows:

Probing: is the process of searching for anomalies by performing *localized* examination at a specific location, or locations, in the object.

Inspection: is the process of *overall* evaluation of the condition of an object or a structure, without providing localized information, i.e. it is a wide-scope probing process.

Monitoring: is a probing or inspection process that does not provide an indication until a significant event takes place.

Gauging: is defined as the process of quantifying a *bulk* property representative of the entire object.

Elemental/Composition Analysis: where information on the elemental content or chemical composition of an object is obtained.

Imaging: is the method of obtaining a detailed portrait of the object, and involves:

- **Radiography:** where an image is obtained by projecting the incident radiation on a plane (typically a film).
- **Tomography:** where a detailed image of a cross-section of the object is acquired.
- **3-D Imaging:** where a complete three-dimensional image is reconstructed.

An examination can encompass simultaneously a number of the above processes. For example, elemental analysis can be accomplished at certain location (probing), to detect overall variation of the composition of the material (inspection), or to determine the spatial distribution of a particular element, or elements, in the object (imaging).

NDE can be performed using different technical methods that rely on acoustic, optical, electric, magnetic, electromagnetic, or other properties. Acoustic techniques encompass ultrasonic testing and passive monitoring of acoustic emissions. Electric and magnetic methods involve the magnetic-flux-leakage and eddy-current methods, as well as techniques involving direct measurement of electric parameters (resistance, capacitance, or inductance). Electromagnetic waves, including x-rays, microwaves, infrared radiation, laser and natural light, are also employed in NDE. The question that naturally emerges is: why use nuclear radiation or isotopes, given the risk of exposure to, or contamination by, potentially harmful radioactivity? In section 1.1, some of the unique features of radiation NDE are stated. However, in practice, radiation methods tend to be employed when there is no other means available for performing the task at hand. This situation arises more often that one would expect, due to the inherent limitations of other NDE techniques. Nevertheless, a potential user of radiation should be aware of the capabilities and limitations of other NDE methods. In this regard, the reader should consult NDE textbooks, such as reference [4], and the Nondestructive Testing Handbooks [5]. However, a brief description of common NDE techniques is give below.

1.3. Conventional NDE Methods

Perhaps the easiest and most practiced from of NDE is visual inspection, which is obviously based on the reflection of light off the surface of the inspected object. However, surface defects are often too small to be seen by the naked eye, or even magnifying glasses. Therefore, a number of NDE techniques have emerged to enhance visual indications of defects.

Penetrant Testing. This is one of the methods used to enhance visual inspection of surface flaws. A colored penetrating dye (often fluorescent) is applied to the surface of the object under examination. The dye seeps into cracks or other surface openings. Subsequently, the surface of the test object is washed and a blotting agent is applied to draw the dye from the surface opening. This produces a two-dimensional visible indicator of the location of flaws. The method is relatively easy to perform, and unlike the other surface enhancement method (magnetic particle inspection), it can be used on the surfaces of both ferrous and non-ferrous materials. However, the method is not sensitive to very small and inaccessible defects. One must also ensure that the chemistry of the penetrant material does not affect the inspected object. The resulting visible indications are not often reproducible, and the interpretation of the results strongly depends on the experience and judgment of the operator. The main limitation is that penetrant-testing is a surface inspection method; i.e. sub-surface flaws, and flaws deep inside the object, cannot be found by the method. If the penetrant cannot reach the flaw, one cannot see it.

Magnetic Particle Testing. Another method to enhance visual inspection is to use small magnetic particles to detect the leakage of magnetic flux from surface and near-to-surface discontinuities in ferromagnetic materials. The test material is first magnetized, with a permanent magnet, or an electrically-induced magnetic field. Then, colored small iron fillings (often fluorescent) are poured onto the surface. The magnetic flux leaking through a crack, or some other defect, will attract the iron particles and provide a visible indication of the defect's location. Some other means can be used to measure flux leakage, such as magnetic rubber, magnetometers, coils, etc. However, the method is limited to ferro-magnetic materials and is only used to inspect surface and sub-surface flaws.

Eddy Current Testing. This is another method that is used to detect surface and near-surface flaws in electrically-conductive materials. Eddy current is a circular electric current created on the surface of an electrically conducting material, by applying a magnetic field in a direction perpendicular to the surface of the material. This is achieved with the aid of a test coil, within which an electric current flows, creating a magnetic field in a direction parallel to the axis of the coil. The presence of a discontinuity on, or near the surface, of a material, disrupts the flow of the eddy current. This, in turn, changes the electrical impedance (or more specifically the inductance) of the test coil. The change in the impedance is then amplified and observed. Since eddy current is affected by metallurgic properties as well as defects, this method is also used for material sorting, and for measurement of hardness and electrical-conductivity. However, the method is limited to the inspection of surface and sub-surface flaws in electrically conducting materials, and is more complex than magnetic-flux leakage methods.

Ultrasonic Testing. Ultrasonic waves are high-frequency (above the audible limit of 20 kHz) vibrational waves that can be used to inspect both internal and surface defects in fine-grained castings, forgings, welds and extrusions as well as non-metallic materials. Sound waves are generated, typically at a frequency of a few MHz, with the aid of a piezoelectric crystal. These waves can penetrate soils and liquids, but their penetrability in gases is very poor. Therefore, an acoustic coupling material, such as grease or glycerin, is used to transmit pulsed waves from the crystal of the transducer to the material of the inspected object. Discontinuities within the material interfere with the sound wave, causing it to be scattered back to the same transducer. The returning wave produces an electrical signal, which can be displayed. The time of arrival of the detected signal, relative to the time of emission of the incident pulse, is indicative of the depth of the discontinuity. However, ultrasonic waves are not very well suited for inspecting objects that contain many interfaces, such as composite materials, or rough crystal grain structures, such as cast iron, and do not effectively penetrate void and gases. Reference [6] provides more information on this method.

Radiography Testing. This is the most common radiation-based method in nondestructive testing. It can be used for detecting internal discontinuities in any solid or liquid object, regardless of its internal structure. X- and gamma-rays are used to penetrate through the test material, to reach a special film placed on the other side of the object. The darkness of the film is indicative of the penetrability of radiation through matter. Therefore, cracks, voids, porosity or inclusions can be detected. While x- and gamma-rays are most suited for detecting flaws in metals, neutrons are used in the radiography of hydrogen-rich materials. Although there are a few limitations on the use of radiography, it is difficult to radiograph thick objects and to interpret structures with intricate geometry. However, it should be recalled that radiography requires access to two opposite sides of the inspected object and that each point on the indication film is indicative of the density of the line along which radiation traverses the material, i.e. flaw-position information are not directly obtainable in radiography. The reader can consult reference [7] on this method. It is also covered in this book.

Acoustic Emission Testing. This is a passive NDE technique in which piezoelectric transducers monitor high-frequency acoustic waves associated with the sudden release of energy during crack propagation, crack growth, plastic deformation or phase transformation in solids. Stress must be applied to the inspected object, in the form of pressure, compression, tension or torsion, to release these waves, which travel to the outside of the test object for detection. The applied stress must exceed the value of previously applied stresses to provide indications. Therefore, the technique is more indicative of the propagation of dislocations within the structure, and is not directly indicative of the mere presence of a flaw. For more information, see reference [8].

Microwave Testing. This technique is complementary to ultrasonic testing. That is, while ultrasonic waves cannot penetrate void, microwaves can, but while ultrasonic waves can easily travel through metals, microwaves are completely reflected off metallic surfaces. However, the phase shift and change in amplitude of microwaves is utilized in NDE for measuring the thickness of metallic sheets. Nevertheless, microwaves are best suited for non-metals, particularly, ceramics where ultrasonic testing is not effective. The affinity of microwaves to water makes them also useful for measuring moisture content. See reference [9] for more details.

Thermography Testing. Thermography produces a temperature profile of a heated object. The thermal conductivity and radiative properties of a material are used to test for thermal insulation, bonding or contact between parts, and overheating of electronic components and transmission lines, and for locating hidden objects. Infrared television camera are used for this purpose.

Many other techniques can be employed for NDE. A coin-operated machine performs a NDE process to determine the value of submitted currency. Liquid leaks from a pipe or a tank are used to detect holes in a container. Vibration and noise analysis is employed for the detection of malfunctions of the rotating parts of machines. Lasers holography is used for surface imaging. Nuclear magnetic resonance is applied to image mainly hydrogen-rich components, but cannot be applied in the presence of metals.

1.4. Elements of NDE

From the discussion in section 1.3 one can infer that in every NDE technique the following basic actions are involved:

- A *source* of some form of energy is sent to the object, or in some cases generated within the object.
- The energy introduced by the source is altered (*modified*) within the object.
- The modified energy exits the object and is *detected* by a sensor.
- The source, object and sensor are arranged so that the detected energy provides a useful *indication* of the condition of the object.
- The measured signal is *interpreted* to explain the nature of the detected change.

Reference [10] introduced these five basic elements of NDE: source, modification, detection, indication, and interpretation. These elements can perhaps be best demonstrated by the common process of photography, where natural or flash light is used as the source. Light reflection off the surface of the object is the process by which physical modification takes place. Detection of the reflected signal is performed with a lens that focuses the light on a film. The indication is the image (picture) obtained after the film is chemically developed. Interpretation becomes the process of identifying the contents of the picture, whether it involves people, places or objects. Obviously, interpretation requires the involvement of someone who is familiar with, or interested in, the subject matter of the picture; otherwise a picture seen by a stranger may have no meaning except for its artistic or aesthetic content. This is also true in industrial NDE, where understanding of both the examination process and the nature of the interrogated object is essential for proper interpretation of the obtained indications.

In any NDE technique, one can identify the above essential elements. For example, in x-ray radiography, the source is provided by an x-ray tube. The intensity of the x-rays is modified (reduced) as the radiation is attenuated by the inspected medium. The amount of attenuation is detected by an x-ray sensitive photographic film. The developed film, as it is being exposed to light, provides an *indication*, as well as a record, of the examination. Trained personnel usually provide an *interpretation* of the radiograph. The lack of any of the above elements renders the NDE process useless. Obviously, radiography cannot be performed without an x-ray source. A radiograph of air may be useless, since air hardly modifies x-rays. Without a film, or an alternative receptor of the transmitted radiation, there will be no indication; and without an indication there is nothing to interpret. Finally, and most importantly, if the obtained image cannot be interpreted into something meaningful, the whole exercise would have been in vain. A radiograph of a piece of machinery means nothing except to a person who is familiar with it.

The coherent logic of the above stated five basic elements of NDE is helpful in the development of a new technique. Therefore, the framework of this book is centered on these elements, albeit with a different view of the last two elements: indication and interpretation. These two elements are combined into one element: technique. A technique is defined here as the method used to arrange the source and the detector, with respect to each other and with respect to the object, so that an interpretable indication is obtained. The use of *technique* as the fourth and last basic element serves to emphasize the importance of the method utilized to set-up an inspection device. In fact, the technique is where innovation and creativity is most often involved.

1.5. Intricacy of Radiation Methods

Atomic and nuclear radiation has long been employed in NDE of both industrial and medical objects. Radiation is attractive for use in NDE mainly because of its ability to penetrate thick objects. Moreover, certain radiation types and interactions can provide information, such as density and composition, that cannot otherwise be obtained using other nondestructive probing techniques.

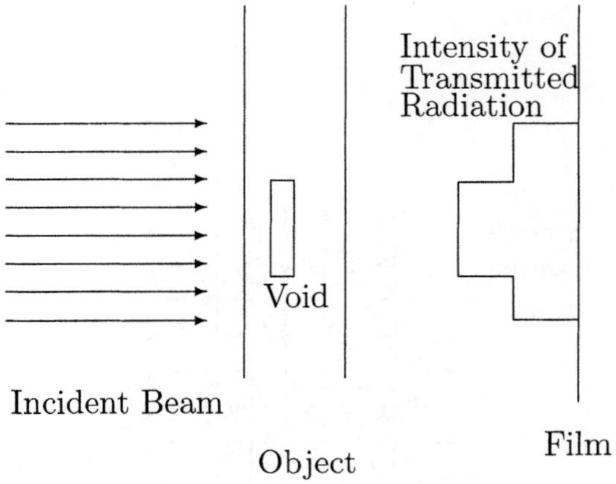

Figure 1.1. A schematic of a an idealized radiography system.

X-ray radiography is one of the most widely used NDE methods. Radiography relies on the well known principle of exponential attenuation of transmitted radiation:

$$I = I_o \exp(-\mu t) \tag{1.1}$$

where I_0 is the intensity of the incident radiation beam and I is the beam's intensity after it travels through a material of thickness t and an attenuation coefficient μ. The attenuated intensity, I, is recorded on a radiographic film that acts as the receiver, as schematically shown in Figure 1.1. The attenuation coefficient depends, among other things, on material density. X-ray radiographs are, therefore, used to infer density information.

In computed tomography (CT), a large number of intersecting transmission maps, along a section of the object, are gathered by detectors, as schematically shown in Figure 1.2. The information is stored in computer memory for subsequent unfolding using an image reconstruction algorithm. The result is a two-dimensional map providing the attenuation coefficient, μ, at each pixel of the image. The μ values, typically normalized to those of water to provide the so called CT numbers, are then utilized to deduce density information.

The above simple physics of x-ray radiography and CT conceals a number of important facts. First, the setup of the imaging system must be such that the exponential law of attenuation, Eq. (1.1), is valid. This requires the use of a narrow well-collimated beam and a detector with a limited view angle; to avoid counting scattered radiation. Even if such

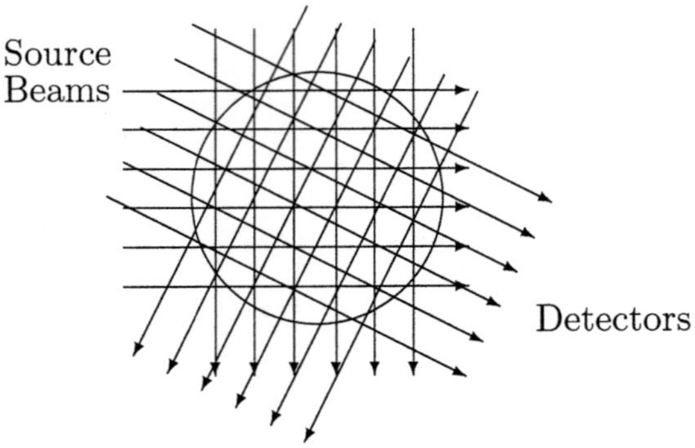

Figure 1.2. A schematic of four projections in computed tomography (CT).

setup is assured, a narrow beam passing through a sufficiently dense object may scatter radiation that contributes to the value of I, thus violating the simple exponential law of attenuation. The attenuation coefficient, μ, depends, not only on the density of the object, but also on the energy of the incident radiation. Since x-rays are produced in multichromatic (many frequencies, energies) fashion, relating measured transmitted intensity to the attenuation coefficient of the material is not a trivial matter, as the value of μ changes with energy. Moreover, since the exponential relationship saturates (reaches a nearly constant value) after the argument (μt) approaches a value of greater than about five, it becomes difficult to obtain any useful information from transmitted radiation for a dense or a thick object. One may then have to increase the energy of the radiation source or seek an alternative technique to obtain the required information; such as using scattering techniques or an alternative type of radiation, e.g. neutrons. Also, in some situations, one may be interested in obtaining bulk or average information, rather than the pictorial indication provided by radiography and tomography. It will then be appropriate to utilize a simpler technique.

The purpose of this book is to introduce the reader to such fundamental concepts. Emphasis is put, not only on both the underlying physics behind a particular technique, but also on that technique's suitability, practicality and limitations. Part I deals with the first, and basic, three elements of NDE, in addition to addressing safety issues associated with the use of radiation. Part II discusses the different technique that can be employed, while Part III presents a variety of applications of the different processes of NDE: probing, inspection, gauging, elemental and content analysis, and imaging. Part IV addresses the design issues faced during the development of new techniques.

PART I: BASICS

This part of the book is devoted to addressing the first three basic elements in a radiation-based NDE method, as identified in section 1.4: source, modification and detection. Chapter 2 discusses the various types of radiation sources that can be employed in NDE. The manner in which radiation interactions with matter modify the incident radiation is explained in chapter 3. The physical principles of radiation measurement and the variety of detectors used are presented in chapter 4. The electronic processes required to convert the detected radiation into a useful electronic signal are discussed at the end of Chapter 4. Chapter 5 provides an overview of the safety aspects of radiation, since this is obviously a concern when dealing with radiation-based NDE techniques.

Chapter 2

RADIATION TYPES AND SOURCES

Atomic radiation is that emitted as a result of transitions of atomic electrons between orbits (energy) states, releasing energy typically in the form of x-rays. Atomic excitation can be produced, for example, externally by bombarding the atoms with energetic electrons. Nuclear radiation, on the other hand, is released as a result of changes in the nuclear state of matter. Such changes occur due to internal, or externally-induced, instabilities in the structure of the nucleus. In order to reach a stable state, the nucleus may release charged-particles, electromagnetic radiation, and in some instances neutrons. The emitted charged-particles are either alpha or beta particles, while the released electromagnetic radiation is known as gamma-rays. One can also induce radioactivity in nuclear interactions, by bombarding stable nuclei with nuclear particles. Then, in addition to alpha and beta radiation, other heavy charged particles, such as protons, deuterons, etc., can be produced. Gamma radiation can be emitted in such nuclear interactions, or as a result of the decay of the excited nuclei in their transition to a stable state. Neutrons can also be produced, usually as a direct product of the interaction. The properties and nature of various types of radiation are discussed below.

2.1. Charged Particles

2.1.1. Alpha Particles

Alpha particles are heavy charged-particles, with a positive charge equal twice the electronic charge, and a mass four times that of a hydrogen nucleus. An alpha-particle is a positive ion of a helium-4 nucleus (^{4}He). For practical utilization in NDE, an alpha-particle must have sufficient energy for it to be able to enter the interrogated object, interact with it, and then exit the object for subsequent detection. The penetration depth of an alpha-particle is defined by its range, which in turn depends on the particle energy, see section 3.3.1.

Alpha particles are produced during the decay of heavy radioisotopes. For an alpha-particle to be able to break away from the binding forces of the nucleus, it must have a sufficiently high energy to penetrate the nuclear barrier. Therefore, the emitted alpha-particles have high energies, from about 3 to 7 MeV[1]. Alpha particles tend to be monoenergetic, or have a narrow energy band. As shown in section 3.3.1, although alpha-particles tend to move in straight lines, their energy loss per unit length (or stopping power) is quite large. Therefore, in practice, alpha-particle sources are made of very thin layers, or contained in thin metallic foils, to retain the monoenergetic nature of the source particles.

There is a variety of heavy isotopes that decay by alpha-particle emission. Table 2.1 shows the properties of some of these isotopes that can be used as particle sources. The Table lists the natural isotopes that emit alpha-particles; namely ^{226}Ra, ^{232}Th, ^{235}U, and ^{238}U. With the exception of ^{226}Ra, the specific activity of these sources is too low to make them of much practical value as external sources; although thorium mixed with zinc sulfide was the radiation source used for coating pointers and figures of clocks and watches for illumination in the dark. Plutonium-238 has a very high specific activity, but it is also a low-energy photon emitter, see Table 2.7. In addition, ^{238}Pu emits some neutrons as its alpha-particles interact with low mass-number elements, such as Si, Al and O, present in the ceramic enamel matrix material in which it is incorporated. Gadolinium-148 has a high specific activity, though a slightly lower particle energy. It is available as a powder in the solid form, or as a solution in HCl with limited activity, 185 kBq lots only [11]. This leaves, ^{241}Am and ^{226}Ra, as the most common sources of alpha-particles.

Radium-226 has been historically the first source of alpha-particles, as it was the source with which radioactivity was discovered by Madam Curie. However, ^{241}Am superseded ^{226}Ra, due to its lower gamma-ray emissions. Americium-241 is also the source employed in ionization chamber smoke detectors, as discussed in section 10.5.2. Americium-241 is produced by the beta-decay (14.29 year half-life[2]) of ^{241}Pu, while the latter isotope is produced by sequential neutron capture in ^{238}U and the

[1] 1 MeV (million electron-volts)= 1.603×10^{-13} J.

[2] Half-life values listed in this book are obtained from references [12] or [13], unless otherwise stated.

resulting plutonium isotopes. Sources of ^{241}Am are formed as foils or circular disks. The foil is made by powdered metallurgy and metal rolling techniques to produce strips up to 1 m in length [14]. The radioactive material is incorporated in a gold matrix ($\approx$ 1 μm thick), as a relatively insoluble compound. The alpha-particle emitting surface is covered by a thin gold or palladium alloy ($\approx$ 2 μm thick), while the other (backing) surface is made of a thicker layer of silver (from 1.5 to 0.2 mm thick) with a thin gold interface ($\approx$ 2 μm thick). This arrangement provides good sealing for the source, while allowing alpha-particle emission through a thin window. The loading of the foil with ^{241}Am varies from 3 to 70 kBq/mm^2, with a typical active width of 12.5 mm, centered across a 20 mm wide strip. Smaller strips and pieces can be cut from the active area of a foil. Both sides of the foil can also be made active, if needed. Disk sources (typically 5 mm in diameter) are formed by mounting a circular foil on a metal holder. The foils can also be shaped into a variety of source configurations.

2.1.2. Beta Particles

Beta particles are electrons emitted as a result of the decay of a radionuclei. They are in practice distinguished from electrons, produced by some other means, by their continuous spectrum of energy. The energy spectrum of beta-particles has a Maxwellian-like distribution, as shown schematically in Figure 2.1. The spectrum starts from zero energy, rises to a maximum and then decreases to a well-defined maximum energy. This upper limit energy, $E_{\max}$, is uniquely defined by the beta-particle emitting nucleus. The majority of beta-particle emitters have a maximum energy limited to about 3 MeV. Although, beta-decay is associated with the decay of a nucleus from an excited state to a stable (or more stable) state, the transition energy between the two states is shared with an anti-neutrino[3]. The energy distribution between these particles varies from emission to emission, resulting in a continuous energy spectrum that peaks at about one-third of the maxim energy. The penetration depth of a beta-particle is determined by its range, which in turn is a function of the maximum energy of the spectrum of the beta-particle, see section 3.3.2.

Neutron bombardment of most nuclei results in beta-active isotopes. However, many beta-particle emitters are of little practical value in NDE, because they either have a too-short half-life (e.g. ^{32}P, 14.28 days), too-

[3]A neutrino is a neutral particle with zero mass, created to conserve momentum and energy. Neutrinos are associated with $\beta+$ (positron) decay, while anti-neutrons accompany beta ($\beta-$) decay, with the two having opposite spin directions.

Table 2.1. Some alpha-particle emitting isotopes [11, 12, 15, 16].

Source	Half-life years	Particle Energy (MeV) (%)	Yield[a]	Specific Activity[b] (GBq/g)
^{241}Am	433	5.489	86	127
		5.443	13	
^{244}Cm	18.1	5.804	76	3×10^3
		5.764	23	
^{148}Gd	75	3.183	100	1.2×10^3
^{210}Po	0.38	5.305	100	1.66×10^5
^{238}Pu	87.7	5.499	71	634
		5.457	29	
^{239}Pu	2.4×10^4	5.155	73	2.3
		5.144	15	
		5.106	11	
^{226}Ra	1.6×10^3	4.785	94	37
		4.602	6	
^{228}Th	1.91	5.423	72	3×10^4
		5.340	28	
^{232}Th	1.4×10^{10}	4.013	77	4×10^{-6}
		3.954	23	
^{235}U	7.1×10^8	4.396	55	8×10^{-5}
		4.368	17	
^{238}U	4.5×10^9	4.199	77	1.2×10^{-5}
		4.149	23	

[a] Percentage of the total decay events.
[b] For calculation of specific activity see Eq. (3.155).

low beta-particle energy (e.g. ^{3}H, 18.6 keV maximum[4]), or are difficult to obtain in a pure form (e.g. ^{14}C). However, there is a number of isotopes that have half-lives and emission energies that makes them suitable in

[4]Emission-energy values listed in this book are obtained from references [12] or [13], unless otherwise mentioned.

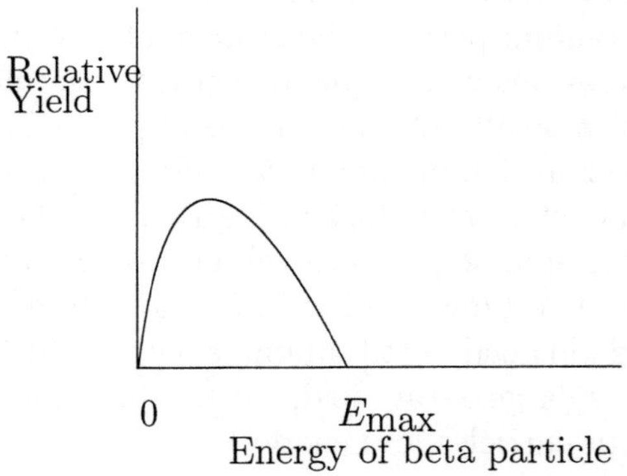

Figure 2.1. Schematic of a typical beta-particle energy spectrum.

industrial applications. Table 2.2 lists some of the commercially available sources and their properties. Since beta-particles have a very small range in metal, to avoid self-absorption within the source's material, beta-particle sources made of solid material are typically deposited on the surface of a thin foil. Various deposition processes are used, depending on the nature of the source material, including electroplating. The radioactive material can also be incorporated in a metal or a compound, in a rolled silver foil, in an enamel that can be mounted on a metallic base with an adhesive material, in a ceramic material, or for point (needle) sources, in a glass bead. The active material is often deposited in a rectangular or a circular shape, to the desired dimensions. The source can also be in the form of an insert within a cylindrical disk equipped with a thin metallic window, usually made of titanium for strength. Point sources are in the form of a small disk or a bead (a mm or so in diameter) mounted at the tip of a cylindrical metallic holder. Below is a brief description of each of the sources listed in Table 2.2.

Krypton-85, unlike other beta-particle sources, is a gas. Therefore, a ^{85}Kr beta-particle sources comes in the form of a capsule in which the gas is sealed. The source capsule is equipped with a thin (25 μm or so) window, to allow the source beta-particles to escape. The window is covered with a cap (protective shield) which can be removed to allow beta-particle exposure. The capsule can be in the form of a disc or a small box. Alternatively, the source's gas can be sealed in a small thin tube, mounted within a protective holder, until it is ready for use.

Steel, nickel and titanium containers are used, for both the capsule's material or the source's windows. Typically the gas is kept under an internal pressure below the atmospheric pressure to prevent leakage, but the strength of titanium permits the storage of gas under pressure, up to a few atmospheres; allowing high concentrations, and hence high source activity within a small volume. For example, a 37 GBq source in a titanium container at 5 atm (about 550 kPa), can be stored in a cylinder 25.4 mm, diameter by a 25.4 m high, while the same source activity under sub-atmospheric pressure will occupy a stainless capsule about 10 times larger in volume ($95.2 \times 23.8 \times 57.3$ mm^3) [17]. When stainless steel is used, instead of titanium, a larger thickness will be required, resulting in a large x-ray field, by the bremsstrahlung effect (x-rays emitted as beta-particles lose energy).

Nickel-63 is a metal that can be electroplated on one face of nickel metal or nickel alloy. The source's holder can be shaped in any desired form, but typically small (30 by 10 mm) slices are used. The maximum practical loading is 3.7 MBq/mm^2 [17]. This source is quite delicate, since when subjected to air or humidity, it gradually tarnishes. Therefore, this source needs to be stored in an inert atmosphere, such as dry argon, prior to use.

Promethium-147, as a carbonate, is incorporated in a rolled silver foil, and shaped as a rectangular or a circular source. For higher activities, the source is embodied in an enamel, mounted on a holder with an adhesive material, and equipped with a thin metallic window. The source is also available as a point source, formed by mounting a small disk (1 mm in diameter) at the end of a metallic holder.

Strontium-90 decays to yttrium-90, which is also a beta-particle emitter. The source as a compound is fused in a roller silver foil. A higher level of source activity is available in the form of a disk in which the source, incorporated in a ceramic insert or rolled silver foil, is inserted in a capsule equipped with a thin window. A point source is obtained by depositing the active material in a small glass bead, sealed at the top of a metallic cylindrical capsule, equipped with a thin window.

Ruthenium-106 also decays to another beta-particle emitter, ^{106}Rh. As a metal, the source is incorporated in a rolled silver foil. Thallium-204, as a chromate compound, is also incorporated in a rolled silver foil, and shaped as a rectangle or a circle.

Two other beta-particle sources are used as calibration sources: ^{3}H (12.35 year, half-life) and ^{14}C (5730 year, half-life), and emit beta-particles with energies up to 18.6 and 156 keV, respectively. Theses sources are typically produced as small disks of thin sheets of a polymer, in which the beta-particle emitting element is embedded [14]. Carbon-

Table 2.2. Common beta-particle sources [11, 16, 17].

Source	Maximum Energy (keV)	Half-life	Working Life[a]	Activity Max.[b], GBq
^{85}Kr	687 (99.57%), 173 (0.43%)	10.73 y	10 y	37
^{63}Ni	66 (100%)	100 y	See note[c]	0.555
^{147}Pm	225 (≈ 100%), 103 (low)	2.623 y	5 y	18.5
^{90}Sr + ^{90}Y	546 (100%) 2280 (99.98%), 519 (0.02%)	28.9 y 64 h	10 y	3.7
^{106}Ru + ^{106}Rh	39 (100%), 3541 (78.6%), 3031 (8.1%), 2407 (10%)	369 d 30 s	5 y	0.37
^{99}Tc	294 (100%%)	2.14×10^5 y	—	—
^{204}Tl	763 (97.4%), e.c.[d] (2.6%)	3.78 y	5 y	1.11

[a] Typical working life that takes into account source construction, half-life, etc. [17].
[b] Typical maximum values that depend on the source design and construction.
[c] Source is sensitive to air and moisture.
[d] e.c.: electron capture, a nuclear decay process in which a nucleus captures an orbital electron, and subsequently releases an x-ray.

14 sources in the form of thin sheets, of up to 0.2 m width and length, are made for use in applications requiring extended sources, such as the radiography of paper, see section 13.3.

2.1.3. Discrete-Energy Electrons

Beta particles, as discussed in section 2.1.2, are electrons with a continuous energy spectrum. It is desirable, however, in some applications, and for calibration purposes to have fast monoenergetic electrons or electrons with known discrete energies. This may be achieved actively by accelerating the electrons with an electric potential, as done in generating the electrons used for operating x-ray machines, see section 2.2.1. However, passive emission of discrete-energy electrons by radioisotopes is also possible by the internal-conversion process or the Auger effect.

Conversion electrons are generated when a decaying nucleus gives its excitation energy to an orbital electron of the atom, instead of releasing it in the form of photons (gamma-rays). If this energy is larger than the binding energy of the electron, the electron will be ejected from the atom, with an energy equal to the excitation energy of the nucleus minus the electron's binding energy. Therefore, a series of electrons with discrete energies, corresponding to their initial orbit, will be released. This process is called internal conversion. The energy spectrum of the conversion electrons consists of discrete lines, in contrast with the continuous spectrum of the electrons arising from the beta-particle decay process. Tables 2.3 lists the properties of some of the radioisotopes that produce conversion electrons. Note that since electron-conversion follows the decay of a nuclei, the Table lists both the parent nuclide and the daughter nuclide that undergoe the internal-conversion process. In order to allow these electrons to leave the source without losing their initial energy, sources are sealed between two layers of low-density polyester films, forming a small (typically 10 mm diameter) active area [14]. The films are supported by a metallic guard ring (washer), for ease of manipulation.

The release of a conversion-electron leaves a vacancy in one of its electron shells, causing the atom to be in an excited state. Typically, the atom releases this energy in the form of x-rays, as an electron from a higher shell fills the vacancy. However, in some cases, these x-rays are not emitted, but their energy is given instead to an electron in a higher orbit, allowing it to escape from the atom. These ejected atomic electrons are called Auger electrons, and usually have lower energy than conversion electrons, but like conversion electrons possess discrete energies, equal to the difference between the atom's excitation energy and the electrons binding energy in the case of Auger electrons. The vacancy left by the release of an Auger electron can lead to the release of a secondary Auger electron, or an x-ray photon. Since x-ray emission is a competing process with the Auger effect, the fraction of electron vacancies, in a given electron shell, that results in the emission of an Auger electron is called the Auger yield. The fraction of filled vacancies that produce x-rays is called the fluoroscopic yield, see section 2.2.2.1. Iron-55 is one of the isotopes that has a good Auger yield (60.7%), and emits electrons with 5.19 keV energy. However, the source still has a significant fluoroscopic yield, as such it is also a source of low-energy photons, see Table 2.7. In order to allow the electrons to escape from the source material, thin sources are formulated, e.g. by electroplating ^{55}Fe on the face of a 50 μm thick nickel alloy foil, normally 24 to 30 mm long and 10 mm wide [14].

2.1.4. Positrons

Positrons are positively charged electrons emitted when a nucleus decays to a nucleus with the same mass-number, but with a lower atomic-

Table 2.3. Some discrete-energy fast-electron sources [11, 12, 14, 16].

Source Isotope	Half-life (Years)	Electron Energy (keV)	Emission Ratio (%)	Specific Activity GBq/g
^{133}Ba	10.53	45	44	9.5×10^3
		75	7	
		266	0.7	
		319	1.3	
^{207}Bi	33	482	1.55	2×10^3
		555	0.45	
		566 - 579	0.15	
		570	0.15	
		976	7.05	
		1049	1.85	
		1060 - 1064	0.60	
^{109}Cd	1.26	62	41.70	9.6×10^4
		84	44.00	
		87	8.900	
^{137}Cs	30.1	624	7.80	3.2×10^3
		656	1.37	
		661	0.37	
^{139}Ce	0.38	127	17.15	2.5×10^5
		160	2.31	
^{57}Co	0.74	7	71	3.1×10^5
		13	7.4	
		115	1.83	
^{55}Fe	2.7	5.2	61	9.3×10^4
^{113}Sn	0.32	364	28.40	3.7×10^5
		387	5.57	
		391	1.11	

number. Many nuclei undergo this decay process, but their half-life is typically too small to be of any practical value. Moreover, as the emitted positron comes to rest within the source material, it annihilates itself with a negative electron. Since both particles have a mass equivalent to 0.511 MeV, the annihilation process results in the release of two photons (into two opposite directions to conserve momentum), each of an energy equal to 0.511 MeV. Therefore, care must be taken in designing

Table 2.4. Some positron sources [12].

Source	Half-life	Energy[a] (keV)	Yield (GBq/μg)	Specific Activity
^{11}C	1223.1 s	960.2	99.759	3.1×10^4
^{13}N	9.965 min	1198.4	99.8036	5.4×10^4
^{14}O	70.606 s	173.01 1808.25 4120.96	0.0190 99.249 0.610	4.2×10^5
^{15}O	122.24 s	1731.9	99.9003	2.3×10^5
^{22}Na	2.6088 y	545.4 1820.0	89.84 0.056	0.23
^{44}Ti	60.1 y	ec[b]		6.4×10^{-3}
^{44}Sc	3.97 h	1474.2	94.27	671
^{58}Co	70.86 d	474.6	14.90	1.2
^{65}Zn	244.06 d	328.8	1.420	0.3
^{68}Ge	270.8	ec[b]		0.26
^{68}Ga	67.61 min	1899.1	87.94	1.5×10^3

[a] Listed energy is the end-point (maximum) energy.
[b] ec: electron capture, producing a positron emitter.

a positron source to allow the particles to escape from the source before being annihilated within its material. This is typically done by electrodepositing the active material over a platinum disk, covered with a thin (10 μm) titanium window. The disk is then mounted on a metallic (aluminum) holder and sealed with an epoxy resin [14]. Cobalt-58 and ^{22}Na are the two most common sealed positron sources, because of their reasonably long half-lives, 70.8 days and 2.60 years, respectively. Other short-lived positron isotopes are also available, as shown in Table 2.4, but are only used when they are produced in an accelerator or a reactor in the close vicinity of the application. Like a beta-particle, a positron is emitted, along with a neutrino, with a continuous energy spectrum similar to that shown in Figure 2.1.

2.1.5. Heavy-Charged Particles

2.1.5.1 Accelerated

Similar to producing fast-electrons by acceleration in an electric field, as discussed in section 2.1.3, positively-charged particles can also be generated at high energies. By stripping an atom of its electrons, positive ions are produced and subjected to a an electric field that accelerates them to the desired energy. Charged-particle acceleration is typically achieved in a linear accelerator (Linac), where charged-particles are accelerated in a straight-line by an alternating voltage. A cyclotron can also be used to accelerate the ions, which are forced to move in a circular orbit of increasing voltage, subjecting them to successive acceleration in an alternating constant-frequency electric field. A synchrotron is another accelerating device, in which a charged-particle is confined to an orbit by a magnetic field that gradually increase in strength as the particle's momentum increases. An alternating electric field that is in synchronism with orbital frequency causes the particles' acceleration. In a synchrotron, the speed of light particles, such as electrons and protons, can approach that of light, producing extremely high energies, since under such conditions, the speed changes only slowly as the energy changes.

Protons are simply hydrogen ions. Once the hydrogen atom is stripped off its electron, it can be accelerated to the required voltage, typically in a linear accelerator. Hydrogen ions can be produced by injecting a hydrogen gas into a metallic cylinder surrounded with an electric field to break down the gas into its elementary constituents, protons and electrons, or by the use of high power microwaves. Negative hydrogen ions can also be created by adding electrons, and stripping the electrons away after acceleration is completed. Protons are also produced by irradiating thin solid targets with a high-powered laser pulsed beam. The protons originate from hydro-carbon impurities located on the surface of the target, so that proton beams are observed even in targets which nominally do not contain hydrogen [18].

Charged particles of any type can be produced and accelerated to the required energy. For example, alpha particles can be produced by ionizing a ^{4}He gas and accelerating the ions. Particles with a heavier mass can also be produced similarly, as long as the material is ionized. Neutron fission (in a nuclear reactor) and the spallation of nuclei (such as Hg, Pb and U) with accelerated charged-particles (such as protons) also result in the production of charged-particles. It is difficult, however, to extract these particles as independent sources in such complex processes.

2.1.5.2 Passively-Emitted

Aside from the active processes discussed above, positively-charged particles, with a mass heavier than that of alpha-particles, can be passively produced by the spontaneous fission of some heavy nuclei, such as ^{252}Cf. Californium-252, though is mainly a neutron source, see section 2.3, decays by alpha-particle emission, and accordingly is also a source of alpha-particles. In addition, the fission process results in the production of two fission fragments, typically one with a lighter mass and the other with a heavier mass. The mass-number of fission fragments varies from about 78 up to 172, with an average mass-number of 106 for the light fragments and 142 for the heavy ones. The total energy of the two fragments is about 181 MeV, with the lighter fragments receiving more kinetic energy (102.5 MeV on average) than the heavier particle (78.5 MeV on average). Since these heavy-charged particles are easily stopped within the material, the source needs to be formed in a very thin layer. This was done, for example, by electrodepositing californium oxide on a platinum foil (0.13 mm thick) in the from of a disk (12.7 mm in diameter) [16]. The source is normally covered with a thin layer of gold to prevent loss of ^{252}Cf by recoiling as it decays.

2.2. Photons

Electromagnetic radiation is released when an atom or a nucleus undergoes a transition from an excited state to a stable, or a more stable, state. The radiation that results from atomic transitions is known as x-rays, while radiation produced by transition in the nucleus is called gamma-rays. Atomic and nuclear radiation has a much shorter wavelength than other forms of electromagnetic rays, such as light, infrared, or radio waves. This short wavelength gives x-rays and gamma-rays the penetration depth that makes them attractive for use in NDE.

Associated with the short wavelength is a high frequency, since electromagnetic radiation propagates at the speed of light, c ($\approx 3\times 10^8$ m/s), so that:

$$c = \lambda\nu \tag{2.1}$$

where λ is the wavelength and ν is the frequency. X-rays that are useful in NDE have a wavelength from a few picometers to about 100 pm, while the wavelength of gamma-rays is in the range of 0.1 to 10 pm; visible light has a wave length between 0.38 to 0.76 micrometer. According to Eq. (2.1), the frequency of this type of radiation is a few exahertz to a few hundred exahertz. It is, therefore, customarily to utilize energy, instead or frequency or wavelength, in describing the quality of atomic and nuclear radiation. The frequency is related to energy according to

Table 2.5. Penetrating depth (in mm) of photons in some materials[a].

Material	Energy, MeV						
(kg/m^3)	0.05	0.1	0.2	0.5	1.0	2.0	5.0
Water (1000)	220	292	365	517	707	1012	1650
Aluminum (2700)	50	108	152	219	302	429	652
Iron (7800)	3.3	17	44	76	107	151	204
Lead (11300)	0.6	0.8	4.5	27	57	97	104

[a] Taken equal to five mean-free-paths (defined by Eq. (3.11), with data from reference [19].

the equation:

$$E = h\nu \tag{2.2}$$

where h is Planck's constant (4.136×10^{-21} MeV s) and E is the energy of a photon of radiation; a photon is a quantum of radiation. The x-ray energy is in the range of a few keV to a few hundreds of keV; while gamma-ray energy is in the few hundred keV range to a few MeV.

A common point of misunderstanding among beginners in the field of radiation is the distinction between energy and intensity. Radiation intensity, as in I in Eq. (1.1), refers to the number of photons, while the energy E denotes the energy of a single photon. Therefore, a radiation beam may be attenuated in intensity, according to Eq. (1.1), as it travels through matter while the photon energy remains unchanged. This is due to the fact that the photons that successfully pass through a material are mainly those which have not suffered any interactions, the so-called uncollided photons, and therefore there is no reason for them to change their energy. Approximate penetration depths of photons at different energies are given in table 2.5.

2.2.1. X-ray Machines

X-rays are photons which are generated as a result of electron transitions between the shells of an excited atom. In practice, this is achieved by bombarding a target of a high atomic-number material with a beam of fast electrons. The incident electrons interact with the atomic electrons of the target, displacing some of them from their stable orbits to orbits of higher energy, i.e. to excited states. As the excited electrons return to their stable orbits, energy is released in the form of x-ray photons. The energy of emitted photons is equal to the energy difference between the excited and stable orbital states. Therefore, x-rays with well characterized energies can be produced. However, the incident beam of electrons is also subjected to acceleration by the electrostatic field of the nucleus. The accelerated electrons, as they are slowed-down, emit elec-

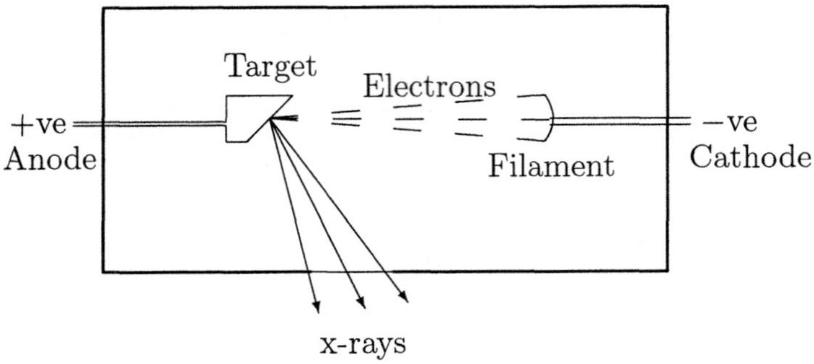

Figure 2.2. A schematic of an x-ray tube.

tromagnetic radiation, called "bremsstrahlung" radiation. The energy of bremsstrahlung photons can carry an energy up to the maximum energy of the original incident electrons. Therefore, the energy distribution of x-ray photons consists of a continuous spectrum, due to bremsstrahlung radiation, with characteristic photopeaks representative of the atomic transitions in the target material. It should be noted here that most of the energy deposited in the target by the incident electrons is dissipated as heat and only a small fraction, typically about three percent or less, is emitted as x-rays.

Figure 2.2 shows a schematic of an x-ray tube. The fast-electron beam is produced by a filament, which is heated by an electric current to liberate electrons from the filament material. The freed electrons are accelerated in a high-voltage field and directed towards the target. Tungsten is usually used as the target material, because of its high atomic-number and high melting-point. The filament and target are enclosed in a vacuum chamber, to avoid energy loss by ionizing the gas that would have been otherwise occupying the chamber.

The intensity, I, of radiation emitted in an x-ray tube, with a target of atomic-number, Z, can be approximately expressed as:

$$I = KZiV_p^n \tag{2.3}$$

where i the filament current, V_p is the operating voltage of the tube and n is an index that is typically equal to about two. The energy of the emitted x-ray, as mentioned above, varies with the tube voltage.

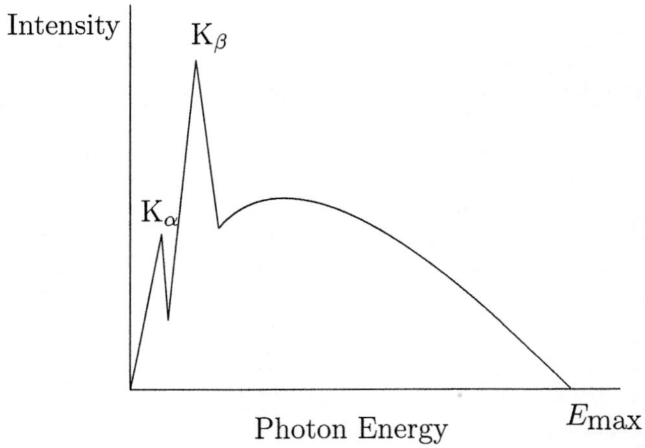

Figure 2.3. A typical x-ray spectrum.

However, emitted x-rays cannot exceed the energy corresponding to V_p. This maximum energy, in eV, is therefore given by:

$$E_{max} = eV_p \tag{2.4}$$

where e is the charge of an electron (the elementary charge). Since the value of e is constant, the maximum or peak energy is commonly expressed as voltage rating in kilovoltage, or kV_p.

The energy spectrum of the continuum of x-rays generated in a machine with a target of atomic-number Z operating at constant potential can be expressed as [20]:

$$I(E) = CZE^2 \left(\frac{E_{max}}{E} - 1\right) + BZ^2E \tag{2.5}$$

where $I(E)$ is the number of photons emitted at energy E and B and C are constants; with the value of B so small that the second term in Eq. (2.5) is usually neglected. The intensity of the characteristic peaks in the x-ray spectrum, see Figure 2.3, result in "jumps" in the intensity spectrum at energies corresponding to their K and L energy levels. Table 2.6 gives the jump ratios for different x-ray machine targets.

The average energy of x-rays emitted from a tube is roughly equal to one third of the energy of the applied electrons, i.e. that corresponding

Table 2.6. Jump ratios at the characteristic lines for various x-ray machine targets [20, 21, 22].

Target	K Line		L Line	
Element	Energy (eV)	Jump Ratio	Energy (eV)	Jump Ratio
Mg	1.30	11.59		
Ti	4.96	8.53		
Cu	8.99	7.96	0.93	2.87
Zn	9.66	7.60	1.02	5.86
Zr	18.00	6.75	2.22	3.98
Mo	20.01	6.97	2.52	3.67
Sn	29.19	6.47	3.93	3.06
Nd	43.57	5.99	6.21	2.66
W	69.51	5.12	10.20	2.62
Pb	89.00	4.99	13.04	2.74
U			17.17	2.82

to kV_p. Therefore:

$$E_{\text{ave}} \approx \frac{1}{3} e V_p \tag{2.6}$$

This average energy may be used to provide an estimate of the attenuation coefficient in some material. However, caution should be exercised in using such an attenuation coefficient, as it does not adequately reflect the actual attenuation characteristic of the complex multichromatic radiation of x-rays, see appendix F.

The penetrability of x-rays depends on the strength and energy of the x-ray source, the penetrated material and the sensitivity of the detection medium. That is, a source of higher energy can penetrate deeper inside an object, without losing much of its intensity, recording a certain intensity at the detector or on a radiographic film. However, the same recorded radiation intensity can be attained, with the same penetrated material, using a source of a higher intensity, a detecting medium of larger sensitivity and efficiency, or simply a longer exposure time. Therefore, in determining the penetrability of x-rays, many factors come into play. Such factors can be summarized as follows: (1) energy of x-ray source, (2) intensity of x-ray source, (3) thickness of object, (4) material and density of object, (5) sensitivity and efficiency of detecting medium, and (6) exposure time of the object, and consequently of the detection medium, to radiation.

The geometric arrangement of the source-object-detector can also affect the apparent (measured) penetrability of x-rays. Emitted x-rays diverge as they are propagated away from their point of origin; typically in inverse proportion to the square of the propagation distance, see section 3.6.2. Therefore, the source-to-film (detector) distance affects the amount of radiation detected. The radiographic film is usually positioned directly against one side of the object. However for thick objects, divergence of radiation with distance can cause some loss of sharpness in the recorded image, see section 6.3.

Owing to the above mentioned intermingled factors, empirical means have emerged for determining the penetrability of x-rays. For example, radiographic equivalent factors (REF's) are often used in industrial practice to estimate the penetrability of x-rays. Such factors are listed in nondestructive testing handbooks, such as reference [5], and provide for each typical x-ray the penetrability of different materials with respect to a reference material, typically aluminum or steel. X-ray machines are, in turn, provided with exposure charts that determine the required exposure for various thickness of the standard material. X-ray exposure is conventionally defined as the multiplication of the intensity of the x-ray machine (expressed in electric current) by the exposure time, see section 6.3. These exposure values for the reference materials can accordingly be adjusted using the REF values to accommodate other materials. It should be noted here that the exposure graphs, provided by a manufacturer, are given for a particular film type, i.e. film sensitivity is taken into account. Also exposure charts are for a typical target-to-film distance.

The above process of determining x-ray exposure for obtaining an acceptable x-ray penetrability takes into account all the six factors listed above. The suitability of particular machine depends mainly on the reasonableness and practicality of the exposure time it requires for a given object. It is essential to ensure that x-rays used in an examination have a sufficiently good penetrability to go through the entire thickness of the object and still provide a detectable signal at the receiving radiographic film. It is also equally important that the same x-rays be capable of detecting small flaws within the object. A practical way to determine sensitivity is to place a thin object (called a pentrameter) containing a small size hole, representing a flaw, in the path of the x-ray beam. If such a hole is distinguishable on the radiographic film, one can then be assured that the setup is suited for detecting at least a flaw of the size of the inspection hole. Naturally, the size of the hole can be decreased until it ceases to appear on the film, then one can determine the smallest flaw size detectable by the employed x-rays. The thickness of this object can also be used to determine the minimum detectable depth, i.e. the thickness below which the thin object is not distinguishable on the film, see section 6.3.1. Standards pentrameters are available for this purpose [7].

2.2.2. Low-Energy Photon Sources

X-rays can also be emitted when an electron is moved out if its orbit, creating a vacancy in the orbit. This vacancy is subsequently filled by an electron residing in a higher energy orbit. The energy difference between the two orbits results in the emission of monoenergetic characteristic x-rays. A transition into the K shell (the lowest orbital shell in the atom) generates so-called K photons. If the transition occurs from the closest orbit to the K shell, (L orbit), it is designated as K_α x-rays; from the second next nearest orbit (M), it is called K_β, and so on. The maximum photon energy occurs when an unbound (free electron) moves to the K shell, called simply K x-rays, and have an energy equal to the binding energy of this shell. Electron transition to other orbits are given similar designations, e.g. L_α refers to x-rays generated by the movement of an electron transition to the L shell from the nearest higher up orbit (M). Naturally, as atomic electrons move downwards to fill in a vacancy in a lower shell, they leave a vacancy in their own shell, which is subsequently filled with electrons from a higher orbit, and so on. Therefore, if an atom is excited, a series of transitions to the K, L, M, ... shells will occur, leading to the emission of corresponding characteristic x-rays. Since the energy levels of atomic electrons varies from one atom to another, these transition are unique to each element and are called fluoroscopic emissions. The question now is how to trigger such an atomic excitation process.

2.2.2.1 Passively-Excited Sources

Atomic excitation can occur either passively (intrinsically) or by external excitation. Sources based on the latter mechanism are discussed in section 2.2.2.2. Passive excitation of an atom can occur either by the electron capture or internal conversion processes. In electron-capture, the nucleus captures an orbital electron, often from the orbit closest to it, the K shell. This creates the vacancy required for triggering a chain of fluoroscopic x-rays. The electron captured by the nucleus decreases its charge by one electronic charge, transmuting it into a new element. The electron orbits of the new atom will have a vacancy left by the captured electron. The capture of the electron by the nucleus can also destabilize it, leading to subsequent decay by gamma-ray emission, at discrete energies corresponding to excitation levels of the nucleus. Therefore, isotopic sources that decay by electron-capture produce both x- and gamma-rays, and accordingly are low-energy photon sources, since the photons emitted in both cases have usually a low energy. The capture of the negatively charged electron by the positively charged nucleus leads to acceleration of the electron, which in turn leads to photon emission by the bremsstrahlung process. The result is a continuous, albeit weak, energy spectrum superimposed on the discrete energy levels of the emitted x-rays and gamma-rays.

In the internal conversion process, an excited nucleus gives its excess energy to one of the orbital electrons of the atom, rather than releasing it in the form of photons. This energy is usually given to electrons in the shell nearest to the nucleus, the K shell, leading to the creation of a vacancy in the shell as the electron gaining the energy is elevated to a higher orbit or is rejected from the atom. This in turn triggers a chain of K-series x-rays. In addition to decay by internal conversion, the nucleus can still decay by gamma-ray emission. Therefore, both x-rays and gamma-rays can be emitted from the same source. In addition, the energy given to the conversion electron leads also to its acceleration, and the subsequent emission of a continuous energy spectrum of photons by the bremsstrahlung process. For example, ^{137}Cs decays by beta-particle emission to $^{137\mathrm{m}}$Ba (in a metastable state of half-life of 2.6 minutes). Internal conversion in Ba leads to the release of K x-rays (32 to 38 keV), while the decay of the nucleus leads to the release of 662 keV photons. However, ^{137}Cs is considered a primary source of photons due to the dominance (81%) and high energy (662 keV) of its gamma-rays, in comparison to the low emission ratio of its x-rays (about 8%).

It should be noted that the atomic excitation process may not always result in the production of x-rays. The excitation energy may instead be given to an outer electron, ejecting it altogether from the atom. This Auger electron, see section 2.1.3, possesses an energy of no more than a few keV. Therefore, it is typically reabsorbed within the source, unless the material is quite thin, one then has a source of electrons. The fluoroscopic yield is the parameter reported to quantify the probability of x-ray emission due to atom de-excitation; and is usually less than 100% due to competition by Auger-electron emission. For example, the fluoroscopic yield of ^{55}Fe is only about 28%, which makes it a good source of Auger neutrons (with an emission rate about 61%), as well as being a source of K x-rays from its electron conversion to Mn.

Table 2.7 lists some of the common isotopic sources of low energy photons and their characteristics. Note that all the listed sources, apart from ^{241}Am, have characteristic x-rays associated with a daughter nucleus that resuls from electron-capture (which reduces the atomic-number of the parent nucleus by one). In the case of ^{241}Am, it decays by alpha-particle emission to ^{237}Np, emitting in the process some gamma-rays. In addition, the atom of the daughter nucleus is left in an excited state and as it rearranges itself it releases a cascade of L x-rays. Curium-244

Table 2.7. Some common low-energy gamma and x-ray sources [11, 16, 17].

Source	Half-life	Photon Energies (keV)	Specific Activity (GBq/g)	Type of Decay
^{241}Am	433 y	12 - 22	127	L x-rays of Np ($\approx$ 40%)
		59.5		γ decay (35.9%)
		26		γ decay (2.5%)
^{109}Cd	462 d	22 - 26	9.6×10^4	K x-rays of Ag (96.2%)
	40 s	88		γ decay of ^{109m}Ag (3.8 %)
^{57}Co	271.7 d	122	3.1×10^5	γ decay (85.5)
		135		Mössbauer effect[a] (10.8%)
		14		γ decay (9.5%)
		6 -7		K x-rays from Fe ($\approx$ 55)
^{244}Cm	18.1 y	12 - 23	2.3×10^3	From L x-rays of Pu ($\approx$8%)
^{153}Gd	241.5 d	41 - 48	7.4×10^3	From Eu X-rays ($\approx$ 110%)
		97		γ (30%) decay
		103		γ decay (20%)
		70		γ decay (2.6%)
^{125}I	60 d	35	6.4×10^5	γ decay (7%)
		27 - 32		K x-rays from Te (138%)
^{55}Fe	2.7 y	5.9 - 6.5	8.9×10^4	K x-rays from Mn
^{238}Pu	87.7 y	11 - 22	633	L x-rays from U ($\approx$ 13)
		43		internal conversion 100
^{153}Sm	47 h	103	1.6×10^7	γ decay (28%)
		75		γ decay (17%)

[a] See section 2.2.2.3.
Percentages in brackets represent emission ratios and may not add up to 100% as minor emissions are ignored; when the sum exceeds 100%, it is due to the involvement of more than type of decay.

also decays by alpha-particle emission, to ^{240}Pu, but with a very low emission rate for gamma-rays, and still weak but detectable, L x-rays from the de-excitation of the daughter nuclei.

Low energy photon sources are produced in a variety of configurations, as disk, annular, line or point sources. Stainless steel capsulation is used, if the low energy x-rays need to be suppressed. However, when the x-rays are also required, a beryllium window is utilized, due to the low absorption properties of this low atomic-number solid material. The radioactive material is incorporated into a ceramic enamel or electrodeposited on a metal backing, and then shaped to the desired form. Iodine-125, and sometimes ^{57}Co, are incorporated as ion exchange resin pellets, as they are separated from other related isotopes in an ion exchange process. Gadolinium-153 in the form of a pressed pellet is recessed into a stainless steel insert and encapsulated in titanium. Point sources are made by incorporating, when possible, the source into a ceramic bead, or pellet, while a line source is produced by packing the beads into a tube. The source sizes vary from a 1 mm in diameter for a pellet to a few tens of mm of diameter, or height, for a disk, annular and line sources, see reference [17] for specific details on each source listed in Table 2.7. The working life of a source depends on its initial activity and the applications in which it is used; but most importantly by the life of its container; since breaching the containment will lead to leak of radioactivity and contamination. The recommended working life for these isotopic source varies from 5 to 15 years.

2.2.2.2 Externally-Excited Sources

External excitation of atomic electrons can also lead to the production of photons. The excitation process is accomplished by bombarding the target material with charged-particles. Electron excitation is the process used to generate x-rays, but the produced x-rays have a wide energy spectrum, as discussed in section 2.2.1. However, low-energy monoenergetic photons can be generated from an x-ray machine by using a filter to cut-off the bremsstrahlung component of the spectrum, while keeping its low energy K_α x-rays. For example, an x-ray generated by a copper target can be filtered with a nickel foil, while those generated from molybdenum can be filtered with zirconium filter, with the filter having an absorption edge slightly higher energy than that of the x-ray target so that it allows the passage of the K_α photons while absorbing most of the others [23]. This method is employed to generate the monochromatic x-rays used in crystallography, which have energies less than about 10 keV.

An x-ray source can be used to generate another source of x-rays, this time with characteristic energies of a target material; called a secondary target to distinguish it from the primary target that produces the primary x-rays. Low atomic-number target materials can be used to generate soft (low energy x-rays), while targets with high atomic-number are used to produce higher energy x-rays, due to the larger orbital electron energy transitions in the latter materials. Obviously, the energy of emitted x-rays cannot exceed the energy of particles used to excite the target material.

Alpha particles can also be used to generate low-energy photons. Alpha-excitation is achieved by the use of an alpha-particle emitter, such as ^{241}Am and ^{244}Cm, see section 2.1.1. Since alpha-particles do not penetrate deep into matter, they only excite electrons of the atoms present near the surface of the target material. Therefore, the target material is usually placed, as a backing material, behind the source. Alpha excitation results in the emission of x-rays as the excited electrons return to a more stable orbit. Note also that in addition to the characteristic x-rays produced by the target, ^{241}Am also emits its own 60 keV photons by nuclear decay and L x-rays from excited atomic state of its daughter, ^{237}Np. As Table 2.1 shows, ^{241}Am emits high energy, 5.489 and 5.443 MeV alpha-particles. This energy far exceeds the binding energy of orbital electrons in most material, making it possible to excite the atoms of a wide variety of target materials. A wide range of targets can be used with ^{241}Am, including Cu (8.04 - 8.91 keV), Rb (13.37 - 14.97 keV), Mo (17.44 - 19.63 keV), Ag (22.10 – 24.09 keV), Ba (32.06 - 36.55 keV) and Tb (44.23 - 50.65 keV), where the shown energies are the mean values for the K_α and K_β energies for the element. These targets can be interchanged to produce various values of photon energies, with the same ^{241}Am source. An interesting observation about ^{241}Am is that it can be used to produce alpha-particles, photons (both x- and gamma-rays), and neutrons. Neutrons can be emitted deliberately by augmenting the source with a proper target, see section 2.3, or inadvertently as the alpha-particles interact with low mass-number materials (such as Si, Al and O) present in the ceramic materials in which the source is often incorporated, or the thin beryllium window used to eliminate the fluoroscopic x-rays. In the latter case, the neutron emission is quite low (0.27 neutrons/s per MBq), but is not negligible at high source activities.

Curium-244 is another alpha-source used for producing ultra-low energy photons. The 5.764 MeV alpha-particles, see Table 2.1, are used to bombard a wide variety of targets, producing photons with principal energies from 0.110 keV, with a beryllium target, to 4.97 keV, with a titanium target [16]. A multi-target assembly can be used to produce a variable-energy source. Typically targets of Mylar (0.523 keV), Teflon (0.667 keV), magnesium (1.254 keV), aluminum (1.487 keV), polyvinyl chloride (2.622 keV) and titanium (4.97 keV) are used to generate the indicated photon energies. Note that as shown in Table 2.7, ^{244}Cm emits higher-energy photons, in the energy range of 12 to 23 keV, as fluoroscopic L x-rays of its Pu daughter.

Beta-particle excitation is achieved with relatively high-energy beta-particle emitters, such as ^{3}H (18.6 keV maximum energy) and ^{147}Pr (225 keV maximum energy). Tritium, a gas, is absorbed in a layer of titanium or zirconium, 0.10 mm thick, which functions as a target material [14]. The source is also equipped with a backing material (tungsten), mounted on a metallic (stainless steel holder), with a frontal window made of plastic. Beta-particle excitation of Ti and Zr lead, respectively, to the generation of Ti K x-rays ($\approx$ 4.5 keV) and Zr L x-rays ($\approx$ 2 keV). The beta particles of ^{147}Pr, are used to excite zirconium (15.7 keV), tin (25.3 keV), barium (32.2 keV) and samarium (40.1 keV) targets, with the values in brackets referring to the energy of the emitted fluoroscopic x-rays. The source is incorporated in a ceramic enamel containing the target material, supported by alumina, and sealed in a monel metal capsule, with a beryllium window. Beta-particle excited sources are typically constructed in cylindrical capsules, 10 to 20 mm in diameter, and about 5 mm in height [14].

In addition to the generation of fluoroscopic x-rays, beta-particle excitation also produces a continuous energy-spectrum of photons by the bremsstrahlung process. Therefore, even in the absence of a target material, photons can be produced by the bremsstrahlung process, if the beta-particle source is encapsulated to prevent beta-particle emission. Hence, a ^{147}Pr source, without the target material, will produce a continuous bremsstrahlung photon spectrum, with an energy not exceeding its maximum beta-particle energy of 225 keV. The same process can be used with other beta-particle sources, as those listed in Table 2.2, by suppressing beta-particle emission by metallic encapsulation. It should be also noted that x-ray machines, discussed in section 2.2.1, also employ electron excitation, but with electrons produced in an external electron accelerator. Table 2.8 lists some of the commonly used bremsstrahlung sources. For example, when photon-excitation is achieved with a ^{55}Fe source, which emits 5.9 and 6.5 keV photons, see Table 2.7. This energy can excite the electrons in targets with an atomic-number lower than that of iron, such as Ti (4.51 - 4.93 keV), Sc (4.09 - 4.46 keV), Ca (3.69 - 4.01 keV), S (2.31 - 2.46 keV), Si (1.74 - 1.84 keV), and Al (1.49 - 1.55 keV), where the values in brackets correspond to the K_α and K_β mean energies for each element [14]. These targets can be interchangeably placed in front a ^{55}Fe source to produce a variable energy x-ray source.

2.2.2.3 Mössbauer Sources

Gamma-ray emission leads the originating nucleus to recoil in the opposite direction to conserve momentum. Therefore, emitted gamma-rays

Table 2.8. Source for low-energy (bremsstrahlung) photons produced by isotopic excitation [14, 24].

Excitation Beta Source	Target Material	Photon Energy (keV)
^{3}H	Ti, Zr	3 – 12
^{55}Fe	Ti, Sc, Ca, S, Si, Al	1 – 5
^{85}Kr	C, Cu, Au, Pb	25 – 80
$^{90}Sr/^{90}Y$	Al, Ag, Zr, Sn, Pb, U	60 – 150
^{147}Pm	Al, Ag, Zr, Cu,Fe Sn, Mo, Ceramics	15 – 50

will have an energy slightly less than the difference between the energy of the initial and final nuclear states, thus preventing the photons from being reabsorbed within the originating nucleus, allowing their emission. However, in some cases, the nucleus may not be free to move. This happens, for instance, in some solids where the atoms are hold together in a crystal lattice structure, that is sufficiently strong that it cannot be broken by the recoil energy given to the nucleus. Then, the entire solid recoils together as one rigid mass, and the recoil energy of a single nucleus becomes negligible. These recoilless gamma-rays have an energy very close to the difference between the two nuclear states that produced them. Therefore, if these photons bombard a target made of the same material as the source they can be resonantly absorbed, as they have an energy that almost matches the energy of two states in the target nucleus. Now, if the energy of the gamma-rays is slightly changed, say by moving the source, which adds energy to or subtracts energy from the photons by the Doppler effect, a mismatch between the photon energy and corresponding nuclear states occurs, and the number of photons absorbed dramatically decreases. The process of emission or absorption of radiation by nuclei without impacting a recoil energy to the nuclei is knows as the Mössbauer effect.

Iron-57 is the isotope with the strongest recoilless absorption of gamma-rays at room temperature. It is produced by the decay of ^{57}Co (270.5 days, half-life, see Table 2.7). Of all the excited ^{57}Fe nuclei, 10.8% will emit a 14.4 keV Mössbauer gamma-ray, i.e. photons with an energy that closely matches the initial and final nuclear states of the nucleus. The photon energy is emitted with a Gaussian-like energy distribution, called a Lorentzian line, with a width defined by the width of the distribution at an intensity corresponding to half of the maximum intensity, the full-width-at-half-maximum. For ^{57}Fe, this width is equal 4.7×10^{-9} eV, and is called the natural line width. Moving the source at

Table 2.9. Some Mössbauer Sources [14]*.

Source	Parent Nuclide	Half-life (years)	Photon Energy Mössbauer (keV)	Other (kev)	Measuremnt Condiction (Temperature)
^{57}Fe	^{57}Co	0.74	14.4	see Table 2.7	room
^{151}Eu	^{151}Sm	87	2.16	5 – 8: Eu L x-rays	room
^{125}Te	^{125}Sb	2.77	35.4	27.4: Te K x-rays	liquid nitrogen or liquid helium
^{119}Sn	^{119m}Sn	0.67	23.9	25 – 29: Sn K x-rays	room or lower
^{133}Cs	^{133}Ba	10.8	81	see Table 2.11	liquid helium
^{195}Pt	^{195}Au	0.50	98.9 - 127.7	65 – 78: Pt K x-rays	liquid helium
^{182}W	^{182}Ta	0.32	100.1	see note below	liquid nitrogen or liquid helium
^{237}Np	^{241}Am	433	59.9	see Table 2.7	liquid nitrogen or liquid helium

* Reference [25] lists additional sources.
The photon energies, in keV, emitted by ^{182}Ta and their emission ratios are: 68 (41.8%), 101 (14.3%), (152 7.2%), 179 (3.2%), 222 (7.7%), 229 (3.8%), 264 (3.7%), 1113 (0.4%), 1212 (35.0%), 1189 (16.5%), 1221 (27.5%), 1231 (11.6%), 1257 (1.5%), 1289 (1.4%), others (< 3% each).

a velocity, v, of 1 mm/s toward, or away from, the target increases the energy of the photons by 4.8×10^{-8} eV for ^{57}Fe, ($= 14.4 \times 10^3 \times \frac{v}{c}$, where c is the speed of light). This energy shift corresponds to about 10.2 line widths, and is sufficient to put the photon energy almost completely off the matching range of the corresponding two nuclear states. Iron-57 is particularly useful as it can be used in an iron, or steel, target, allowing the examination of some interesting properties of this widely used material, see section 12.4.6. However, there are other Mössbauer sources as shown in Tables 2.9. Note that the Table shows the conditions under which the Mössbauer effect is observed, since this effect is not always observable at room temperature where nuclear recoil is often possible.

Cobalt-57, the parent nuclei of ^{57}Fe , a Mössbauer source, is typically loaded in a metal matrix, such as natural iron, stainless steel, rhodium, palladium or copper [14]. The matrix material provides the rigid body that takes the recoil momentum following gamma-ray emission. Therefore, each target material has its own Mössbauer emission characteristic, in terms of the recoilless fraction and the spread it introduces into the energy of the Mössbauer photons. Note that one natural line width for ^{57}Co corresponds to a velocity of 0.097 mm/s, and in turn the line width for different targets is measured in mm/s. Natural iron splits the spectrum into many lines, by the effect of its ferromagnetism. All other loading materials, for ^{57}Co, listed above, give a high recoilless fraction of 0.62 to 0.78, and line widths not exceeding 0.12 mm/s. The active area of a source is typically a circle of about 10 mm in diameter, 6 to 25 μm thick foils, loaded up to 18.5 TBq of the active material. Highly concentrated sources tend to produce line-broadening due to the electrostatic and magnetic effects between neighboring cobalt and iron atoms. Samarium is usually loaded in its own oxide or fluoride matrix, while ^{125}Sb is loaded in rhodium, and ^{119m}Sn is loaded in calcium stannate. Mössbauer sources are mounted on holders equipped with threads to facilitate their installation on a source moving mechanism. Sources that require nitrogen or helium cooling are supplied in a cryogenic holders that allow the flow of the cooling material around the source.

2.2.3. Primary Gamma Rays

Gamma-rays originate from within the nucleus, as an excited nucleus decays to a more stable state. Gamma-ray emission usually follows the beta-particle decay of the parent nuclide, which leaves the daughter nucleus in an excited state. The emitted gamma-rays carry the difference in energy between the initial and final states of the nucleus, and consequently gamma-rays are distinguished by their well-defined discrete energies. Since gamma photons are generated from the nucleus, they have higher energy than the x-ray photons generated by the transitions of atomic electrons. This gives gamma-rays a larger penetration depth than x-rays, since the photon penetration depth increases with energy as shown in Table 2.5.

Table 2.10 lists some of the most common gamma-ray sources. Note that the specific activities of these sources are quite high, in comparison with those of other radioactive listed in Tables 2.1 and 2.2. This high specific activity results in a high emission rate from small-volume sources, typically less than 10 mm in dimensions. Sources are usually contained in stainless capsules, often with double encapsulation. The capsules are usually cylindrical in shape, with a non-active extension part to enable easy handling. The working life of a source depends on its half-life and the required source intensity, in addition to the expected life of the encapsulation. However, a working life of 15 years is usually recommended. The short half lives of ^{203}Hg and ^{192}Ir limits their use, which in addition to the liquid state of mercury, limits the amount of

Table 2.10. Some gamma-ray sources [11, 12, 14, 16, 17].

Source	Half-life	Photon Energy (keV)	Emission Rate (%)[a]	Specfic Activity (GBq/g)
^{198}Au	2.7 day	412	95.58	9×10^6
	2.27 day	215	77.0	1.1×10^7
^{133}Ba	10.53 year	81	34.1	9.5×10^3
		356	62.2	
^{207}Bi	32.9 year	570	98	1.9×10^3
		1064	75	
^{60}Co	5.27 year	1333	99.98	4.2×10^3
		1173	99.85	
^{137}Cs	30.1 year	662	85.1	3.1×10^3
^{203}Hg	47 day	279	81.5	5.1×10^5
^{192}Ir	74 day	317	83	3.4×10^5
		470	48	
^{226}Ra	1600 year	609[b]	41.7	3.7×10^1
		352[c]	32.0	
^{75}Se	120 day	265	60	5.4×10^5
		136	57	
^{170}Tm	129 day	84	3.4	2.2×10^5
^{169}Yb	32 day	63	44	8.9×10^5
		198	35	

[a] The two most significant emissions are given, for a complete list see reference [12].
[b] From decay of daughter ^{214}B.
[c] From decay of daughter ^{214}Pb.

total activity obtainable in a small volume. However, ^{192}Ir is one of the sources utilized in industrial radiography, along with ^{60}Co and ^{137}Cs. Iridium-192 is favored in radiography as it enables the imaging of thin metallic objects, in the range covered by x-ray machines, but without the need to have access to an electric power source to operate the machine.

The distinctiveness of photon energies produced by gamma-ray sources enables their use in energy calibration of gamma-ray detectors. Small activity sources, less than 1 MBq, in the form of small beads encased in plastic disks or positioned at the tip of cylindrical rods are utilized, and supplied in the form of a reference calibration set. Table 2.11 provides a list of the sources used in a typical set of gamma-ray reference sources. Note that ^{22}Na emits 511 keV photons, as it is also a positron emitter. The positively charged positron is combined with one of the atomic electrons in an annihilation process that results in the release of two photons (in two opposite directions), each carrying an energy equivalent to that of the rest-mass of the electron (or positron). Yttrium-88 also emits positrons, but at a very small emission rate ($\approx 0.2\%$), hence its 511 keV annihilation gamma-rays are negligible in comparison to those resulting from the competing process of electron conversion.

2.2.4. Indirect Gamma Rays

Gamma-rays emission almost always accompanies neutron emission. Therefore, the neutron sources discussed in section 2.3 can also be considered as gamma-ray sources. High-energy gamma-rays can be produced by the decay of excited nuclear states of metastable nuclei created in an accelerator. For example, the decay of ^{16}O* leads to the emission of 6.13 and 7.12 MeV photons, creating a quasi monoenergetic gamma-ray source. The excited state, ^{16}O*, can be formed in a number of ways [26]. The 14 MeV neutrons emitted from the (d,t) reaction, see section 2.3.1.3, can be used to activate ^{16}O (say in water) via the reaction: ^{16}O(n,p)^{16}N; the latter decays by beta-particle emission, with a 7.13 s half-life, to ^{16}O*. Alternatively, the proton bombardment of ^{19}F (e.g. in calcium fluoride) at energies above 2 MeV results in the reaction: ^{19}F(p,α)^{16}O*.

2.3. Neutrons

The neutron is a neutral particle, i.e., it carries no electric charge, with a mass of 1.6745×10^{-27} kg. Being one of the constituents of the nucleus, its generation requires a process that liberates it from the forces that hold the nucleus together. This process could be fission that breaks down a heavy nucleus. This happens in a nuclear reactor, where induced fission occurs, or in some unstable nuclei that decay by spontaneously fission. Alternatively, neutrons can be produced by bombarding a nucleus with a charged-particle, that is absorbed by the nucleus, liberating a neutron in the process. For example, when an alpha-particle bombards a beryllium target, a nuclear reaction occurs which releases neutrons. The

Table 2.11. Gamma-ray reference sources [14].

Source	Half-life (Years)	Photon Energy (keV)	Emission Ratio (%)	Decay Mechanism
^{241}Am	433	14	13.3	alpha decay
		18	18.9	
		21	4.9	
		26	2.5	
		59.95	35.3	
^{133}Ba	10.53	30 - 36	≈123	K x-rays from Cs
		81	33.8	Mössbauer photons
		80	2.4	electron capture
		276	7.1	
		303	18.4	
		356	62.2	
		384	8.9	
^{137}Cs	30.1	32 - 38	8.0	K x-rays from Ba
		662	85.1	beta decay of ^{137m}Ba
^{57}Co	0.741	6-7	≈55	K x-rays from Fe
		14	9.4	electron capture
		122	85.2	
		136	11.1	
		692	16	
^{60}Co	5.27	1173	99.86	beta decay
		1333	99.98	
^{54}Mn	0.856	835	100	electron capture
^{203}Hg	0.128	71 – 85	12.8	K x-rays from Tl
		279	81.5	beta decay
^{22}Na	2.60	511	181	positron emission
		1275	99.95	electron capture
^{88}Y	0.292	14 - 16	≈60	K x-rays from Sr
		898	93.2	electron capture
		1836	99.4	
		2734	0.60	

alpha-particles can be produced by the decay of heavy isotopes, such as ^{241}Am or ^{239}Pu. Similarly, energetic charged-particles, produced in an accelerator, generate neutrons when interacting with different types of particles. In summary, neutrons can be produced in a reactor, with a spontaneous-fission radioisotope, with beryllium bombarded by alpha-particles emanating from isotopes, or by the interaction of with various types of accelerated charged-particles with a variety of nuclei. More details on these types of neutron sources are given below. It should be kept in mind that almost all the above reactions produce fast-neutrons. Neutrons of engineering interest have an upper limit energy of about 14 to 17 MeV. This is about the maximum energy of neutrons produced by fission or via the ^{3}H(d,n) reaction in a neutron generator.

Neutrons are classified according to their energy in a number of overlapping categories. Table 2.12 summarizes this classification. A neutron is considered to be fast, if its kinetic energy is in the MeV range. With 1 eV being equal to 1.602×10^{-19} J, a 1 MeV neutron has a velocity of about 13 832 km/s. An intermediate-energy neutron has an energy in the keV range; while a slow-neutron has energy in the eV to a few keV range. Thermal neutrons are those in thermal equilibrium with the medium, with an energy designated to be equal to kT, where k is Boltzmann's constant (1.3806×10^{-23} J K^{-1}) and T is the medium's temperature in degree Kelvin. This is the kinetic energy corresponding to the most probable velocity, based on the Maxwell-Boltzmann distribution, see Eq. (3.89). At room temperature, this energy is equal to 0.025 eV. Slow neutrons that have an energy above the thermal energy are called epithermal neutrons. On the other hand, neutrons with energy well below the thermal energy (say in the meV range) are known as cold neutrons. Cadmium has the ability to absorb neutrons with energy below 0.5 eV; subsequently neutrons with an energy below 0.5 eV are often called sub-cadmium neutrons; while those with higher energy are called epicadmium neutrons. The above classification is somewhat arbitrarily, and the reader should pay attention to the definitions used by different authors. Approximate penetration depths of neutrons at different energies are given in Table 2.13. Table 2.14 provide a summary of most common sources used in NDE. These sources are discussed in more detail in the following subsections.

2.3.1. Fast Neutrons

Fast-Neutron sources are readily available, either in the form of radioisotopes or generators. Radioisotopes are very small in volume, maintenance-free, and do not require a power source. However, they continuously emit neutrons and thus must be provided with permanent

Table 2.12. Classification of Neutrons.

	Energy	Notes
Fast	MeV	maximum energy of 20 MeV, for all practical purposes.
Intermediate	keV	
Slow	≤ 1 keV	
Fission	50 keV to 17 MeV	see section 3.5.4.
Resonance	1 eV – 100 eV to 1 keV	depends on nucleus with which the neutrons interact, see section 3.5.6.2.
Thermal	0.025 eV	at room temperature, see section 3.5.6.1.
Epithermal	above thermal, below resonace	slow, but not in equilibrium with medium
Subthermal	below thermal energy	
Epicadmium	above 0.5 ev	above cut-off energy of cadmium absorption cross-section.
Subcadmium	≤ 0.5eV	
Cold	< 5 meV	well below thermal energy, see section 2.3.4

Table 2.13. Penetrating depth (in mm) of neutrons in some materials[a].

Material (kg/m^3)	Energy		
	Thermal[b]	Fission[b]	14 MeV
Light Water (1000)	14	141	498
Heavy Water (1100)	111	193	470
Graphite (1600)	127	462	255
Aluminum (2700)	514	255	482
Iron (7860)	43	185	231
Lead (11250)	135	236	280

[a] Taken equal to five mean-free-paths (defined by Eq. (3.11), with data from reference [27], except for thermal-neutrons where reference [28] is used.
[b] Values averaged over the spectrum of thermal-neutrons at room temperature, and for neutrons produced by fission neutrons.

shielding. On the other hand, neutron generators are more intense, can be operated in as pulsed or continuous sources, and can be turned off by disconnecting the power source. Also, the producing reaction can be selected to generate monoenergetic neutrons. Accelerator-based sources tend, however, to be delicate, operate for a limited number of hours before requiring maintenance and/or replacement of target and can be

Table 2.14. Common neutron sources.

Source	Energy	Half-life	Specific Activity	Working Life	Neutron Yield
^{241}Am/Be	MeV range ≈ 4.5 (ave.)	433 y	110 Bq/g $(AmO)_2$	15 y [a]	6×10^4 (n/s)/GBq
^{252}Cf	Fission spect. ≈ 2.3 MeV (ave.)	2.65 y	20 GBq/mg	8 y[b]	2.3×10^9 (n/s)/mg[c]
d-^{3}H[d]	14 MeV	12.33 y	9.64×10^3 (^{3}H)	100 h 10^5 pulses	10^{11} n/s 10^8/pulse
^{124}Sb	23 keV	60.2 d		180 d[b]	2.1×10^4 (n/s)/GBq

[a] Typical working life that takes into account source construction, half-life, etc. [17]
[b] Working life is determined by half-life (assumed to be 3 half-lives).
[c] 1.15×10^8 (n/s)/GBq.
[d] Given working life and neutron yield are for a typical sealed tube.

more expensive. Nevertheless, both source types have been employed in field applications. A brief description of some of the most common isotopic neutron sources and generators is given here; concentrating on those readily usable in an NDE instrument. More details on neutron sources can be found in references [29, 30].

2.3.1.1 Alpha-particle Reaction Sources

This is a family of sources that utilize an alpha-particle emitting radioisotope mixed with a low atomic-number material to produce neutrons via the (α,n) reaction. Isotopes with a reasonably long half-life that are alpha-particle emitters with a good yield include: ^{239}Pu (24000 y), ^{226}Ra (1602 y), ^{241}Am (433 y), ^{238}Pu (87.4 y), ^{244}Cm (18.1 y), ^{227}Ac (21.6 y) and ^{228}Th (1.91 y). Shorter half-life alpha-particle emitting isotopes, such as ^{242}Cm (162 days) and ^{210}Po (138 days), are also used. Low mass-number targets include: natural boron (^{10}B and ^{11}B), fluorine (^{19}F), carbon (^{13}C), natural lithium (^{7}Li) and beryllium (^{9}Be). Other low mass-number materials, such as silicon, aluminum and oxygen, are also susceptible to the $(\alpha$, n) reaction. However, among these elements, beryllium gives by far the highest neutron yield per alpha-particle, about 70 per million [29]. Among the alpha-particle emitting isotopes, ^{241}Am is the most widely used isotope, because of the much lower specific activity of the longer-lived isotope, ^{239}Pu, and the limited availability of this and other isotopes. Therefore, ^{241}Am/Be is one of the two most common (α,n) sources, and indeed of the two most widely used isotopic neutron sources, the other is ^{252}Cf. These two sources are the preferred sources in NDE, since they are reliable, compact, relatively inexpensive and have been used for many years.

Americium-241 produces alpha-particles at an energy of 5.484 MeV, which interact with beryllium in accordance with the reaction:

$$^{9}\mathrm{Be} + ^{4}\mathrm{He} \longrightarrow ^{12}\mathrm{C} + \mathrm{n}$$

The spectrum of neutrons emitted by this source is quite broad, extending to about 10 MeV [30]. However, the source also emits a significant number of low-energy neutrons, about 23% below 1 MeV with a mean energy of 400 keV [17]. The neutron emission rate of this source per second is 6×10^4 per GBq [17], i.e. only 6×10^{-5} neutrons per second per disintegration of ^{241}Am. This low rate of production and the long half-life of the source (433 years), consequently its low specific activity, make it necessary to have a relatively large source to produce a sufficient number of neutrons. For instance a source producing about 10^6 neutrons per second requires a capsule 22.1 mm in diameter and 31 mm in height, compared to a capsule 1/25th smaller in volume (of 7.8 mm in diameter and 10 mm in height) for a ^{252}Cf source [17]. Although that size for such an ^{241}Am/Be source is manageable, the significant effect of the low specific activity of the source is in the associated gamma-ray dose rate. The excited ^{12}C nuclei produced in the reaction subsequently decay by emitting 4.43 MeV gamma-rays, with a yield between 0.50 to 0.75 photons per neutron [29]. Low energy photons, at 25.3 and 59.5 keV, are also emitted. As a result, the gamma-ray exposure rate for an unshielded source at 1 m is about 10 μGy/h per 10^6 neutrons; while the comparable value for ^{252}Cf is only 0.6 μGy/h [17]. Consequently, the gamma-ray shielding for an ^{241}Am/Be source is much more considerable than that for a ^{252}Cf source, for the same neutron yield. The Am/Be source is typically in the form of a compacted mixture of americium oxide (AmO_2) with beryllium metal, doubly encapsulated in wielded stainless steel to prevent the leakage of alpha-particles that can present a serious radiological contamination hazard. The recommended half-life for such capsules is 15 years [17]. For sake of comparison, Table 2.15 lists a number of alpha-particle sources, including ^{241}Am, that can be used along with a target like beryllium, to from a neutron source.

2.3.1.2 Californium-252

This radioisotope is produced in a high-flux reactor as a result of a series of neutron captures by ^{242}Pu or ^{243}Am (two isotopes obtained from reactor fuel reprocessing). Californium-252 decays by alpha-particle

Table 2.15. Alpha-particle reaction neutron sources.

Source	Half-life	Energy, MeV Neutrons (Average[b])	Gamma	Specific Activity[a] GBq/g	Neutron Yield (per s per g)
^{241}Am/Be	433 y	4	0.06, 4 .43[c]	127	1×10^9
^{210}Po/Be	138 d	4.3	0.8, 4.43[c]	1.7×10^5	1.28×10^9
^{238}Pu/Be	88 y	4	0.1, 4.43[c]	633	4.7×10^7
^{241}Am-^{242}Cm/Be	163 d	4	0.04, 0.06, 4.43[c]	– (80% Am, 20% Cm)	1.2×10^9
^{242}Cm/Be	163 d	4	0.04, 4.43[c]	1.1×10^5	1.46×10^{10}
^{244}Cm/Be	18 y	4	0.04, 4.43[c]	3×10^3	2.4×10^8
^{226}Ra/Be	1600 y	5	0.186, 4.43[c]	37	–
^{227}Ac/Be	21 d	5	0.015, 4.43[c]	10^6	–
^{228}Th/Be	1.9 y	6	0.084, 4.43[c]	3×10^4	–
^{241}Am/F	433 y	1.5	0.06	127	–
^{241}Am/B	433 y	3	0.06	127	–
^{239}Pu/Be	24,100 y	4	0.052	2.3	–
^{239}Pu/^{13}C	89 y	4	0.052	2.3	–

[a] Specific activity per gram of radioactive element, unless otherwise stated [11].
[b] Approximate values.
[c] Energy of de-excitation of $^{12}C^*$ resulting from the Be(α,n)C reaction.

emission (97%) and spontaneous fission (3%), yielding an average of 3.75 fast-neutrons per fission. The half-life for alpha-particle decay is 2.73 years, which although much shorter than that of an ^{241}Am/Be source, is sufficiently long for industrial uses. The half-life for the fission process (85.5 y) is also sufficiently low to allow a high production rate of neutrons, i.e., high specific activity. The effective half-life of the source is then 2.65 years. The source produces 2.3×10^9 neutron/s per mg, and has a specific activity of about 20 GBq/m [17]. Therefore the source emits 1.16×10^8 neutrons/s per GBq (compared with only 6×10^4 neutrons/s per GBq for an ^{241}Am/Be source). This high specific-activity results in a much lower gamma-ray exposure, and a smaller volume sources, than in the case of the ^{241}Am/Be source, as the discussion in section 2.3.1.1 indicated. The energy spectrum of a ^{252}Cf source, being a fission spectrum, is well-characterized and can be described by a Maxwellian distribution as [30]:

$$N(E) = \sqrt{E} \exp\left[-\frac{E}{T}\right] \tag{2.7}$$

where $N(E)$ is the number of neutrons emitted at energy E, per unit energy interval, and T is a calibration constant estimated to be between 1.3 MeV [30] and 1.424 [31], giving an average energy between 2.136 and 2.348 MeV. The difference in these values is likely due to the method of measuring the neutron spectrum and of fitting data. This analytical expression of the source's energy spectrum does not only come handy in numerical simulations, but it also has the potential of enabling analytical characterization of the behavior of the response of a particular device. In addition, the fission fragments, in the form of charged-particles, can be monitored, allowing time-tagging of neutron emission [32]. This time-tagging can be achieved by incorporating the source in a gas scintillator or by depositing it on a parallel-plate of ionization chamber, or by timing the emission rate of the prompt gamma-rays associated with the fission process with a scintillation detector placed adjacent to the source.

The decay of alpha-particles, the fission process, and the fission products, generate gamma radiation, at a rate of 1.3 $\times 10^{10}$ photons/s per mg [33]. However, the energy spectrum of the emitted gamma radiation is a broad monotonically deceasing spectrum up to about 6 MeV, with about 78% of the photons with an energy below 1 MeV. While alpha, beta and charged particles, are stopped within the source's capsule, they produce a heat output of about 39 W/g; which is small amount for most applications that usually employ a few tens of micrograms of the source. Similar to ^{241}Am/Be, ^{252}Cf is doubly-encapsulated in welded stainless to provide protection against the leakage of alpha-particles and fission fragments, typically designed to a working life of 15 years. The source is usually in the form of wire or pellet of californium-oxide, platinum/iridium alloy, or palladium alloy, depending on its activity [17].

2.3.1.3 Neutron Generators

Neutrons can be produced by bombarding a target with accelerated charged particles. By proper selection of the charged-particles and target, monoenergetic neutrons can be produced. The bombarding particles are typically protons (p), deuterons (d) or , tritons (t), since their small mass allows their acceleration to high energies in an electrostatic generator (such as a Van de Graff) or a cyclotron. High neutron-yield targets are light-weight nuclei such as deuterium (^{2}H), tritium (^{3}H) and lithium (^{7}Li), typically in a solid form. The reaction ^{3}H(d,n)^{4}He, known as the d-T reaction, provides a high yield of neutrons with low accelerated-particle energy [29]. Therefore, the d-T reaction has become the most widely-used neutron-generating mechanism in field applications.

In a d-T neutron generator, deuterons are accelerated to an energy between 100 and 400 keV. The target is usually a metal (Zr, W, Ti, etc.) in which tritium is absorbed. Neutrons are generated through the following exothermic reaction:

$$^2\mathrm{d} + ^3\mathrm{H} \longrightarrow ^4\mathrm{He} + \mathrm{n} + 17.59\ \mathrm{MeV}$$

Inexpensive and compact high-voltage power supply systems are commercially available at ratings up to 150 kV. Therefore, the deuterons are typically accelerated to 150 keV in energy (note that the charge of a deuteron is equivalent to that of a proton, and in turn equal in magnitude to that of an electron). Therefore, the resulting reaction particles, the neutron and alpha-particle, carry a total energy of 17.74 MeV (the incident deuteron energy plus the reaction energy). The emerging alpha-particle gains an energy of about 3.7 MeV, leaving the neutron with an energy of about 14 MeV. Therefore, sources employing this type of reaction are nominally known as 14 MeV neutron generators; although the neutron energy slightly varies depending on the angle of emission. With a deuteron current of 1 mA, 150 keV deuterons produce about 1.5×10^{11} neutrons/s, for a fresh tritium target [29]. However, the target often deteriorates rapidly with deuteron bombardment, resulting in lower yields in most generators. The deuteron ions are produced by ionizing a deuterium gas, which is consequently accelerated within a vacuum chamber (to avoid ionization of air and the associated sparking and waste of energy). Vacuum can be achieved in an evacuated (with the aid of a vacuum pump) acceleration tubes or in sealed (already-vacuumed) tubes. The latter are preferred for field use, as they eliminate the need for a vacuum system and are easily replaceable. However, the neutron yield for sealed tubes is much lower, for a given tube current, than that in evacuated accelerator tubes. Either systems can operate in a continuous mode of neutron emission, or in a pulsed mode. The pulse length can vary from a few nanoseconds to a few tenths of a second, at a repetition rate from 1 Hz to a few hundreds of kHz. Sealed d-T tubes typically produce an output in the order of 10^{11} neutrons/s in the continuous mode, or 10^8 neutrons/pulse. The life-time of a tube is in the order of 100 hours of continuous operation, or 10^5 pulses [29]. The sealed tube

Table 2.16. Some fast-neutron generation by accelerated charged-particles [29, 35].

Reaction	Energy, MeV Bombarding	Neutron	Comments
$^{2}H(d,n)^{3}He$	0.1 - 0.3	2 - 3	prolific reaction
$^{3}H(d,n)^{4}He$	0.1 -0.3	14	prolific reaction
$^{9}Be(d,n)^{10}B$	1	1 - 5	relatively weak
$^{3}H(p,n)^{3}He$	5	0.7 - 3	
$^{7}Li(p,n)^{7}Be$	5	1.5 - 3	widely used
$^{9}Be(p,n)^{9}B$	16	1 - 10	

size can be reasonably small, and can be used along with a compact high-voltage power supply, thus enabling portability in field use.

It should be noted that neutron generation can be produced by many other types of interactions, usually in a linear accelerator. Table 2.16 lists some of these reactions. Reference [34] demonstrated that a 5 to 6.6 MeV pulsed neutron beam incident on a thick beryllium target generates an almost uniform ("white") neutron energy-spectrum between about 0.75 to 4 MeV.

2.3.2. Intermediate-Energy Neutrons

Sources producing neutrons in the keV energy are not as readily available as those of fast-neutrons. Although, fast-neutron sources can be slowed-down by a suitable moderator to soften (reduce) the source energy, a large number of the source neutrons are wasted by divergence (spread) and leakage from, and absorption by, the moderating material, see section 15.3. Filters also can be placed in front of a source to allow neutrons to be transmitted over a limited range of energy, see section 15.2. For example, one can use filters composed of ^{238}U, ^{45}Sc, Fe, S and Si to obtain intensified neutrons at 0.186, 2.0, 24.3, 55 and 144 keV, respectively [29], see section 15.2.2.1. However, filtering is employed for neutron beams extracted from a reactor, as it requires a strong neutron source to produce a reasonably intense radiation at the desired energy. A third option to produce intermediate energy neutrons is to employ a photoneutron source.

Photoneutron sources take advantage of the fact that high-energy photons can liberate neutrons in nuclei with a low neutron binding-energy, such as deuterium or beryllium. A photon of energy of at least 2.226 MeV is required to liberate a neutron from deuterium, while a 1.666 MeV photon is needed for producing neutrons from beryllium. These photon energies are within the range of isotopic gamma-ray sources. For example, ^{24}Na emits 2.7541 MeV photons, which can produce neutrons in deuterium and beryllium, at 263 and 967 keV, respectively [29]. The production of these monoenergetic neutrons, from monoenergetic photons, are dictated by the kinematics of the reaction. This is a unique feature of photoneutron sources. However, the short half-life of ^{24}Na, 15 hours, limits its use to the proximity of an isotope-producing facility. An alternative source, ^{124}Sb, with a 60.2 day half-life, emits photons at 1.691 MeV, making it useful for use with a beryllium target to produce 23 keV neutrons. This antimony-beryllium source is, therefore, the most widely used photoneutron source in industrial applications. The neutron yield for this source is 2.1×10^4 neutrons/s per GBq. A strong gamma-ray field accompanies neutron production. This is not only problematic in terms of radiation shielding, but gamma-rays can also interfere with neutron counting, since most neutron detectors are also sensitive to gamma-rays. A photoneutron source is constructed by placing a suitable gamma-ray source typically within a spherical target of heavy water, deuterated polyethylene (CD_2) or beryllium metal. Although with a monoenergetic gamma-ray source, monoenergetic neutrons are produced, some of these neutrons will be slowed-down by the low mass-number target nuclei, resulting in a low-energy tail in the spectrum of the emitted neutrons. The sources can be activated and deactivated by designing the source so that the photon emitter can be inserted in, and removed from, the target assembly. Some photoneutron sources employ isotopes with longer half-lives, as shown in Table 2.17. Low-energy neutrons can also be produced in an accelerator by bombarding a heavy target with protons, as shown in Table 2.18.

2.3.3. Slow Neutrons

Slow-neutrons, particularly thermal neutrons, are quite useful in many applications because of their high interaction probability with many elements. However, they are the most difficult neutrons to produce for use in industrial applications, since there is no readily available sources that produce directly neutrons in the eV range or below. Therefore, one has to resort to moderating neutrons emitted from other readily available sources, such as those listed in Table 2.14. This is obviously not a very efficient process, since most of the source neutrons are wasted in

Table 2.17. Photoneutron sources [29].

Source	Half-life	Energy, MeV Neutron	Energy, MeV Gamma	Neutron Yield (per GBq)
^{24}Na/Be	15 h	0.967	2.754	3.4×10^4
^{24}Na/D	15 h	0.263	2.754	3.3×10^4
^{28}Al/Be	2.24 min	0.101	1.779	3.3×10^3
^{38}Cl/Be	37.3 min	0.446	2.168	4.3×10^3
^{56}Mn/Be	2.58 h	0.129, 0.398, 1.149	1.811, 2.113, 2.960	9.2×10^3
^{56}Mn/D	2.58 h	0.365	2.9598	16
^{72}Ga/Be	14.1 h	0.174, 0.476, 0.748	1.861, 2.202, 2.508	6.5×10^3
^{72}Ga/D	14.1 h	0.140	2.508	2.5×10^3
^{76}As/Be	26.3 h	0.109, 0.383	1.788, 2.096	305
^{88}Y/Be	107 d	0.152, 0.949	1.836, 2.734	2.3×10^4
^{88}Y/D	107 d	0.253	2.734	16
^{116m}In/Be	54.1 min	0.397	2.112	1.6×10^3
^{124}Sb/Be	60.2 d	0.023	1.691	2.2×10^4
^{140}La/Be	40.3 h	0.760	2.522	10^3
^{140}La/D	40.3 h	0.147	2.522	660
^{144}Pr/Be	17.3 min	0.462	2.186	69

the moderation process. During moderation, neutrons are spread over a large area, some are not fully slowed-down, and some of the fully slowed-down neutrons are absorbed within the moderating material before being extracted for use. Nevertheless, the high interaction probability of slow-neutrons tends to compensate for neutron loss during the moderation process. Therefore, considerable effort has been devoted to developing efficient moderating assembles, see section 15.3.

Table 2.18. Some reactions for slow-neutron generation by accelerated charged-particles [29].

Reaction	Energy, MeV	
	Bombarding	Neutron
$^{45}Sc(p,n)^{45}Ti$	2.9	0.056 - 0.052
$^{51}V(p,n)^{51}Cr$	1.6	0.002 - 0.786
$^{57}Fe(p,n)^{57}Co$	1.65	0.002 - 1.43

In order to enhance the use of slow-neutrons, some moderating assemblies are devised so that thermal-neutrons are produced within the assembly itself, and the interrogated material has to be inserted, or pass through, the moderating assembly. Other moderating assemblies allow the extraction of neutron beams, and are often called neutron howitzers. Some designs allow both bulk neutron production within the moderating assembly and neutron-beam extraction. Refer to section 15.3 for the design of such assemblies.

2.3.4. Cold Neutrons

Cold neutrons, as the name indicates, are produced by slowing-down neutrons to energies below that of the thermal-energy corresponding to room temperature (0.025 eV). At these low energies, neutrons possess the characteristics of waves, see section 3.5.5.1. Cooled neutron moderators, such as water, polyethylene, methane, hydrogen or beryllium, can be used to further slow the neutrons down. This cooling process can take place inside a reactor. It is also possible to produce cold-neutrons from accelerator-based sources [36]. However, in order to accommodate the high intensity required in most applications, a number of specialized highly intense neutron sources are emerging, see reference [37].

The cooled neutrons are typically extracted using a collimator. The distribution of neutrons inside a reactor is governed by the Maxwell-Boltzmann distribution of Eq. (3.89). However, the distribution of neutrons extracted from a reactor via a long collimator can be described by the following distribution [38]:

$$\frac{n(\lambda)d\lambda}{n} = \left(\frac{h^2}{2mkT}\right)^2 \frac{2}{\lambda^5} \exp\left[-\frac{h^2}{2mkT\lambda^2}\right] \tag{2.8}$$

where $\frac{n(\lambda)d\lambda}{n}$ is the fraction of neutrons emerging from a collimator with a wavelength between λ and $\lambda + d\lambda$, k is Boltzmann's constant, T is the temperature of the moderating medium, and m is the neutron mass. Note that λ is related to the neutrons momentum, mv, according to the duality principle, Eq. (3.84), as $\lambda = \frac{h}{mv}$. The neutron energy, $E = mv^2$, is equal to $E = \frac{h^2}{2m\lambda^2}$. Using the distribution of Eq. (2.8), it can be shown that the maximum intensity occurs at a wavelength equal to $\frac{h}{\sqrt{5mkT}}$, corresponding to a neutron energy of $\frac{5}{2}kT$. Using the above relationship, the reader can prove that $\frac{dE}{E} = 2\frac{dv}{v} = 2\frac{d\lambda}{\lambda}$, and drive the distribution of neutrons extracted from the collimator as function of velocity, $\frac{n(v)dv}{n}$, or energy, $\frac{n(E)dE}{n}$, keeping in mind that $n(\lambda)d\lambda = n(v)dv = n(E)dE$. It can be then shown, for instance, that the velocity corresponding to the peak of the distribution, $n(v)dv$ is given by $v^2 = \frac{3kT}{m}$, i.e. at an energy of $\frac{3}{2}kT$. Note that the corresponding energy for a Maxwell-Boltzmann distribution is equal to kT.

A monochromatic beam of neutrons can be obtained by Bragg reflection in a crystal, such as calcium fluoride. Collimated neutrons reflected at a particular angle θ by a crystal of interplanar (lattice) spacing d will emerge at a wavelength centered around a value λ that satisfies Bragg's law, Eq. (3.66), i.e. such that $\lambda = 2d \sin\theta$ [38].

2.4. Natural Sources

Naturally radioactive materials are readily available sources of radiation. Natural activity is the result of either the decay of unstable elements created upon the formation of the Earth (about 4.5 billion years ago), or nuclear interactions by cosmic radiation[5] in the upper atmosphere. Those created upon the formation of the Earth have very long half-lives, in the order of 10^9 years, while terrestrial isotopes can have shorter half-lives as they are continuously produced. A list of naturally occurring radioisotopes is given in Table 2.19. While some of these isotopes decay to elements that are stable, such as ^{40}K which decays to the stable isotope ^{40}Ar, others heavier isotopes have complex decay schemes, which in turn in produce another set of naturally occurring radioisotopes. The decay schemes of Earth-originated radioisotopes are as follows [39]:

[5]Cosmic radiation consists of highly energetic (10^{15} to 10^{20} eV) electrons, positrons, protons, mesons, muons and various heavy nuclei. Muons easily penetrate most materials producing neutrons and activation products in their wake.

Thorium Series.

$$^{232}\mathrm{Th} \xrightarrow{\alpha} {}^{228}\mathrm{Ra} \xrightarrow{\beta-} {}^{228}\mathrm{Ac} \xrightarrow{\beta-} {}^{228}\mathrm{Th} \xrightarrow{\alpha}$$
$$^{224}\mathrm{Ra} \xrightarrow{\alpha} {}^{220}\mathrm{Rn} \xrightarrow{\alpha} {}^{216}\mathrm{Po} \xrightarrow{\alpha} {}^{212}\mathrm{Pb} \xrightarrow{\beta-}$$
$$^{212}\mathrm{Bi} \longrightarrow \left\{ \begin{array}{l} (36\%) \xrightarrow{\alpha} {}^{208}\mathrm{Tl} \xrightarrow{\beta-} \\ (64\%) \xrightarrow{\beta-} {}^{212}\mathrm{Po} \xrightarrow{\alpha} \end{array} \right\} \longrightarrow {}^{208}\mathrm{Pb}\ \text{(stable)}.$$

Actinium ^{235}U Series.

$$^{235}\mathrm{U} \xrightarrow{\alpha} {}^{231}\mathrm{Th} \xrightarrow{\beta-} {}^{231}\mathrm{Pa} \xrightarrow{\alpha} {}^{227}\mathrm{Ac} \xrightarrow{\beta-}$$
$$^{227}\mathrm{Th} \xrightarrow{\alpha} {}^{223}\mathrm{Ra} \xrightarrow{\alpha} {}^{219}\mathrm{Rn} \xrightarrow{\alpha} {}^{215}\mathrm{Po} \xrightarrow{\alpha}$$
$$^{211}\mathrm{Pb} \xrightarrow{\beta-} {}^{211}\mathrm{Bi} \xrightarrow{\alpha} {}^{207}\mathrm{Tl} \xrightarrow{\beta} {}^{207}\mathrm{Pb}\ \text{(stable)}.$$

Uranium Series.

$$^{238}\mathrm{U} \xrightarrow{\alpha} {}^{234}\mathrm{Th} \xrightarrow{\beta-} {}^{234}\mathrm{Pa} \xrightarrow{\beta-} {}^{234}\mathrm{U} \xrightarrow{\alpha} {}^{230}\mathrm{Th} \xrightarrow{\alpha}$$
$$^{226}\mathrm{Ra} \xrightarrow{\alpha} {}^{222}\mathrm{Rn} \xrightarrow{\alpha} {}^{218}\mathrm{Po} \xrightarrow{\alpha} {}^{214}\mathrm{Pb} \xrightarrow{\beta-} {}^{214}\mathrm{Bi} \xrightarrow{\beta-}$$
$$^{214}\mathrm{Po} \xrightarrow{\alpha} {}^{210}\mathrm{Pb} \xrightarrow{\beta-} {}^{210}\mathrm{Bi} \xrightarrow{\beta-} {}^{210}\mathrm{Po} \xrightarrow{\alpha} {}^{206}\mathrm{Pb}\ \text{(stable)}.$$

Each of the above series will reach an equilibrium state at which the rate of production of the decay products becomes equal to their rate of decay. Equilibrium conditions are reached after the passing of a period of time equal to about six (6) half-lives of the longest-lived daughter of the parent element, provided that the parent nucleus and its daughter are left undisturbed during that period, see section 3.7. For ^{232}Th, the longest lived daughter is ^{228}Ra, with a half-life of 5.76 years [13], leading to an equilibrium period of about 35 years. The longest-lived daughter of ^{235}U is ^{231}Pa, with a half-life of 3.28×10^4 years, resulting it a very long equilibrium period of about 200 thousand years. The time to reach equilibrium in the ^{238}U decay series is even longer, about 1.5 million years, since its longest-lived daughter, ^{234}U, has a half-life of 2.46×10^5. These long equilibrium times makes it possible to assume that when found in nature, the daughters of ^{232}Th, ^{235}U and ^{238}U have reached equilibrium conditions, unless one of the decay products have been previously tampered with (e.g. extracted).

Natural activity is encountered in the Earth crust, which contains about 2.35% K, 1.2×10^{-3}% Th and 3×10^{-4}% U [40]. Granite rocks and basaltic rocks, clay, shales and certain beach sands can have measurable levels of radioactivity. Although the concentration of thorium and uranium is low in basal, it is a dominant rock; the ocean's floor consists almost entirely of basalt, covered and interlayered with thin pelagic

Table 2.19. Some sources of natural radioactivity [13].

Isotope	Abundance in Element (Atom %)	Radiation Emitted (Prominent, or Maximum, Energies, MeV)	Half-life (year)
^{3}H	10^{-18}	$\beta-$ (0.0186)	12.32
^{14}C	10^{-10}	$\beta-$ (0.157)	5.730×10^{3}
^{40}K	0.0117	$\beta-$ (1.33) γ (1.461) $\beta+^{a}$	1.28×10^{9}
^{232}Th	100	α (4.013, 3.947), γ (0.059^{a}, ...)	14×10^{9}
^{235}U	0.720	α (4.395, 4.365), γ (0.186, 0.144, ...), neutrons (spontaneous-fission^{a})	7.04×10^{8}
^{238}U	99.2745	α (4.197, 4.147), γ (0.0496^{a}, ...), neutrons (spontaneous-fission^{a})	4.47×10^{9}

[a] Weak emissions.

sediments, and large areas of the Continent are covered by basaltic material [41]. Potassium-40 is also found in food stuff, such as Brazil nuts (210 Bq/kg), banana (130 Bq/kg), raw lima-bean (170 Bq/kg,) carrot and white potatoes (125 Bq/kg), and red meat (110 Bq/kg) [42, 43]. Many minerals contain ^{40}K, but its main mineral sources are the alkali potassic feldspars (found in sandstones) and micas (such as muscovite). Under favorable temperature and humidity conditions, potassium is liberated from muscovite, in the form of positive ions that find their way to water streams and rivers. These potassium ions are then absorbed in clay or by plants. Being one of the basic minerals in fertilizers (in the compound potash), potassium is a constituent of all plant and animal tissue. Uranium mineral sources are igneous rocks of acid origin, which includes granite and basalt [40]. However, since uranium is soluble, it can also find its way to spring water and can be absorbed by plants (appearing in peat, wood and cellulose). Uranium can also be absorbed by some coal deposits leached out from uranium-bearing rocks. Igneous rock also bear thorium. However, unlike uranium, thorium is an insoluble rock, and is transported by water mainly in suspension. Nevertheless, a small amount of thorium is absorbed by clay minerals. The sand beaches of Brazil, India and Egypt are known to contain thorium minerals. Uranium and thorium tend also to accumulate in the thyroid glands of animals, the leaves of plants and the bones of fish.

Cosmogenic Nuclides. Cosmic-rays continually produce isotopes, known as cosmogenic nuclides. The isotopes are produced in the Earth's atmosphere or on its surface, but can also be produced in meteorites and other extraterrestrial materials. Cosmogenic isotopes include: ^{3}H (12.32 year), ^{7}Be (53.28 day), ^{10}Be (1.5 $\times 10^{6}$ years), ^{14}C (5715 year), ^{24}Na (14.95 hr), ^{26}Al(7.1 $\times 10^{5}$ year), ^{36}Cl (3.01 $\times 10^{5}$ year), ^{38}Cl (37.24 min), ^{39}Cl (55.6 min), ^{41}Ca (1.03 $\times 10^{5}$), ^{129}I (1.57 $\times 10^{7}$ year), where the values in brackets are their half-lives. The radiation emitted from these nuclides enables the studying of the material in which these isotopes are affiliated. The low concentration of long-lived isotopes can be measured with accelerator mass spectrometry, see reference [44].

Chapter 3

MODIFYING PHYSICS

3.1. General

Radiation can interact with atomic electrons and nuclei, or their associated electromagnetic fields. It is, therefore, necessary to understand some of the basic characteristics of the atom and its constituents. The interaction of radiation with matter can be classified into two broad categories: the first is for charged-particles (such as alpha and beta particles), and the second deals with uncharged (neutral) radiation (neutrons and photons). When charged-particles enter matter they are continuously affected by the electromagnetic field created by the charge and motion of the atomic electrons and the nucleus. On the other hand, neutral particles interact with the nucleus only when they reach its proximity. That is, a neutral particle can travel a long distance without encountering any interaction at all, since a very small portion of the space within matter is occupied by nuclei.

The radius, R, of a nucleus can be approximately estimated as:

$$R = R_0 A^{\frac{1}{3}} \tag{3.1}$$

where A is the mass-number of the nucleus and R_0 is about 1.25×10^{-15} m. This leads to a nuclear density of 2.03×10^{17}kg/m^3, assuming a spherical nucleus with a uniform density and using a nucleon (proton or neutrons) mass of 1.6605×10^{-27}kg. The latter mass is known as the atomic mass unit, u, and is equal to 1/12th of the mass of a ^{12}C atom. Note that when u is expressed in grams, $\frac{1}{\mathrm{u}}$ $(= 0.602 \times 10^{-24})$ becomes numerically equal to Avogadro's number[1], N_A.

[1] Avogadro's number is the number of atoms in 12 g of ^{12}C, or the number of atoms per mole.

The number of atoms (nuclei) per unit volume is given by:

$$N = \frac{\text{mass of material per unit volume (density)}}{\text{mass of a single atom}} = \frac{\rho}{A\text{u}} \tag{3.2}$$

where ρ is the material density (mass per unit volume) and Au is the mass of an atom of mass-number A. Eq. (3.2) is often expressed as $N = \frac{\rho N_A}{A}$, with ρ in g/cm^3.

Let us consider the case of water at the normal density of 1000 kg/m^3. There are

$$N = \frac{1000}{18\text{u}} = 3.346 \times 10^{28} \text{ H}_2\text{O molecules in one m}^3$$

where 18 is the molecular weight of water. The volume occupied by the nuclei of one H_2O molecule, V_{nucl}, is estimated, with the aid of Eq. (3.1) assuming a spherical nucleus, as:

$$V_{nucl} = \frac{4\pi R_0^3}{3}(2 + 16) = 1.473 \times 10^{-43} \text{ m}^3$$

The total condensed volume occupied by H_2O nuclei in one m^3 of water is:

$$NV_{nucl} = 3.346 \times 10^{28} \times 1.473 \times 10^{-43} \approx 5 \times 10^{-15}\text{m}^3$$

Therefore, in one m^3, nuclei occupy only 5×10^{-15} of the total space. The rest of the space is simply vacant, aside from the small volume occupied by the atomic electrons, As such, a neutral particle can travel a long distance in that vacant space before encountering one of the nuclei constituting water.

The mean radius of a water molecule, or in other words, the mean separation distance between the O and H in the H-O-H bond (Van der Waal's radius) is 1.41×10^{-10} m. This leads to a molecular volume of 1.174×10^{-29} m^3, assuming a spherical molecule. The total condensed volume occupied by the H_2O atoms in 1 m^3 of water is, therefore, 1.174×10^{-29} m^3 $\times 3.346 \times 10^{28}$ or about 0.393 m^3. This condensed volume is much lower than the nominal 1 m^3 volume which the molecules occupy in total. This is a testimony to the low packing efficiency of the water molecules. Note that the maximum number of molecules that can be packed in a 1 m^3 based on the above molecular size is $\frac{1}{1.174} \times 10^{-29}$, or 8.518×10^{28} molecules, i.e. 2.544 the number of molecules in 1 m^3 at a density of 1000 kg/m^3.

Another interesting observation is to consider the total condensed mass occupied by the water nuclei. The condensed mass of the nuclei of an H_2O molecule, M_{nucl}, can be estimated as:

$$M_{nucl} = V_{nucl} \times 2.03 \times 10^{17} = 2.989 \times 10^{-26} \text{ kg}$$

where 2.03×10^{17} is taken as the condensed nuclear density in kg/m^3. Therefore, the total condensed mass of the H_2O nuclei in one m^3 of water is:

$$NM_{nucl} = 3.346 \times 10^{28} \times 2.989 \times 10^{-26} = 1000.1 \text{ kg}$$

The total mass of the electrons is $10m_0N$, where m_0 is the electron's mass (9.109×10^{-31} kg). This amounts to a total of about 0.037 kg of electron mass in one m^3 of water, a negligible amount. Therefore, the total mass of an H_2O molecule estimated by this method is very close to the expected value of 1000 kg in one m^3 of water.

Another way of calculating the mass of an H_2O molecule is to consider the mass of individual nucleons and electrons. This leads to:

$$\begin{array}{ll} & = [10m_p + 10m_0 + 8m_n] \\ \text{Total mass of nucleons} & = [10 \times (1.672623 \times 10^{-27})+ \\ \text{in an } H_2O \text{ molecule} & \quad 10 \times (9.109389 \times 10^{-31})+ \\ & \quad 8 \times 1.674928 \times 10^{-27}] \\ & = 3.013476 \times 10^{-26} \text{ kg} \end{array}$$

where m_p, m_0 and m_n refer, respectively, to the mass of a proton, an electron and a neutron, and their values are taken from reference [13]. This leads to a total mass of 1008.2 kg in one m^3, slightly higher than the assumed 1000 kg.

One other approach is to consider the mass of the H and O atoms, instead of the mass of individual nuclei. Then, the mass of an H_2O molecule is given as:

$$\begin{array}{ll} \text{Total mass of atoms} & = 2m_H + m_O \\ \text{in an } H_2O \text{ molecule} & = [2 \times 1.00794 + 15.9994] \times 1.660540 \times 10^{-27} \\ & = 2.991509 \times 10^{-26} \text{ kg} \end{array}$$

where m_H and m_O refer, respectively, to the mass of the hydrogen and oxygen atoms, with their values calculated using their atomic-number multiplied by u. The above molecular mass leads to a total mass of 1000.85 kg. Taking a molecular mass of 18.021528, rather than the nominal value of 18 used in calculating the value of N above, introduces a correction factor of $\frac{18}{18.021528} = 0.99915$. This correction brings the 1000.85 kg value to 1000 kg, i.e. to the expected mass of one m^3 of water.

Correcting similarly the above obtained mass values, one can summarize the results as follows:

$$\text{Mass of an } H_2O \text{ molecule} = \begin{cases} 2.986 \times 10^{-26}\text{kg} & \text{using nuclear size and density} \\ 3.011 \times 10^{-26}\text{kg} & \text{using mass of nucleons} \\ 2.989 \times 10^{-26}\text{kg} & \text{using atomic masses} \end{cases}$$

Although the three values are very close to each other, the middle value is slightly higher. In fact, the mass of any atom (nucleus) is always lower than the sum of its individual constituents of nucleons and electrons. The difference is known as the mass excess. This mass excess, when converted to energy, with the atom at rest, using the relationship $E = mc^2$, see appendix C, is equivalent to the binding energy of the atom. In other words, the excess mass is in effect consumed into energy that keeps the atom (mainly the nucleus) together. This binding energy is released when the nucleus is broken down into smaller species, as in nuclear fission. Also this excess energy determines whether a nuclear reaction is exoergic (releases energy) or endoergic, requires external energy.

3.2. Cross Sections

Classical mechanics deals with the kinetics of particle collision, i.e. conservation of momentum and energy, but does not address the question of the probability of occurrence of the collision in the first place. The interaction probability of elementary particles with an atom, or its nucleus, is addressed by quantum mechanics, see appendix D. The probability of interaction of radiation with an atom (or a nucleus) is expressed by a quantity called the microscopic cross-section, σ, while the probability of interaction, per unit volume, with *all* atoms in a material, is called the macroscopic cross-section, denoted by Σ. The probability of scattering into a particular solid-angle is defined by the differential cross-section. These concepts are discussed below.

3.2.1. Microscopic Cross-Section

The microscopic cross-section is expressed in a unit called "barn" (b), which is a unit of area, with one barn equals to 10^{-28} m^2 or 10^{-24} cm^2. The microscopic cross-section measures, in effect, the probability of radiation interacting with an atom or a nucleus, as radiation may pass the atom/nucleus without interacting with it, see appendix D. The terminology cross-section signifies that the atom/nucleus projects itself to the incident radiation as an area; the larger the area the more likely that radiation will be affected by the atom/nucleus and interact with it. The microscopic cross-section depends not only on the type of atom/nucleus, but also on the type and energy of the interacting radiation. Each type of interaction is assigned its own cross-section, with the total cross-section, σ_t, being the summation of all these interaction cross-sections and presents the probability of interaction in one form or another. Note, however, that no matter how large the cross-section is, there is a finite probability that radiation may not interact at all with the nucleus; since σ_t has always a finite value.

3.2.2. Differential Cross-Section

For the scattering interaction, the scattered radiation will emerge in various directions. The probability of scattering towards a certain direction, $\vec{\Omega}$, within a solid angle $d\Omega$, around $\vec{\Omega}$, is defined by the differential cross-section, also called the angular cross-section, $\frac{d\sigma}{d\Omega}$, so that:

$$\sigma_s = \int_0^{4\pi} \frac{d\sigma}{d\Omega} d\Omega \tag{3.3}$$

where σ_s is the scattering cross-section at all angles. Note that for isotropic scattering, $\frac{d\sigma}{d\Omega} = \frac{\sigma_s}{4\pi}$. Defining the direction $\vec{\Omega}$ by the polar angle, θ and the azimuthal angle ϕ, as schematically shown in Figure 3.1, one can express $d\Omega$ as:

$$d\Omega = \sin\theta d\theta \; d\phi = d\cos\theta \; d\phi \tag{3.4}$$

Therefore, the differential cross-section can also be expressed as:

$$\frac{d\sigma}{d\Omega} = \frac{d\sigma}{d\cos\theta d\phi} \tag{3.5}$$

For isotropic scattering, $\frac{d\sigma}{d\cos\theta d\phi} = \frac{1}{2}\frac{1}{2\pi}$, noting that $\cos\theta$ varies in value from $-1 \longrightarrow 1$, while ϕ covers the range from $0 \longrightarrow 2\pi$.

3.2.3. Macroscopic Cross-Section

The microscopic cross-section deals with the interaction probability of radiation with individual atoms/nuclei. On the other hand, the macroscopic cross-section relates to the probability of interaction with the aggregate of atoms/nuclei that compose a medium. Since the microscopic cross-section, σ, defines the interaction probability per atom/nuclei, the macroscopic cross-section, Σ, is related to σ by the number of atoms/nuclei per unit volume, N, given by Eq. (3.2), so that:

$$\Sigma = N\sigma \tag{3.6}$$

The macroscopic cross-section has, therefore, units of reciprocal length (usually cm^{-1}), since σ has units of area. The units of Σ can also be

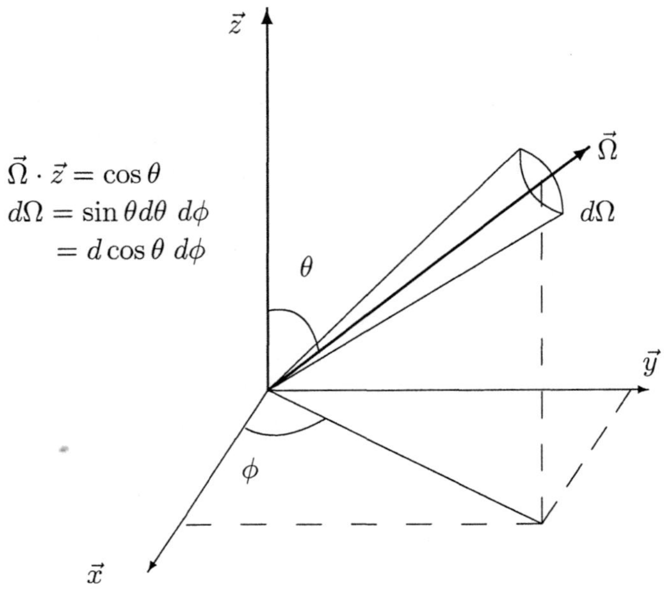

Figure 3.1. A schematic of a change in direction from the z-direction, $\vec{z}$, to the direction $\vec{\Omega}$ within a solid angle, $d\Omega$ around $\vec{\Omega}$; with the latter defined by the polar angle, θ, and the azimuthal angle, ϕ, $\vec{x}$, $\vec{y}$, $\vec{z}$ and $\vec{\Omega}$ are unit vectors.

expressed as barns per unit volume, i.e. the macroscopic cross-section per unit volume. This concept facilitates the calculation of the macroscopic cross-section of mixtures, by weighting the macroscopic cross-section of each component in the mixture by its volume fraction. Appendix E shows how the macroscopic cross-section of a chemical compound, or a material consisting of a mixture of nuclides of different types, can be calculated from the microscopic cross-sections of its constituents.

In order to provide a physical meaning for the macroscopic cross-section, consider a radiation beam incident on a material with a total macroscopic cross-section Σ_t. The incident radiation will interact with the medium's material, in all possible mechanisms, since the total cross-section is considered. The intensity of the radiation beam after traveling from a distance x to $x + dx$ within the medium will decrease from an intensity I at distance x, to $I - dI$ at $x + dx$. The amount of beam attenuation (decrease in intensity) will depend naturally on the how large dx is, as well as on the original value of I. That is:

$$-dI \propto Idx, \text{ or, } \quad -dI = \Sigma_t Idx \tag{3.7}$$

The negative sign signifies a decrease in the value of I, and Σ_t is used as a proportionality constant as it accounts for all the other physical factors (radiation energy, the nature of the medium's nuclei and their number per unit volume). Integrating Eq. (3.7) results in:

$$\int \frac{dI}{I} = -\int \Sigma_t dx, \text{ or, } \frac{I}{I_0} = \exp(-\Sigma_t x) \tag{3.8}$$

where I_0 is the intensity of the radiation beam at $x = 0$. Based on Eq. (3.8), one can state that the probability $p(x)$ that a radiation beam will to travel a distance x within a medium without interacting, then interact at a distance between x and $x + dx$ is equal to $\Sigma \exp(-\Sigma_t x)$; with Σ_t introduced in front of the exponential term for normalization purposes as the integration of Eq. (3.10) below shows. That is,

$$p(x)dx = \Sigma_t \exp(-\Sigma_t x)dx \tag{3.9}$$

This relates the concept of cross-section to the conventional concept of probability. After traveling an infinite distance, one would expect that the radiation beam to have been completed attenuated, that is:

$$\int_0^\infty p(x)dx = \int_0^\infty \Sigma_t \exp(-\Sigma_t x)dx = 1 \tag{3.10}$$

which can be easily mathematically proven.

Eq. (3.9) is also useful for introducing the concept of "mean-free-path". The average distance a radiation beam will travel between interactions is known as the mean-free-path, λ, and can be obtained from Eq. (3.9), (dropping the subscript as the concept applies to all types of interactions), as:

$$\lambda = \int_0^\infty xp(x)dx = \int_0^\infty x\Sigma \exp(-\Sigma x)dx = \frac{1}{\Sigma_t} \tag{3.11}$$

That is, the mean-free-path is equal to the reciprocal of the macroscopic cross-section. Therefore, the mean distance a radiation beam travels before encountering any type of interaction with the medium is equal to $\frac{1}{\Sigma_t}$. In other words, within a distance d, the radiation beam will on average encounter $\Sigma_t d$ interactions of any type. Therefore, the macroscopic cross-section can also be seen as the average number of interactions per unit length. The distance $\frac{1}{\Sigma_t}$ is referred to as the *relaxation distance*, because of the exponential nature of beam attenuation, since according to Eq. (3.8) $\frac{1}{\Sigma_t}$ is the distance required to attenuate the beam to e^{-1} (= 0.3679) of its initial value.

When dealing with photons, the macroscopic cross-section is often referred to as the attenuation coefficient, since its value determines the degree of attenuation a radiation beam is subjected to after traveling a certain distance, as indicated by Eq. (3.8). For photons, the mass attenuation coefficients, or the attenuation coefficient per unit mass, $\frac{\Sigma}{\rho}$, is typically reported, as it provides a density-independent value.

3.3. Charged Particles

Charged particles are generally classified into "light" particles (electrons, positrons) and "heavy" particles (mesons, protons, α particles, etc.). Particles employed in NDE are typically those with velocities that are much larger than the velocity of thermal agitation, called "swift" or fast particles. These particles can interact with atomic electrons of the atom, with the nucleus, or with other free charged-particles, producing detectable NDE indications.

Inelastic collision with atomic electrons is the predominant mechanism by which a fast charged-particle loses energy as it travels through matter. The energy given to the atomic electron can cause it to experience a transition to an excited state (excitation) or to an unbound state (causing ionization). The kinetic energy of a liberated electron in the ionization process can be large, since its mass is small. Hence, the ejected electron may itself be a swift charged-particle, (called a delta-ray), usually of energy in the order of 1 keV or more. These ejected electrons may in turn produce more electrons, and so on. If a charged-particle approaches a nucleus, it invariably will be deflected; resulting in some cases in the emission of a quantum of radiation, called bremsstrahlung radiation. The photons of the bremsstrahlung affect also produce secondary electrons, resulting in a cascade shower of electrons, as explained in section 3.6.5. Therefore, a single incident charged-particle can produce a cascade of electrons, as it is being brought to rest.

Two free charged-particles, such as those resulting from the electron shower, can also interact with each other. Then, the only force between them is the coulomb force between their electrostatic charges. Elastic collisions between such charged-particles is not necessarily a "head-on" collision, but occurs at some distance between the two particles. As two particles approach one another, the coulomb forces gradually begin to take effect and the velocities of the two particles change in direction and magnitude, according to the laws of conservation of energy and momentum; taking into account the effect of the potential energy of the coulomb force.

There are computer codes that can simulate in detail, using the Monte Carlo method, the complex kinematics and transport of charged-particles; such as the EGS4 code [45] and the TIGER series of codes [46]. However, for most applications, the concepts of stopping-power and particle's range are utilized to describe, respectively, the amount of energy lost per unit length and the total distance a charged-particle can travel within a material before it delivers all its energy, and is subsequently absorbed. The stopping power, S, is defined as:

$$S = -\frac{dE}{ds} \tag{3.12}$$

where ds refers to an incremental distance and dE is the energy lost during that interval. The range, R, of a particle with an initial energy, E_0 is given by:

$$R(E_0) = \int_0^{R(E_0)} ds = \int_{E_0}^{0} \frac{ds}{dE} dE = \int_0^{E_0} \frac{1}{\frac{-dE}{ds}} dE = \int_0^{E_0} \frac{dE}{S} \tag{3.13}$$

The range, $R(E_0)$, in a material is, therefore, the thickness of the material required to absorb an incident charged-particle of energy E_0. A computer code, called SRIM, is available to calculate the stopping power and range of positively charged-particles (ions) in solids, liquids and gases [47], and mathematical expressions giving approximate estimates of these quantities are given in the following sections for various charged-particles. It should be noted, however, that since charged-particles do not travel in straight lines, the range is an "average" distance a particle will travel before losing all its energy. Charged-particles do not move in straight lines; in fact they have a curved path due to the influence of the electrostatic field of the atom. Therefore, some charged-particles may survive for a distance larger than that corresponding to the range value, and some other may be absorbed before reaching a distance equal to the particle's range. This phenomenon is known as straggling.

3.3.1. Alpha Particles

When a heavy charged-particle, such as an alpha-particle, travels through a medium it loses energy through one or more of the scattering mechanisms discussed below. As the energy of a particle is reduced, this positively charged-particle starts to pick up electrons from the surrounding, until it is eventually neutralized and in effect stopped. The range of alpha-particles, R_α, is approximately given by:

$$R_\alpha(\text{mm}) = \frac{E_0^{1.5}(\text{MeV})}{\rho(\text{kg/m}^3)}\sqrt{A} \tag{3.14}$$

where A is the mass-number of the material, ρ is its density and E_0 is the energy of the alpha-particle as it enters the material. Therefore, the range of a 4 MeV alpha-particle in air (at standard temperature and pressure, $\rho = 1.2$ kg/m^3, $A \approx 14$) is about 25 mm, while in aluminum ($A = 27$, $\rho = 2700$ kg/m^3) it is about 0.015 mm. This shows that, within the energy range of alpha-particles emitted from radionuclides, alpha-particles will not deeply penetrate the surface of most materials. The information deduced from the interaction of alpha-particles is mainly superficial, barely underneath the surface of the material within the range of the alpha-particle.

For an alpha-particle to penetrate a medium, its energy must be larger than the thermal-equilibrium energy of the medium, otherwise the particle will simply absorb an electron and become neutral. Particles with energy much larger than that of the thermal energy, 0.025 eV at room temperature, are called "swift" particles. Alpha particles of significance in NDE must have energy much higher than the thermal energy, so that they can penetrate a medium to some depth. Therefore, the discussion on the modification of alpha-particles is limited here to swift particles. Such energetic particles are mainly subjected to the coulomb forces of the matter, which are caused by either the atomic electron (attraction forces) or the nucleus (repulsion forces). The information embedded in alpha-particles emerging from a medium is mainly related to the strength of the coulomb forces of the atoms making up the medium, which depends in turn on the atomic-number. Therefore, some useful chemical composition information can be obtained using alpha-particles. The coulomb interactions can be either elastic (where kinetic energy is conserved) or inelastic (where some of the energy is absorbed by the atom or the nucleus). These modes of interactions are discussed below.

3.3.1.1 Elastic Scattering[2]

An elastic collision of an alpha-particle with a nucleus simply results in a deflection of the alpha-particle, with a corresponding loss in kinetic energy in accordance to the conservation laws of momentum and energy between two colliding particles. Elastic scattering of alpha-particles may also be caused by the electric field of the atomic electrons, but the energy transferred to the atomic electrons is generally too small to cause a significant effect on the status of the atom or its electrons.

The elastic scattering of charged-particles by the nucleus is known as Rutherford scattering. If the nucleus causing the scattering is so heavy that its motion during the interaction is neglected, the probability of scattering of a charged-particle into a solid angle, $d\Omega$, around the direction of a scattering angle, θ, is given by the differential cross-section [48]:

$$\frac{d\sigma}{d\Omega} = \frac{(Zze^2)^2}{16E^2} \sin^{-4}\frac{\theta}{2} \tag{3.15}$$

[2]The discussion in this section is written so that it is applicable to different types of charged-particles.

where E is the initial energy of the incident particle. It is obvious from relationship (3.15) that the angle of scattering is much higher in the forward direction, i.e., at a small angle of scattering. However, forward-scattering can be observed only if the particle can penetrate the target. This can only occur if the target is very thin, due to the short range of alpha-particles.

With the nuclei of light elements, elastic scattering of alpha-particles is affected by the mass of the target nucleus, due to the weakening of the electrostatic field of the nucleus. The probability of scattering of a charged-particle (of mass m) by the nucleus of a light elements of mass M can be approximated by the differential cross-section [49]:

$$\frac{d\sigma}{d\Omega} = \frac{(Zze^2)^2}{16E^2 \sin^4 \theta} \frac{\left[\cos\theta \pm \sqrt{1 - \left(\frac{m}{M}\right)^2 \sin^2\theta}\right]^2}{\sqrt{1 - \left(\frac{m}{M}\right)^2 \sin^2\theta}} \tag{3.16}$$

The positive sign applies when $m < M$ and the negative sign is taken when $m > M$. Not that in atomic mass units, $\frac{M}{m} \approx \frac{A}{4}$ for alpha-particles, where A is the mass-number of the target nucleus.

The kinematics of elastic scattering requires conservation of momentum and energy, which leads to the relationship [48]:

$$E_s = E\left[1 - 2(1 - \cos\Theta)\frac{mM}{M+m}\right] \tag{3.17}$$

where E_s is the energy of an alpha-particle scattered by an angle Θ in the center-of-mass-system. The angle θ in the laboratory frame of reference is related to the angle Θ in the center-of-mass system by:

$$\tan\theta = \frac{\sin\Theta}{\frac{m}{M} + \cos\Theta} \tag{3.18}$$

Combining relationships (3.17) and (3.18) results in the following expression:

$$E_s = E\left[\frac{m\cos\theta + (M^2 - m^2\sin^2\theta)^{\frac{1}{2}}}{M+m}\right]^2 \tag{3.19}$$

Both Eqs. (3.17) and (3.19) convey the same physical indication, that is, the energy of scattering is uniquely related to the angle of scattering, for a given incident energy. Moreover, the energy of the scattered particle along a particular direction is dependent on the mass of the target nucleus. This, along with the dependence of the probability of scattering on the atomic number of the target, Eqs. (3.15) or (Eq. 3.16), makes elastic scattering of alpha-particles by the nucleus a useful indicator for elemental composition of material.

3.3.1.2 Inelastic Scattering with Electrons

Inelastic scattering by bound atomic electrons is the most dominant scattering modality of alpha-particles. In such collision, a bound electron is moved to a higher orbit in the atom, bringing the atom to an "excited state". The affected electron can also become unbound, causing ionization of the atom. The energy gained by the electron is an energy lost by the incident alpha-particle. However, this loss in kinetic energy hardly changes the path-line of an alpha particle through the material, due to the large mass of the particle in comparison to that of the electron. The stopping power, i.e., the energy loss per unit length, due to this inelastic scattering interaction, and the associated excitation/ionization process, can be approximated by [48]:

$$-\frac{dE}{ds} = \left(\frac{2\pi e^4 z^2 m}{m_0 E}\right) N\ Z\ \ln\left(\frac{m_0 E}{mI}\right) \tag{3.20}$$

where e is the elementary (electron) charge, m_0 is the electron's rest mass, m is the mass of an alpha-particle, z is its atomic-number ($z = 2$), Z is the atomic-number of the target nucleus, N is the number of material atoms per unit volume, E is the energy of the alpha-particle and I is the ionization potential of the material. For most elements, I is linearly proportional to Z. The strong dependence of energy-loss on Z is evident in Eq. (3.20), and this can in turn be used to monitor changes in the chemical composition of a material via changes in Z. The density of the material, through N, also affects the amount of energy loss. Therefore, by measuring the energy of alpha-particles after passing through a material, one can obtain some relevant density and composition information.

3.3.1.3 Inelastic Scattering with a Nucleus

The possibility that an alpha-particle can cause nuclear excitation, by inelastically depositing energy in the nucleus, is quite remote, due to the repulsion effect of the electrostatic force between the two positively charged bodies. However, the same repulsive force can cause deflection of an alpha-particle, as it approaches the nucleus. In this process, the alpha-particle may lose a quantum of energy, which is emitted as photons (bremsstrahlung radiation). This makes the process, in effect, an inelastic interaction, because of this radiative emission. However, the amount of bremsstrahlung radiation produced by an alpha-particle is very small, due to its large mass, and this effect is usually neglected; and only the elastic Rutherford scattering, discussed in section 3.3.1.1, is the dominant nuclear interaction process of alpha-particles.

3.3.1.4 Nuclear Absorption

Energetic alpha-particles, of a few MeV, can overcome the coulomb field of light nuclei and interact directly with the nucleus. This produces a temporary nucleus, called the compound nucleus, which is richer in neutrons, in comparison to protons, and typically emits a neutron to restore the proton/neutron balance in the nucleus. This is known as the (α,n) reaction. Only a small fraction of alpha-particles, about 1 in 10^4, interacts in this fashion, the rest absorbs electrons and becomes neutralized. This small fraction is, however, sufficient to be utilized in forming neutrons sources, see section 2.3.1.1. Nuclei affected by this reaction include, ^{6}Li, ^{7}Li, ^{9}Be, ^{10}B, ^{11}B, ^{13}C, ^{17}O, ^{18}O, ^{19}F, ^{27}Al and ^{28}Si. Obviously, this reaction can be used to detect the presence of these isotopes, if they exist in a sufficient concentration to produce a measurable neutron yield.

3.3.2. Beta Particles

The continuous distribution of the energy of beta-particles, see Figure 2.1, results in varying stopping distance. The range of a beta-particle is usually taken as that corresponding to the maximum beta-particle energy, E_{max}, as it results in the largest penetration distance. This range, R_β, can be approximated by:

$$R_\beta(\text{mm}) = 4 \times 10^3 \frac{E_{max}^{1.4}(\text{MeV})}{\rho(\text{kg/m}^3)} \tag{3.21}$$

where ρ is the material density. For 2 MeV particles, the range in air is about 8.8 m and in aluminum is 3.9 mm. These range values are much larger than those for alpha-particles, reported following Eq. (3.14). This is what makes beta-particles useful for application in NDE. Another approximate expression relating R to E_{max} is given by:

$$E_{max}(\text{MeV}) = 184 \frac{R_\beta(\text{mm})}{\rho(\text{kg/m}^3)} + 0.294 \tag{3.22}$$

Eq. (3.22) is known as the Feather's equation [50]. This expression is obviously valid only at high energies, since at zero range, Eq. (3.22) gives an energy of 0.294 MeV.

The wide energy spectrum of emitted beta-particles allows expressing empirically the attenuation of the intensity of beta-particles in an exponential fashion, permitting the definition of an attenuation coefficient. That is, the intensity of a beta beam, with an initial intensity I_0, after traversing a distance x ($\leq R$) in matter can be approximated as:

$$I = I_0 \exp[-\mu x] \quad \text{for } x \leq R \tag{3.23}$$

where μ is the attenuation coefficient, which can be approximated by:

$$\mu(\text{mm}) = 2.2 \times 10^{-3} \frac{\rho(\text{kg/m}^3)}{E_{max}^{\frac{4}{3}}(\text{MeV})} \tag{3.24}$$

Taking a beta-particle source with E_{max} of 4 MeV, for sake of comparison with alpha-particles, the attenuation coefficients for aluminum and air are, respectively, 0.94 and 4×10^{-4} mm^{-1}.

The utilization of beta-particles in NDE is more complicated than in the case of alpha-particles. This is because beta-particles change significantly their direction after scattering, and are more likely to re-scatter before leaving the object. This straggling effect makes it also difficult to define a precise range of a beta-particle through a straight line. Moreover, having a complex energy spectrum makes extracting useful information from the interaction of beta-particles a difficult task. Adding to the complexity is the relativistic effect of the change in the mass of the beta-particle as its velocity approaches that of the speed of light, see appendix C. Nevertheless, beta-particles have found some very useful applications in NDE, as Part III of this book shows.

3.3.2.1 Scattering with Atomic Electrons

When a beta-particle interacts with the electrons of the atom, the energy transferred to the atomic electrons causes excitation or ionization of the atom. However, when the energy transferred to the atomic electron is much larger that its binding energy, the interaction can be viewed as a collision between two particles of equal mass, with the atomic electron considered to be at rest with respect to the incident electron. Then, Eq. (3.16), with $m = M = m_0$ and $z = Z = 1$, becomes:

$$\frac{d\sigma}{d\Omega} = \frac{e^4}{4E^2} \cos\theta \sin^{-4}\theta \tag{3.25}$$

Since the two colliding particles have equal mass, backscattering ($\theta > \frac{\pi}{2}$) is not possible. The scattered electron from the beta-particle and the recoil atomic electron are experimentally indistinguishable. Therefore, the probability of scattering, or the differential angular cross-section, per atom, of electrons into a solid angle $d\Omega$ around the direction $\vec{\Omega}$ at an angle θ, becomes:

$$\frac{d\sigma}{d\Omega} = \frac{Ze^4}{4E^2} \left(\sin^{-4}\theta + \cos\theta^{-4}\right) \cos\theta \tag{3.26}$$

The atomic-number, Z, of the target material is incorporated into the above equation in order to provide a per atom, rather than per electron, probability of scattering.

The stopping power, or energy loss per unit length, due to ionization and excitation is given by the so-called Bethe formula as [48]:

$$\left.\frac{dE}{ds}\right|_c = -\frac{2\pi e^4 NZ}{m_0 v^2} \times \left[\ln \frac{m_0 v^2 E}{2I^2(1-\beta^2)} - (\ln 2)\left(2\sqrt{1-\beta^2} - 1 + \beta^2\right) + (1-\beta^2) + \frac{\left(1-\sqrt{1-\beta^2}\right)^2}{8}\right] \quad (3.27)$$

where the subscript c signifies energy losses by collisions and all other notations are as those of Eq. (3.20), with $\beta = \frac{v}{c}$.

3.3.2.2 Elastic Scattering with the Nucleus

The elastic scattering of beta-particles by the nucleus follows also the law of Rutherford scattering, discussed for alpha-particles in section 3.3.1. Therefore, Eq. (3.15), with $z = 1$, becomes:

$$\frac{d\sigma}{d\Omega} = \frac{(Ze^2)^2}{16E^2} \sin^{-4}\frac{\theta}{2} \quad (3.28)$$

Since the mass of the electron is much lower than that of the nucleus, elastically scattered electrons are deflected by a large angle after one collision, particularly when the electron energy is low and the atom has a high atomic-number. Such large deflection angles take place only when the electron passes near the nucleus. Usually, due to the small nucleus volume, electrons travel far away from the nucleus, and are scattered only by a small-angle, making small-angle deflection more probable than large angle-scattering.

Comparison of the probability of nuclear scattering of beta-particles, Eq. (3.28), to that with the atomic-electrons, Eq. (3.26), indicates that nuclear scattering increases with Z^2 while electronic scattering increases only with Z. Therefore, in general, nuclear scattering predominates over the scattering by electrons by a factor of Z. The two mechanisms of scattering have the same probability of occurrence only with hydrogen, $Z = 1$.

3.3.2.3 Inelastic Scattering with the Nucleus

Inelastic interactions of beta-particles with the nucleus results in the production of a continuous x-ray spectrum, or bremsstrahlung. The coulomb field of the nucleus accelerates the incident electron, which consequently emits a pulse of radiation (a photon) as it slows-down. In such radiative collision, the photon carries a very small momentum and can be emitted in any direction. The problem can then be considered as a collision of an electron with the nucleus, but with some energy loss to the emitted photon. The angular distribution of the electron is then expected to be similar to that described by Eq. (3.28). The probability of occurrence of this radiative collision process is, however, much lower than that of other scattering forms. However, energy loses by this process is significant when the beta-particle energy is high and the material has a high atomic-number, as indicated by the stopping power through radiative collisions, given by [48]:

$$\left.\frac{dE}{ds}\right|_r = -\frac{NEZ(Z+1)e^4}{137m_0^2c^4}\left[4\ln\frac{2E}{m_0c^2}-\frac{3}{4}\right] \tag{3.29}$$

where the subscript r refers to capture and all other notations are as those of Eq. (3.20). Note that the total stopping power is the total of the collision and capture losses, i.e. Eq. (3.27) plus Eq. (3.29).

3.4. Photons

A photon is a packet, a quantum, of energy, in the form of electromagnetic radiation and as such travels at the speed of light. The energy of a photon, E, is equal to $h\nu$, where h is Planck's constant (6.626 $\times^{-34}$J s) and ν is the frequency of the electromagnetic wave. The interest here is in photons that have a sufficient energy to penetrate matter and be affected by it, to provide a useful signal for an NDE technique. This includes x-rays and gamma rays. As discussed in section 2.2, x-rays are photons emitted in atomic transitions of bound electrons in the atom, while gamma-rays consist of photons emerging from nuclear decay. Gamma-ray sources are typically radioisotopes, such as ^{60}Co and ^{137}Cs, that have certain half-lives and emit photons with discrete energies. X-rays, however, are generated electronically and have a continuous energy spectrum that spans from zero up to an energy corresponding to the maximum voltage applied to the triggering electrons. This difference in the energy distribution of gamma-ray and x-ray photons affects the way they are modified by matter. An x-ray beam experiences the so-called beam-hardening as it travels through matter, due to the absorption of low-energy photons in the early stages of its travel, so that only photons with higher energy remain in the beam. This makes the beam "harder" in energy, and affects its overall interaction characteristics. Obviously, monoenergetic photons, such as those of gamma-rays, do not suffer from this beam hardening effect, since as a photon is absorbed it simply disappears from the beam, leaving the other photons in the beam unaltered. Nevertheless, the interaction of a photon with matter depends on its energy, not its origin.

When a photon approaches an atom, it is exposed to the electrons of the atom and to the nucleus; as well as to the electromagnetic field generated by the movement of the atomic electrons and the spin of the nucleus. This can lead to a number of interactions. In the energy range of commonly used x- and gamma-ray photon sources, 0.01 to 10 MeV, there is a number of dominant interactions:

Photoelectric absorption. This is the process in which a photon is completely absorbed in the atom, ejecting an orbital electron. The process is known as the photoelectric effect, and dominates at low photon energies and atoms with large atomic numbers.

Compton scattering. When the corpuscular properties of a photon dominates, it behaves like a particle. When this particle-like photon collides with another particle, such as an atomic electron, the laws of conservation of momentum and energy apply to the kinematics of this collision-like process. When the energy of the incident photon is higher than the binding-energy of the atomic electron, the electron is considered to be practically free and at rest, in comparison to the incoming photon. The collision process is then called Compton scattering. This is an inelastic scattering process, as it involves delivering some energy to the atom, via its electrons. The process is also called incoherent scattering, since the classical interference between waves, constructively or destructively, between waves, does not apply[3]. Compton scattering is the most dominant mode of interaction for most materials in the photon energy range from 50 keV to 1.5 MeV.

Coherent scattering. When the wave nature of photons takes precedence, photons are affected by the electromagnetic field of the atom as a whole. In this coherent scattering process, wave amplitude-addition and phase-shift take effect. As far as the atom is concerned, this is an elastic scattering process, since the integrity of the atom is not affected. Three different modes of wave scattering can occur:

Rayleigh scattering. In this process, the photon energy does not change, but its direction is deflected by a small angle. This small-angle deflection makes it difficult to distinguish the scattered radiation from that of an incident beam. Because of this slight-variation in direction,

[3]See appendix D for a more precise definition of coherent and incoherent scattering.

the effect of this elastic-scattering process is often referred to as deflection. Rayleigh scattering is dominant at low photon energies and for high atomic-number atoms..

Bragg Diffraction. In materials with ordered crystalline structure, photons coherently scattered by adjacent electrons can interfere with one another as waves. In a crystalline structure, the atoms are structured in array form, the layers of which present themselves to the photon as scattering centers. Waves scattered from such centers would have generally traveled slightly different distances, and would consequently be out of phase of each other. When the phase shift is a whole number of wavelengths, constructive interference would occur, resulting in a high intensity scattered wave, otherwise destructive interference occurs. Therefore, a pattern emerges consisting of high and low intensity reflected waves. This phenomenon is know as Bragg diffraction, and occurs when the photon wavelength is about equal to the spacing of atoms in a crystal, i.e. a about 100 pm (at about 12 keV energy).

Refraction. At photon energies below that required to excite an atomic electron, less than 20 keV or so, photons experience refraction when they cross an interface between two media, in the same manner light is refracted by an optical lens or a prism.

Pair production. In the presence of the strong electromagnetic field of nuclei of a high atomic-number, a photon can be annihilated and its energy is converted into mass, generating an electron-positron pair. This process requires the photon to have an energy of at least 1.022 MeV, since the mass of an electron, or a positron, is equivalent to 0.511 MeV. The coulomb field of an electron can also lead to pair production, with the target electron acquiring recoil energy that ejects it from the atom. This gives the appearance of "triplet production", which requires threshold photon-energy of 2.044 MeV. For materials of high atomic-number, pair-production dominates all other reactions at high energy (a few MeV or higher).

The above interactions are discussed in more detail in the following sections. It should be noted, however, that since chemical bonds are of the order of a few electron volts, and thus have no significant effect on photons of energies of practical interest, the cross-sections of photons for mixtures of elements, or a chemical compound, is usually assumed to depend only upon the sum of the cross-sections of individual atoms in the mixture. Therefore, they can be combined by the methods discussed in appendix E.

3.4.1. Photoelectric Absorption

In this process, a photon transfers all of its energy to an orbital (bound) atomic electron. The electron is subsequently ejected from its orbit, carrying a kinetic energy of $E - B$, where E is the energy of the incident photon and B is the binding energy of the electron. The orbital vacancy left by the ejected electron can be subsequently filled by another electron from a higher orbit, producing a low-energy photon, with an energy equal to the difference in the binding energy of the two orbits. This is in effect an x-ray photon, also called a fluorescent photon, which can in turn be absorbed in outer electronic shells of the atom, in a photoelectric-like process, ejecting an electron. The latter process is called fluorescence, and the ejected electron usually escapes the atom and gradually loses its energy in the surroundings. In some instances, especially in heavy elements, the fluorescent photon itself may escape the atom, but it will have an energy much lower than that of the original incident photon, and one can usually consider the initial photon to be completely absorbed. It is easier to liberate an electron from a low atomic-number (Z) material, while the probability of x-ray emission from the K shell of the atom is nearly unity in high-Z elements, and is nearly zero in low-Z elements. For elements with an intermediate Z number, both transitions are possible.

Tightly bound electrons have the highest probability of absorbing a photon. In fact, 85% of photoelectric absorption takes place in the K shell, outer shell, of the atom. However, the photoelectric effect is usually considered as an interaction with the entire atom, and its probability of interaction is expressed in terms of the atomic cross-section, τ. This cross-section is strongly dependent on both the atomic-number of the target atom and the incident energy of the photon, so that:

$$\tau \propto \frac{Z^n}{E^m} \tag{3.30}$$

where n varies from 3 to 5, and m is about 3, depending on Z and the photon energy, E. The strong dependence of τ on Z can be used in NDE to provide composition, or more accurately atomic-number, information. An approximate value of τ is given by the relationship [40]:

$$\tau = 12.1\frac{Z^{4.6}}{E^{3.15}} \tag{3.31}$$

with E in keV and τ in barns. A more accurate expression for the photoelectric cross-section for the dominant absorption by K-shell electrons,

τ_K, is given by [51] for $E > 0.2$ MeV:

$$\tau_K = Z^5 \left[\frac{1.6268 \times 10^{-9} - 2.683 \times 10^{-12} Z}{(1 + 4.173 \times 10^{-2} Z) E} + \frac{1.5274 \times 10^{-9} - 5.110 \times 10^{-13} Z}{(1 + 1.027 \times 10^{-2} Z) E^2} + \frac{1.1330 \times 10^{-9} - 2.177 \times 10^{-12} Z}{(1 + 2.013 \times 10^{-2} Z) E^{3.5}} - \frac{9.12 \times 10^{-11}}{E^4} \right] \text{barn/atom} \tag{3.32}$$

Note that when the photon energy is below the binding energy of a given electron shell, the cross-section sharply drops in value since the electron cannot be ejected from the shell. This leads to the "absorption edges" observed in the photoelectric cross-sections, which are characteristic of the elements involved, since each element has its own unique binding energies. Cross-sections in the range of characteristic energies from 4.509 keV (Ti K_α) to 24.942 keV (Ag K_β) are tabulated in reference [52].

Although the photoelectric effect is due to an interaction with an electron in a particular orbit in the atom, an average cross-section per electron can be defined as:

$$\tau_e = \frac{\tau}{Z} \tag{3.33}$$

3.4.2. Incoherent/Inelastic (Compton) Scattering

3.4.2.1 Kinematics

The kinematics of Compton scattering is governed by the conservation of momentum and energy in this particle-like collision process. A photon of energy E has a momentum equal to $\frac{E}{c}$, where c is the speed of light, see appendix C. The target atomic electron can be assumed to be at rest, if E exceeds the electron's binding energy. At rest, the electron is considered to have an energy equivalent to its rest-mass energy of m_0c^2, equal to 0.511 MeV, where m_0 is the mass of the electron. Note that when the photon energy is comparable in value to the binding energy of the atomic electron, the photoelectric effect has a larger probability of occurrence than Compton scattering.

As schematically shown in Figure 3.2, if a photon is scattered by an angle θ, the electron will gain kinetic energy and move at an angle ϕ with respect to the initial direction of the photon. Energy conservation requires that:

$$E = E' + T \tag{3.34}$$

where E' is the kinetic energy of the scattered photon and T is that of the recoiled electron. Conservation of momentum in the direction of the incident photon gives:

$$\frac{E}{c} = \frac{E'}{c}\cos\theta + q\cos\phi \tag{3.35}$$

where q is the momentum of the electron, and is related to its kinetic energy, T, by the relativistic relationship $q^2c^2 = T^2 + 2m_0c^2T$, see Eq. (C.11). Conserving the momentum in the direction normal to that of the incident photon leads to:

$$0 = \frac{E'}{c}\sin\theta - q\sin\phi \tag{3.36}$$

Combining Eqs. (3.34), (3.35) and (3.36) to eliminate T and ϕ, leads to the well-known Compton relationship:

$$E' = \frac{E}{1 + \frac{E}{m_0c^2}(1 - \cos\theta)} \tag{3.37}$$

The momentum transfer to the electron, obtained from Eqs. (3.35) and (3.36), is:

$$q = \frac{1}{c}(E^2 + E'^2 - 2EE'\cos\theta)^{\frac{1}{2}} \tag{3.38}$$

Therefore, the maximum value of q is attained at $\cos\theta = -1$, and is equal to $\frac{E+E'}{c}$.

In NDE applications, relationship (3.37) can be used to determine the direction from which photons are scattered, if the energy of the incident photons is known and the energy of the scattered photons is measured. It should be noted, however, that Eq. (3.37) applies only to a single scattering event of the photon; subsequent scattering will further reduce the photon energy and entangle its direction of scattering.

3.4.2.2 Cross Sections

Another modifying effect of interest in NDE is the fact that the probability of Compton scattering is proportional to the number of available electrons, which in turn depends on the atomic-number of the target atom and the density of the material. This is due to the fact that when the photon-energy is higher than the electron's binding energy, all atomic electrons are available for interaction with the incident photons.

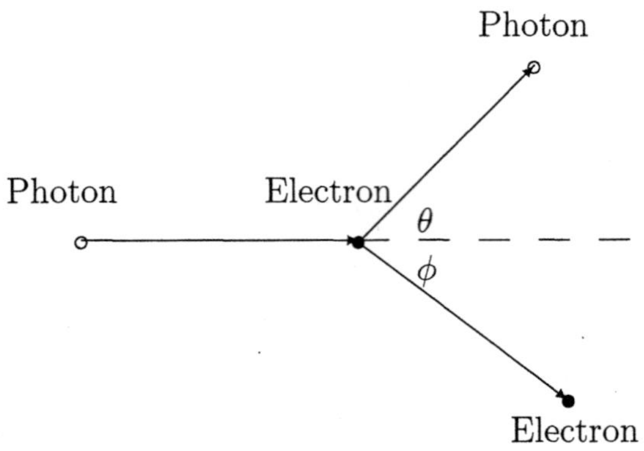

Figure 3.2. Schematic of Compton scattering.

The probability of Compton scattering per a single electron depends only on energy, and increases with increasing photon energy, so that:

$$\sigma_e \propto \frac{1}{E} \tag{3.39}$$

where σ_e is the cross-section per electron. Therefore,

$$\sigma \propto \frac{Z}{E} \tag{3.40}$$

where σ is the cross-section per atom, and Z is the atomic-number of the target atom.

Differential Cross-Section. The differential cross-section, or the probability per electron that a photon will Compton scatter into a unit solid angle, Ω, about a scattering angle, θ, is given, to a very good approximation, by the Klein-Nishina relationship:

$$\begin{aligned} \frac{d\sigma_e(E,\Omega)}{d\Omega} &= r_0^2 \left(\frac{1+\cos^2\theta}{2}\right)\left[\frac{1}{1+\alpha(1-\cos\theta)}\right]^2 \times \\ & \left\{1+\frac{\alpha^2(1-\cos\theta)^2}{[1+\alpha(1-\cos\theta)](1+\cos^2\theta)}\right\} \end{aligned} \tag{3.41}$$

where $\alpha = \frac{E}{m_0c^2}$ and r_0 is a constant called the classical electron radius ($r_0 = \frac{e^2}{m_0c^2} = 2.818 \times 10^{-15}$ m, where e is the elementary charge). The scattering in the azimuthal direction, see Figure 3.1, is isotropic, i.e. with an equal probability of $\frac{1}{2\pi}$. Therefore, relationship (3.41) can also be expressed in terms of the cosine of the polar angle, $\cos\theta$ in Figure 3.2, as:

$$\frac{d\sigma_e(E,\theta)}{d\cos\theta} = 2\pi\frac{d\sigma_e(E,\theta)}{d\Omega}$$

Eq. (3.41) can also be expressed as:

$$\frac{d\sigma_e(E,\Omega)}{d\Omega} = \frac{r_0^2}{2}\left(\frac{\alpha'}{\alpha}\right)^2\left\{\frac{\alpha}{\alpha'}+\frac{\alpha'}{\alpha}+\cos^2\theta-1\right\} \quad (3.42)$$

where $\alpha' = \frac{E'}{m_0c^2} = \frac{\alpha}{1+\alpha(1-\cos\theta)}$ is used in the above equation for simplification. The expression of Eq. (3.42) resembles the physical explanation of Compton scattering by quantum electrodynamics, see section D.4.

Figure 3.3 plots the Klein-Nishina cross-section, $\frac{d\sigma_e(E,\theta)}{d\cos\theta}$, for different incident photon energies. The Figure shows that at high energies, photons favor scattering in the forward direction, while backscattering becomes pronounced at low-photon energies. Note that at low photon energies, E' and E are about equal even at maximum energy loss, $\cos\theta = -1$, as Eq. (3.37) indicates. Then, $\alpha \approx \alpha'$, $\frac{\alpha}{\alpha'}+\frac{\alpha'}{\alpha} \approx 2$, and relationship (3.42) is reduced to:

$$\begin{aligned}\frac{d\sigma_e(E,\Omega)}{d\Omega} &= \frac{r_0^2}{2}\left\{\cos^2\theta+1\right\} \\ &= \frac{3\sigma_T}{16\pi}\left\{\cos^2\theta+1\right\} \quad (3.43)\end{aligned}$$

where $\sigma_T = \frac{8\pi r_0^2}{3}$. Relationship Eq. (3.43) is valid only at low incident energies ($E < m_0c^2$) and describes the differential Thomson cross-section[4].

Cross-Section per Electron. Integration of Eq. (3.41), over all scattering angles, leads to the total cross-section for Compton scattering per electron, which is given by:

$$\begin{aligned}\sigma_e(E) &= 2\pi r_0^2\left\{\frac{1+\alpha}{\alpha^2}\left[\frac{2(1+\alpha)}{(1+2\alpha)}-\frac{1}{\alpha}\ln(1+2\alpha)\right]+\right. \\ &\left.\frac{1}{2\alpha}\ln(1+2\alpha)-\frac{1+3\alpha}{(1+2\alpha)^2}\right\} \quad (3.44)\end{aligned}$$

[4]Thomson scattering is the scattering of electromagnetic radiation (of energy much less that m_0c^2) by a free electron. The scattering cross-section, integrated over all angles, is given by σ_T, see Eq. (3.47)

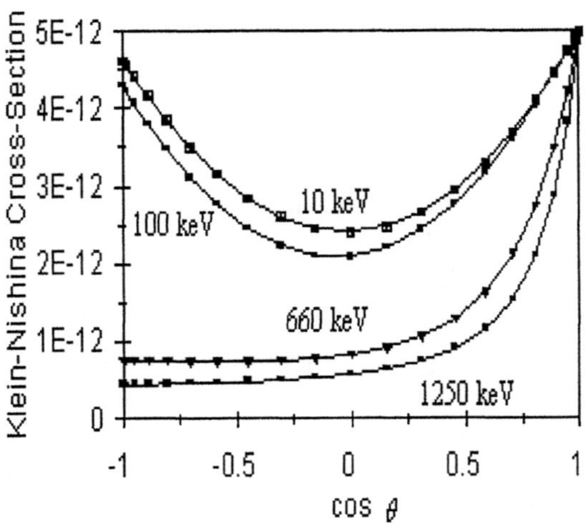

Figure 3.3. The Klein-Nishina cross-section, $\frac{d\sigma_e(E,\theta)}{d\cos\theta}$, plotted at different incident photon energies as a function of the scattering angle, θ.

Another useful approximation or σ_e is given by the Hastings approximation [51]:

$$\sigma_e(E) = \pi r_0^2 \frac{c_1\eta^2 + c_2\eta + c_3}{\eta^3 + d1\eta^2 + d_2\eta + d_3} \tag{3.45}$$

where $\eta = 1+0.222037\alpha$, $c_1 = 1.651035$, $c_2 = 9.340220$, $c_3 = -8.325004$, $d_1 = 12.501332$, $d_2 = -14.200407$, and $d_3 = 1.699075$. At photon energy below 100 keV, the logarithmic and algebraic terms in the expression of Eq. (3.45) nearly cancel out. The following expression can be used below 100 keV photon energy [51]:

$$\begin{aligned}\sigma_e(E) &= \frac{8\pi}{3} r_0^2 \frac{1}{1+2\alpha^2}\left(1 + 2\alpha + \frac{6}{5}\alpha^2 - \frac{1}{2}\alpha^3 + \frac{2}{7}\alpha^4\right.\\ &\left.-\frac{6}{35}\alpha^5 + \frac{8}{105}\alpha^6 + \frac{4}{105}\alpha^7 - \cdots\right)\end{aligned} \tag{3.46}$$

Note that at very low energy, the above relationship leads to (as $\alpha \longrightarrow 0$):

$$\sigma_T = \frac{8\pi}{3} r_0^2 = 0.6653 \text{ barn/electron} \tag{3.47}$$

which is the classical Thomson cross-section[5]. This limit is, however, valid for a free electron; for bound electrons the $\sigma_e \longrightarrow 0$ as $\alpha \longrightarrow 0$ [51].

For very large values of α (>>1), i.e. at high photon energy [51]:

$$\sigma_e(E) = \frac{\pi r_0^2}{\alpha}(\ln 2\alpha + \frac{1}{2}) \tag{3.48}$$

The reader can prove, from the approximations of Eqs. (3.45), (3.46) and (3.48), that σ_e is reasonably proportional to $\frac{1}{\alpha}$, as relationship (3.39) indicates.

Attenuation Coefficients. The linear attenuation coefficient, or macroscopic cross-section, Σ, is obtained, according to Eq. (3.6), by multiplying the atomic, or microscopic cross-section, by the number of atoms per unit volume, N. Therefore,

$$\Sigma = \sigma N = NZ\sigma_e = \rho_e \sigma_e \tag{3.49}$$

where ρ_e is the electron's density, which is related to the number of atoms per unit volume, N, defined according to Eq. (3.2), by the relationship:

$$\rho_e = ZN \tag{3.50}$$

since each atom contains Z electrons. At a given source energy, Eq. (3.49) shows that the Compton cross-section is directly proportional to the electron density, ρ_e. Using Eqs. (3.50) and (3.2), Eq. (3.49) becomes:

$$\Sigma = [Z\sigma_e]\left\{\frac{\rho}{A\mathrm{u}}\right\} = \frac{Z}{A}\frac{\rho\sigma_e}{\mathrm{u}} \tag{3.51}$$

For most atoms, except hydrogen, $\frac{Z}{A} \approx 0.5$. Therefore, one can conclude that Σ is generally proportional to the mass density, ρ. This makes Compton scattering attractive for use in NDE, as it can provide density information, independent of the nature or the physical state of the inspected material.

The mass-attenuation coefficient for Compton scattering, or the attenuation coefficient per unit mass, $\frac{\Sigma}{\rho}$, becomes, as one can deduce from Eq. (3.51), nearly independent of the nature of the medium, (with $\frac{Z}{A} \approx 0.50$), and is also independent of the actual density and physical state (gas, liquid, or solid) of the material.

[5] The Thomson cross-section of Eq. (3.47) can be obtained from the differential Thomson cross-section of Eq. (3.43) by integrating the latter over all angles.

3.4.2.3 Electron Binding-Energy and Momentum

The basic assumption in arriving at the above described kinematics of Compton scattering is that the electron with which the photon interacts is free and at rest, so that the electron's binding energy and its initial momentum are neglected. However, the effect of electron's binding energy and its momentum (motion) can be taken into account as shown below.

Scattering with Bound Electron. Binding energy effects are accounted for by assuming that a sudden impact action is applied to the atomic electron, transferring to it a momentum, q, given by Eq. (3.38), and in turn causing atomic excitation and ionization. The probability of scattering (differential cross-section) per electron, the Klein-Nishina cross-section, Eq. (3.41) or Eq. (3.42), is affected by the electron's binding energy, decreasing it at small scattering angles (i.e. in the forward direction), and increasing it at large angles. This change in the differential cross-section is analytically accommodated by a multiplicative correction factor, $S(q, Z)$, called the incoherent scattering function, applied to the Klein-Nishina cross-section, so that:

$$\frac{d\sigma_e^B(E,\Omega)}{d\Omega} = \frac{d\sigma_e(E,\Omega)}{d\Omega} S(q,Z) \tag{3.52}$$

where the superscript B refers to the modified cross-section due to the inclusion of the function $S(q, Z)$. This function represents the probability of exciting, or ionizing, the atom as a result of a sudden impulsive action which imparts a recoil momentum q to the atom [51]. At a given Z value, $S(q, Z)$ varies from a minimum of zero, i.e. causing a complete absorption of the incident photon, to a maximum value equal to Z. Tabulated values for $S(q, Z)$ are given in reference [53]

Doppler Broadening. The motion (initial momentum) of the atomic electrons causes a broadening of the energy spectrum of scattered photons. If the electron is at rest, then a monoenergetic beam of photons of energy E scattered at a certain angel θ will emerge at a unique energy E', given by Eq. (3.37). Therefore, a photon emerging at angle θ will have a single energy, exhibiting an energy spectrum in the shape of a delta function at E', i.e. with a zero intensity at all energies except at E'. The spectrum is known as the Compton profile. This spectrum can, however, be broadened by the Doppler effect, i.e. the relative motion of the incoming photon with respect to the moving electron. The relative energy of an incoming photon, with respect to that of the electron, will be lower, or larger, than E, depending on whether the electron is moving towards, or away, from the photon, respectively. The extent of this Doppler broadening depends on both the incident photon energy and the structure of the atom with which the photon interacts. The Compton profile reveals, therefore, some information on the nature of the target atom.

The differential cross-section of Eq. (3.52) takes into account only the effect of the binding energy of the atomic electron, by changing the angular probability of scattering. The effect of electron motion and its associated spread (broadening) requires the introduction of a double-differential cross-section that takes into account the change in photon energy due to the electron's motion [54]. The result is the cross-section given by [55]:

$$\frac{d\sigma^2}{dE'd\Omega} = \frac{r_0^2}{2}\frac{E'E'_0}{E^2}\frac{m_0c^2}{E_e}\frac{dp_z}{dE'}XJ(p_z) \tag{3.53}$$

where the variables are as given below Eq. (3.41), with the added variables defined as follows:

- E_e is the relativistic total energy (see appendix C) of the atomic electron before the collision, given, using Eq. (C.13), by:

$$E_e^2 = p_z^2c^2 + (m_0c^2)^2 \tag{3.54}$$

  Note that E_e is equal to the electron's rest mass, when $p_z = 0$.

- p_z is the projection of the initial momentum of the electron on the direction of scattering:

$$p_z = \frac{1}{c^2q}[EE'(1-\cos\theta) - m_0c^2(E-E')] \tag{3.55}$$

- E'_0 is the energy of the photon scattered by a free electron, and thus satisfies Eq. (3.37).

- q is the momentum transfer to the electron after collision, defined by Eq. (3.38). The parameter X in Eq. (3.53) is given by:

$$X = \frac{R}{R'} + \frac{R'}{R} - 2\left(\frac{1}{R'} - \frac{1}{R}\right) + \left(\frac{1}{R'} - \frac{1}{R}\right)^2 \tag{3.56}$$

$$R = \frac{E}{m_0c^2}\left\{\left[1 + \left(\frac{p_z}{m_0c}\right)^2\right]^{\frac{1}{2}} + \frac{E - E'\cos\theta}{cq}\frac{p_z}{m_0c}\right\} \tag{3.57}$$

$$R' = R - \frac{EE'}{(m_0c^2)^2}(1-\cos\theta) \tag{3.58}$$

- The derivative $\frac{dp_z}{dE'}$ is defined using Eqs. (3.55) and (3.37) as:

$$\frac{dp_z}{dE'} = \frac{m_0 E}{qE'_0}$$

- Due to Compton Broadening, the energy of the scattered photon, E' in Eq. (3.53), does not follow the Compton kinematics of Eq. (3.37), but is given by:

$$E' = \left[\frac{p_z cq}{Em_0 c} + 1\right]\left[\frac{E}{1 + \frac{E}{m_0 c^2}(1 - \cos\theta)}\right] \qquad (3.59)$$

- $J(p_z)$ is the Compton profile, and $J(p_z)dp_z$ is dimensionless. $J(p_z)$ is normalized to unity so that:

$$\int_{\infty}^{\infty} J(p)dp = 1 \quad \text{per electron}$$

Note that, when $p_z = 0$, i.e. the electron is at rest, Eqs. (3.59) and (3.37) become identical. Under these conditions, the Compton profile $J(p_z)$ becomes equal to $\delta(0)$, i.e. a dirac delta function at E'_0. Moreover, at $p_z = 0$, one can show that $R = \alpha$, $R' = \alpha'$, and $\frac{1}{\alpha'} - \frac{1}{\alpha} = 1 - \cos\theta$, where α and α' are as defined in Eq. (3.42). Then, $X = \frac{\alpha}{\alpha'} + \frac{\alpha'}{\alpha} + \cos^2\theta - 1$. Also, $E_e = m_0 c^2$ at $p_z = 0$. Therefore, integrating Eq. (3.53) with respect to E', keeping in mind the nature of the delta function, leads to Eq. (3.42), with the latter inherently incorporating the unique energy-angle relationship of Eq. (3.37).

The relativistic effects, before collision, can be neglected when $p_z^2 c^2 << (m_0 c^2)$, so that $E_e = m_0 c^2$, as can be deduced from Eq. (3.54). This is, in effect, equivalent to setting $p_z = 0$ in Eq. (3.53). The non-relativistic formulation is valid for atoms of atomic-number, Z, less than 36 [56], where the electron moment p_z is quite small. Applying relativistic effects after collision depends on whether the energy acquired by the recoil electron is sufficiently high to bring its velocity to close to that of the speed of light. This, in turn. depends on the energy of the incident photon, as Eqs. (3.37) and (3.59) indicate. Generally, when $E << m_0 c^2$, relativistic effects can be neglected. Then, the maximum value of p_z (at $\cos\theta = -1$), according to Eq. (3.55), is equal to $\frac{2E^2}{c^2 q}$, and $\frac{p_z cq}{Em_0 c}$ is equal to $\frac{2E}{m_0 c^2}$ at most, which is much less than one, given that $E << m_0 c^2$. This has the effect of eliminating the contribution of p_z, as is the case with the before-collision non-relativistic formulation. Note also that when $E << m_0c^2$, the Thomson approximation, Eq. (3.43), can also be applied.

The function $J(p_z)$ is bell-shaped and symmetrical about $p_z = 0$. It has been measured by many workers, see for example references [57], [58] and [59]. Tabulated values of the Compton profile, $J(p_z)$, are given in reference [56], for individual atomic orbital and for the entire atom, for elements of atomic-number from 1 to 102. For a compound material, the Compton profile can be estimated as the sum of the atomic profiles of individuals atom in the molecule, in a manner similar to that used for calculating the macroscopic cross-section for a compound (see appendix E) [55]. The dependence of the Compton profile on the nature of the scattering element offers the possibility of providing elemental (composition) information. It can also be used to study the atomic structure of compounds, see for example reference [60].

Doppler Broadening with Binding Effects. For Compton scattering to take place, the energy deposited to the atomic electron, $E - E'$ must be greater than its binding energy, B. Therefore, the double-scattering cross-section that incorporates both Doppler broadening and binding-energy effects, is expressed as [55]:

$$\frac{d\sigma^2}{dE'd\Omega} = \frac{r_0^2}{2}\frac{E'E_0'}{E^2}\frac{m_0c^2}{E_e}\frac{dp_z}{dE'}XJ(p_z)\Theta(E - E' - B) \tag{3.60}$$

where the function $\Theta(x)$ is the Heaviside step function:

$$\Theta(x) = \begin{cases} 1 & \text{if } x > 0 \\ 0 & \text{otherwise} \end{cases}$$

Note by integrating the double-scattering cross-section of Eq. (3.53) over all photon scattering energies, dE', the single differential cross-section of Eq. (3.52) is obtained. Therefore, one can state that the incoherent scattering function, $S(q, Z)$, used Eq. (3.52), corresponds to the integration of the Compton profile $J(p_z)$ over all possible photon scattering energies.

3.4.2.4 Anti-Compton Effect

Photons may also exhibit the reverse of Compton scattering. A recoil electrons that result from Compton scattering can collide with other incident photons, transferring a part or all of its kinetic energy to the photon. The incident photon can then gain energy, shortening its wavelength. This process, called anti-Compton effect, was observed in x-ray interactions with low density polyethylene and polyethylene terephthalate [61]. It is considered useful for studying the properties of electrons in solids, such as the electron band structure in a Tc superconductor. The phenomenon was also proposed as a means of increasing the frequency of an x-ray source [62].

3.4.3. Coherent/Elastic Scattering

3.4.3.1 Rayleigh scattering

In this process, the atom presents itself to the incoming photon as one entity, while the photons behaves like a wave. As a result, the scattered photons emerge with almost no energy loss, and is scattered only by a very small angle. This is because the atom's integrity is maintained, i.e. no excitation or ionization (electron rejection) takes place. The practical effect of this type of scattering is to scatter photons in the forward direction, with no reduction in energy. Therefore, in many cases, coherent scattering is regarded as if no interaction has taken place, since the direction and angle of the incident neutron hardly change. On the other hand, if this effect can be detected, it provides some useful information that can be exploited in NDE, since its probability of occurrence, as discussed below, is highly dependent on the atomic number, and thus can be used for distinguishing different compositions.

This mode of scattering is known as Rayleigh scattering and occurs mostly at low photon energies and for high atomic-number materials. The differential scattering cross-section of Rayleigh scattering is equal to the classical Thomson cross-section by an electron, Eq. (3.43), multiplied by the probability that the Z electrons of an atom acquire a recoil moment, q, without absorbing energy. This probability is expressed by the square of the atomic form factor, $F(q, Z)$. Therefore, one obtains [51]:

$$\begin{aligned} \frac{d\sigma_r(E,\Omega)}{d\Omega} &= \frac{r_0^2}{2}(1+\cos^2\theta)[F(q,Z)]^2 \\ &= \frac{3\sigma_T}{16}(1+\cos^2\theta)[F(q,Z)]^2 \end{aligned} \tag{3.61}$$

where r_0 is the classical electron radius, θ is the angle of scattering, Z is the atomic-number of the target atom, q is the momentum transfer ($q = 2\alpha \sin\frac{\theta}{2}$, with $\alpha = \frac{E}{m_0c^2}$, E is the incident photon energy), and σ_T is given by Eq. (3.47). Tabulations of $F(q, Z)$ are given in reference [53]. The Rayleigh scattering cross-sections as function of energy and angle are also available on-line [63]. The differential scattering cross-section for Rayleigh scattering can be approximately expressed as [64]:

$$\frac{d\sigma_r(E,\Omega)}{d\Omega} = k\frac{\pi}{4}r_0^2 f(\theta)\frac{Z^3}{E^3} \tag{3.62}$$

$$\sigma_r(E) \quad \propto \quad \frac{Z^3}{E^3} \tag{3.63}$$

where k is a constant equal to 9.491 when E is in units of keV, and $f(\theta)$ is a function that determines the angle dependence of Rayleigh scattering. At least 75 % of the Rayleigh scattering angles are confined to scattering angles which are smaller than the angle [51]:

$$\theta_0 = 2\sin^{-1}\frac{0.0133Z^{\frac{1}{3}}}{E} \tag{3.64}$$

with E in MeV. This is due to the fact that recoil energy imparted into the atom must not produce excitation or ionization. For example, for Fe ($Z = 26$), 75% of Rayleigh scattering occurs within angles smaller than 4.6°, 4.5° and 0.45°, for incident photon energies of 0.1, 1 and 10 MeV, respectively.

The cross-section of Rayleigh scattering is calculated by integrating Eq. (3.61) to obtain:

$$\begin{aligned} \sigma_r(E) &= \pi r_0^2 \int_{-1}^{+1} \frac{1}{2}(1+\cos^2\theta)[F(q,Z)]^2 d\cos\theta \\ &= \frac{3\sigma_T}{8} \int_{-1}^{+1} \frac{1}{2}(1+\cos^2\theta)[F(q,Z)]^2 d\cos\theta \end{aligned} \tag{3.65}$$

where σ_T is given by Eq. (3.47).

The strong dependence of Rayleigh scattering on the atomic-number, indicated by Eq. (3.63), makes it an attractive method for composition analysis, particularly, for atoms of high Z value. Rayleigh scattering is easily distinguished from Compton scattering by the fact that Rayleigh scattering is detected at the same energy as the incident radiation, while Compton scattering causes energy degradation in accordance to Eq. (3.37).

3.4.3.2 Diffraction

The atoms in a crystal are placed at equal distance from each other in a structure of parallel lines. Hence, they present themselves as a "grating" that can diffract photons, similar to the grating used to diffract light. Photons, are diffracted by a crystal when their wavelength is comparable in value to the lattice spacing (distance between atoms in the crystal). The well-known Bragg's law of diffraction states that constructive interference between waves scattered by a crystalline structure occurs when:

$$n\lambda = 2d\sin\frac{\theta}{2} \tag{3.66}$$

where λ is the wavelength of the incident radiation, θ is the scattering angle, d is the is the atomic lattice spacing, n is an integer indicating the cycles of constructive interference. Setting in Eq. (3.66), $d = 100$ pm $= 10^{-10}$ m, $n = 1$, the minimum value for constructive interference, is then:

$$\frac{1}{\lambda}\sin\frac{\theta}{2} = \frac{1}{2\times 10^{-10}}\ \mathrm{m}^{-1} = 5\ \mathrm{nm}^{-1}$$

Bragg diffraction takes place when $\lambda \geq d$, and as such it requires the satisfaction of the following condition:

$$\frac{1}{\lambda}\sin\frac{\theta}{2} \leq 5\ \mathrm{nm}^{-1} \tag{3.67}$$

or equivalently,

$$E(\mathrm{keV})\sin\frac{\theta}{2} \leq 6.2 \tag{3.68}$$

where E is the photon energy corresponding to λ. For example, for a 10° scattering angle, E should be less than about 70 keV for diffraction to occur in a crystalline materials, while $\theta = 45°$ requires an energy less than 16 keV. When Bragg scattering occurs, its pattern can also be used to characterize small crystalline materials [65]. A diffraction pattern is produced when a narrow beam of x-rays passes through a thin specimen of single-crystal material, giving rise to a set of diffraction rings (fringes) that can be observed on a photographic plate. This method is known as the Laue diffraction technique, first observed by Max von Laue in 1912.

3.4.3.3 Refraction

Photon can be refracted in a manner similar to light refraction at photon energies below less than 20 keV or so, i.e. below the energy required to excite an atomic electron. Refraction occurs when photons cross an interface between two media. This refraction is described, like the refraction of light, by Snell's law, so that:

$$\eta = \frac{\lambda}{\lambda'} = \frac{\sin\phi}{\sin\phi'} \tag{3.69}$$

where, λ and λ' are respectively, the wavelength of the incident and refracted rays, the angles, ϕ and ϕ' are the diffraction angles, see Figure 3.4, and the ratio η becomes equal to the refractive index when the incident x-ray is in vacuum. Unlike ordinary light, where the refractive index is greater than one, the refractive index of x-rays is less than unity, and is approximately given by [66]:

$$\eta = 1 - \frac{1}{4\pi\epsilon_0}\frac{N_e e^2\lambda^2}{2\pi m_0 c^2} \tag{3.70}$$

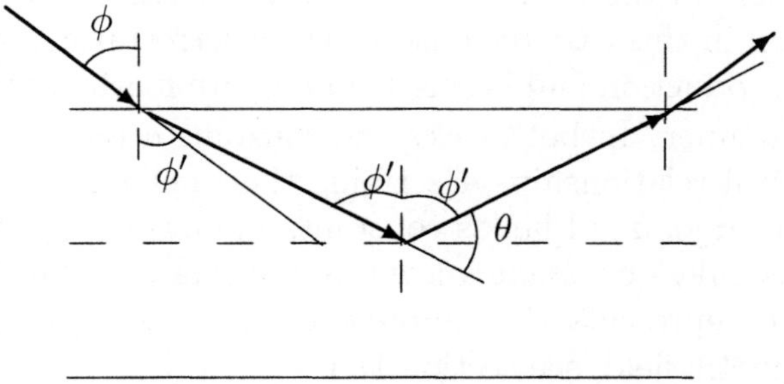

Figure 3.4. A schematic showing combined refraction and diffraction of an x-ray beam.

where N_e is the electron density, m_0c^2 is the electron's rest-mass energy, and ϵ_0 is the electric perimittivity of vacuum[6] ($= 8.854 \times 10^{-12}$ C^2/J m). For example, η for glass at 8 keV x-ray energy is about $1 - 10^{-5}$, i.e. very close to unity [67]. Note that the refractive index has also an imaginary component that accounts for the phase-shift of the refracted waves. With $\eta < 1$, x-rays tend to diverge when incident from vacuum into a material, unlike light which converges when incident on a convex lens because its refractive index is greater than one. Eq. (3.69) indicates that refraction is a density-dependent phenomenon, and thus can be used to obtain density information; in small light samples because of the limited penetration depth of the low-energy x-rays at which diffraction is observed.

Note that, because of refraction, Bragg's law of diffraction, Eq. (3.66), should be rewritten as:

$$n\lambda' = 2d \sin \frac{\theta}{2} \tag{3.71}$$

with λ replacing λ', to reflect the fact that diffraction follows refraction, as schematically shown in Figure 3.4. However, since η is usually very close to unity, the effect of refraction on diffraction is often ignored.

3.4.4. Pair Production

In the bremsstrahlung process discussed in section 3.3, charged-particles moving in the electromagnetic field of the nucleus emit pho-

[6]Note that $\epsilon_0 = \frac{1}{\mu_0 c^2}$, where μ_0 is the permeability of vacuum ($= 4\pi \times 10^{-7}$ H/m), and c is the speed of light.

tons. The pair production process can be viewed as the reverse of the bremsstrahlung effect, that is photons in the presence of the electromagnetic field of the nucleus can generate charged-particles, a positron and electron in this case to conserve the electrostatic charge. Therefore, both pair-production and bremsstrahlung are mathematically treated in a similar manner. In both cases the constant a $(= \frac{\pi e^2}{hc})$ appears in the mathematical relationships describing their behavior, where e is the elementary charge, h is Plank's constant, and c is the speed of light. This is a dimensionless constant known as the *fine-structure constant*, equal to $\frac{1}{137}$, and represents the coupling between charged-particles and an electromagnetic field, see section D.4.

In the presence of the electric field of the nucleus, and to a lesser extent that of the outer orbital atomic electron, a photon can be completely absorbed and its energy is distributed equally between a positron and an electron, such that

$$h\nu = (E_- + m_0c^2) + (E_+ + m_0c^2) \tag{3.72}$$

where $h\nu$ is the incident photon energy, E_- and E_+ are the kinetic energy of the electron and the positron, respectively, and $m_0c^2 = 0.511$ MeV is the rest-mass energy of an electron, see appendix C. A positron is essentially an electron with a positive electric charge, and a rest-mass energy equal to that of the electron, each being equal to 0.511 MeV. Therefore, as Eq. (3.72) shows, the incident photon must possess an energy greater than 1.022 MeV to create an electron-positron pair. If the incident photon has exactly an energy of 1.022 MeV, the emerging pair of particles will have no kinetic energy and will annihilate each other, reproducing the incident photon, as if no reaction has taken place. Therefore, a photon of an energy greater than 1.022 MeV is required, with the excess energy shared as kinetic energy between the electron and the positron, and to a less extent with the nucleus. The positron and electron tend to emerge in the forward direction, same direction as the incident photon.

The energy of the created pair is affected by "coulomb potential" of the positively charged nucleus, causing repulsion of the positron, and attraction of the electron towards the nucleus. A pair created at a distance r from the center of a nucleus of charge Ze will possess a potential $\frac{Ze^2}{r}$ for each member of the pair. Thus, this electrostatic field increases the original difference in kinetic energy of the positron and the electron by $\frac{2Ze^2}{r}$. Pair production occurs in a region of the nuclear field for which r lies between $\frac{h}{2\pi m_0 c}$ and $\frac{h}{2\pi m_0 c} \times \frac{h\nu}{2m_0c^2}$ [48]. The average positron, therefore, acquires a maximum of about $\frac{2Ze^2}{\frac{h}{2\pi m_0 c}} = 0.0075Z$ MeV more kinetic energy than the average electron, for small values of $h\nu$. At very high photon energies, pair production may take place at a distance greater than the radius of the K-electron shell. Then the positive electrostatic potential of the nucleus is "screened" (reduced), by the negative potential of the atomic electrons. This reduces the potential, from $\frac{Ze^2}{r}$ to $\frac{Ze^2}{r}e^{-\frac{r}{a}}$ where a is equivalent radius of the cloud of atomic electrons that screens the nucleus $\left(a = \frac{(\frac{h}{2\pi})^2}{m_0 e^2 Z^{\frac{1}{3}}}\right)$ [48]. This screening effect causes the cross-section to attain asymptotically a constant value at high photon-energies (i.e. when complete screening occurs) [68].

The angular distribution of the positron and electron is mainly forward peaked for incident quanta of high energy. The average angle between the created particle and the incident photon direction is given by $\theta = \frac{m_0 c^2}{h\nu}$ [45]. The two members of the pair are emitted at two opposites azimuthal directions with respect to the incident photon to conserve momentum.

The differential cross-section, $\frac{d\kappa_n}{d\Omega}$ per nucleus, for the production of a positron of kinetic energy E_+ (and an electron of kinetic energy $h\nu - 2m_0c^2 - E_+$), can be expressed in a simple form as [48]:

$$\frac{d\kappa_n}{dE_+} = \frac{\sigma_0 Z^2 P(h\nu, Z)}{h\nu - 2m_0c^2} \tag{3.73}$$

where $\sigma_0 = \frac{r_0^2}{137} = 0.579$ mb, with r_0 being the classical electron radius. and the function $P(h\nu, Z)$ is a complicated function that takes into account the effect of electrostatic (coulomb) potential of the nucleus, the screening of the field by atomic electrons, and radiative effects, see section D.4.

The total pair-production cross-section per nucleus (or atom) can be found by integrating Eq. (3.73) over all possible positron energies. The resulting integral can be approximately expressed as [48]:

$$\kappa_n = \sigma_0 Z^2 \bar{P} \tag{3.74}$$

where $\bar{P}$ is analogous to the function $P(h\nu, Z)$ in Eq. (3.73), that is, it incorporates the coulomb, screening and radiative corrections. Reference [68] gives detailed expressions for these corrections. However, some understanding of the behavior of κ_n can be gained by considering special cases. For $2m_0c^2 << h\nu << 137m_0c^2Z^{-\frac{1}{3}}$ (about 16 MeV for Pb, $Z = 82$), neglecting the screening effect [48]:

$$\kappa_n \approx \sigma_0 Z^2 \left(\frac{28}{9}\ln\frac{2h\nu}{m_0c^2} - \frac{218}{27}\right) \tag{3.75}$$

which indicates by, comparison with Eq. (3.74), that $\bar{P}$ changes approximately logarithmically with $h\nu$. In the case of complete screening, for $h\nu >> 137 m_0 c^2 Z^{-\frac{1}{3}}$, κ_n is approximately given by [48]:

$$\kappa_n = \sigma_0 Z^2 \left(\frac{28}{9} \ln \frac{183}{Z^{\frac{1}{3}}} - \frac{2}{27} \right) \tag{3.76}$$

which shows that at very high energy, the pair-production cross-section depends only on screening and is independent of the photon energy.

Pair production can also take place in the field of an atomic electron, with the electron acquiring recoil energy that can either cause excitation or ionization of the atom. Therefore, the process is called incoherent pair production, as opposed to the coherent pair production caused by the field of the nucleus which does not affect the state of the atom. An energy slightly greater than $2m_0c^2$ (1.022 MeV) is required for the excitation process, but it is normally ignored because of its low probability of occurrence. In the case of ionization, the recoiled electron cannot be distinguished from the electron created in pair production, and the process is then called the "triplet production". Conservation of momentum and energy dictate that the incident photon energy be greater than $4m_oc^2$ (2.044 MeV) for triplet production to take place [48]. The cross-section for triplet production is designated as κ_e, to distinguish it from the cross-section for pair production in the nuclear field, κ_n. Detailed expressions for the triplet cross-section are given in reference [68]. However, in a simple form, κ_a can be related to κ_e by the relationship:

$$\frac{\kappa_e}{\kappa_n} = \frac{1}{CZ} \tag{3.77}$$

where C is parameter that is nearly independent of Z, but varies with energy from as high as 15 near the threshold energy of triplet production to about unity at 100 MeV, as can be deduced from the tables of reference [68]. Eq. (3.77) indicates that triplet production is particularly important for low-Z elements, and at high photon energy (which has low C values).

At high energies, the pair production and triplet processes dominate, and the total cross-section becomes equal to the sum of the cross-section of the two processes. This results in a minimum in the total cross-section, reached at photon energies of 55 MeV in carbon, 8 Mev in copper, and 3.5 MeV in lead [69].

The generated positrons in pair and triplet production eventually recombine with the emerging electrons or the electrons of the atom, generating two photons, each having 0.511 MeV energy and moving in opposite directions to preserve momentum. However, these annihilation photons can emerge in any two opposite directions, with equal probability, i.e. their angular distribution is isotropic. NDE can exploit these annihilations as evidence of the occurrence of pair production, since their energy is unique and is lower than that of the incident photons. The strong dependence of κ_n on the atomic-number makes pair-production also a good indicator of material composition. However, pair-production requires a high photon energy, usually produced by particle accelerators.

3.4.5. Photo-nuclear Interactions

At high energies, typically above about 10 MeV, photons can interact directly with the nucleus, resulting in the emission of neutrons, charged-particles, or gamma rays. The cross-sections for such photo-nuclear interaction exhibits a broad peak ("giant " resonance) at about 24 MeV for light nuclei, decreasing with energy to about 12 MeV for heavy nuclei [51]. The width of this resonance (full-width at half-maximum[7]) varies from 3 to 9 MeV, depending on the structure of the nucleus. In spite of this resonance, the photo-nuclear cross-section represents at most 10% of the total cross-section, which includes all other competing interactions with the atomic electrons.

Nuclei with low neutron binding-energy, such as deuterium and beryllium, are affected by photo-nuclear interactions at lower energies, releasing in the process neutrons. These neutrons are liberated from deuterium at a photon energy of 2.226 MeV or above, while a 1.666 MeV photon energy, or higher, is required to release neutrons from beryllium.

3.5. Neutrons

The neutron is one of the fundamental nuclear particles, having a mass and size about equal to that of the a proton (another nuclear particle). Unlike the proton, which has a positive charge, the neutron is a neutral particle, i.e. has a zero electric charge. Therefore, the electromagnetic field of the nucleus, and for that matter any external field, have a very small effect on its behavior[8]. Consequently, neutrons easily penetrate the cloud of electrons surrounding the nucleus and directly interact with the nucleus. This makes the neutron one of the few probes that can directly and easily reach the nucleus, at any kinetic energy. That is to say, neutrons interact with all nuclei of all types of isotopes. This

[7]The concept of full-width at half-maximum is discussed in section 4.1.

[8]In spite of its neutral electric charge, the neutron possesses a magnetic dipole moment ($|\mu| = 60.3$ neV/T), the rotation of which is analogous to that of a negative charge of about $0.4 \times 10^{-21} e$, where e is the elementary charge.

makes the neutron a very attractive probe for determining elemental (or isotopic) composition.

A nucleus can "modify" the physical attributes of a neutron in a number of ways:

- loss of neutron kinetic energy, and momentum, by colliding with the nucleus; in a particle-to-particle *elastic scattering* process, referred to as the (n,n) reaction;
- disappearance of the incident neutron and the emergence of another one, with a different energy; this is usually accompanied as well by the emission of a gamma-ray photon. This *inelastic scattering* process is designated by (n,n$'\gamma$);
- *absorption* of the incident neutron. This may accompanied by the emission of gamma rays, or secondary particles, from within the nucleus in a process known as *radiative capture*, expressed as (n,γ) or (n,ch), where "ch" designates a charged-particle;
- disappearance of the incident neutron and the emergence of many other neutrons in a neutron multiplication process. The emerging neutrons have energies that are different from that of the incident neutron. This type of interaction may result in a splitting of the target nucleus in a *fission* process, (n,f), accompanied by the emergence of other forms of radiation.
- at very low energies (meV range), neutrons assume wave properties, subjecting themselves to *diffraction* and *reflection* in a manner similar to that experienced by light, in addition to magnetic effects due to the neutron magnetic moment.

The occurrence of any of the above interactions depends on the energy of the incident neutron, as well as on the nature of the target nucleus (i.e. the type of isotope). Therefore, each of the above modification processes can be exploited in NDE, as each one can reveal information on the identify of the target with which the neutron interacts. The above modifying interactions are discussed in more details below, in view of their role of providing indications for an NDE technique.

3.5.1. Elastic Scattering

In this process, a neutron loses kinetic energy by a simple hard-ball-like collision process with the nucleus. The (n,n) designation of this reaction emphasizes that the same incident neutron emerges from the reaction, though after losing some of its energy. The loss in neutron kinetic energy is governed by the conventional laws of conservation of momentum and energy.

The kinematics of neutron elastic scattering are described in most nuclear engineering textbooks, see for example references [28] and [70], where the conservation of momentum and kinetic-energy are applied to a simple two-particles collision process between the neutron and a nucleus. These conservation laws are usually applied in a frame of reference whose origin is at the center-of-mass of the neutron and the target nucleus. This is done to simplify mathematical manipulation, since the total momentum, before and after collision, is zero in the center-of-mass system. The results of this analysis are converted to the laboratory frame of reference, where the nucleus is at rest (not moving). Figure 3.5 schematically shows the neutron collision process in both the laboratory ($\mathcal{L}$) and center-of-mass ($\mathcal{C}$) frames of reference. The energy of the outgoing neutron, after colliding with a nucleus with a mass-number A, is related to that of the energy of incident neutron by the relationship:

$$\frac{E'}{E} = \frac{A^2 + 2A\cos\psi + 1}{(A+1)^2} = 1 - \frac{1-\alpha}{2}(1 - \cos\psi) \tag{3.78}$$

where E' is the outgoing neutron, E is the energy of the incoming neutron, ψ is the angle of scattering in the $\mathcal{C}$ system, A is the mass-number of the target nucleus and α is given by:

$$\alpha = \left(\frac{A-1}{A+1}\right)^2$$

The angles θ and ψ are related by the relationship:

$$\cos\theta = \frac{A\cos\psi + 1}{(A^2 + 2A\cos\psi + 1)^{\frac{1}{2}}} \tag{3.79}$$

Eqs. (3.78) and (3.79) can be combined into the following single equation:

$$\frac{E'}{E} = \left[\frac{\cos\theta + (\cos^2\theta + A^2 - 1)^{\frac{1}{2}}}{(A+1)}\right]^2 \tag{3.80}$$

Eq. (3.80) reveals some interesting characteristics that can be used as indicators in NDE. These are:

- The energy of the outgoing neutron, E', is uniquely defined at an angle θ for scattering by a nucleus of mass-number A at a given monoenergetic incident energy E. Therefore, if the scattering energy is measured at a specific angle for a given source energy, E, one can

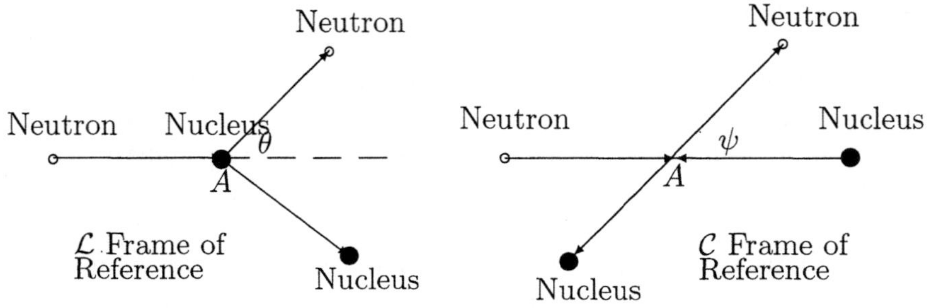

Figure 3.5. Schematic of neutron scattering with a nucleus of mass-number A in the laboratory ($\mathcal{L}$) and center-of-mass ($\mathcal{C}$) frames of references.

determine whether a nucleus of mass-number A is present or not, by monitoring its energy of scattering at a certain scattering angle θ. If it is present, the intensity of the neutron signal, at that angle and energy of scattering, will be an indicator of the density (concentration) of that nucleus.

- The minimum energy, E_{min}, that a neutron can have following a collision with a nucleus of mass-number A is attained when $\cos\theta = -1$. Therefore, using Eq. (3.80):

$$E_{min} = E\left(\frac{A-1}{A+1}\right)^2 = \alpha E \tag{3.81}$$

  It is obvious that neutrons lose more energy when they interact with a lighter nucleus, small A. When scattered by a high mass-number nucleus, the neutron hardly loses energy. Therefore, neutron slowing-down can be used as an indicator of the presence of light nuclei. On the other hand, since metals have a high mass-number, they do not affect the neutron energy loss. Therefore, neutrons can be used to examine lighter materials enclosed in metallic walls, such as water in a metallic pipe. This is one of the unique features of neutrons.

- The mass of a hydrogen nucleus, $A = 1$, is about equal to that of the neutrons. Therefore, when a neutron collides with a hydrogen nucleus it can only scatter in the forward direction, i.e. no backscattering is allowed. This becomes evident by setting $A = 1$ and using $\theta > \frac{\pi}{2}$ in Eq. (3.80), the energy of scattering is always zero, as $\cos\theta$ becomes negative. Therefore, a neutron can scatter with a hydrogen nucleus (a proton) only at angles from $\theta = 0$ to $\frac{\pi}{2}$. Moreover, using Eq. (3.80), with $A = 1$, one can show that $E' = \cos^2\theta$. That is, a neutron can lose all of its energy in a collision with a proton (at $\theta = \frac{\pi}{2}$), or lose no energy at all (at $\theta = 0$). The average energy loss of neutron colliding with hydrogen is, therefore, equal to $\frac{1}{2}E$. This is the highest energy loss per neutron-collision with any type of nuclei, which makes neutron slowing-down an ideal indicator of hydrogen content, or for that matter water content, as fast neutrons are slowed-down in a few collisions to the thermal energy. Moreover, in a single-scattering process, backscattering of neutrons can be used as an indicator of elements other than hydrogen, since hydrogen can only singly-scatter neutrons in the forward direction.

It should be noted that Eq. (3.80) is applicable only when a neutron scatters once; i.e. in single scattering. More collisions will complicate the above indications. On the other hand, the effect of light nuclei will be compounded when multiple scattering occurs, thus showing a large degree of neutron slowing-down.

Neutron Compton Scattering. At high momentum-transfer (between $200h$ fm^{-1} and $1500h$ fm^{-1}, where h is Plank's constant) and high energy-transfer (> 1 eV), an effect similar to that of Compton broadening of photons, see section 3.4.2.3, occurs in neutron scattering [71]. That is, the energy of scattered neutrons will possess a distribution around the energy determined by Eq. (3.80). The profile of neutron Compton scattering can be used to determine single-particle kinetic energies and single-particle momentum distributions. This process is also called deep inelastic neutron scattering to distinguish it from the Bragg scattering of neutron waves (see section 3.5.5.1). In order to observe such behavior, monoenergetic neutrons (in the eV range) are required. However, with a wide-spectrum beam, a filter-difference technique (see section 15.2.2.2) can be used; by observing scattering in the presence and absence of filters that selectively absorb incident and/or scattered neutrons [72]. This method has been used, as indicated in reference [72], to determine the mean atomic kinetic energies in condensed matter (liquids and solids), such as those of helium, noble gases, glasses, graphite, molecular solids and solid hydrides.

3.5.2. Inelastic Interactions

In this process, the nucleus appears to the neutron as a "soft" ball. The result is that the nucleus is internally strained, with some of its nucleons elevated to an excited state. As the nucleus reaches a more stable state it can release individual nucleons, or combination of, in the form of more complex particles, such as alpha-particles.

When a neutron is emitted, the reaction is called inelastic scattering, since in effect one cannot distinguish between the incoming and the emitted neutrons, and the process appears to an observer in the laboratory frame of reference as an apparent scattering process. However, in this process the momentum and kinetic energy are not conserved in the classical sense. The process is usually accompanied by gamma-ray (photon) emission, as the excited nucleus relaxes from its excited state to a more stable state. This reaction is designated as (n,n$'\gamma$).

Since in inelastic scattering, a portion of the incident-neutron energy is deposited in the target nucleus, the relationship between energy and angle found in elastic scattering, Eq. (3.80), no longer applies. However, the kinematics of inelastic scattering can be expressed in a manner similar to that of elastic scattering by accounting for the excitation energy of the target nucleus, Q, i.e. the energy absorbed by the nucleus due to its "softness". Then, the energy of a scattered neutron can be expressed using the elastic-scattering relationship, Eq. (3.80), but with A replaced by an effective-mass, $\bar{A}$, defined as:

$$\bar{A} = A\left[1 - (\frac{A+1}{A})\frac{Q}{E}\right]^{\frac{1}{2}} \tag{3.82}$$

The mass $\bar{A}$ can be viewed as the mass that makes the "soft " nucleus equivalent to a "hard " nucleus.

Unlike elastic scattering, inelastic scattering is not possible with all nuclei. However, the value of Q depends on the type of the nucleus causing inelastic scattering. Therefore, the gamma-ray emitted during inelastic scattering is used as a nucleus identifier; since the energy it carries depends on the Q-value, which in turn is nucleus dependent. Thus, inelastic scattering can be viewed as an activation (secondary-particle producing) reaction.

The inelastic scattering (n,n$'$) reaction is only one of many other reactions in which neutrons are re-emitted but the kinetic energy of the system is not conserved. For example, the reactions (n,n$'\alpha$), (n,n$'$p) produce, respectively, an alpha-particle and a proton, as the nucleus returns from its excited state to a more stable state. However, such reactions occur in certain nuclei and are not as common as inelastic scattering. It should be kept in mind that all non-elastic scattering reactions are not genuine scattering processes, since the original incident neutrons disappear and are replaced by a new neutron emitted from the target nucleus.

Some neutron multiplicity reactions, such as (n,2n),(n,3n) and (n,fission), can also be seen as non-elastic scattering processes, as they involve change in neutron energy and direction. However, the most revealing indication of these modifying reactions is that they result in the multiplication of neutron population. Therefore, they are discussed here separately in section 3.5.4.

3.5.3. Absorption

Reactions in which a neutron is absorbed by the nucleus are quite common, particularly at low neutron energy. In some cases, the neutron is absorbed within the nucleus, converting it from a nucleus with mass-number A to an isotope of a mass-number $A + 1$, keeping the same atomic-number, Z. In other cases, neutron absorption leads to a change in Z, as in the case of the reactions that involve the production of charged-particles: (n,p), (n,d), (n,α), where "p" designates a proton, "d" a deuteron, and α an alpha-particle.

The (n,γ) reaction is known as radiative capture, as it is accompanied by the release of gamma radiation. It is one of the common reactions for thermal neutrons. The energy of the emitted gamma rays is uniquely related to the emitting nucleus. Therefore, these gamma-rays are often called "activation gamma" and are used as indicators to characterize the remitting nucleus, in thermal-neutron activation analysis.

There are some isotopes that are known for their high ability to absorb thermal neutrons; such as ^{3}He, ^{10}B, ^{6}Li and the isotopes of cadmium and gadolinium. The first three isotopes are exploited in neutron detection, see chapter 4. On the other hand, the presence of these isotopes in some other materials can be measured, by the amount of neutron absorption they induce. For example, the concentration of boron in glass, used to control the viscosity of molten glass, can be measured by neutron absorption. The presence of a strong neutron absorber can also interfere with indications provided by other neutron interactions, by suppressing the neutron flux.

3.5.4. Fission and Multiplicity Reactions

Reactions such as fission, (n,2n) and (n,3n), are reactions in which more neutrons are produced than those bombarding the nucleus. For example, a thermal-neutron absorbed by a ^{235}U nucleus can cause a fission that results in the production of 2.54 neutrons, on average. While neutron fission is widely exploited in nuclear reactors to produce energy, the other reactions are less common and their probability of occurrence is quite low. In NDE, neutron multiplication is seen as an indicator of the presence of a neutron-multiplying nuclei; often a fissile material. Therefore, it can be used to the detect the presence, or absence of uranium and plutonium, in techniques known as nuclear assaying.

One of the main characteristics of multiplicity reactions is that the energy of the emerging neutrons is independent of the incident energy, and is in general higher than that of the incident neutrons. For example, a thermal neutron absorbed by a ^{235}U nucleus produces fission neutrons with an energy from 0.075 to 17 MeV. The energy of fission neutrons is usually expressed by empirical relationships obtained from experimental data. For example, the energy distribution of ^{235}U fission neutrons is given by the Watt's formula [70]:

$$N(E) = 0.48\, e^{-E} \sinh \sqrt{2E} \quad \text{neutrons/MeV/fission} \tag{3.83}$$

where $N(E)$ is the fraction of neutrons per unit energy emitted per fission and E is neutron energy in MeV. This increase in neutron energy in a fission process can also be seen as a definite indicator of the presence of a fissile material. Fission is also used in neutron detection, see Chapter 4.

For other multiplicity reactions, such as (n,2n) and (n,3n), the energy of the emitted neutrons depends on both the energy of the incident neutron and the type of the target nucleus, as well as on the manner on which the nucleus is excited. In these reactions, unlike fission, the nucleus is not completely destroyed, rather the neutron usually excites the nucleus. The excited nucleus subsequently decays and releases two or more neutrons. It is, therefore, reasonable to use an approach similar to that employed in inelastic scattering, Eq. (3.82), assigning a proper value for the excitation energy Q. However, the excitation energy can also be made dependent on the energy of the incident neutron.

3.5.5. Coherent Scattering

3.5.5.1 Diffraction

Neutrons at sufficiently low energy can, like photons (see section 3.4.3.2), experience coherent (Bragg) diffraction. While photons coherently scatter by the atom, neutrons are scattered by the nucleus. This enables the use of neutron diffraction to distinguish between two isotopes of the same element. While the probability of photon coherent scattering for materials of low atomic-number is quite low, see Eq. (3.63), coherent scattering of neutrons is not dependent on the atomic-number, and can occur with a high probability in low mass-number elements. Like photons, neutron diffraction takes place when the distance between two adjacent nuclei in a crystal is equal to the wavelength of the incident radiation. For example, in a carbon crystal, the carbon-to-carbon single bond is around 154 pm [23]. The neutron wavelength, λ, is related to its momentum by the de Broglie relationship (see appendix C):

$$\lambda = \frac{h}{mv} \tag{3.84}$$

where m is the mass of the neutron (1.675×10^{-27} kg), v is the velocity of the neutron and h is Plank's constant (6.625×10^{-34} Js). Therefore, for the carbon-to-carbon spacing of 154 pm, Bragg diffraction occurs, according to Eq. (3.84), when $v = 2568$ m/s, which is equivalent to a kinetic energy of about 0.034 eV; a value that is close to that of thermal neutrons at room temperature, see discussion following Eq. (3.89). For crystals of longer lattice spacing, neutrons of lower velocity, or lower energy, are required. Cold neutrons, see section 2.3.4, can provide such lower energy neutrons. The neutron wavelength can be approximated by [73]:

$$\lambda(\text{pm}) = \frac{904}{\sqrt{E(\text{meV})}} \tag{3.85}$$

That is, a neutron at an energy of 1 meV has a wavelength of 904 pm.

Small-Angle Diffraction. Diffraction of neutrons at small-angle by small particles is affected by the particle size, shape and concentration. A sphere of radius r causes a small deflection of a cold-neutron beam by an angle: $\Delta\phi = \frac{\sqrt{3}\lambda}{\pi r}$, where λ is the neutron wavelength [38]. Therefore, the diffraction angle is inversely proportional to the particle size, offering the possibility of detecting precipitates and corrosion products in alloys [74]. The cross-section per solid angle for neutron scattering by particles is given by the Guinier approximation [73]:

$$\frac{d\sigma}{d\Omega} = N_p V_p^2 (b_p - b_m)^2 \exp\left[-\frac{Q^2 r^2}{3}\right] \tag{3.86}$$

where N_p is the number of particles per unit volume, V_p is the particle volume, $(n_p - n_m)$ is the difference in the scattering length[9] between the particles and the surrounding medium, and Q is the momentum transferred from the neutron to the particle. Therefore, according to Eq. (3.86), small-angle diffraction offers the possibility of measuring par-

[9]The scattering energy is distributed over the surface of a sphere of a minimum surface area of $4\pi b^2$. The quantity b is known as the neutron scattering length, since neutrons are scattered by the nucleus of the atom, b is of the order of 10^{-14} m for all nuclei, see section 3.5.6.

ticle size and concentration; and even shape since r in Eq. (3.86) can be replaced by the gyration radius.

According to Bragg's law, Eq. (3.66), λ reaches a maximum value, $\lambda_{max} = 2d$, when $\theta = 0$, $n = 1$. In a polycrystalline medium, with many values of d, $\lambda_{max} = 2d_{max}$, where d_{max} is the largest spacing in the crystal. Corresponding to this maximum value of λ is a minimum energy that can be calculated using Eq. (3.85). Below this energy, Bragg scattering ceases to exist and the scattering cross-section sharply decreases, see reference [75]. This is called Bragg cutoff phenomenon.

3.5.5.2 Reflection and Refraction

Cold neutrons are also reflected and refracted off a flat surface in a manner similar to the reflection of light by a mirror. The fraction, R, of the beam reflected by a flat surface is governed by the Fresnel law of reflection, that is [73]:

$$R = \left[\frac{\sin\phi - (\eta^2 - \cos^2\phi)^{\frac{1}{2}}}{\sin\phi + (\eta^2 - \cos^2\phi)^{\frac{1}{2}}}\right]^2 \tag{3.87}$$

where ϕ is the angle of incidence measured with respect to the normal to the surface and η is the neutron refractive index of the material. The real-part of the complex index of refraction is given by [76]:

$$\eta = 1 - \lambda^2 \frac{Nb}{2\pi} \tag{3.88}$$

where λ is the neutron wavelength, given by Eq. (3.85), N is the atomic density, and b is the bound coherent scattering-length, see definition following Eq. (3.86) and reference [77] for values. The refractive index for neutrons is generally less than unity (about $1 - 10^{-6}$ or so). Therefore, total (specular) reflection takes places at very small glancing angles. Reference [37] provides a handy source for information on the physics of the scattering of cold neutrons. Changes in the density or flatness of a surface cause interference between the reflected waves, which can be used to detect and characterize surface irregularities [73].

3.5.6. Cross Sections

Different types of neutron interactions occur with different probability (microscopic cross-section, σ). The value of σ for a particular interaction depends on the type of the target nucleus, as well as the energy of the incident neutron, E. That is, σ is a function of E, and is denoted as $\sigma(E)$. A subscript is also associated with σ, to identify the reaction type. For example, σ_{in} refers to inelastic scattering, σ_f designates the fission cross-section, while σ_t is the total cross-section for all reactions. Measured and calculated cross-sections for different nuclides (isotopes) are compiled in computer files. A number of such libraries are available; such as the Evaluated Nuclear Data Files, ENDF/B [78]. These files were complied over the years from reported laboratory measurements and are continuously revised. The on-line program ENDFPLOT [27] provides handy plots for various elements showing how interaction cross-sections change with energy for different elements.

Appendix E discusses the methods used for calculating the cross-section of a chemical compound, or a material consisting of a mixture of nuclides of different types. Note, however, that at thermal-neutron energies, the thermal motion of the molecules affect the neutron cross-section under thermal-equilibrium conditions, as discussed below.

3.5.6.1 Thermal Neutrons

Although the neutron interacts with the nucleus, without any regard to chemical bonds, at low neutron energies, chemical bonds have some effect on neutrons. When the neutron energy is greater than the molecular binding energy (a few eV), the kinetic energy associated with the thermal motion of the molecules of the medium can be neglected, in comparison to that of the incident neutron. That is to say, as seen by the neutron, the nucleus appears to be at rest. However, at low neutron energies, particularly when the neutron is in thermal equilibrium with the medium, its kinematic behavior becomes comparable with that of the molecules of the medium. The neutron can in turn gain energy from the molecules, or lose energy to them. Therefore, for thermal neutrons, the physical state of the medium can affect the value of the nuclear cross-section. In other words, the state of the nucleus as being bound in a crystal lattice or in a chemical compound will have an effect on the neutron's interaction probability. Water is an example of a material where the cross-section formulated by adding the contribution of its individual elements is different from that measured at the thermal neutron energy.

The distribution of thermal-neutrons in a medium is similar to that of a free gas, provided that the medium is not a strong neutron absorber. Therefore, the Maxwell-Boltzmann distribution, derived from the kinetic theory of gases, is used to describe the kinetic energy distribution of thermal neutrons. The probability of neutrons having an energy E is expressed in the form [28]:

$$\frac{n(E)}{n} = \frac{2\pi}{(\pi kT)^{\frac{3}{2}}} \exp\left[-\frac{E}{kT}\right] E^{\frac{1}{2}} \tag{3.89}$$

where $n(E)$ is the number of neutrons having energies in the range from E to $E + dE$, n is the total number of neutrons in the system, k is the Boltzmann constant, and T is the absolute temperature. It can be shown from Eq. (3.89) that the kinetic energy corresponding to the most probable velocity interval is given by kT, where T is the temperature of the medium in degree Kelvin, with $k = 1.381 \times 10^{-23}$J/K $= 8.617 \times 10^{-5}$ eV/K. At room temperature, 300 K, this energy is equal to 0.025 eV.

While chemical binding tends to increase the scattering cross-section of thermal neutrons, for the reasons discussed below, it hardly affects the value of the absorption cross-section [79]. The absorption cross-section for free nuclei can, therefore, be equally used for bound nuclei. In scattering, chemical binding causes the atom, rather than the nucleus alone, to recoil as a whole unit. Since the atom is chemically bound to a molecule, neutron scattering has to be seen as an interaction with the whole molecule. With the molecule's mass typically larger than that of the neutron, the center-of-mass of the nuclear system (neutron and the molecule) coincides with that of the molecule, and in turn with the $\mathcal{L}$ system. The result is that scattering with bound atoms becomes isotropic in the $\mathcal{L}$ system, while scattering with free nuclei is usually isotropic in the $\mathcal{C}$ system [79], see section 3.5.6.3. With these two separate cases of isotopic scattering, one can express the scattering cross-section for bound and free atoms as:

$$\begin{aligned} \sigma_f(\cos\psi) &= \frac{\sigma_f^s}{4\pi} \\ \sigma_b(\cos\theta) &= \frac{\sigma_f^s}{4\pi} \\ \sigma_f(\cos\psi)d\cos\psi &= \sigma_f(\cos\theta)d\cos\theta \end{aligned} \tag{3.90}$$

where σ_f is the scattering cross-sections for free atoms and σ_b is that for bound atoms, which are respectively isotropic in the $\mathcal{C}$ and $\mathcal{L}$ systems, the superscript s refers to scattering at all angles, and ψ and θ are the scattering angles in the $\mathcal{C}$ and $\mathcal{L}$ systems, respectively, see Figure 3.5. However, when $\psi = 0$, θ is also equal to zero, see Eq. (3.79), and there is no energy loss, according to Eq. (3.78). Then, neither the free atom nor the molecule recoil, in effect nothing happens. Then $\sigma_f(\cos 0) = \sigma_b(\cos 0)$. Consequently, one can write Eqs. (3.90) as:

$$\begin{aligned} \sigma_b^s &= 4\pi\sigma_b(\cos 0) = 4\pi\sigma_f(\cos 0) \\ &= \sigma_f(\cos\psi)\left.\frac{d\cos\psi}{d\cos\theta}\right|_{\cos\psi=\cos\theta=0} \end{aligned}$$

$$= \sigma_b(\cos\theta)\left.\frac{d\cos\psi}{d\cos\theta}\right|_{\cos\psi=\cos\theta=0}$$

Now using the relationship between θ and ψ, Eq. (3.79), one obtains:

$$\sigma_b = \left(1+\frac{1}{A}\right)^2 \sigma_f \tag{3.91}$$

This equation demonstrates that the bound cross-section of scattering for thermal neutrons is higher than that for free atoms. In the case of hydrogen, $\sigma_b = 4\sigma_f$, but for heavy atoms, the two values are approximately equal.

The differential cross-section for thermal neutrons is often described by the free-gas approximation, known as the $S(\alpha,\beta)$ treatment. Assuming isotropic scattering in the azimuthal direction, see Figure 3.1, the double-differential cross-section with an atom of mass-number, A, in terms of the polar angle of scattering, θ, is given by [79, 80]:

$$\begin{aligned}\frac{d\Omega}{d\cos\theta dE} &= \sigma(E_i \longrightarrow E_f;\cos\theta)\\ &= \frac{\sigma_0}{\pi kT}\sqrt{\frac{E_f}{E_i}}\exp\left[-0.5\beta\right]S(\alpha,\beta) \qquad (3.92)\\ \text{where,}\ \alpha &= \frac{E_i+E_f-2\sqrt{E_iE_f}\cos\theta}{AkT} \quad \text{and} \quad \beta=\frac{E_f-E_i}{kT}\end{aligned}$$

where E_i and E_f are, respectively, the initial and final energies of the neutron at a thermal temperature T, σ_0 is the scattering cross-section integrated over all values of the final energy, E_f. In Eq. (3.92), β is obviously related to the change in the energy of the neutron, while α is proportional to the square of the change in its momentum[10], during the scattering process. Under thermal equilibrium, the cross-section for energy gain is related to that of energy loss as:

$$\sigma(E_i \longrightarrow E_f;\cos\theta)E_i\exp\left[\frac{-E_i}{kT}\right] = \sigma(E_f \longrightarrow E_i;\cos\theta)E_i\exp\left[\frac{-E_f}{kT}\right] \tag{3.93}$$

Therefore, $S(\alpha,\beta)$ is invariant under the transformations $E_i \longrightarrow E_f$ and $\beta \longrightarrow -\beta$. The $S(\alpha,\beta)$ values are available in the ENDF/B files [78] for

[10]From Figure 3.5 in the $\mathcal{L}$ system, momentum conservation in the direction of the incident neutron leads to: $p-p'_z = \sqrt{2m_n}(\sqrt{E}-\sqrt{E'})$, where p and p'_z are, respectively, the momentum of the neutron before and after scattering in the direction of the incident neutron, with $p=\sqrt{2m_nE}$, and $p'=\sqrt{2m_nE'}$; m_n being the mass of the neutron and p' the momentum of the scattered neutron. Therefore, $(p-p'_z)^2 = 2m[E+E'-2\sqrt{EE'}\cos\theta]$.

some materials, such as light water, heavy water, graphite and polyethylene, at some selected temperatures. The integrated cross-section, σ_0, at energy E can be expressed mathematically as [81]:

$$\sigma_0(E) = 4\pi R^2 \left[\frac{2}{\gamma\sqrt{\pi}} \exp[-\gamma^2] + \left(1 + \frac{1}{2\gamma^2}\right) \mathrm{Erf}(\gamma)\right] \quad (3.94)$$

$$\text{with} \quad \gamma = \sqrt{A\frac{E}{kT}} \quad \text{and} \quad \mathrm{Erf}(\gamma) = \frac{2}{\sqrt{\pi}} \int_0^{\gamma} \exp[-x^2]\, dx$$

where R is a constant known as the scattering length, and A is its mass-number. The error function (Erf) is present due to the integration of the Maxwell-Boltzmann distribution, Eq. (3.89) over energy. There are two limiting values to σ_0:

$$\sigma_0 = \begin{cases} 4\pi R^2 & \text{for } E >> kT \quad \text{where } \mathrm{Erf}(\gamma) \to 1 \\ 4\pi R^2 \frac{3}{\sqrt{\pi}}\sqrt{\frac{kT}{AE}} & \text{for } E << kT \quad \text{where } \mathrm{Erf}(\gamma) \to \frac{2\gamma}{\sqrt{\pi}} \end{cases} \quad (3.95)$$

3.5.6.2 Energy Dependence

The cross-sections for various reactions and isotopes are available, at different neutron energies, in cross-section libraries, such ENDF/B [78]. It is important, however, to be aware of the behavior of cross-sections with neutron energy for different nuclides, as it affects the modification of the neutron signal by the material of interest. A summary is provided here for the major trends of cross-section change with energy.

The reason the cross-section changes from one nuclide to another is that, similar to the electron orbits of the atom, each nuclide has its own unique discrete nuclear energy levels where nucleons (protons and neutrons) are allowed to reside. A nucleon can move from one level to another, if it acquires an exact energy (a quantum leap) that enables it to reach the next allowed energy level, or levels. The nucleus then becomes excited (in a metastable quasi-stationary state), forming a compound nucleus. Each nucleus has its own energy-dependent excitation function. Neutrons with energy below or above the excitation energy have a lower probability of forming a compound nucleus.

At high energies, the excitation levels in most nuclei are closely spaced, broad and partially overlapped. This, in turn, results in a continuous excitation function. Therefore, when a high-energy neutron bombards a nucleus, the nuclear levels appear to it as continuous levels; thus resulting in a continuous probability of interaction, i.e. smooth cross-sections. This trend appears at high neutron-energy, as shown in Figure 3.6 and 3.7 for neutrons of energy above a few keV .

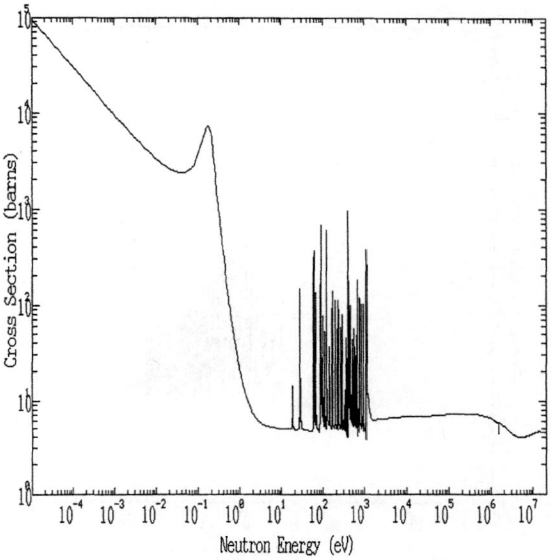

Figure 3.6. Fine resonances in neutron cross-sections, σ_t for natural Cd, plotted using ENDFPLOT [27].

At lower energies, in the eV to near the keV range, the excited levels of the nucleus become discrete. Therefore, an incident neutron with an energy exactly equal to the excitation energy has the highest level of interaction with the nucleus, as it can move a nucleon from one level to another. This results in what is known as the resonance cross-sections; as for example in the case for natural cadmium, shown in Figure 3.6. If the nuclear levels are widely spaced, then the resonance can be easily observed, and are called resolved resonances. This is in contrast to the *unresolved* resonances, where the resonances take a smoother appearance due to the narrowing of the gap between neighboring energy levels, and the subsequent overlapping of the resonances. This typically occurs in heavy elements, as shown in Figure 3.7 for ^{235}U.

The terminology of "resonances" arises from acoustics and corresponds with a constructive interference of waves. Therefore, another explanation of the behavior of neutron cross-sections arises from wave mechanics, see appendix D. The neutron is viewed as a wave, with the nucleus representing itself to the neutron as a "potential" field that can interfere with the incoming neutron wave. When the neutron is away from the nucleus, so that it does not feel its presence, it can be described as a plane wave. The neutron wavelength, λ, depends on the momen-

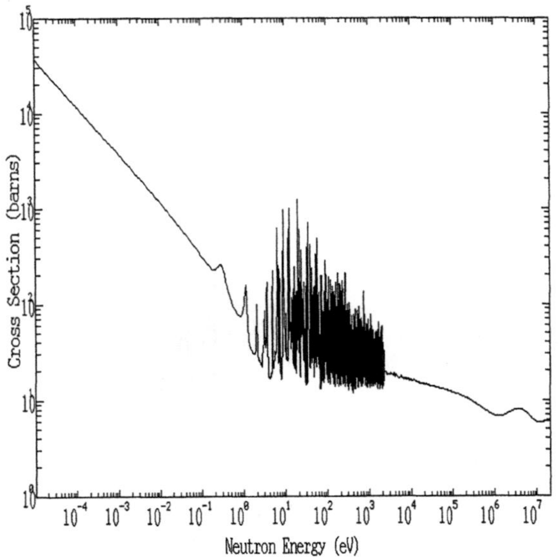

Figure 3.7. Broad (unresolved) resonances in neutron cross-sections, σ_t for ^{235}U, plotted using ENDFPLOT [27].

tum of the neutron, mv, and is given by the de Broglie relationship, Eq. (3.84).

The neutron plane wave is in effect the result of many spherical partial waves (harmonics). In the absence of external effects, these partial waves are all in phase, thus constructively interfere with each other to form the plane wave. On the other hand, when the plane wave approaches a nucleus, the field of the nucleus affects the partial waves differently, resulting in patterns of constructive and destructive interferences between them. A constructive interference corresponds to a resonance, hence a maximum probability of interaction.

Potential Scattering. Simple reflection of the neutron plane wave, by the surface of the nucleus, is analogous to elastic scattering, and is called potential scattering. In this process, when the neutron wave hits a nucleus, it reverses its direction, and the incoming and outgoing waves combine into a single wave with an amplitude double that of the incident wave. The probability of potential scattering (microscopic cross-section) is proportional to the square of the amplitude of the neutron wave. As indicated in appendix D, the corresponding maximum possible scattering cross-section, for the ℓth partial wave, is equal to [48]:

$$\sigma^{\ell}_{s,max} = 4\pi\frac{(2\ell+1)}{k^2} \tag{3.96}$$

where k is the wave number ($k = \frac{2\pi}{\lambda}$), and $\ell = 0, 1, 2, \cdots$. When $\ell = 0$, the wave is called the "s" wave, the first (undisturbed) harmonic of the wave. The potential scattering cross-section should also be a function of the size (or surface area) of the nucleus, since the scattering energy is distributed over the surface of a sphere of surface area $4\pi R^2$, where R is the radius of the nucleus. For an s-wave neutron, $\ell = 0$, the cross-section for potential scattering can be expressed as [48]:

$$\sigma_{pot} = \frac{4\pi}{k^2}\sin^2(kR) \tag{3.97}$$

where kR represents the phase shift of the scattered neutron wave. Note that the maximum value of σ_{pot}, at $\sin(2kR) = 1$, is consistent with that obtained from Eq. (3.96), for $\ell = 0$. At low neutron energies, $kR << 1$, Eq. (3.97) is reduced to:

$$\sigma_{pot} \simeq 4\pi R^2 \quad \text{at low energy} \tag{3.98}$$

Figure 3.8 shows the smoothly varying scattering cross-section of hydrogen. The constant value, of about 20 b, appearing at most of the energy range, may be explained by potential scattering. However, using Eq. (3.1), R for hydrogen ($A = 1$) is equal to 1.25×10^{-15}m, which leads to a value of about 0.2 b for σ_{pot}, according to Eq. (3.98). The discrepancy between the two values is due to the fact that Eq. (3.98) represents only s-wave scattering, while the cross-section includes all partial waves, see Eq. (D.40). Neutron scattering must be represented by an infinite number or partial waves, as the comparison of Eqs. (3.113) and (3.114) shows, to accommodate the fact that backscattering, in a single neutron collision, is not possible in the laboratory frame of reference.

The radius corresponding to the 20 b value is about ten times larger than that obtained using Eq. (3.1) for hydrogen. This indicates that the effect of the potential field of a nucleus can extend far beyond its physical dimensions. Therefore, the value of R obtained from cross-section measurements is often called the scattering length. Note also that in Figure 3.8, at very low energies, the cross-section decreases with energy, with a $\frac{1}{\sqrt{E}}$ trend, due to thermal-equilibrium effects, see Eq. (3.95). At high energies, the trend can be explained as discussed following Eq. (3.102) below. The scattering cross-section for hydrogen, $\sigma_s^H(E)$, for energies

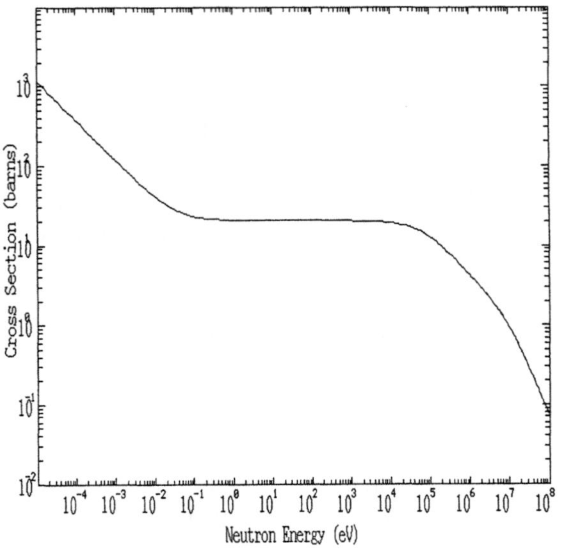

Figure 3.8. Neutron-scattering cross-sections, σ_s for ^{1}H, plotted using ENDF-PLOT [27].

above 0.1 eV is approximately given by [82]:

$$\sigma_s^H(E) = 4\pi \left[0.75 \times \frac{0.2884}{1 + 0.2427E + 0.0028E^2} + 0.25 \times \frac{5.603}{1 + 7.417E + 0.1105E^2}\right] \quad (3.99)$$

with E in MeV and σ_s is in barns. This is a spin-averaged cross-section. The first term representing 25% of the spin-singlet state and the second term 75% of the spin triplet state of a neutron-proton nuclear system, see appendix D. Note that at low energies, $E << 1$ MeV, $\sigma_s^H(E) \approx 20$ b, as shown in most of the range of Figure 3.8.

When the incoming wave enters the potential field of the nucleus, some absorption of the incoming wave can occur, reducing the amplitude of the outgoing partial wave. When complete (maximum) absorption occurs, there will be no outgoing waves. Then one observes only the presence of the incident wave around the nucleus; unlike in elastic scattering where the wave amplitude is doubled. Since the probability of scattering is proportional to the square of the wave's amplitude, the maximum cross-section for neutron capture, $\sigma_{c,max}$, is one-fourth that of potential scattering, using Eq. (3.97) at $\ell = 0$:

$$\sigma_{c,max} = \frac{\pi}{k^2} \tag{3.100}$$

Note that the same conclusion was arrived at in appendix D, Eq. (D.25), using the complete absorption of the incident plane wave. The maximum cross-section of Eq. (3.100) is also the cross-section for the formation of a new nucleus by the absorption of the incident neutron. As Eq. (3.100) indicates, for low neutron energies, $kR << 1$, the value of $\sigma_{c,max}$ can far exceed the geometric cross-section of the target nucleus, πR^2.

At large energies, where $\frac{1}{k} << R$, both the capture and potential scattering cross-sections for neutrons approach the same asymptotic value:

$$\sigma_c = \sigma_s = \pi \left(R + \frac{1}{k} \right)^2 \quad \text{at high energy} \tag{3.101}$$

Therefore, the total cross-section, for high energy neutrons, is twice the value given by the above equation.

Resonances. When the incident neutron wave has a wavelength that resonates with the nucleus, or in other words if the neutron energy matches one of the excitation levels of the nucleus, a compound nucleus will be formed. Then the cross-section, σ_{com}, takes the shape of one of the resonances, shown in Figure 3.6. At a neutron energy, E, in the neighborhood of a resonance energy E_0, the shape of the resonance is described by the Breit-Wigner formula [48]:

$$\sigma_{com} = \sigma_a \sqrt{\frac{E_0}{E}} \frac{1}{\frac{(E_0 - E)^2}{(\frac{\Gamma}{2})^2} + 1} \tag{3.102}$$

where σ_a is the potential (smooth) absorption cross-section and Γ is the full width of the resonance at half of its maximum value. If the compound nucleus returns to a stable state by the emission of gamma-rays, σ_{comp}, corresponds to the capture cross-section, $\sigma_{(n,\gamma)}$. Whereas, σ_{comp} corresponds to the inelastic scattering cross-section, $\sigma_{(n,n'\gamma)}$, if a neutron and a photon are emitted, as the compound nucleus stabilizes. For broad resonances, $\Gamma >> (E - E_0)$, and one can show from Eq. (3.102) that σ_{comp} becomes proportional to $\frac{1}{\sqrt{E}}$, i.e. proportional to $\frac{1}{v}$, where v is the velocity of the incident neutron. The same trend of change in the cross-section is observed when $E << E_0$, or $E >> E_0$, i.e. at the far ends of a resonance. This $\frac{1}{\sqrt{E}}$ behavior commonly governs the absorption cross-section of slow-neutron, away from the resonances, as can be seen in Figures 3.6 and 3.7.

Doppler Broadening. The thermal motion of nuclei affects the relative velocity of the neutron with respect to the nucleus, which changes the energy of the incident neutron as seen by the nucleus. The result is a broadened resonance with a reduced peak value. The effect is known as Doppler-broadening. This effect is particularly important when the temperature of the medium is sufficiently high so that the resonance energy is affected by the Maxwell-Boltzmann energy distribution of thermal neutrons, Eq. (3.89).

Resonance Integral. Within the resonance energy range (a few eV to a few keV), neutrons can be considered to be, more or less, moderated, but have not yet reached thermal equilibrium. The neutron flux can then be considered to be proportional to $\frac{1}{E}$, where E is the neutron energy. The rate of occurrence of a particular interaction in the resonance region is proportional to the integral of the cross-section, as a function of energy, multiplied by the neutron flux. With the neutron flux assumed to be proportional to $\frac{1}{E}$, one can define the resonance integral, I_r, as:

$$I_r = \int \sigma(E)\frac{dE}{E} = \int \sigma(E)\, d\ln E \tag{3.103}$$

The lower limit of the integral is usually taken as the cadmium cutoff energy of 0.5 eV. The value of I_r for a given reaction is determined by measuring the reaction rate of a foil of the material of interest when it is covered with cadmium, and when it is not covered [83]. The resonance integral is widely reported as a measure of the interaction rate within the resonance range of neutrons [84, 85].

3.5.6.3 Differential Cross-Sections

The discussion here focuses only on elastic scattering, since inelastic and nonelastic scattering processes are usually assumed to be isotropic in the laboratory ($\mathcal{L}$) frame of reference. Scattering in the azimuthal direction, see Figure 3.1, is also assumed to be isotropic. The angular distribution of scattering in the polar angle is represented by the differential cross-section:

$$\frac{\partial \sigma}{\partial \cos\theta} = \sigma(E \rightarrow E'; \cos\theta) = \sigma_s(E)p(E \rightarrow E'; \cos\theta) \tag{3.104}$$

where $\sigma_s(E)$ is the scattering cross-section at energy E for all angles, E' is the energy of scattering at an angle of scattering θ, and p refers to the probability of scattering.

The probability of elastic scattering from an energy E to an energy E' with an angle θ in the $\mathcal{L}$ system is customarily reported in terms of

Legendre coefficients (see appendix D) in the center-of-mass ($\mathcal{C}$) system (see Figure 3.5) as [86]:

$$\begin{aligned} p(E \rightarrow E'; \cos\theta) &= \frac{1}{4\pi}\sum_{j=0}^{J}(2j+1)f_j(E)P_j(\cos\psi) \times \\ & \delta\left[E' - E + E\frac{1-\alpha}{2}(1-\cos\psi)\right] \end{aligned} \quad (3.105)$$

where $\alpha = (\frac{A-1}{A+1})^2$, A is the ratio of the target mass to the rest mass of the neutron (A is very close in value to the mass-number of the nucleus with which the neutron is colliding), ψ is the angle of scattering in the $\mathcal{C}$ system and δ is the dirac delta function, P_j is an ordinary Legendre polynomial of order j, and f_j is called the jth Legendre coefficient in the $\mathcal{C}$ system. ENDF/B files [78] contain these Legendre coefficients and all the constants necessary for the evaluation of the cross-sections.

The dirac delta function in Eq. (3.105) imposes the unique relationship between the energy and angle of scattering in elastic scattering, in accordance to Eq. (3.80). This makes the probability of scattering an inseparable function of energy and angle; i.e. it cannot be expressed as the multiplication of two functions, one in energy and the other in angle.

The choice of the number of coefficients, J, in the Legendre expansion of Eq. (3.105) depends on the mass of the target nucleus and the neutron energy. However, some rough criteria are available to estimate the required order of expansion in the $\mathcal{C}$ system. Elastic scattering is isotropic (also called spherically symmetric, or s-scattering) when $J = 0$, in the $\mathcal{C}$ system, if the neutron de Broglie wavelength, see Eq. (3.84), is greater than its radius, or equivalently, if [87]:

$$E < 10A^{-\frac{2}{3}} \quad (3.106)$$

where E is the incident neutron energy in MeV and A is the mass-number of the nucleus. When scattering is isotropic, all scattering energies become equally probable. Therefore, using Eq. (3.78):

$$p(E \rightarrow E')dE' = \begin{cases} \frac{\sigma_s(E)dE'}{(1-\alpha)E} & \text{for } \alpha E \leq E \leq E' \\ 0 & \text{for } \alpha E > E > E' \end{cases} \quad (3.107)$$

The cosine distribution (called the p-scattering approximation) occurs at $J = 1$, in the $\mathcal{C}$ system, and is valid when [87]:

$$10A^{-\frac{2}{3}} < E < 40A^{-\frac{2}{3}} \quad (3.108)$$

provided that resonance scattering is ignored.

In general, the required number of coefficients may be chosen such that it satisfies the inequality [87]:

$$E < 10(J+1)^2 A^{-\frac{2}{3}} \tag{3.109}$$

The use of s-scattering or p-scattering approximations, in the $\mathcal{C}$ system, is adequate for a large number of isotopes and over a wide range of incident-neutron energies; as relationships (3.106) and (3.108) indicate.

For hydrogen, the assumption of isotropic scattering in the $\mathcal{C}$ system is valid up to 10 MeV neutron energy, as relationship (3.106) indicates. Moreover, hydrogen is an effective slowing-down element, and thus brings high-energy neutrons to below the above energy limit after a few collisions. Isotropic scattering in the $\mathcal{C}$ system can, therefore, be applied to hydrogen-containing materials without much loss of accuracy. This is, of course, provided that there is a sufficient amount of hydrogen to rapidly reduce the energy to the few MeV energy range. It should, however, be mentioned that this assumption breaks down in the thermal energy range, where chemical binding becomes important.

The p-scattering approximation can accommodate situations where the angular distribution shows peaks in the forward or backward directions. The coefficient f_1 in Eq. (3.105) takes a negative value when scattering is predominantly backward, and a positive value when it is predominantly forward. The bias in the forward direction is of the order of $0.07A^{\frac{2}{3}}E$, with E measured in MeV [83].

In the $\mathcal{L}$ system, the probability of scattering must be given in terms of the angle θ (see Figure 3.5). For large mass-number nuclei, the $\mathcal{C}$ and $\mathcal{L}$ systems almost coincide with each other and θ becomes approximately equal to ψ. The Legendre expansion of Eq. (3.105), and the criteria of Eqs. (3.106) to (3.109), become equally applicable to both frames of reference.

For elements with a low mass-number, the transformation to the $\mathcal{L}$ system is not as straightforward. In the $\mathcal{L}$ system, the probability of scattering can be expressed as [86]:

$$p(E \to E'; \cos\theta) = \frac{1}{4\pi\sigma_s(E)} \sum_{\ell=0}^{L} S_\ell(E \to E') P_\ell(\cos\theta) \tag{3.110}$$

where S_ℓ is the ℓth Legendre coefficient in the $\mathcal{L}$ system. The relationship between the Legendre coefficients in the $\mathcal{L}$ and $\mathcal{C}$ systems is given by [86]:

$$S_{ell}(E \to E') \quad = \quad \frac{\sigma_s(E)}{(1-\alpha)E}(2\ell+1)P_\ell[\cos\theta_o(E \to E')] \times$$

$$\sum_{j=0}^{J}(2j+1)f_j(\cos\psi)P_j(\cos\psi) \tag{3.111}$$

The angle θ, in the $\mathcal{L}$ system, is related to the angle ψ, in the $\mathcal{C}$ system, by Eq. (3.79). The angle $\theta_o(E \to E')$ expresses the energy-momentum conservation in the $\mathcal{L}$ system and is given, using Eq. (3.80), as:

$$\cos\theta_o(E \to E') = \frac{A+1}{2}\sqrt{\frac{E'}{E}} - \frac{A-1}{2}\sqrt{\frac{E}{E'}} \tag{3.112}$$

In order to demonstrate the difficulty of determining the number of Legendre coefficients required in the $\mathcal{L}$ system, let us consider the simple case of s-scattering, which requires only the first term of the expansion ($J = 0$) in the $\mathcal{C}$ system. Utilizing Eqs. (3.110) and (3.111):

$$p(E \to E'; \cos\theta) = \frac{1}{4\pi(1-\alpha)E} \times$$

$$\sum_{\ell=0}^{L}(2\ell+1)P_l(\cos\theta)P_l[\cos\theta_o(E \to E')] \tag{3.113}$$

This is while the same probability expressed in the $\mathcal{C}$ system is given by:

$$p(E \to E'; \cos\theta) = \frac{1}{4\pi}\delta\left[E' - E + E\frac{1-\alpha}{2}(1-\cos\psi)\right] \tag{3.114}$$

It can be shown that the right-hand-sides of Eqs. (3.113) and (3.114) can only be equal when L in the former approaches infinity; then the summation in Eq. (3.113) approaches a delta function, $\delta(\cos\theta - \cos\theta_o)$ [79].

Therefore, low mass-number nuclei require a very large number of coefficients to approach isotropic scattering in $\mathcal{C}$ system. Since, in practice, only a limited number of coefficients can be employed, scattering of neutrons in light elements is always dealt with in an approximate manner, if Legendre expansion in the $\mathcal{L}$ system is employed. Failure to use a large enough number of coefficients may lead to a negative value for the probability of scattering, a physically unacceptable condition. The situation improves as the mass of the target nucleus increases, leading to a decreased distance between the points of reference in the two systems. Eventually, for large mass-numbers, the two systems coincide and the angles ψ and θ become equal, enabling the expression of S_ℓ, using Eqs. (3.105) and (3.110), as:

$$S_\ell = \Sigma_s(E)(2\ell+1)\delta\left[E' - E + E\frac{1-\alpha}{2}(1-\cos\psi)\right] \tag{3.115}$$

3.6. Radiation Transport

As radiation travels in matter, it is affected by the reactions discussed in the above sections. Radiation direction and energy change by scattering, some of it is absorbed, and some new radiation may be generated. Even in the absence of a medium to interact with, i.e. in void, the intensity of radiation changes, simply as it spreads over a wider volume and angle. As indicated in section 3.1, charged-particles interact continuously with matter, thus their transport is limited to within their short range; although their secondary effects can complicate the radiation transport process as discussed in section 3.6.5. On the other hand, neutral entities, such as photons and neutrons, can move a considerable distance within a material. The intensity of radiation is "modified" as it is transported within matter, as discussed below.

The population of photons and/or neutrons transported in a material is usually sufficiently large that one can neglect the effect of statistical fluctuation in their number. However, in comparison to the number of atoms (or nuclei) of matter, the number of photons and neutrons is so small that it is reasonable to assume that such particles do not interact with each other. Consequently, neutrons or photons can be viewed as a "gas" seeping through a porous medium filled with atoms. This "gas", unlike normal gases, can penetrate boundaries, thus the medium of boundaries is within the domain of radiation transport. As discussed in section 3.1, most of the medium with which the radiation "gas" travels, is vacant, as the nuclei and electrons occupy only a small portion of available space. Therefore, a neutron or a photon can travel a long distance before encountering an atom or a nucleus to interact with. Being neutral particles, neutrons and photons travel in straight-lines and without changing energy, until they suffer an interaction. It is reasonable to assume that the medium properties are not altered at a rate faster than the rate of interaction. It is also reasonable to assume that external fields, such as gravity and electromagnetic fields, have no effect on the transported particles. The above considerations enable the use of a bookkeeping process similar to that formulated by the Austrian physicist Ludwig Boltzmann (1844-1906) in the kinetic theory of gases.

The Boltzmann transport equation involves a systematic cataloging of all the possible ways radiation "gas", or photons or neutrons, leave, be created, or destroyed, within an infinitesimal volume. On the limit, this process leads to an integro-differential equation for the number of particles per unit volume, or the particle density, $F(\vec{r},\vec{v},t)$, of the form [88]:

$$\frac{\partial F(\vec{r},\vec{v},t)}{\partial t} = -\vec{v}\cdot\nabla F(\vec{r},\vec{v},t) - v\Sigma_{total}(\vec{r},\vec{v})\ F(\vec{r},\vec{v},t)$$

$$
\begin{aligned}
&+ \int v' \Sigma_{scatter}(\vec{r}, \vec{v}' \to \vec{v}) \; F(\vec{r}, \vec{v}', t) \; d\vec{v}' \\
&+ \int v' \nu \Sigma_{generation}(\vec{r}, \vec{v}' \to \vec{v}) \; F(\vec{r}, \vec{v}', t) \; d\vec{v}' \\
&+ Q(\vec{r}, \vec{v}, t) \qquad (3.116)
\end{aligned}
$$

where $\vec{r}$ is a position vector, $\vec{v}$ is the particle velocity, t refers to time, Σ is the total cross-section of the interaction defined in its subscript and Q is the intensity of particles supplied by an external source. The terms in the right-hand-side of Eq. (3.116), represent, respectively, the number of particles per unit time:

1 entering, minus those leaving, the volume element at $\vec{r}$, with velocity $\vec{v}$, at time t, (called the streaming or divergence term);

2 removed from volume by absorption or scattering;

3 scattered into volume element, with velocity $\vec{v}$;

4 generated within element (say by neutron fission or positron-electron annihilation or photon emission), at a velocity $\vec{v}$, with ν being the average number of neutrons produced per generation;

5 generated by an external source, with velocity $\vec{v}$.

The transport equation spans a seven-dimensional space, with three variables for the spatial coordinates, one variable for the magnitude of velocity (or equivalently energy), two variables for the vector defining the direction of velocity, and one variable for time.

Since photons move at a constant speed, equal to the speed of light, energy replaces the scalar independent variable representing the velocity in the transport equation, Eq. (3.116). Therefore, the transport equation is often written in a form suitable for both photons and neutrons, by replacing the velocity vector $\vec{v}$, with the particle energy, E, and a vector $\vec{\Omega}$, representing the direction of travel. The velocity vector $\vec{v}$ can be expressed as $\vec{v} = v\vec{\Omega} = \sqrt{\frac{2E}{m_n}}\vec{\Omega}$ for neutrons, and simply as $\vec{v} = c\vec{\Omega}$ for photons, where m_n is the neutron mass and c is the speed of light. The number of particles per unit volume having an energy in dE about E and direction in $d\Omega$ about $\vec{\Omega}$, is defined in terms of the particle density function, $n(\vec{r},E,\vec{\Omega},t)$, as $n(\vec{r},E,\vec{\Omega},t) \; dE \; d\Omega \; dt$. The transport equation can then be rewritten as:

$$
\begin{aligned}
\frac{\partial n(\vec{r}, E, \vec{\Omega}, t)}{\partial t} &= -\vec{\Omega} \cdot \nabla v n(\vec{r}, E, \vec{\Omega}, t) - \Sigma_{total}(\vec{r}, E) \; v n(\vec{r}, E, \vec{\Omega}, t) \\
&+ \int \Sigma_{scatter}(\vec{r}, E' \to E; \vec{\Omega}' \to \vec{\Omega}) \; v' n(\vec{r}, \vec{v}', t) \; dE' \; d\Omega'
\end{aligned}
$$

$$+ \int \nu\Sigma_{generation}(\vec{r}, E' \to E; \vec{\Omega}' \to \vec{\Omega})\, v' n(\vec{r}, \vec{v}', t)\, dE'\, d\Omega'$$
$$+ Q(\vec{r}, E, \vec{\Omega}, t) \qquad (3.117)$$

Using the angular flux density, $\phi(\vec{r}, E, \vec{\Omega}, t) = vn(\vec{r}, E, \vec{\Omega}, t)$, the transport equation becomes:

$$\begin{aligned} \frac{1}{v}\frac{\partial\phi(\vec{r}, E, \vec{\Omega}, t)}{\partial t} &= -\vec{\Omega}\cdot\nabla\phi(\vec{r}, E, \vec{\Omega}, t) - \Sigma_{total}(\vec{r}, E)\phi(\vec{r}, E, \vec{\Omega}, t) \\ &+ \int \Sigma_{scatter}(\vec{r}, E' \to E; \vec{\Omega}' \to \vec{\Omega})\phi(\vec{r}, \vec{v}', t)\, dE'\, d\Omega' \\ &+ \int \nu\Sigma_{generation}(\vec{r}, E' \to E; \vec{\Omega}' \to \vec{\Omega})\, \phi(\vec{r}, \vec{v}', t)\, dE'\, d\Omega' \\ &+ Q(\vec{r}, E, \vec{\Omega}, t) \end{aligned} \qquad (3.118)$$

The introduction of the angular flux density enables direct evaluation of the interaction rate, R. A particle i traversing a distance l_i will interact with matter at a rate proportional to its path length l_i. Therefore, for all particles, the interaction rate, R, i.e. the number of interactions per unit volume, per unit time, in an infinitesimal volume, ΔV:

$$R \propto \frac{d}{dt}\left(\frac{\sum_i l_i}{\Delta V}\right) \propto \frac{\sum_i v_i}{\Delta V} \qquad (3.119)$$

where v_i is the speed of particle i; $v_i = \frac{dl_i}{dt}$. For all particles of the same velocity, v, or energy, E:

$$R(E) = \Sigma(E)\, nv(E) \qquad (3.120)$$

where $n(E)$ is the number of particles per unit volume that have a velocity v, or equivalently energy E, and the proportionality constant is replaced by the macroscopic cross-section, $\Sigma(E)$ as it provides a measure of the probability of interaction with matter. Eqs. (3.119) and (3.120) demonstrate that the interaction rate is proportional to nv, and equivalently to the sum of all track lengths for all particles per unit time per unit volume. The latter quantity is called the particle flux density, or simply flux, $\phi(v) = nv$. The flux is also defined as the number of particles entering a sphere of unit cross-sectional area per unit time.

The flux at position $\vec{r}$ can be decomposed into a number of components, the most elementary of which is $\phi(\vec{r}, E, \vec{\Omega}, t)$, the flux of particles of energy in dE about E and direction in $d\Omega$ about $\vec{\Omega}$. Therefore, the flux of particles in all directions having an energy in dE about E is given by:

$$\phi(\vec{r}, E, t)\, dE = \int_{4\pi} \phi(\vec{r}, E, \vec{\Omega}, t)\, d\Omega\, dE \qquad (3.121)$$

The flux at all particle energies, at position $\vec{r}$, is:

$$\phi(\vec{r},t) = \int \phi(\vec{r},E,t)\, dE \tag{3.122}$$

The integration of the above quantity over time defines the particle fluence:

$$\Phi(\vec{r},E) = \int \phi(\vec{r},E,t)\, dt \tag{3.123}$$

This quantity is also called the amount of "irradiation" or "exposure", as it defines the amount of radiation a material is exposed to over time.

Another quantity of interest is the current density, i.e. the directed flow of particles in a given direction $\vec{\Omega}$, per unit area per unit time:

$$\begin{aligned} \vec{J}(\vec{r},E,\vec{\Omega},t)\, d\Omega\, dE &= \vec{v}\, n(\vec{r},E,\vec{\Omega},t)\, d\Omega\, dE \\ &= \vec{\Omega}\, vn(\vec{r},E,\vec{\Omega},t)\, d\Omega\, dE \\ &= \vec{\Omega}\, \phi(\vec{r},E,\vec{\Omega},t)\, d\Omega\, dE \end{aligned} \tag{3.124}$$

Also:

$$\begin{aligned} \vec{J}(\vec{r},E,t)dE &= \int_{4\pi} J(\vec{r},E,\vec{\Omega},t)\, d\Omega\, dE, \quad \text{and} \\ \vec{J}(\vec{r},t) &= \int \vec{J}(\vec{r},E,t)\, dE \end{aligned}$$

Note that the vector quantity, $\vec{J}$, measures the passage of particles through a surface and it includes directional characteristics, while ϕ is simply a scalar quantity. The two quantities are related as:

$$|\vec{J}(\vec{r},E,\vec{\Omega},t)| = |\vec{\Omega}\phi(\vec{r},E,\vec{\Omega},t)| = \phi(\vec{r},E,\vec{\Omega},t) \tag{3.125}$$

Hence, the flux is often defined as the flow of particles per unit area.

3.6.1. Classical Laws of Conservation

The solution of the transport equation in void shows how particle transport is related to classical laws of mechanics and light propagation. In void, and in the absence of an external source, all terms in the right-hand-side of the transport equation, Eq. (3.117), disappear, except for the divergence term, leading to:

$$\frac{\partial n(\vec{r},E,\vec{\Omega},t)}{\partial t} = -\vec{v}\cdot\nabla n(\vec{r},E,\vec{\Omega},t) \tag{3.126}$$

The above equation can be arrived at by applying Newton's First Law; namely, if a body in motion is not acted upon by external forces, its momentum remains unchanged. The particle flux density is actually equal to the momentum of particles per unit area per unit mass (per unit time). Thus, if one considers a small volume element, $\Delta x\ \Delta y\ \Delta z$, then:

$$\text{Rate of momentum loss through volume} = \tag{3.127}$$
$$\Delta y\Delta z[\phi(x) - \phi(x+\Delta x)] + \Delta x\Delta z[\phi(y) - \phi(y+\Delta y)] + \Delta x\Delta y[\phi(z) - \phi(z+\Delta z)] \tag{3.128}$$

According to Newton's first law, this momentum loss is not permitted and must be accounted for by a change in the particle mass within the volume:

$$\text{Rate of mass change} = \Delta x\Delta y\Delta z\frac{\partial n}{\partial t} \tag{3.129}$$

where n is the particle density within the volume. Equating the above two equations, replacing the flux density by the product of the velocity and the particle density, dividing by the volume and taking the limit, one obtains:

$$\begin{aligned}\frac{\partial n}{\partial t} &= -\left(\frac{\partial}{\partial x}nv_x + \frac{\partial}{\partial y}nv_y + \frac{\partial}{\partial z}nv_z\right)\\ &= -\nabla\cdot n\vec{v} = -\vec{v}\cdot\nabla n - n\nabla\cdot\vec{v} = -\vec{v}\cdot\nabla n\end{aligned} \tag{3.130}$$

Since there is no reason for the particle's velocity to change in void, then $\nabla\cdot\vec{v}$ must be equal to zero and the above equation becomes identical to that obtained from the Boltzmann's equation, Eq. (3.126). The term on the end of Eq. (3.130) is called the streaming term, as it represents the spread of particles in space, even in the absence of material.

The classical law of mass conservation leads also to Eq. (3.126). The conservation of mass as seen by an observer moving with the particle "gas", at speed $\vec{v}$ requires that:

$$\frac{dn}{dt} = 0 \tag{3.131}$$

In fluid mechanics, this is known as the "material derivative". This derivative is related to the partial derivative, estimated at a fixed position in space, and is given by:

$$\frac{dn}{dt} = \frac{\partial n}{\partial t} + \vec{v}.n \tag{3.132}$$

It can be easily shown that Eqs. (3.131) and (3.132) lead to Eq. (3.126). The above analysis indicates that the left-hand-side and the first term of the right-hand-side of the Boltzmann particle transport equation,

Eq. (3.117), are due to the conservation of mass, as seen by an observer moving at the same speed as the particles. Equivalently these two terms can be considered as being the result of conservation of momentum, as interpreted by an observer in the laboratory frame of reference.

3.6.2. Divergence Law

Like light, the intensity of particles decreases in inverse proportion to the distance of travel, as the same amount of radiation spreads over a larger area. In steady state, the transport equation, Eq. (3.118), in void, where v is constant, is reduced to:

$$\vec{\Omega} \cdot \nabla \phi(\vec{r}, E, \vec{\Omega}) = Q(\vec{r}, E, \vec{\Omega}) \tag{3.133}$$

Since energy does not change as the particles travel in void, one can use the equivalent equation:

$$\vec{\Omega} \cdot \nabla \phi(\vec{r}, \vec{\Omega}) = Q(\vec{r}, \vec{\Omega}) \tag{3.134}$$

The "dot" product can be resolved along the direction in which the particles are traveling, $\vec{\Omega}$, as:

$$\vec{\Omega} \cdot \nabla \phi(\vec{r}, \vec{\Omega}) = |\vec{\Omega}| \frac{\partial}{\partial R}\, \phi(\vec{r}, \vec{\Omega}) = \frac{\partial}{\partial R}\, \phi(\vec{r}, \vec{\Omega}) \tag{3.135}$$

where R is some distance along $\vec{\Omega}$. Therefore,

$$\frac{\partial}{\partial R}\, \phi(\vec{r}, \vec{\Omega}) = Q(\vec{r}, \vec{\Omega}) \tag{3.136}$$

Integrating the above equation along R, in an infinite space, results in:

$$\int_{-\infty}^{\infty} \frac{\partial}{\partial R} \phi(\vec{r}, \vec{\Omega})\, dR = \int_{-\infty}^{\infty} Q(\vec{r}, \vec{\Omega})\, dR \tag{3.137}$$

Therefore, the value of the flux along the direction $\vec{\Omega}$ is given by:

$$\phi(\vec{r}, \vec{\Omega}) = \int_{-\infty}^{\infty} Q(\vec{r}\,', \vec{\Omega})\, dR \tag{3.138}$$

Note that the integral in the above equation represents the angular flux along $\vec{\Omega}$, only at some position $\vec{r}$, since any other non-zero value of the flux along $\vec{\Omega}$ has a different value of $\vec{r}$, and therefore does not enter the integration process. In the right-hand-side of Eq. (3.138), the dummy variable $\vec{r}\,'$ was introduced to reflect the fact that a source anywhere along $\vec{r}'$ with neutrons directed in $\vec{\Omega}$ will contribute to the value of the flux at $\vec{r}$, since these particles will eventually reach $\vec{r}$, see Figure 3.9.

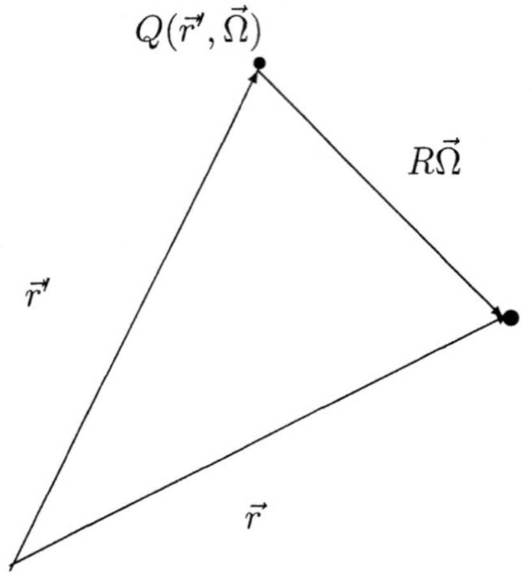

Figure 3.9. Relationship between $\vec{r}$, $\vec{r}'$ and $R\vec{\Omega}$.

The relationship between $\vec{r}$ and $\vec{r}\,'$ can be expressed by the vector relationship [79]:

$$\vec{r}\,' = \vec{r} - R\vec{\Omega} \tag{3.139}$$

Introducing the above relationship and integrating Eq. (3.137) over all possible values of $\vec{\Omega}$ results in:

$$\phi(\vec{r}) = \int_{-\infty}^{\infty} \int_{4\pi} Q(\vec{r} - R\vec{\Omega}, \vec{\Omega})\; dR\; d\Omega \tag{3.140}$$

Making use of the fact that $d\Omega = \frac{dA}{R^2}$ and $dV = dR\, dA$, where A and V refer, respectively, to area and volume. Then,

$$\phi(\vec{r}) = \int \frac{Q(\vec{r} - R\vec{\Omega}, \vec{\Omega})}{R^2}\; dV \tag{3.141}$$

For an isotropic point source at some location $\vec{r}_0$, Eq. (3.141) can be expressed as:

$$Q(\vec{r}, \vec{\Omega}) = \frac{Q_0}{4\pi} \delta(r_0) \tag{3.142}$$

where Q_0 is the source strength (number of particles per unit volume) and the delta function is equal zero everywhere except at $\vec{r} = \vec{r}_0$ where it is equal to unity. The resulting flux then becomes:

$$\phi(\vec{r}) = \frac{Q_0}{4\pi} \int \frac{\delta(\vec{r} - R_o\vec{\Omega})}{R^2} dV = \frac{Q_0 V}{4\pi R_0^2} \tag{3.143}$$

Note that in performing the above integral, advantage was taken of the fact that the delta function is equal to zero everywhere, except at $\vec{r} = \vec{r}_0$, i.e. at $R = R_o$, where it is equal to unity. The above equation is a representation of the well-known $\frac{1}{R^2}$ law of radiation propagation, keeping in mind that $Q_0 V$ is equal to the total number of particles emanating from the point source. Eq. (3.143) represents also the streaming effect of a point source.

3.6.3. Attenuation Law

A beam of radiation decreases exponentially in intensity in accordance to Eq. (3.8). This relationship can also be derived from the particle transport equation. For a well-collimated, narrow, beam of radiation, at steady state and in the absence of an external source (in other words without including the physical volume of the source within the transport domain), and assuming that no particles enter the beam from outside the region traversed by the beam nor are generated within, the particle transport, Eq. (3.118) reduces to:

$$-\vec{\Omega} \cdot \nabla\phi(\vec{r}, E, \vec{\Omega}, t) - \Sigma_{total}(\vec{r}, E)\phi(\vec{r}, E, \vec{\Omega}, t) = 0 \tag{3.144}$$

The ∇ operator affects the flux density in only distance. Time as a variable can be dropped from the above equation. In addition, all particles within the narrow beam maintain the same direction and energy, thus the energy and directions can also be dropped as variables. Within a narrow beam, particle transport can be seen as occurring in a one-dimensional space, say in the x-direction, with $\vec{\Omega}$ becoming equal to the unit vector $\hat{x}$ in the x-direction. Then the divergence term in Eq. (3.144) reduces, after performing the dot product, to $-\hat{x} \cdot \frac{d}{dx}\hat{x} = -\frac{d}{dx}$. Eq. (3.144) then becomes:

$$\begin{aligned} -\frac{d\phi(x)}{dx} &= \Sigma_{total}(x)\phi(x) \\ \frac{d\phi(x)}{\phi(x)} &= -\Sigma_{total}(x)dx \end{aligned} \tag{3.145}$$

Integrating the above equation from $x = 0$ to some value x along the beam, leads to:

$$-\int \frac{d\phi(x)}{dx} = -\int_0^0 \Sigma_{total}(x)\phi(x)$$

$$
\begin{aligned}
-\ln \frac{\phi(x)}{\phi(0)} &= -\int_o^x \Sigma_{total}(x)dx \\
\phi(x) &= \phi(0) \exp\left[-\int_o^x \Sigma_{total}(x)dx\right] \qquad (3.146)
\end{aligned}
$$

If the cross-section Σ_{total} remains constant along the distance of integration, then:

$$\phi(x) = \phi(0) \exp[-\Sigma_{total} x] \qquad (3.147)$$

Eq. (3.146) and (3.147) represent two different forms of the attenuation law of radiation, which is traditionally derived using the approach followed in arriving at Eq. (3.8). The above derivation from the transport equation emphasizes the many assumptions involved in the derivation of this attenuation law. The most important of these assumptions is that of a narrow beam, which ignores radiation divergence and the introduction of particles from outside the beam to within the beam. Nevertheless, the simple form of the attenuation beam makes it attractive for use in many NDE applications, such as in computed tomography, see section 6.7.

3.6.4. Diffusion Theory

Diffusion is a particle transport process in which the transported quantity does not alter its properties. In radiation, this implies that the cross-sections do not change as the particles are diffused, or equivalently the particle energy stays constant. The rate of diffusion is assumed to obey a simple linear law: the diffusion of molecular species is governed by Fick's law; heat transfer by conduction obeys Fourier's law; diffusivity of momentum is described by Newton's law of viscosity; while the conduction of electricity in metal follows Ohm's law.

In radiation, Fick's law takes the form:

$$\vec{J} = -D\nabla\phi \qquad (3.148)$$

where D is the diffusion coefficient. This law states that the directional flow of particles, i.e. the current, is driven by the gradient of their concentration (flux), with the negative sign indicating that the flow is from high to low concentration. This simply indicates that at regions of high flux, there will be more radiation scattering (at all directions) than in regions of low flux, and consequently more particles will be directed from the high to the low flux region rather than the other way around. In fact, the diffusion coefficient is approximately related to the macroscopic scattering cross-section, $\Sigma_{scatter}$, by [28]:

$$D = \frac{1}{3\Sigma_{scatter}(1 - \overline{\cos\theta})} = \frac{1}{\Sigma_{transport}} \qquad (3.149)$$

where $\overline{\cos\theta}$ is the average value of the cosine of the angle of scattering of the particles in the medium, for neutrons $\overline{\cos\theta} = \frac{2}{3A}$, where A is the mass-number of the scattering nucleus. The transport cross-section, $\Sigma_{transport}$, in Eq. (3.149), is used to elucidate the fact that the diffusion coefficient incorporates the scattering process, and in effect removes the need to consider the scattering term, third term in the right-hand-side of the transport equation, Eqs. (3.116), (3.117) or (3.118). The appearance of $\cos\theta$ in the definition of D is a reflection of the fact that diffusion is in effect a p-scattering approximation [88], see section 3.5.6.3.

Fick's law addresses only the spatial gradient of the flux, and does not address energy or angular changes. Therefore, diffusion analysis is applied to particles of the same energy (or by lumping all energies into a single energy group), and when directional preference is not important, i.e. when isotropic scattering is predominant. The last three terms in Eq. (3.118) can be lumped into one source term, S, with the understanding that the scattering term is a source term in the sense that it can bring particles from a different energy group to the energy group analyzed by diffusion. Since there is no energy or angle dependence in diffusion analysis, the flux becomes a function of space and time only and is designated as ϕ_g, with the subscript referring to the energy group considered. With these assumptions, Eq. (3.118) can be rewritten as:

$$\frac{1}{v_g}\frac{\partial\phi_g(\vec{r},t)}{\partial t} = -\vec{\Omega}\cdot\nabla\phi_g(\vec{r},t) - \Sigma_{g,total}(\vec{r}) + S_g(\vec{r},t) \tag{3.150}$$

To eliminate the appearance of $\vec{\Omega}$ in the first term of the right-hand-side of the above equation, the following relationship in introduced:

$$\nabla\cdot(\vec{\Omega}\phi) = \vec{\Omega}\cdot\nabla\phi + \phi\nabla\cdot\vec{\Omega} = \vec{\Omega}\cdot\nabla\phi \tag{3.151}$$

where $\phi\nabla\cdot\vec{\Omega}$ is eliminated, since diffusion is an isotropic process. Incorporating in Eq. (3.151) the definition of current, Eq. (3.124), as $\vec{J} = \vec{\Omega}\phi$, along with Fick's law, Eq. (3.148), one gets:

$$\vec{\Omega}\cdot\nabla\phi = \nabla\cdot(\vec{\Omega}\phi) = \nabla\cdot J = -\nabla\cdot(D\nabla\phi) \tag{3.152}$$

Substituting Eq. (3.152) in Eq. (3.150), for a medium of a constant diffusion coefficient, D, leads to the familiar diffusion equation:

$$\frac{1}{v_g}\frac{\partial\phi_g(\vec{r},t)}{\partial t} = D_g\nabla^2\phi_g(\vec{r},t) - \Sigma_{g,total}(\vec{r}) + S_g(\vec{r},t) \tag{3.153}$$

The subscript g is also given to the diffusion coefficient, since it is a function of the scattering cross-sections, Eq. (3.149), which is in turn a function of particle energy.

Reference [89] derives the diffusion equation in a number of different ways. These derivations emphasize the limitations of the diffusion theory. Diffusion is an isotropic or near isotropic dispersive process, applicable only in regions away from system boundaries and strong sources or sinks (absorbers), and is unsuitable for describing particle transport within a void. Therefore, the diffusion equation is mainly used for depicting the overall distribution of neutrons in a nuclear reactor, where such restrictions can be reasonably tolerated [28]. In radiation NDE applications, the diffusion equation, if used at all, should be applied with great care due to its limited applicability. Since in most NDE applications, a strong radiation source is usually present, radiation has to travel in air (void space) from the source to the object and from the object to the source, and radiation is detected near boundaries, the conditions for the use of the diffusion conditions are not usually met.

3.6.5. Transport of Charged-Particles

The main difference between the transport of charged-particles and that of neutral radiation (such as neutrons and photons), is that the interaction cross-section of charged-particles is quite large, and approaches infinity as the amount of transferred energy approaches zero. In other words, the mean-free-path tends to be very small, and the charged-particles appear to interact continuously with matter. As a result, the exact values of the cross-sections of charged-particles are not well known. However, during interactions, charged-particles do not significantly change their direction, and do not travel very far before being absorbed. Charged-particle interactions are, therefore, low momentum transfer events. As a result, the concepts of stopping power and range, discussed in section 3.3, are often used in charged-particle transport calculations.

Charged-particle interactions can produce electrons. Electron transport, whether arising from the source itself, or as the result of the interaction of other radiation, is a complex process, due to the generation of secondary electrons. Electron interactions can result in the production of secondary electrons in a number of ways. The secondary electrons can in turn produce more electrons, and so on. The result is an electron cascade that is triggered by the original electron [45]. A secondary electron may be produced by collision with an orbital electron. If the incident electron has a sufficient kinetic energy to liberate an electron from the atom, it will cause ionization. The liberated electrons are often referred to as "delta" rays. However, the main mechanism of secondary-electron generation is associated with radiative energy losses (bremsstrahlung), a process that tends to occur at high energies. The photons produced by bremsstrahlung produce more electrons as they interact with matter, via Compton scattering, or pair production. The resulting electrons themselves may lose energy by the bremsstrahlung process, producing more photons, and so on. This leads to the generation of a cascade of electrons. The photoelectric absorption of the bremsstrahlung photons and the coulomb scattering of the electron by the atom act against the formation of this cascade. This, along with Compton scattering, produce a lateral spread, thus the electron cascade is referred to as a "shower". Electron energy losses are caused by collisions or radiative (bremsstrahlung) emission. At low electron energies, energy loses by collision are more than the cascade-producing radiative loses. Then the shower tends to "tail off " as the energy of the electrons dissipate into excitation and ionization of the atoms of the medium. Due to the complexity of this electron shower process, the Monte Carlo method is used in computer codes such as the EGS4 code [45] and the TIGER series of codes [46]. Reference [90] provides a brief description of the process involved in the simulation of electron interactions; namely, energy loss by ionization, straggling, bremsstrahlung, and multiple collisions with the atomic nuclei.

3.7. Radioactive Decay

3.7.1. Kinetics of Decay

Another radiation modifying process with time is radioactive decay, which not only affects radioisotopic sources but also the radioactivity induced by radiation activation. Radioactive decay is a random process described by the Poisson statistical distribution discussed in appendix G. However, like most statistical process, radioactive decay can be expressed in terms of the expected-value. The kinetics of radioactive decay is described by the following first-order equation:

$$\frac{dN}{dt} = -\lambda N \tag{3.154}$$

where N is the number of nuclei per unit volume of a certain radioactive nuclide available at time t, and λ is a constant, called the decay constant. The value of λ is a property of the decaying nucleus, and does not depend on any other external factors, such as chemical bonding, temperature, pressure or the presence of any other nuclei. Eq. (3.154), therefore, reflects a simple physical fact that the rate of decay depends on the nature of the decaying nucleus and the number of nuclei available. The right-hand-side of Eq. (3.154) defines, therefore, the activity of the decaying nuclei. This activity is measured in disintegrations per second, or becquerels (Bq). The old unit called curie (Ci), with 1 Ci equal to 3.7×10^{10} Bq; the curie was originally supposed to represent the activity of 1 g of ^{226}Ra. The specific activity (activity per gram) can be calculated by replacing N in Eq. (3.154) by the number of atoms per gram ($= \frac{1}{A\mathrm{u}}$, where A is the mass-number of the decay nucleus and u is the atomic mass unit in gram), that is:

$$\text{Specific activity (Bq)} = \frac{\lambda}{A\mathrm{u}} \tag{3.155}$$

Integrating Eq. (3.154) over time leads to the familiar exponential decay equation:

$$N = N(0)\exp(-\lambda t) \tag{3.156}$$

where $N(0)$ is the value of N at $t = 0$. The half-life of a nuclide is the time it takes for a nuclide to lose half its activity, or in other words it is the time at which $N = 0.5N_0$. Using Eq. (3.156), the half-life, $t_{\frac{1}{2}}$, is given by:

$$t_{\frac{1}{2}} = \frac{\ln 2}{\lambda} \approx \frac{0.693}{\lambda} \tag{3.157}$$

3.7.2. Parent / Daughter Decay

When the daughter of a parent nuclide is also radioactive, its rate of accumulation depends, not only on the decay constant of the parent nuclide, but also on that of the daughter. The rate of accumulation of a daughter nuclei is then equal to its rate of formation minus its rate of decay. Therefore, in accordance to Eqs. (3.154) and (3.156), one can express the balance of the daughter nuclei as:

$$\frac{dN_2}{dt} = \lambda_1 N_1 - \lambda_2 N_2 = \lambda_1 N_1(0)\exp(-\lambda_1 t) - \lambda_2 N_2 \tag{3.158}$$

where the subscript 1 refers to the parent nuclei and 2 to the daughter nuclei. The right-hand-side term in Eq. (3.158) is obtained from Eq. (3.156), assuming that the parent nuclei are not being produced by some other means. Eq. (3.158) can be have a solution in the from $N_2 = f(t)\exp[-\lambda_2 t]$, where $f(t)$ is some function of time, leading to:

$$\begin{aligned} N_2 &= f(t)\exp(-\lambda_2 t) = \left[N_2(0) - \frac{\lambda_1 N_1(0)}{\lambda_2 - \lambda_1}\right]\exp(-\lambda_2 t) \\ &\quad + \frac{\lambda_1 N_1(0)}{\lambda_2 - \lambda_1}\exp(-\lambda_1 t) \end{aligned} \tag{3.159}$$

If $N_1(0) = 0$, i.e. no daughter nuclei are available at $t = 0$, then:

$$N_2 = \frac{\lambda_1 N_1(0)}{\lambda_2 - \lambda_1}\left[\exp(-\lambda_1 t) - \exp(-\lambda_2 t)\right] \qquad (3.160)$$

3.7.3. Equilibrium

Conventionally equilibrium is viewed as a balance between the number of nuclei of a daughter, N_2, relative to that of the parent, N_1. Such a balance is manifested by a constant relative content, i.e., by $\frac{N_2}{N_1}$ reaching a constant value. If both N_1 and N_2 change with time while their ratio remains constant, the equilibrium is defined as transient equilibrium. However, when the values of each of N_1 and N_2 stabilize with time, a lasting and continuing type of equilibrium, referred to as secular equilibrium, is attained. Another way is to define the states of equilibrium between the daughter and the parent as a balance in either activity or relative change in content, between the daughter and the parent. As Table 3.1 and the discussion below show, these types of equilibrium lead to the conventionally defined types of equilibrium under certain conditions. Table 3.1 also indicates that equilibrium is only attainable when $\lambda_1 < \lambda_2$. In other words when the half-life of the parent is larger than that of the daughter, with secular equilibrium reached only when $\lambda_1 << \lambda_2$. When $\lambda_1 > \lambda_2$, i.e. when the parent's half-life is shorter than that of the daughter, equilibrium can never be attained. This is best demonstrated, using Eq. (3.160), when $\lambda_1 >> \lambda_2$, Eq. (3.160) gives $N_2 \approx N_1(0)\exp(-\lambda_2 t)$, indicating that N_2 will reach a maximum value and then will continue to decay at a rate determined by its own time constant, independent of that of the parent.

3.7.3.1 Equilibrium in Activity

Equilibrium between a parent and a daughter can be attained when the activity of the parent nuclei is equal to that of the daughter nuclei, which according to Eq. (3.158) requires that:

$$\lambda_1 N_1^{equ} = \lambda_2 N_2^{equ} = \frac{\lambda_1 \lambda_2 N_1(0)}{\lambda_2 - \lambda_1}\left[\exp(-\lambda_1 t) - \exp(-\lambda_2 t_{equ})\right] \qquad (3.161)$$

When $\lambda_1 << \lambda_2$, or according to Eq. (3.157) when the half-life of the parent is much longer than that of the daughter, Eq. (3.161) is simplified to:

$$\lambda_2 N_2^{equ} \quad = \quad \lambda_1 N_1^{equ}$$

Table 3.1. Balance between parent (N_1) and daughter (N_2) in radioactive decay.

Equilibrium	Condition	Relative Content
In Activity:	$\lambda_1 N_1 = \lambda_2 N_2$;	$\frac{N_2^{equ}}{N_1^{equ}} = \frac{\lambda_1}{\lambda_2}$
Secular	$\lambda_1 << \lambda_2$	$\frac{N_2^{equ}}{N_1^{equ}(0)} = \frac{\lambda_1}{\lambda_2}$
Example:	^{226}Ra ($t_{\frac{1}{2}} = 1599$ years) $\xrightarrow{\alpha}$ ^{222}Rn ($t_{\frac{1}{2}} = 3.8235$ days)	
In Relative-Rate:	$\frac{1}{N_1}\frac{dN_1}{dt} = \frac{1}{N_2}\frac{dN_2}{dt}$;	$\frac{N_2^{equ}}{N_1^{equ}} = \frac{\lambda_1}{\lambda_2 - \lambda_1}$
Transient	$\lambda_1 < \lambda_2$	
Example:	^{211}Pb ($t_{\frac{1}{2}} = 2.14$ minutes) $\xrightarrow{\beta}$ ^{211}Bi ($t_{\frac{1}{2}} = 2.14$ minutes)	
Never	$\lambda_1 > \lambda_2$	
Example:	^{218}Po ($t_{\frac{1}{2}} = 3.10$ minutes) $\xrightarrow{\alpha}$ ^{214}Pb ($t_{\frac{1}{2}} = 27$ minutes)	

$$\begin{aligned} \lambda_1 N_1(0)\,[1 - \exp(-\lambda_2 t_{equ})] &= \lambda_1 N_1(0) \exp[-\lambda_1 t_{equ}] \\ 1 - \exp(-\lambda_2 t_{equ}) &\approx 1 \end{aligned} \tag{3.162}$$

where use is made of Eq. (3.156) for N_1 and Eq. (3.160) for N_2, setting $\exp(-\lambda_1 t) \approx 1$, due to the assumed small value of λ_1. The last relationship in (3.162) indicates that the time to reach equilibrium is when $\exp(-\lambda_2 t_{equ}) \approx 0$, which is practically attained when $\lambda_2 t \geq 5$, or so. Secular equilibrium is attained, according to Eq. (3.157), when $t \geq 7t_{\frac{1}{2}}$ or so, where $t_{\frac{1}{2}}$ is the half-life of the daughter nuclei. From the relationships of Eq. (3.162), one can also conclude that:

$$N_2^{equ} \approx N_1(0)\frac{\lambda_1}{\lambda_2} \tag{3.163}$$

When $\lambda_1 << \lambda_2$, this activity equilibrium is referred to as secular equilibrium.

3.7.3.2 Equilibrium in Relative-Rate

When the relative rates of decay of the parent and its daughter are equal, a state of equilibrium is reached. Mathematically this requires that:

$$\frac{1}{N_1}\frac{dN_1}{dt} = \frac{1}{N_2}\frac{dN_2}{dt} \tag{3.164}$$

Not that the relative rate of decay is also equal to the natural logarithmic rate of decay, since $\frac{d\ln N}{dt} = \frac{1}{N}\frac{dN}{dt}$. Using Eqs. (3.154) and (3.158), the condition of Eq. (3.164) can be expressed as:

$$
\begin{aligned}
-\lambda_1 &= \frac{\lambda_1 N_1^{equ} - \lambda_2 N_2^{equ}}{N_2^{equ}} \\
N_2^{equ} &= \frac{\lambda_1 N_1^{equ}}{\lambda_2 - \lambda_1}
\end{aligned}
\tag{3.165}
$$

This is the condition for transient equilibrium. Note that when $\lambda_1 << \lambda_2$, Eqs. (3.162) and (3.165) become identical, i.e. transient equilibrium approaches secular equilibrium, as $\frac{dN_2}{dt} \to 0$.

Substituting the solutions for N_1, and N_2, Eqs. (3.156) and (3.160), in Eq. (3.165), one can show that transient equilibrium is attained when:

$$
\exp(-\lambda_2 t_{equ}) = 0 \tag{3.166}
$$

That is, like secular equilibrium, Eq. (3.162), is practically reached when $\lambda_2 t \geq 5$, or so.

3.7.4. Decay Chains

For a decay sequence that involves many progenies in a simple decay chain:

$$
N_1 \longrightarrow N_2 \longrightarrow N_3 \cdots\cdots\cdots N_i - 1 \longrightarrow N_i \longrightarrow N_{i+1} \cdots\cdots\cdots
$$

the activity of the ith type of nuclide can be expressed as [39]:

$$
\begin{aligned}
\lambda_i N_i &= \lambda_i N_i(0) \exp(-\lambda_i t) + \\
\sum_{l=1}^{i-1} & \left[N_l(0) \left(\prod_{j=l}^{i} \lambda_j \right) \sum_{j=1}^{i} \left\{ \exp(-\lambda_j t) \prod_{k=l, k \neq j}^{i} (\lambda_k - \lambda_j)^{-1} \right\} \right]
\end{aligned}
\tag{3.167}
$$

For more complex decay chains see reference [39].

If the production of a nuclide is caused, not by the decay of a parent nuclide, but by a nuclear reaction, such as neutron activation, the following overall "removal" and "production" constants for the $i$the nuclide, Λ_i and Γ_i, respectively, can be defined:

$$
\begin{aligned}
\Lambda_i &= \lambda_i + \sigma_{r,i} \phi_i \\
\Gamma_i &= \sigma_{p,i} \phi_i
\end{aligned}
\tag{3.168}
$$

where σ_r and σ_p represent the microscopic cross-sections for the removal and production of the concerned nuclides, respectively, at a flux, ϕ of the activating radiation. Note that $\sigma\phi$ is used here, instead of $\Sigma\phi$ of Eq. (3.120), since the latter is the interaction rate per unit volume, while the former is for all available nuclei, as N is defined here as the number of nuclei (not their number per unit volume). Then using the definitions of Eqs. (3.168) [39]:

$$\begin{aligned}\lambda_i N_i &= \Lambda_i N_i(0)\exp(-\Lambda_i t) + \\ \sum_{l=1}^{i-1} &\left[N_l(0) \left(\prod_{j=l}^{i} \Gamma_j \right) \sum_{j=1}^{i} \left\{ \exp(-\Lambda_j t) \prod_{k=l, k\neq j}^{i} (\Lambda_k - \Lambda_j)^{-1} \right\} \right] \quad (3.169)\end{aligned}$$

Note that in the above equation, the parent, N_1, is assumed to have been already existing, i.e. not produced by activation. When the parent is a stable nuclide, $\lambda_1 = 0$, but is the first activated nuclide.

Chapter 4

DETECTION METHODS

4.1. Introduction

Detection aims at measuring the intensity of radiation (i.e. the number of particles[1] per unit time) and/or its energy. The distribution of the number of particles with energy gives the energy spectrum. In other words, the energy spectrum provides the number of particles at each energy (or more accurately energy interval), $N(E)dE$, where $N(E)$ is the number of particles per unit energy having an energy from E to $E + dE$. Therefore, $\int N(E)dE$ provides the total number of particles, or the radiation intensity.

The essence of radiation detection is to produce electronic signals, or pulses, that can be further processed by a suitable electronic device to provide an indication of the radiation intensity, and/or particle energy. Charged-particles lend themselves naturally to the detection process, since charged-particles can directly produce an electrical signal as they travel through matter, by ionizing the atom or exciting its electrons, as discussed in section 3.3. On the other hand, photons and neutrons are charge-neutral and require an intermediary medium that can induce an electric charge from the incident neutral radiation. Consequently, one can view the detection process itself as an NDE process with a specific purpose. Thus, one can use the five basic elements of the NDE, described in Chapter 1, to describe the detection process: where a *source* of radiation is made incident on a *modifying* material that produces a measurable (*detectable*) signal which gives an *indication* that can be *interpreted* to give the strength of the signal and or the energy of the incident particles.

[1]For x- and gamma-rays, photons should be used instead of particles.

The following three sections describe the different methods (modifying media) used to detect various types of radiation, and discuses some of the main characteristics of the detectors. Section 4.5 presents the electronic devices used to process the electric signal produced by the detectors. Some concepts common to all types of detectors are first introduced below.

Detection Efficiency. The intrinsic efficiency of a detector is a measure of how many of the incident radiation particles will produce an electric signal in a detector. However, the number of radiation particles incident on a detector is often difficult to know, since what is usually known is the source strength (the number of particles emitted, in all directions, from the source within a certain time period). Therefore, the detector efficiency is sometimes defined as the ratio between the detector count rate (pulses per unit time) and the source strength, and is then called the *absolute* efficiency as it incorporates the source strength. Another practical measure of a detector's efficiency is to relate its efficiency to that of some other reference detector, to obtain a *relative* efficiency. Given these different definitions of efficiency, the reader should pay attention to reported detector efficiencies.

For a stream of monoenergetic particles bombarding a detector, one would expect a train of pulses of the same amplitude (pulse height), as particles should deposit the same amount of energy in the detector. However, a distribution of pulse-heights is usually observed, centered around a peak, as schematically shown in Figure 4.1. This reflects the fact that incident particles interact in variety of ways, and at different locations within the detector, and thus affect the detector in slightly different ways. The peak of the registered signal represents the most probable event within the detector, and the width of the peak represents its energy resolution; the narrower the peak the better the resolution. The peak width is measured by the full-width-half maximum (FWHM), i.e., the full width of the peak at half its intensity, as schematically shown in Figure 4.1. Since the energy resolution of a detector usually depends on the energy of the incident particle, it is often reported as the ratio between the FWHM and the energy of the incident particle. It should be emphasized that resolution is a measure of the width of a single energy peak at a specific energy. Note also that if the pulse distribution is described by a normal (Gaussian) distribution, the area of the beak covered by the FWHM will correspond to 2.35σ, where σ is the standard deviation of the distribution [15]. The FWHM, therefore, defines an area that covers about 76% of the peak, if it were normally distributed. The Gaussian distribution is indicative of a large number of events occurring

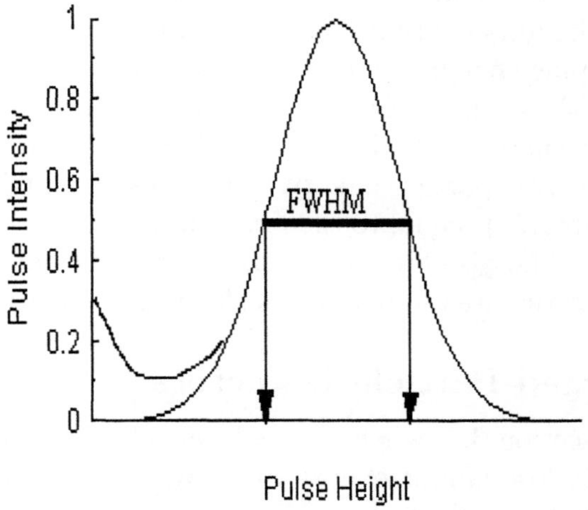

Figure 4.1. Schematic of pulse-height Gaussian distribution with a low-energy tail, showing the full-width at half-maximum (FWHM).

independently, which should be the case in a radiation detector. However since the number of events occurring in a detector is usually finite in number, within a finite measurement time, the theoretical shape of the Gaussian distribution is not usually attained. Nevertheless, having such a normal shape is desirable as it facilitates analysis of multiple peaks, say by curve fitting. An easy way to determine how close a distribution is to a Gaussian distribution is to measure the ratio $\frac{\mathrm{FW}x\mathrm{M}}{\mathrm{FWHM}}$, where FW$x$M is the full width at some fraction x of the maximum, and compare it to the calculated value for the Gaussian distribution. Typically, the ratios $\frac{\mathrm{FW0.1M}}{\mathrm{FWHM}}$ and $\frac{\mathrm{FW.02M}}{\mathrm{FWHM}}$ are measured, where FW.1M and FW.02M are, respectively the full widths at one-tenth and at 0.02 of the maximum. The closer these ratio are to the calculated values for Gaussian distribution, ($\frac{\mathrm{FW0.1M}}{\mathrm{FWHM}} = 1.83$ and $\frac{\mathrm{FW0.2M}}{\mathrm{FWHM}} = 2.38$), the closer the measured distribution to a normal distribution [91].

Some incident radiation particles can leave the detector before depositing their full energy within the detector, giving rise to 'lower' energy signals. This leads to a low-energy tail accompanying the detected distribution around the peak of monoenergetic radiation. Particle scattering on the wall of the detector can also contribute to this tail. Moreover, in some detectors, the incident particles give their energy to secondary particles, which are in turn transported in the detectors depositing lower energies that also contribute to the signal's tail. Obviously, the tail end tagging to the main peak deteriorates the quality of the pulse-height-distribution; particularly when it overlaps with peaks produced by other lower energy particles incident on the detector. The quality of the signal is measured by the full-energy peak efficiency, which is the ratio between the area in the pulse-height Gaussian (or a Gaussian-like) distribution under the peak, shown in Figure 4.1, by the total area of the signal including its lower energy tail. The higher this efficiency, the better the detector is in resolving particle energy. This is also an intrinsic efficiency, since the total integral is equal to the total number of particles measured by the detector. The absolute full-energy peak efficiency is defined as the ratio between the area under the peak to the source intensity.

4.2. Charged-Particle Detectors

As shown in section 3.3, when charged-particles interact with matter they continuously lose kinetic energy to the atom, causing its ionization or the excitation of its electrons. These physical processes can be directly utilized in detecting charged-particles, by employing a material that is highly sensitive to any of these interactions. Such material needs to be confined within an enclosure. However, charged particles also interact with the enclosure's material, and if that material is not carefully chosen, not all charged-particles incident on the detector may be able to reach its sensitive material. Therefore, special attention needs to be given to the detector's window, to allow charged-particles to reach the detector's sensitive material. This challenge is similar to that encountered with charged-particle sources, where an enclosure is required to contain the radioactive material, while a window is needed to allow the source particles to leave the source, see section 2.1.

A number of processes can be used to obtain detectable indications from the charged-particle kinetic energy loss and the associated ionization and excitation. The following sections discus the various processes used in charged-particle detection.

4.2.1. Detection by Chemical Reactions

Chemical reactions result from rearrangements of the electrons of adjacent atoms. Since charged-particles interact with atomic electrons, they can induce chemical reactions, the impact of which can be used as indication of the effect of the incident charged-particles. However, the effect of chemical interactions is not directly convertible to an electric signal, and therefore is difficult to quantify. On the other hand, chemical reactions provide a simple, passive, and equipment-free means for the detection of radiation, and can also produce a permanent record of detected events. Photographic emulsion and track-etching are two of these passive detection methods.

4.2.1.1 Photographic Emulsion

A photographic film is a cellulose film (or a plate) coated with an emulsion of finely divided silver-halide crystals in a viscous liquid (e.g. gelatin). When charged-particles pass through a film, they, like light, interact with the electrons of the silver halide molecules, destroying their chemical bonding and making them susceptible to chemical reduction into a metallic form in a subsequent development process. The unaffected silver salts are removed chemically with an oxygen-deficient chemical, called hypo, that forms a water solvable complex with the silver salts. Washing the film with water leaves, in the emulsion, metallic silver particles, which carry a visible (black) indication image of the tracks of the incident particles. Note that film badges used to monitor radiation exposure, see Chapter 5, employ radiographic emulsion to record the accumulated radiation exposure.

The process of detecting radiation with photographic films is associated with the process commonly know as radiography. The film thickness, or more accurately the thickness of the emulsion in radiographic films, is usually small (typically 10 to 20 μm), much smaller than the range of most charged-particles. Therefore, an incident particle may not deposit its entire energy within the emulsion. Moreover, subsequent particles can add to the impact of previous particles. Therefore, the indication left on the film reflects the cumulative indication of many individual particles incident on the film. Consequently the obtained indication is an overall indication, not the indication of a particular particle. However, if the film is made thicker and the concentration of the silver salt is increased, the track of individual particles can be recorded on the film. These are known as nuclear emulsions, and can be as thick as 0.5 mm, allowing the recording of entire individual tracks of charged-particles. The tracks can be seen under a microscope. The depth of the track within the emulsion reflects the particle energy, while its darkness is indicative of the particle's stopping power (energy loss per unit length, as defined in section 3.3), which can be related to its type.

4.2.1.2 Track Etching

Charged particles can also engrave tracks on some dielectric materials (such as quartz or mica) as the energy transferred by charged-particles destroys the chemical bonds between the molecules of the material. The trail of broken molecules can be made visible by etching in a strong acid, which affects more the damaged molecules than the unbroken ones. The formed tracks become visible under a microscope. However, forming tracks on a dielectric plate requires more energy than in the case with emulsion films. Therefore, track-etching is used with particles that has a large stopping power, making them unusable for electrons, and more suited for heavy charged-particles (ions of elements of high atomic number). Tracks formed by this method carry information on the particle type and energy. However, the etching process destroys many of the features of individual tracks, making it difficult to obtain information on singles particles, as is the case in nuclear emulsion.

4.2.2. Detection by Direct Ionization

As a charged-particle moves through matter it can, as discussed in section 3.3, ionize an atom by freeing one of its electrons. The ionized atom becomes positively charged. If an electric field is applied across the ionized material, the positive ions will move towards one direction (towards the cathode), while the electrons will move in the opposite direction (towards the anode). This creates an electric current, thus providing an electronic signal that can be amplified, shaped, measured, etc. Gases are employed in this detection process as they are easy to ionize and the resulting current is readily directed towards the electrodes (cathode and anode). This is the concept used in gas-filled detectors, such as ionization chambers, Geiger-Müller (GM) counters, proportional counters, multiwire counters, and position-sensitive counters. The concepts behind the operation of each of these types of detectors are described below, after briefly describing the behavior of the ionization process in gases as the strength of the applied electric-field (voltage gradient) is increased.

Ionization Plateau. Each charged-particle incident on a detector containing a gas causes ionization along its track within the detector. The electrons produced by ionization are collected in the form of a pulse. The pulse duration is determined by the time it takes the electrons generated by each charged-particle to reach the anode, since the ionization process itself is almost instantaneous. Therefore, each incident charged-particle produces a pulse. Obviously, if no voltage is applied between the electrodes of the detector, all the formed ions will recombine with the free electrons and no signal will be obtained. As the magnitude of the applied voltage increases, the voltage gradient between the two electrodes increases, and more electrons are accelerated towards the anode, damping the amount of ion recombination, and increasing the strength (amplitude) of the signal. This is almost a linear processes, as shown in

Figure 4.2, that is, as the voltage increases, the amplitude of the signal is proportionally increased. However, when all the generated electrons are collected, ion recombination ceases. Further increase in the voltage will simply increase the energy of the electrons, without affecting the amplitude of the signal, and thus the pulse-height reaches a plateau. When ion recombination ceases, the number of collected electrons will directly reflect the number of ionizations caused by the charged-particle incident on the detector. Then the detector will function as an ionization chamber, operating typically at less than 1 kV applied voltage.

Proportionality. If the applied voltage is allowed to increase, electrons will gain sufficient energy to begin themselves to ionize the gas. This breaks the plateau of constant count rate, which is the domain of operation of ionization chambers, and marks the beginning of the region of operation of proportional counters. This typically requires an applied voltage from 1 to 2 kV. Position-sensitive and multiwire detectors also operate at this proportionality range as discussed later in this section. In this range, the electrons produced in this secondary ionization process are also accelerated by the same applied voltage, causing further ionization, and so on. The result is an avalanche of ions, which is further enhanced by increasing the applied voltage, causing the steep increase in the amplitude of the pulse observed Figure 4.2. However, the amplitude of the signal will still be proportional to the number of electrons generated in the original ionization process caused by the incident radiation, albeit with a greatly enhanced multiplication. Increasing the voltage further increases proportionally the amplitude of the signal, by increasing the amount of secondary electron-induced ionization. This proportional increase remains linear with voltage until the amount of ionization becomes so large that it creates a large cloud of positive ions within the detector chamber. Positive ions migrate towards the cathode more slowly, because of their higher mass, than the electrons moving towards the anode. As ionization increases they will begin to present a significant cumulative positive charge that can distort the voltage gradient within the detector, and hence its linear response. Therefore, it is not desirable to operate a detector in this so-called limited linearity region.

Avalanche. If the applied voltage increases further, the intensity of the avalanche increases, to the point that its electrons can trigger another avalanche. The electrons of the new avalanche can also trigger a new one, and so on, creating a propagating chain reaction of avalanches within the tube. This process can continue to grow indefinitely except for the fact that the cloud of the slowly moving positive ions associated with these avalanches also grows and begins to gradually mask the effect of the applied field, and dampen the avalanche propagation process until they ultimately terminate the chain reaction. During the propagation of avalanches, the produced electrons ultimately lose any relationship to the electrons generated in the initiating ionization process caused by the incident particles. However, each incident particle, at this high voltage, produces its own burst of avalanches. The intensity, or amplitude, of this burst will depend only on the magnitude of the applied voltage and the nature of the gas, but not on energy, or even type, of the incident particle. Under these conditions, the detector operates as a simple counter of intensity of the incident radiation, and is called a Geiger-Müller, or simply a GM, counter. In order to avoid using an excessively high voltage, gases used in GM tubes are kept at sub-atmospheric pressure, thus requiring an operating applied-voltage from only 0.5 to 2 kV, depending on the type of gas used. If the applied voltage is further increased beyond the domain of GM counters, the increased voltage can overcome the masking effect of the positive ions, and a continuous discharge of the gas will be produced. Then even information on the number of incident particles will be lost and the gas will cease to be of any value as a detection medium.

Detector Window. In order for any of the above described ionization processes to occur, it is imperative that the incident charged-particles reach the gas contained within the detector. It is also equally important to contain the gas within a well-sealed container, and at high pressure. The container wall may need to be constructed from a strong metal to withstand the pressure. Therefore, for charged-particles, gas detectors are equipped with thin windows, made of glass, mica, or ultra-thin foils of metal (aluminum, stainless steel or titanium). Beryllium windows are also used, as it is a metal that is sufficiently strong but is not a good stopper of charged-particles because of its low atomic-number. Detectors equipped with a window at one of their sides (or ends) are called end-window detectors.

4.2.2.1 Ionization Chambers

As indicated above, an ionization chamber is the simplest form of an ionization-based detector. The electric charge collected at the electrodes of such detector is equal to the number of the charges produced by the incident ionizing particle. Therefore, in an ionization chamber, no magnification of the charge is introduced. Moreover, as shown in Figure 4.2, ionization chambers are stable counters, since their amplitude does not

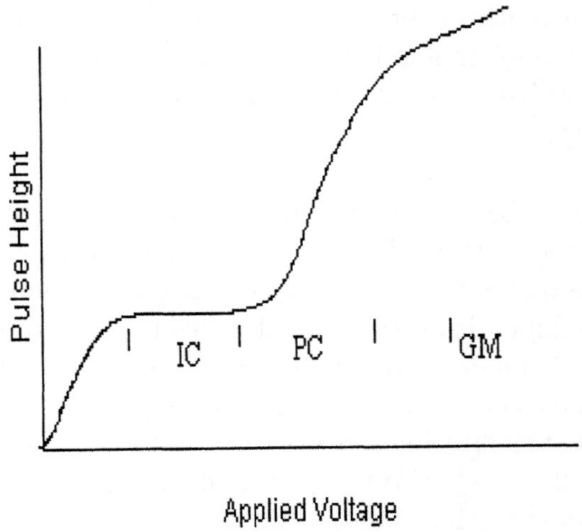

Figure 4.2. Pulse height versus voltage in a gas-filled detector, showing range of operation of ionization chambers (IC), proportional counters (PC), and GM tubes (GM).

change with variations in the applied voltage, due to power instabilities, and depends only on the amount of ionization introduced by the incident particle. Ionization chambers may be used to measure the spectrum of charged-particles, although semiconductors are usually employed for this purpose, as discussed in section 4.2.4.

Particle Energy. When a charged-particle passes through an ionization chamber, it loses an energy, ΔE, by ionizing the gas. It takes on average about 30 eV to produce an ion pair (positive ion and a free electron) in most gases [15]. Therefore, the number of ion pairs generated by the particle would be equal to $\frac{\Delta E(\text{eV})}{30}$. Each ion pair carries a total charge (positive and negative) of $2e$, where e is the elementary charge (1.602×10^{-19} C). At the plateau where an ionization chamber operates, see Figure 4.2, no ion recombination occurs, and all the charge deposited by the particle will produce a current. Therefore, this particle will produce a pulse of a charge (amplitude) equal to $\frac{2e\ \Delta E(\text{eV})}{30}$. The energy loss ΔE depends on the type and energy of particles, and the type and pressure of the gas. Therefore, for a given type of charged-particles, and a given detector, the amplitude of the pulse will be proportional to the particle's energy. This of course is only valid if the incident particle deposits all its energy in the gas, which can be made possible by increasing the detector's size and/or its pressure (to increase its density). It is inevitable, however, that some particles will escape the detector before losing their entire energy, and others, if negatively charged, may be attracted to one of its anode's causing distortions in the pulse. Nevertheless, the amplitude of the pulse can still give a good indication of the particle-energy.

Pulse Duration. The immediate charge collected by the applied voltage is caused by the drifting electrons, while the slow motion of the much heavier ionized atoms produce a delayed response. Therefore, the duration of the pulse depends on the time it takes the generated electrons to reach the anode of the detector. This time in turn depends on the velocity with which the electrons drift to the anode of the detector. The value of electron's drift velocity increases with the magnitude of the gradient of the applied voltage, $\frac{dV}{dx}$ (Volt/m), due to the increased particle acceleration effect. On the other hand, a higher gas pressure, p, increases the density of the gas and decreases the drift velocity, by offering resistance to the effect of the applied voltage. Therefore, the drift velocity, v, is proportional to $\frac{1}{p}\frac{dV}{dx}$; the proportionality constant is called the mobility factor, μ_b. Then, $v = \frac{\mu_b}{p}\frac{dV}{dx}$. The movement of the electrons is slowed-down by inelastic collisions with the atoms of the gas, a process that depends on how tightly bound are the atomic electrons. Electrons of high energy will resist drifting under the effect of the applied voltage, as their high energy enables them to move away from the anode. Therefore, gases with atoms that can slow-down the electrons by inelastic scattering will have a higher electron mobility, and the value of μ_b depends on the composition of the gas. However, for gases of medium atomic number, the value of μ_b for electrons is in the order of 10^4 Newtons/V s [15]. Therefore, at a typical voltage gradient of one kV/m, and gas pressure of 100 kPa (1 atm), the electron's drift velocity will be in order of 1 km/s. In a detector where the density between the electrodes is 50 mm, and assuming that ionization occurs on average at a distance half-way between the anode and the cathode, then the collection time of the electrons, and hence the pulse's duration, would be 25 μs ($\frac{50\times0.5\times10^{-3}}{10^3}$ s). Therefore, a detector would be ready to receive another pulse, without overlapping with a previous one, only after 25 μs, thus allowing a maximum count rate of 40,000 counts per second. This is a high count rate for many applications; although scintillators can accommodate higher count rates as explained in section 4.2.3. Obviously, the gas pressure can be used to control the pulse duration, since as the pressure increases, the drift velocity decreases, and the charge collection time also decreases. Almost any level of gas pressure can be used in an ionization chamber, but gas pressures in the range of 100 to 600 kPa (1 to 5 atm) are commonly employed.

Ionization Current. In addition to the pulse mode of operation, which provides a count rate, and enables processing of the pulse amplitude to obtain energy information, see section 4.5, the collected charge can be accumulated, i.e., integrated, over time. The charge collected per unit time defines the electric current, hence the terminology *current mode*, see section 4.5.3. This mode of operation is simple to implement. The resulting current can be measured directly with an ammeter (or more realistically a picoammeter), or by means of a high-impedance voltmeter, connected across a resistance through which the current flows. In either case, a direct indication of the current is obtained. This is the most common mode of operating ionization chambers and is best suited for high count rates, where time between successive pulses is too short to allow pulse-by-pulse processing.

Ionization Gas. Almost any gas can be used in an ionization chamber, even atmospheric air. However, in some gases, the electrons, as they move towards the anode, can become attached to the neutral atoms of the gas, forming slowly moving negative ions. This is undesirable, as it will hinder the movement of the electrons and tremendously elongate the duration of the signal. Therefore, attention is given to the choice of the ionizing gas to minimize this effect. An inert (noble) gas, typically argon because of its low cost, is used. However, noble gases have large ionization potential due to the nature of their tightly bound atoms. Therefore, noble gases have a low value of μ_b, as they allow electrons to maintain their original energy, giving them the freedom to move in any direction and resist drifting by the applied potential towards the anode. Therefore, a hydrocarbon gas (about 10% of methane or carbon dioxide) is often added to the inert gas used in ionization chambers, to increase their electron mobility. This in turn increases their drift velocity and reduces the charge collection time, or pulse duration.

Applied Voltage. Ionization chambers operate at the plateau voltage, shown in Figure 4.2. The value of this voltage depends on the size and shape of the detector and the type of gas used. Manufacturers provide a recommended operating voltage for each detector. The user can verify the recommended value, by observing the change in the pulse's amplitude with increasing voltage, essentially to reproduce Figure 4.2, until the operation of the detector just begins to depart from the range of operation of the ionization chamber and enter the region for proportional counters.

Geometry. The geometry of the electrodes in an ionization chamber has a direct impact on the distribution of the electric voltage within the chamber, and in turn on its gradient, $\frac{dV}{dx}$. If the anode and the cathode are arranged as two parallel plates, then $\frac{dV}{dx}$ becomes uniform, except near the walls of the chamber, and is equal to the applied voltage divided by the separation distance between the plates. This uniform field enables a relatively easy prediction of the shape and duration of the pulse, see references [92] and [15]. Cylindrical ionization chambers are also used, with the cylindrical body grounded and a thin (25 to 100 μm diameter) tungsten or platinum wire employed as the anode. In such cylindrical geometry, the gradient of the electric field varies inversely with radius, generating a strong gradient near the anode wire, slightly complicating the shape of the pulse. However, cylindrical ionization chambers, equipped with thin windows on one of the flat surfaces of the cylinder, has been used for measuring beta-particles, see for example reference [93].

The main disadvantage of ionization chambers is the small amplitude of the pulse, a few mV. Therefore, pulse analysis with them requires high gain amplifiers, carefully designed to avoid amplifying the electronic noise that always exist in electronic systems. Ionization chambers are, therefore, more suited for operation in the current mode.

Gridded Chambers. Information pertaining to the energy of a charged-particle is usually lost in an ionization chamber. Although electrons produced by ionization in the chamber are immediately removed from the tracks and collected by the anode, they leave behind slowly-moving positive ions. These ions mask the effect of the externally applied voltage, as they create their own positively charged barrier. The amount of masking depends on the orientation of this barrier with respect to the electrodes of the chamber. This masking affects the number of ions collected, so that even for the same incident particle-energy, producing the same amount of ion pairs, the amplitude of the pulse will vary with the direction of incidence, as it affects the orientation of ionization tracks. To overcome this problem, a grid of fine wires spaced millimeters apart, called the Frisch grid, is placed between the anode and the cathode. These wires are just slightly negatively charged so that electrons are not attracted to them, but are allowed to move in between them. The effect of this grid is to create a fixed masking barrier, and thus produce a pulse amplitude that is dependent only on the number of ion pairs produced, and in turn is dependent on the particle energy. However, provisions have to be made to prevent ionization in the region between the grid and the anode, otherwise the grid loses its fixed masking effect. The source of the particles needs also to be confined so that they only reach the zone between the cathode and the grid. This is done, either by collimating the source, or by physically placing it on the cathode (if possible). Instead of using anode-cathode wires, ultra-fine layers of conductive strips are used in gas-microstrip detectors [94]. These strips are deposited on an insulating, or partially insulating, glass substrate and arranged in an anode cathode pattern. The electric field necessary for the multiplication process is produced between adjacent strips, not by the voltage difference between the strips and the detector's cathode. The fine size of the strips and high gain enable the use of moderate gas pressures, thus providing high spatial-resolution, high contrast resolution and excellent mechanical stability [95]. Other advantages of microstrip wire chambers over multi-wire proportional counters include gain uniformity over a large area, operation at lower voltages, faster ion collection time, counting at higher rates without gain deterioration and better ability to reject background pulses using pulse shaping [96]. Recently a new position-sensitive <u>microm</u>esh <u>ga</u>seous <u>s</u>tructure (Mircomegas) chamber has been developed [97]. It consists of planar electrodes forming a conversion gap in which incident radiation causes ionization and a thin amplification gap. The two gaps are separated by a few mircometers. The particles drifting into the amplification gap are collected using electrodes on a printed circuit. This concept of detection leads to inexpensive high-precision position-sensitive detectors that can function in high radiation fields.

4.2.2.2 Proportional Counters

As Figure 4.2 shows, proportional counters require a higher applied voltage, or more precisely a higher voltage gradient (electric-field strength), than ionization chambers. However, in order to produce a high-voltage gradient, without applying an excessively high voltage, proportional counters are made in the form of cylinders with a thin wire (25 to 100 μm in diameter) stretched along its axis. A positive voltage is applied across the wire, and the detector's walls are grounded. Then the voltage gradient becomes proportional to $\frac{1}{r}$, where r is the radial distance from the wire. Therefore, an extremely high-voltage gradient is produced at the vicinity of the wire. As mentioned earlier, the avalanche process associated with charge multiplication is the result of the ionization of the gas by the resulting secondary electrons. The electrons are accelerated as they are attracted to the anode. The multiplication process associ-

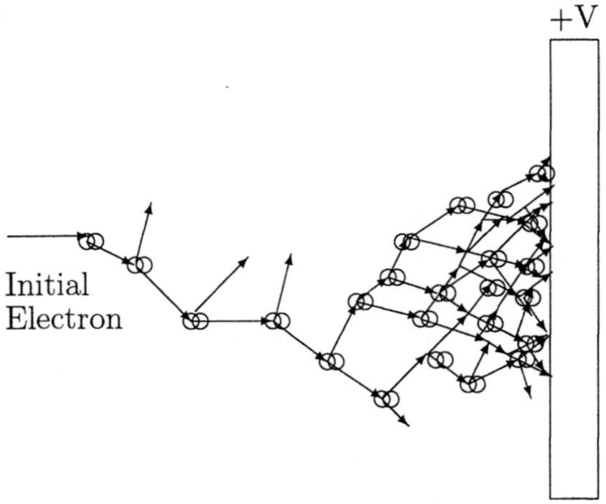

Figure 4.3. Schematic of initiation of ionization and propagation near the wire anode of a proportional counter, based on reference [98].

ated with the avalanches of proportional counters occurs only in a very small region around the anode's wire, where the electrons acquire the energy needed to cause ionization. Confining charge multiplication into a small volume has the added advantage of providing almost a uniform multiplication process. Therefore, while the initial ionization process by the incident charged-particles occurs everywhere in the detector, charge multiplication takes place only within a small volume around the anode wire, as schematically shown in Figure 4.3. When an electron from the initial ionization event drifts towards the anode's wire, it creates an avalanche extending along the length of the wire, and away from it, by a distance equivalent to no more than a few times the wire's diameter (less than a fraction of a mm). This localization of the avalanche enables the design of position-sensitive proportional counters, as discussed later in this section.

Multiplication Factor. The total charge produced in a proportional counter has the same dependence on particle energy, gas pressure and electron mobility, as in an ionization chamber, see section 4.2.2.1. However, in proportional counters, the charge is further magnified by the detector's multiplication factor, M. In a cylindrical detector, the multiplication factor is often estimated by Diethorn's model as [15]:

$$\ln M = \frac{V}{\Delta V} \frac{\ln 2}{\ln \frac{b}{a}} \ln \left\{ \frac{V}{K\, p\, a\, \ln \frac{b}{a}} \right\} \tag{4.1}$$

where V is the applied voltage, ΔV is the potential driving an electron's movement between successive ionization events, a is the radius of the anode's wire, while b is the inside radius of the cylindrical detector and p is the gas pressure. The value of the parameter K is obtained by operating the detector as an ionization chamber, i.e. with $M = 1$, which requires reducing the voltage to the plateau value shown in Figure 4.2, say V_0. Then from expression (4.1), with $M = 1$, $K = \frac{V_0}{p\, a\, \ln \frac{b}{a}}$. Note that both the values of K and ΔV depend on the type of gas used to fill the detector. However, the value of ΔV is typically about 25 V, while K is around 50 V/m Pa [15]. Expression (4.1) indicates that, for a given gas pressure, M increases rapidly, almost exponentially, as the applied voltage is increased. This makes proportional counters quite susceptible to changes in the applied voltage, and therefore very stable high-voltage suppliers are required for their operation. The magnification factor can be quite high as can be demonstrated by considering the case for $a = 12.5$ μm, $b = 25$ mm, $V = 1500$ V, $p = 100$ kPa (about 1 atm), with the typical values of ΔV and K reported above, then using Eq. (4.1), $M = 540$. Magnification by a few decades to a few tens of thousands were obtained in commercially available proportional counters, see reference [15].

Pulse Shape. The shape of a pulse generated by a single charged-particle interacting within a proportional counter is governed by the drift of the positive ions of the avalanche formed near the anode towards the cathode. This is because the charge produced by the initial ionization event is very small compared to that produced by the electrons. Moreover, the generated electrons are collected almost immediately by the anode's wire, due to the higher voltage gradient attained near the anode where avalanches occur. However, the same high-voltage gradient pushes positive ions away toward the cathode at a high velocity, resulting in a fast rise in the amplitude of the signal, as schematically shown in Figure 4.4. As ions move away from the anode, the voltage gradient decreases, the drift velocity of the atoms also decreases, slowing the rise time of the pulse. The pulse amplitude eventually saturates when the entire charge of the positive ions are all collected at the cathode. However, by electronic shaping of the pulse by a differentiating circuit (see section 4.5) the slow portion of the pulse is removed, as schematically shown in Figure 4.4. The effective (useful) duration of the pulse

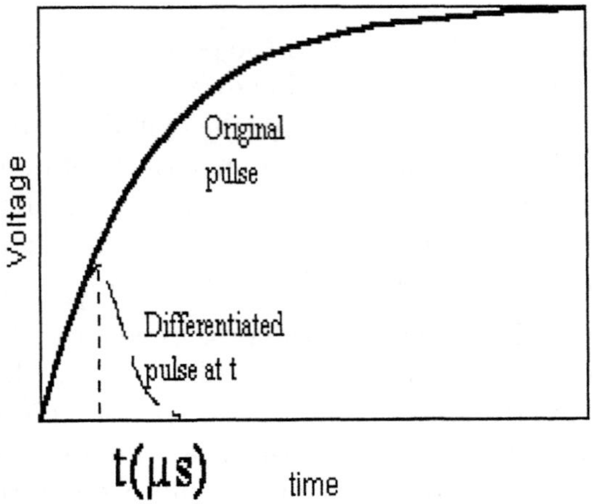

Figure 4.4. Shape of a pulse produced by a proportional counter, before and after differentiation at time, t, based on reference [92].

is, therefore, determined by its rise time, which is in the order of a few microseconds.

Filling Gas. Gases to which the moving electrons can stick to atoms, creating heavy negative charged-particles, should be avoided. Such gases can reduce the multiplication factor by eliminating the fast-electrons that trigger the avalanche. Since air is one of these gases, tight sealing of the detector is essential for sustaining its high multiplication factor. Since electron attachment to atoms of noble gases is not strong, they are often used in proportional counters, as in the case of ionization chambers, section 4.2.2.1. Argon is the preferred gas because of its low cost. However, it is more difficult to ionize an inert gas, because of its tightly packed atoms, electrons lose their high energy and cease to trigger avalanches. These lower-energy electrons, although incapable of causing ionization, can introduce excitation of the atomic electrons of the metallic walls of the detector. In addition, when the heavy positive ions reach the cathode, they can also excite its atoms, and release electrons from its metallic material. Although, such additional secondary electrons will also drift towards the anode and can trigger additional charge multiplication, they are not directly related to the initiating ionization event by the incident charged-particle. Thus, these secondary electrons distort the proportionality between the energy of the incident particle and the amplitude of the pulse, and introduce spurious signals. Due to these reasons, proportional counters employing inert gases are operated at a low multiplication factors, less than 100, to avoid the above mentioned complicating effect. Operating such counters at higher multiplications becomes possible if these undesirable electrons are suppressed by a polyatomic gas, such as methane (CH_4). Then the energy of both the secondary electrons, and other low-energy electrons are consumed in disintegrating the weakly bound molecules of the added gas. Moreover, the added gas will stand in the way of the positive ions moving towards the cathode, and absorb their energy before they can bombard the cathode and release some of its atomic electrons. Because of their extinguishing effect, the added gas is called the quench gas. An added advantage of a hydrocarbon gas, such as methane, is that it increases the mobility, and consequently drift velocity, of the electrons, as explained in section 4.2.2.1. This is beneficial as it helps maintain a high-voltage gradient near the anode, by quickly attracting electrons before they begin to accumulate around the anode forming a negative charge cloud that can mask the positive charge of the anode and reduce the voltage drop. The so-called P-10 gas (10% methane, 90% argon) is one of the most widely used gas mixtures in commercial proportional counters. The gas pressure is usually kept below atmospheric pressure.

Applied Voltage. Proportional counters can operate under applied voltages from 300 to 3000 V, depending on the size of the detector and its filling gas. Manufacturers of detectors usually provide a recommend voltage under which the performance of a certain detector is stable. The operating voltage of a proportional counter is based on the so-called counting plateau, which is different from the amplitude plateau of Figure 4.2, used in determining the operating voltage of ionization chambers. The concept of counting plateau is based on the fact that rarely a single avalanche is relied upon, as the detector would be susceptible to any distortions of this single pulse by one or more of the many processes occurring within the detector. Instead, a number of ion-pairs produced by various incident particles are utilized. But even if all the incident particles are of the same type and energy, they would generate pulses with slightly different amplitudes due to the randomness of their emission, the location and length of their track path within the detector, etc. Therefore, one observes a pulse with a peak corresponding to the most probable events, and a distribution around both sides of the peak, as shown in Figure 4.1. The distribution around the peak reflects low and high energy depositions inside the detector. If the applied voltage is too low, the portion of the signal corresponds to low-energy events can be mixed with the other low-energy events, arriving for example from unmagnified initial ionization and even from electronic noise. Therefore, a low-energy discrimination level is usually set in counting experiments to remove these low energy events. However, this discrimination level will also remove a portion of the desired signal. The applied voltage can increase to further magnify the signal beyond the noise level, by giving more energy to the collected ions to permit more of them to exceed the discrimination level. As the applied voltage is further increased, more of the produced ions will be collected. When all ions are gathered, elevating the applied voltage will no longer increase the detector counts, and a counting plateau is reached. Increasing the applied voltage within this plateau region should not affect the count rate. However, if the applied voltage is allowed to significantly increase, overlapping and spurious pulses will start to contribute to the detector counts, and the count rate will start to increase again. Therefore, proportional counters have a region of a stable count-rate plateau, at certain range of applied voltage, within which they should be operated. It should be noted, however, that the count rate does not remain exactly constant within this plateau, due to the slight increase in amplitude with voltage in most detectors [92]. This is attributed to the increase in the detector's efficiency with increasing voltage, by attracting more ions towards the cathode.

Position-Sensitive Detectors. Avalanche generation is a localized event, around a point in the anode at the point of arrival of the inducing charged-particle. This fact is used to obtain information on the position of particle-incidence on the detector, enabling for example the determination of the spatial distribution of incident particles. Spatial information is either obtained using a single anode with a high resistance per unit length (position-sensitive detectors), or using multiple anode wires (multiwire counters). In the first arrangement, the signal is extracted from both ends of the wire. Then because of the increasing resistance, a signal produced by an avalanche will generate a signal with a higher amplitude at the end close to its occurrence and a lower-amplitude signal at the other end farther away. Further processing of the signal will then enable determination of the position of the signal along the wire. The same concept can be generalized by employing multiple anodes. This is typically done in an arrangement that utilizes two parallel plates as cathodes, with a grid of anode wires placed between them [99]. Then two-dimensional position information are obtained; one from the position of the wire near which the avalanche takes place, and the other along the axis of the anode using the amplitude of the pulses recorded at each of its two ends.

Drift Chambers. These are multiwire proportional counters but with the wires widely spaced (by a few centimeters). The electrons produced by ionization tend to drift in the low electric field away from the wires, before reaching the high field near a wire. The drift-time becomes then indicative of the position of the charged-particle originating ionization. When a drift chamber has a large drift volume, and the drift direction is perpendicular to the plane of its wires, the chamber is called a time-projection chamber. For more information on drift chambers see reference [100].

4.2.2.3 Geiger-Müller Counters

While ionization chambers can accommodate high rates of radiation incidence on the counter by operating in the current mode, proportional counters operate in the pulse mode providing both count rate and particle-energy information. For low count rates and low-energy particles, a high multiplication factor is required, otherwise the signal will be lost in the noise. This is the domain of Geiger-Müller (GM) counters, which operate at a high voltage, as Figure 4.2 shows, but are similar in design to proportional counters. The increase in magnification aims at spreading the occurrence of avalanches along the entire length of the anode wire in a cylindrical chamber. The amplitude of the pulse is then increased to the 20 to 100 V range. This comes, however, at the expense of losing any energy information, as chain of avalanches eventually lose memory of their initiating ion-pair formation. Moreover, the chain of avalanches produced by a single incident particle takes a considerable time to discharge to the cathode (about 1 ms). Within this time, avalanches from a newly arriving particle will overlap with those generated by the preceding particle. This is another reason for the counter to operate at a low count rate (less than 1000 counts per second). However, because of the high amplitude of the signal, no amplification of the single is required, as it can directly operate in the current mode, see section 4.5.3. This simplifies the device and reduces its cost. Since the produced pulse does not depend on the nature of the initiating event, the counter is also independent of the nature of the incident radiation. Therefore, GM counters are mainly used as survey meters for detection of contamination or measuring low-level radiation background, such as that found in underground mines.

4.2.2.4 Other Chambers

Spark and Flash Chambers. Spark chambers are gas chambers employing a set of wires or plates with a large voltage difference between them. A charged-particle passing through the chamber generates a spark between the wires, producing a visible track. Increasing the applied potential difference enhances the visibility of the spark. At even higher voltage, the chamber becomes "saturated" and the recombination of electrons and positive ions leads to the production of visible streamers. Further increase in the applied voltage can lead to an electrical breakdown of the gas, as the ionization would spread to both electrodes. However, by applying short (a few nanoseconds) voltage pulses across the wires or the plates, the streamers are localized along the track of the charged-particle. The chamber is then known as a streamer chamber. Flashes, rather than sparks, can be produced by placing the wire, or the plates, inside plastic wraps filled with a gas (typically a neon-helium mixture). The applied voltage then produces flashes, instead of sparks. Both spark and flash chambers have a slow response time, and thus are suited only for use at very low count rates [99]. Sparks, streamers and flashes can be photographed through transparent electrodes to show pictures of particle tracks, or electrically counted as pulses of electric charges.

Cloud Chambers. A cloud chamber is a chamber filled with humid air, that is suddenly cooled when a charged-particle enters the chamber. This brings the air to a supersaturated state in water vapor. Charged-particles entering the chambers cause ionization in the air. The formed ions act as nucleation centers around which water droplets are formed, marking the tracks of ions. If the chamber is placed inside a strong magnetic field, the tracks of the particles can be observed and their momentum can be measured [101]. The chamber is also known as Wilson chamber, and can employ any suitable mixture of vapor in equilibrium with liquid, and a non-condensable gas that can be brought into a supersaturated state by expansion [102].

Bubble Chambers. These chambers, like cloud chambers, function on phase transformation. Unlike cloud chambers which rely on liquid droplet formation in a supersaturated gas, bubble chambers, as the name indicate, rely on the formation of bubbles in a superheated liquid. The liquid in the chamber is brought into the superheated state by sudden lowering of its pressure. The ions generated by a charged-particle act as nucleation centers for bubble formation. The bubbles formed along the path of a charged-particle can be made visible, allowing the detection of particle tracks [101]. Typically, the chamber uses liquid hydrogen, kept under pressure at temperature slightly above its boiling point.

4.2.3. Detection by Scintillation

The continuous energy loss of charged-particles as they travel in matter can be converted in some solid and liquid materials into visible light (scintillation) that can be easily measured with a photomultiplier tube, or a photodiode. Scintillation is an attractive mode of operation as it enables the use of liquid and solid materials, which are denser than gases, thus allowing more efficient detection of more energetic particles that may pass through a gas detector without causing much ionization. The scintillation process is similar in nature to the fluorescence processes, discussed in sections 2.1.3 and 2.2.2 for producing fast-electrons and low-energy photons, respectively. That is, the process is associated with electron transition between the shells (excitation states) of atomic electrons. However, since the scintillation process produces visible photons, it is associated with transitions between very close atomic shells, eV energy difference, rather than the keV transitions in conventional fluoroscopic transitions. In order to achieve such low-energy transitions, closely packed atomic-energy levels are required. These levels are generated in some inorganic materials by adding some impurities, called activators, with electron excitation states that can fill in the gaps between the wide states of the parent material. In organic materials, use is made of the fact that a compound material can have many excitation states, belonging to each of its elements, that can overlap each other, creating the small gaps required for scintillation. This usually occurs in a class of organic materials known as aromatic compounds. When a charged-particle enters one of these scintillating materials, the energy it loses is used to excite electrons to a higher level, and light is emitted as these electrons return to a more stable state.

Scintillation detectors must be transparent to light, to allow the generated light to leave the detector without being absorbed within. This is facilitated in these materials by the fact that the energy of the emitted light photon is usually lower than the energy needed to excite atomic electrons. In inorganic scintillators, the excited electrons tend to settle in one of the many shells that exist between their original states and the excitation states created by the impurity in the inorganic material. This reduces the energy of the emitted photon to below the initial excitation energy, and lowers the chance for reabsorption of the photon within the material. A similar effect occurs in organic scintillators, where lattice vibrations cause the excited electrons to lose energy and land at higher energy levels after de-excitation.

Scintillation Efficiency. Not all excited states created by the incident radiation are necessarily converted into light, as some simply result in the emission of heat by giving kinetic energy to the molecules of the detector's material. The fraction of particle-energy that results in the emission of light defines the scintillation efficiency. For most inorganic scintillators, this efficiency is less than 12%, and is even lower (by about one third to one fifth) for organic scintillators. The scintillation efficiency is usually reported relative to the absolute efficiency of a common scintillator. For inorganic scintillators, the efficiency is reported with respect to that of NaI(Tl) (which has an absolute efficiency of 11.3% [15]), and for organic scintillators anthracene (with a conversion efficiency of about one third of that of NaI(Tl) [92]) is used as a reference.

Inorganic Scintillators. Although they produce pulses with amplitudes (pulse-heights) that are linearly proportional to the particle energy, up to 1 MeV for protons and deuterons and 15 MeV for alpha-particles [92], inorganic scintillators are not usually employed for the detection of charged-particles. Owing to their high density (typically above 3500 kg/m^3), the energy loss per unit length, $\frac{dE}{ds}$, is quite high in inorganic scintillators. This highly localized energy damages the molecules and quenches the scintillation process, thus reducing both the efficiency and resolution of inorganic scintillators to charged-particles, in particular heavier ones. Nevertheless, silver-activated zinc sulfide, ZnS(Ag), a scintillator available only in powder form, has been historically used for detecting alpha-particles and other heavy ions, in the form of thin screens. Another organic scintillator suitable for detecting alpha and beta particles is YAP:Ce (Yttrium Aluminum Perovskite doped with Ce, $YAlO_3$:Ce), as it can be used with a very thin entrance window, see section 4.3.2.

Organic Scintillators. The less dense (close to 1000 kg/m^3) organic scintillators are more suited for detecting charged-particles. Table 4.1 lists some of the organic scintillators commonly used for detecting charged-particles. Liquid scintillators can be used in large quantities to increase the detection of low-energy beta-particles. Plastic scintillators are particularly suited for detecting charged-particles, since they do not need a container, thus eliminating the need for detector windows. Plastics can also be shaped into any geometry or size to accommodate the requirements of a given application. The response of organic scintillators tends, however, to be sensitive to the direction of incidence of heavy charged-particles [92], due to their short range. Therefore, attention should be paid to the direction-dependent response of these detec-

Table 4.1. Some common scintillator for detecting charged-particles [103].

Scinitllator	Type	Scintillation Efficiency (relative)	Decay Time (ns)	Alternative Name
Anthracene	Crystal	100	30	
BC-400	Plastic	65	2.4	NE-102
Bc-404	Plastic	64	2.1	PILOT B
B-412	Plastic	60	3.3	NE-110
BC-517L	Liquid	30 - 39	2	NE-235A
BC-517H	Liquid	40 -52	2	NE-235H
Stilbene	Crystal	60	4 -8	

tors, when using such scintillators with particles other than electrons or beta-particles.

Pulse Shape. Atomic excitation occurs almost instantaneously, followed by a rapid prompt decay. This results in a very fast increase in the rise time of the light output pulse. The prompt scintillation response then decays exponentially. Therefore, the pulse produced by scintillation can be described by the exponential relationship [15]:

$$L(t) = L_0 \left[\exp(-\frac{t}{\tau}) - \exp(-\frac{t}{\tau'}) \right] \tag{4.2}$$

where $L(t)$ is the number of light-photons emitted at time t, following the energy deposition by the indent particle, L_0 is the maximum number of light-photons emitted, and τ is the decay time-constant (or simply decay time) of the scintillator, while τ' is the rise time. Since τ' is an order of magnitude less than τ, the second term of Eq. (4.2) is usually ignored and the pulse shape is simply expressed as $L(t) = L_0 \exp(-\frac{t}{\tau})$. Inorganic scintillators have decay-times in the order of microseconds, while organic scintillators are much faster, with decay times of a few nanoseconds, making the latter well-suited for fast-timing measurements.

Since the stopping power, $\frac{dE}{ds}$, is higher for a heavy charged-particle than for a lighter particle with the same energy, see Eq. (3.27), a heavy particle will tend to lose most of its energy over a short distance, thus increasing the probability of causing thermal excitation without light emission. Therefore, a fast-electron, or a beta-particle, will give more light output (intensity) than an alpha-particle of the same energy. On the hand, heavier particles tend to excite the atomic electrons to longer-lived states, because of the larger amount of energy they impact into the atoms. The result is that atoms excited by heavy charged-particles will continue to produce scintillation light for a longer period than lighter particles. Therefore, the tail end of a pulse tends to change slope as delayed emission begins to take place, from decaying with a time constant τ to decaying with a time constant that can be higher by an order of magnitude or two, depending on the particle type. The decay-time for an alpha-particle is longer than that for an electron, or a beta-particle. This delayed emission effectively changes the shape of the pulse, and thus offers the ability to distinguish between different types of particles by their pulse shape, see section 4.5.5.4. Note that the decay-times given in Table 4.1 are for prompt scintillation, which tends to be independent of the particle type.

Phosphorescence & Afterglow. In some scintillators the lifetime of some excited states can be as long as 1 ms, specially in low-energy transitions from thermal excitation. This slow component of light is called phosphorescence, and causes the detector to continue to emit light awhile longer after the termination of exposure to radiation. This phenomena is know as afterglow and presents problems in systems that operate on continuous basis, as a newly measured signal can be contaminated by the effect of a previous signal.

Photomultiplier Tubes. The light emanating from a scintillator is converted into an electric pulse using a photomultiplier (PM) tube or a photodiode. A PM tube is a vacuum tube equipped from one end with a semi-transparent photocathode, that absorbs the incident light and converts it into electrons. The light from the scintillator is often transmitted to the PM tube via *light pipes.* These are transparent solids, with a high index of refraction, (such as Lucite) that internally reflect the incident light, guiding it from on end of the pipe to the other. Typically the photocathode in a PM tube is made of a compound of cesium and antimony, K_2CsSb, called *bialkali.* Since its sensitivity to light matches the frequency of the light emitted from most scintillators, bialkali has a high efficiency for producing electrons from light, and its electron production by heat (thermal emission rate) is low [15]. An applied electric field accelerates these electrons and directs them to an intermediate electrode, called a dynode, usually coated with either cesium and antimony or silver and magnesium. The electrons impending on the dynode liberate more electrons from its metallic material, causing a multiplication effect in the number of electrons. These electrons are in turn accelerated with another electric field and directed towards another dynode, and so on. Typically ten, or more, dynodes are used, to provide a high multiplication level of electrons, in the order of a million. Finally, the electrons are collected by an electrode. The electric field, of about 80 to 120 V between each pair of dynodes, is introduced with the aid of a series of voltage dividers (resistors) located in a plug on the tube base assembly. This assembly divides the total applied voltage, typically in the order of 1 kV, equally between the dynodes and the anode. Note that thermal heating of the cathode can also produce electrons, which are subsequently amplified in the PM tube, producing the so-called *dark current*; a noise signal that should be minimized. In addition, the presence of any external magnetic field can influence the direction of motion of electrons between the dynodes and affect the performance of the tube. However, the effect of these fields can be reduced by wrapping the PM tube with an electromagnetic insulation material, such as the μ-metal.

While the light signal is amplified in a PM tube, its shape remains exponential, but its amplitude is greatly enhanced. The signal is subsequently fed into an integrating RC electric circuit with a large time constant, see section 4.5, where it is transferred into a voltage pulse that continues to rise with time to its saturation value within a span of a few decay times.

Photodiodes. Photodiodes offer an attractive alternative to PM tubes, because of their small size, ruggedness, higher light-to-electron conversion ratio, insensitivity to electric fields, and their lower power consumption. This, however, comes at the expense of low electron-multiplication ratios, since conventional photodiodes simply convert light into an electron signal without any internal amplification; in a process similar to that of radiation detection by semiconductors, discussed in section 4.2.4. Therefore, conventional photodiodes produce low-amplitude pulses that become susceptible to background noise and dark current. Conventional photodiodes require the application of a modest voltage of only 10 to 100 V. However, if the applied voltage is elevated, an internal multiplication of the charge carriers within the anode material can take place, producing an avalanche effect, similar to that of proportional counters, see section 4.2.2.2. The photodiode is then called an avalanche diode, and while it offers higher multiplication than a conventional photodiode, the multiplication effect is quite sensitive to the value of the applied voltage and thus a very stable high-voltage power supply is needed. One general disadvantage of photodiodes is their susceptibility to radiation, which introduces a signal that can compete with that produced by the scintillation process of the detector. Nevertheless, photodiodes can be useful in some applications.

Fibers. Fiber optics offer another means of collecting scintillation light by embedding a scintillating material with a high refractive index within a thin fiber made of a lower refractive index. The scintillation light, emitted when a charged-particle deposits energy in the scintillating material, will be reflected, back and forth, at the interface between the scintillating and non-scintillating materials, because of the difference in the refractive index between the two. As a result, the emitted light will be entrapped within the fiber. The optical fiber guides this light to a light detector located at one end of the fiber. Note, however, that the non-scintillating clad of the fiber represents a dead layer, as far as the incident radiation is concerned. Therefore, the very small size of the fiber makes use of a single fiber very inefficient for detecting radiation. To increases the size of the sensitive scintillating volume of the detector, these fibers are bundled together. Nevertheless, the overall efficiency of such detectors is still low, in comparison to conventional scintillation detectors. Therefore, their application is limited to use in high fields of radiation, or for the detection of very low-energy charged-particles.

Cherenkov Detectors. Light emission from a medium can also be produced by the Cherenkov effect. When fast charged-particles passes through an optically transparent medium, light is emitted when the particle's speed exceeds the speed of light in the medium, i.e. when $vn > c$, where c is the speed of light in vacuum, v is the particles's velocity and n is the medium's refractive index ($n > 1$). The emitted light is bluish white in color and is confined in a cone with a vertex angle θ, so that $\cos\theta = \frac{c}{vn}$. For this condition to occur, the particle has to have a very high energy, or be small in mass. For charged-particles, heavier than electrons, that energy is in the order of hundreds of MeV, while for electrons the same effect can take place at an energy of the order of a few hundreds of keV. Therefore, detectors based on this effect are mainly used for detecting electrons, though their light output is about one-hundredth of that produced in a scintillator. Since only high speed particles emit Cherenkov radiation, the detector is not sensitive to low-energy particles, and as such it provides a natural threshold against detecting low energy particles. Materials with a refraction index between about 1.1 to 1.8 are used in such detectors, including pressurized and liquified gases, glycerin, water, Lucite, glasses and various crystalline materials [15].

Gas Scintillation Detectors. One other interesting variation of scintillation detectors is the gas proportional scintillation detector, which takes advantages of the scintillation caused by the free electrons from ion-pairs in proportional counters, see section 4.2.2.2. In a proportional counter filled with a pure noble gas, most of the energy of the accelerated electrons excite the atoms of the gas rather than ionize it, producing visible or ultraviolet light during the de-excitation process. In conventional counters, this light can generate signal noise by producing electrons when it interacts with the metallic walls of the counter. However, if this light is collected with a PM tube, it will produce a pulse indicative of the number of photons incident on the tube. Although these detectors offer better energy resolution than conventional counters, and have a very short response time, their light yield is quite low [15].

Energy Resolution. It should be noted that scintillation detectors, in general, offer the poorest energy resolution among detectors commonly used to measure particle energy. This poor resolution is mainly due to the fluctuations in the PM tube from one scintillation event to another, and to the statistical variation in the number of electrons emitted from the dynodes of the tube and collected by its anode. In photodiodes the situation is even worse, due to their small size which makes them vulnerable to fluctuations from thermally generated charge carriers (dark current).

4.2.4. Semiconductor Detectors

Semiconductors materials are used to detect radiation because of their ability to provide good energy resolution, due to the large number of charge carriers generated in such materials. Semiconductors, as the name implies, are materials with an electric conductivity that is lower than that of conductors and higher than that of insulators. Silicon, germanium, selenium, copper oxide and cadmium sulfide, are some of the semiconductor materials. In order to explain how these materials are used in radiation detection, let us examine the silicon atom [98].

Valence Band. The silicon atom has 14 electrons, occupying the K, L and M shells, with two electrons completely filling the K shell and eight electrons filling the L shell entirely. This leaves four electrons in the M shell, which requires 18 electrons for complete filling. Electrons in this incomplete shell are called *valence electrons*, and are the ones that are responsible for the physical and chemical behavior of the element, since the other shells are closed and cannot receive any additional electrons. In a silicon crystal, these four valence electrons of individual atoms are used to build a chemical bond formation. Although each of these valence electrons is still tied to its own atom, in a solid material, the atoms are closely packed together in lattice formations. Consequently, the normal single energy levels of isolated atoms are disturbed by the neighboring atoms, resulting in the formation of a wide energy band called the *valence band.*

Conduction Band. The unoccupied electronic shells of the atom cluster together, in solid materials, forming the so-called *conduction band.* There is an energy gap of 1.1 eV in Si between the upper energy level of the valence band and the bottom level of the conduction band. Electrons in the conduction band, if they exist, are not bound to any specific atom, and are free to move. Electric conductivity is due to the ability to mobilize these free electrons by the application of an external electric field. In a silicon crystal, except for electrons resulting from lattice defects or thermal agitation, no electrons are present in the conduction band, and thus silicon is a poor conductor of electricity.

Doping. If an atom of phosphorus, which has five valence electrons, is allowed to replace one of the silicon atoms in the lattice, four of its electrons will be involved in chemical bonding and will reside in the valence band, while the extra fifth one will attempt to reach the conduction band, but will occupy a level slightly below it. Phosphorus becomes then a *donor* of electrons, and the material is called n-type (for negative, due to the presence of an extra electron). On the other hand, if an atom of boron, which has only three valence electrons, replaces a silicon atom, a "hole" will be created in the valence band. This hole can trap an electron from the valence band, making it participate in a weaker chemical link, as it involves a foreign atom, the boron, and leaving a hole somewhere else in the crystal. The electron filling the hole, though is still tied to a specific location in the lattice, is slightly less bound than ordinary valence electrons, and appear as if it occupies an energy level just above the top of the valence band. Boron is called an *acceptor*, as it creates holes that can receive other valence electrons, and the material is called p-type (due to the positive charge created by the hole). The process of adding impurities is known as *doping*, and is essential for making the material a semiconductor of electricity. By adding impurities, and with a slight addition of energy, it becomes easier to move the extra electron from the donor to the conduction band, or allow the "hole" in the valence band to be passed along from one silicon atom to another, allowing electric conduction to take place.

Junctions/ Depletion Layer. If a p-type semiconductor is brought into contact with n-type material, a p-n junction is created. Then, electrons from the donor in the "n" region can move down in energy to fill in the vacancies of the acceptors in the "p"region. This creates a potential difference between the two materials. The p-type material acquires a negative potential with respect to the n-type material, due to the movement of electrons from the latter to the former. The result is that, in this region of the junction, all the acceptors will become neutralized by the donors and the junction area ceases to be conducting, creating the so-called *depletion layer.* If an external voltage is applied across the junction, making the "n" side more positive, further migration of electrons from the donor to the acceptor will occur and the width of the depletion layer will increase.

Charged-Particles Detection. With the reverse bias voltage applied, a charged-particle passing through the depletion layer will deposit energy in the material, creating free electrons in the n-type side, and "holes" in the p-type material, by allowing some of its electrons to gain energy and move to the energy level just above the valence band. The electrons will subsequently move towards the positive (n) side, and the holes will migrate towards the negative (p) side, thus creating a flow of current. This process is quite fast and occurs with about 0.1 ns. The creation of holes in the parent material makes some of its atoms positively charged. When a positive ion captures an electron from a neighboring neutral atom, the atom becomes a positive ion, and can in turn capture an electron from another neighboring neutral atom, and so on. This in effect moves the negative ions, creating a current flow of "holes" with a mirror-image current of electrons generated in the opposite direction. Although these ions do not physically move, they contribute to the flow of current by capturing electrons from adjacent atoms. This process enables positive ions in a semiconductor to carry current at a speed much higher than the positive ions of a gas, where the ions have to physically move.

High-Purity Semiconductors. Since the depletion layer is the part that produces the current as radiation passes through the junction, it is desirable to increase the depth of the depletion layer, in effect to increase the sensitive size of the detector. The thickness of the depletion layer increases as the applied voltage increases, as more acceptors become more neutralized by the donors. There is, however, a limit to the voltage that can be applied before damaging the junction by heating and sparking effects. The doping concentration also affects the thickness of the depletion layer, since as the concentration of the donor and acceptor decreases, the flowing electrons will have to travel a longer distance until they encounter an electron trap (a "hole"). Therefore, very low levels of impurities, i.e. high purity semiconductors, are used in junctions employed in radiation detectors.

Energy Resolution. Although the energy gap between the valence band and the conduction band is only 1.1 eV, it takes about 3.6 eV to create an electron-hole pair in silicon, since part of the energy is dissipated into heat. Recalling that it takes about 30 eV to ionize a gas, see section 4.2.2, a semiconductor junction would produce about eight times ($\frac{30}{3.6}$) more charge-carriers (electron-hole pairs) than in a gas (electron-ion pairs). The increase in the generation of charge-carriers reduces their statistical fluctuations. This fluctuation follows the Poisson statistics of counting, where the relative standard deviation is inversely proportional to the square root of the counts, see section 4.5.6.1. Thus, an increases in the number of charge-carriers by eight times would produce an increase in the precision of the signal by about 2.8 times. With the number of charge-carriers being proportional to the energy of the incident radiation, and the amplitude, or pulse height of the signal, a semiconductor detector can inherently provide an energy resolution that is about three times better than that of a gas detector. Note that the multiplication effect in a proportional counter simply magnifies the pulse, thus making it easier to measure, but it does not affect the inherent statistical fluctuations of the initiating ionizing event, and as such magnification does not provide improvement in the detector's energy resolution. As indicated in section 4.2.3, the energy resolution of scintillators is poorest among common detectors, due to the inherent statistical fluctuations in the photon-electron conversion process of photomultipliers tubes and photodiodes. Therefore, semiconductors are the best type of detectors for performing energy-spectrum measurements.

Applied Voltage. The collection of electrons and holes created by charge-particles in a semiconductor detector results in the formation of a pulse. Since, as described above, the motion of holes does not involve physical movement of the atoms themselves, the mobility of the electrons and the carriers of the positive charge ("holes ") are about equal; unlike the case with gases where electrons have a much higher mobility than the positive ions, see section 4.2.2.1. This means that both electrons and positive ions contribute at about the same time to the collected charge forming the detected signal. The drift velocity, as in gases, increases as the gradient of the applied electric field is increased. However, increasing the applied field also increases the amount of collected charge-carriers. Therefore, carriers of opposite charge traveling towards opposite ends of the detector's electrodes can come in close proximity of each other, creating their own internal electron field that opposes the applied external field. This in turn reduces the drift velocity. Therefore, there is a value of applied voltage beyond which the drift velocity of electrons and holes cannot increase any further. This velocity is called the *saturation velocity*, and is in order 10^5 m/s [15]. Semiconductor detectors usually operate at sufficiently high-voltage gradients, in the order of 100 kV/m, so that the saturation drift velocity is usually reached.

Response Time. For a typical charge-particle semiconductor detector with a thickness not exceeding 1 mm, the charge collection time would be about 10 ns. Since the pulse duration depends on the time it takes the charge-carriers to reach the electrodes of the applied external field, such thin semiconductors have a very fast charge-collection time, making these detectors one of the fastest responding detectors for charged-particles [15]. This short pulse duration reduces the dead-time, see section 4.5.7.4, and makes these detectors able to accommodate higher count rates.

Surface-Barrier Detectors. A p-n junction can be manufactured in a variety of ways, defining the type of detector. In a diffused-junction detector, the junction is created on a thin silicon wafer, usually of the p-type, by diffusing a donor material on its surface, thus creating a junction just underneath the surface. In this arrangement, the surface layer is saturated with the donor material (i.e. is not neutral) and becomes a dead layer, as far as radiation detection is concerned. This is obviously a disadvantage in the detection of charged-particles, since a significant portion of their energy can be lost while passing through the dead layer before reaching the sensitive area of the detector. In surface-barrier detectors, the thickness of the dead layer is considerably reduced by a slight oxidation of the surface of an n-type wafer, creating electron traps (holes) on the surface and neutralizing the material directly underneath it. Although, such detectors reduce energy loses in the dead layer, they are vulnerable to the accumulation of foreign material on the surface and are sensitive to the photons of visible light that reach the sensitive layer. Therefore, for charged-particle detection, surface-barrier detectors are housed in vacuum enclosures, equipped with a thin entrance window, to reduce these effects, but the window itself is susceptible to damage and dirt accumulation.

Table 4.2. Maximum energy of charged-particles measurable at different depletion depth in a PIPS detector [104].

Depletion Depth (μm)	Energy of charged-particles (MeV)		
	Electron	Proton	Alpha
100	0.15	7	15
300	0.31	15	55
500	0.45	21	85
700	0.52	27	105
1000	0.73	33	130

PIPS. Surface doping can also be done by ion implantation, where an energized beam of the ions of the donor, or acceptor, impurities, bombards a wafer of the opposite type of material. This creates a fine surface layer that is less vulnerable to ambient conditions. This type of detector is known as PIPS, an acronym for "Passivated Implanted Planar Silicon", and is the modern form of surface-barrier detectors. If the depletion layer is extended throughout the entire thickness of the wafer, the detector becomes *fully depleted.* The advantage of this arrangement is that, if a particle has sufficient energy to penetrate through the wafer, or if the wafer is sufficiently thin, the amplitude of the pulse will be indicative of the energy loss of the particle within the wafer. In all the above detector types, the depth of the sensitive layer is limited to about 1 mm. However, this thickness exceeds the range in Si, of alpha-particles, beta-particles and electrons emitted from radioisotopes. Note, however, that semiconductor detectors with thicker depletion regions are available and are used for photon detection, see section 4.3.

In order to be able to measure the energy of a charged-particle, the depth of the depletion layer has to be at least equal to the range of the particle, so that the incident particle can fully deposit its energy within the detector. The maximum particle energy that can be measured by a given thickness of depletion layer in a PIPS detector is shown in Table 4.2 for electrons, protons and alpha-particles. As the Table shows, even the thinnest detector is capable of measuring the few MeV alpha particles emitted from radioisotopes, see section 2.1.1 for particle energies. However, only electrons of maximum energies less than 1 MeV are fully absorbed, even in the thickest detector.

4.3. Photon Detectors

Photopeak. Although photons do not carry an electric charge, they produce electrons when they interact at almost any energy with the atomic electrons of any material. As discussed in section 3.4, electrons produced by photons are either the result of the photoelectric effect or Compton scattering. The photoelectric effect results in complete absorption of a photon, while liberating an atomic-electron that carries almost the entire photon energy (minus the binding energy of the electron). This process produces the full-energy peak, also called the photopeak, observed in photon detectors measuring monoenergetic photons, as schematically shown in Figure 4.5.

Compton Continuum and Edge. In Compton scattering, photons give only a portion of their energy to atomic electrons. The amount of energy lost by a photon, hence gained by an electron, depends on the angle of scattering, as Eq. (3.37) indicates. The incomplete energy deposition by Compton scattering adds to the photopeak, a continuum of pulses at lower energy, called the Compton continuum. For a photon of energy, E, incident on a detector, the maximum energy it can give an electron is when it scatters by an angle of π. Therefore, the maximum energy an electron can gain, T_{max}, according to Eq. (3.37), is given by:

$$T_{max} = E - \frac{E}{1 + \frac{2E}{m_0c^2}} = E\left[\frac{2E}{2E + m_0c^2}\right] \tag{4.3}$$

where m_0c^2 is the electron's rest-mass energy (511 keV). The pulse-height corresponding to this energy is known as the Compton edge. The Compton-continuum extends from the pulse-height corresponding to the Compton edge to a zero pulse height (where the photon loses no energy, at zero scattering angle).

Backscattering Peak. A peak, called the backscattering peak, is often observed at a pulse-height corresponding to an energy below $\frac{1}{2}m_0c^2$. As the name implies, it is the result of photons that are reflected back into the detector by scattering from walls of the detector's containers or by the surrounding shielding or structures. Backscattered photons have the highest probability of reaching the detector's sensitive volume, and the longest path length within the detector. The energy of a backscattered photon, using Eq. (3.37), is:

$$E_{back} = \frac{E}{1 + \frac{2E}{m_0c^2}} \tag{4.4}$$

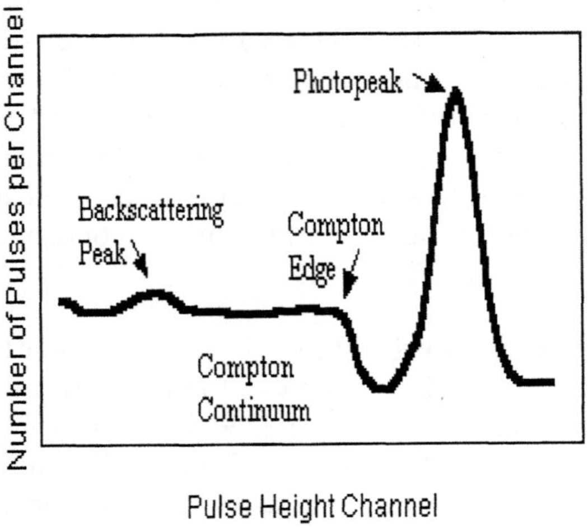

Figure 4.5. A schematic of a typical pulse-height distribution of a detector to a monoenergetic source of photons.

The detector pulse-height corresponding to the full-absorption of E_{back} produces the backscattering peak. The largest possible value of E_{back} is when $2E >> m_0c^2$, and, using Eq. (4.4) is equal to $\frac{1}{2}m_0c^2 = 255.5$ keV. Therefore, the backscattering peak appears at pulse-heights corresponding to energies of 255.5 keV or less.

Escape Peaks. Pair-production distorts the energy-spectrum of photons, when the photon energy is above about 1.5 MeV, i.e. when the probability of this interaction becomes relevant (see section 3.4). If the electron-positron pair resulting in this reaction escapes the detector, then the photon-energy absorbed in the detector would in effect be reduced by $2m_0c^2$ (1.022 MeV). This produces the so-called double-escape peak in the photon spectrum at an energy of $E - 2m_0c^2$. If only one of the particles of the pair escapes the detector, a single-escape peak appears at a photon energy of $E - m_0c^2$. In between these two situations, the pair can deposit part of their energy before escaping, thus producing a continuum between the two peaks.

Peak-to-Compton ratio. In many applications, photons arrive at the detector carrying different energies. Each of these energies produces in the detector its own photopeak and the lower pulse-height Compton continuum shown in Figure 4.5. A photopeak associated with lower-energy photons can overlap with the Compton continuum associated with the photopeak of higher energy photons. It is, therefore, desirable for a detector that can resolve many energies to have a not-too-strong Compton continuum. The parameter that defines this quality is called the peak-to-Compton ratio, i.e. the ratio between the intensity of the photopeak to the mean intensity of the Compton continuum. This ratio should be high for a detector to be able to distinguish between low-energy photopeaks in the presence of high-energy peaks.

The material from which a photon detector is made acts as the intermediary medium that induces the electronic charge, required for the detection of the incident photons. Consequently, any of the detectors utilized for charged-particles, see section 4.2, can be employed as well for detecting photons. However, since the penetration depth of photons in matter is much larger than that of charged-particles, the detector need not be equipped with windows for the radiation to reach the detector's sensitive volume. On the other hand, the same large penetration depth of photons, necessitates the use of a larger volume of detection material, since photons can travel some distance without interacting with the material at all; unlike charged-particles which are continuously affected by the electric fields of the atom and the nucleus and lose energy continuously. However, the same mechanisms used for detecting charged-particles, discussed in section 4.2, are employed for detection photons. Since the physics behind these detection mechanisms were discussed in section 4.2, they are not repeated here, instead emphasis is given to how the methods are used for detecting photons.

4.3.1. Gas-Ionization Detectors

Photons can ionize the atoms of a gas by giving sufficient energy to its atomic electrons to liberate them from their shells. The voltage applied across the electrodes of the detector is what determines whether a gas-filled detector will function as an ionization chamber, a proportional counter or a Geiger-Müller counter, as discussed in section 4.2.2. However, the lack of multiplication, and consequently the low signal amplitude of ionization chambers makes them unsuited for detecting photons. Nevertheless, when used in the current mode, see section 4.2.2.1, ionization chambers can be employed to detect the intensity of photons in a strong radiation field, such as that of a nuclear reactor. On the other hand, for the same reasons explained for charged-particles, section 4.2.2.3, the Geiger-Müller counter is useful for applications involving low count-rates, but offers no energy information.

Proportional Counters. The multiplication process of proportional counters, explained in section 4.2.2.2, makes them useful for photon detection. Because of the low-density of gases, these counters are applied mainly for the detection of low-energy x-rays or gamma-rays, well below 100 keV. While argon is the gas commonly used in proportional counters, xenon and krypton are often employed in gamma detectors because of their higher atomic-number, which increases the probability of photon interaction with the gas, as indicated in section 3.4. The use of either of these gases can, therefore, enhance the detection efficiency, or can make the detector more sensitive to higher-energy photons. Neon is also used in some counters. Although its atomic-number is lower than that of argon, neon offers the ability to preferentially detect low-energy photons in the presence of high-energy ones. As indicated in section 4.2.2.1, noble gases are used in gas detectors, because unlike other gases, electrons do not become easily attached to their atoms. A polyatomic quench gas, such as methane, is often added to the noble gas, for the reasons explained in section 4.2.2.2. The popular mixture of 90% argon and 10% methane, the P-10 gas, is also utilized in photon detectors. Most commercial detectors are pressurized to about 100 kPa (1 atm). In order to avoid using an excessively high voltage, a proportional counter in the form of a cylinder with a thin wire stretched along its axis is often applied, see section 4.2.2.2 for more details. Then an applied voltage from 1 to 2 kV, depending on the size of detector and the gas pressure, is used. The pulse shape and duration characteristics for photons are similar to those of charged-particles, see section 4.2.2.2, since both types of radiation have the same ionization effect on the gas. However, the pulse rise-time for a proportional counters detecting photons tends to be longer than that of heavy charged-particles. This is due to the fact that photons deposit their energy over a longer distance than heavy charged-particles which have a short range. This leads to some ion-pair formations away from the anode, and a subsequent spread in drift-time and the rise-time of the pulse. For the same reasons, photon with higher energy have longer rise-times than photons with lower energy. This differences in rise-times between different particles and photons energies can be used in a pulse-shape discrimination scheme to distinguish between different types of radiation within the detector, see section 4.5.5.4.

The pulse-height of the signal obtained in a proportional counter can be used to measure the energy deposited by a photon, and in turn the photon energy if all its energy is deposited within the detector. However, for the low-energy photons for which proportional counters are most efficient, energy resolution is limited. As indicated in section 4.2.2, about 30 eV is required to produce an electron-ion pair, thus a relatively small number of charge-carries is produced at low-photon energies. It is also likely that some photons will escape without fully depositing their energy in the detector, while some others will lose energy without producing ionization by interacting with the metallic walls of the detector. Moreover, some of the incident photons may not ionize the gas, but instead excite its atomic electrons. The excited states can release x-rays with energies that are relatively high, so that the resulting x-rays may escape the counter without further interaction. These factors limit the photon-energy resolution of proportional counters. Typical energy resolutions are about 16 to 20% full-width at half maximum (FWHM) [104]. In spite of this, measuring the energy spectrum of low-energy photons is perhaps the most useful application of proportional counters.

4.3.2. Scintillation Detectors

Both organic scintillators and inorganic scintillators are used for photon detection. However, because of their higher density, inorganic scintillators provide a higher interaction probability with photons, and thus have a higher detection efficiency. A variety of detector types are reviewed in this section, after a brief description of the shape of pulses produced in this type of photon detectors.

Pulse Shape. The shape of pulses from a scintillation detector is determined not only by the decay-time of the light emission caused by an incident photon, but also by the time-constant of the anode circuit of the PM tube. The anode is in effect an electric circuit consisting of the capacitance of the anode itself and the resistance across which the voltage is applied to the PM tube. This circuit acts as an integrator, see section 4.5, of the exponentially decaying pulse of electrons resulting from the scintillation process. If the time-constant of the anode is set to be much larger than the decay-time of the scintillator, a long-duration pulse with a large magnitude, proportional to the charge of the electron pulse, will be produced. This will enable good pulse-height analysis and consequently good energy resolution. On the other hand, the duration of the signal can be made shorter, for use in applications requiring fast timing, by setting the anode's time-constant at a value lower than the decay-time of the scintillator, at the expense of reduced amplitude [15].

Position-Sensitive Detectors. The scintillation process is localized along the track of radiation incident on the detector. Therefore, particle-position information, within a position-sensitive scintillator, can be obtained from a large size detector. This can be achieved in one dimension by placing two PM tubes at the end of a scintillation rod and monitoring either the light intensity or the signal time-of-arrival at both ends, as both can be related to the position of the particle [15]. An array of PM tubes covering the back of a thin layer of scintillating material can be used to provide a two-dimensional spatial image of the radiation incident on the scintillator. Directional confinement of the incident radiation can further be provided by placing collimators in front of the scintillating plate, as in the case of gamma cameras [15].

Organic Scintillators. Plastic and organic-crystal scintillators have a poor energy resolution, but have a fast decay-time, see section 4.2.3. This makes them useful in timing applications. Plastic scintillators are more robust than organic crystals (compared to NaI), while organic liquids can provide detectors of larger sizes (if needed). Being composed of light elements, organic scintillators are not very efficient for detecting photons, which can easily go through the detector without depositing much of their energy. However, the rise-time of their pulses is in the nanosecond range, and thus they are well-suited in applications demanding fast timing.

Inorganic Scintillators. Because of their higher detection efficiency and their ability to provide particle-energy information, inorganic scintillators are more suited for photon detection. Indeed the most popular photon detector is the thallium-activated sodium-iodide crystal, NaI(Tl), as it provides good performance in terms of efficiency and resolution at reasonably low cost. The amplitude of the scintillation pulse is nearly proportional to the energy deposited by a photon within the crystal, and as such can be used for energy measurements, as well as photon counting. Although, the energy resolution for scintillation detectors is not as high as that of proportional counters and semiconductors, it offers good resolution at higher photon energies. The highest obtainable energy resolution achievable is about 7.5% for 76 mm diameter by a 76 mm length NaI(Tl) crystal at 662 keV, the energy of a ^{137}Cs gamma source [104]. Small sizes provide even a lower resolution due to the incomplete energy deposition of the photons, as they can leave the crystal before losing their energy. Increasing the crystal size tends to lower the energy resolution, due to variation in the efficiency of collecting light from a large volume [15]. Therefore, lower photon energies require a smaller detector size to optimize detector resolution, but this comes at the expense of reduced detector efficiency as more photons can pass through the detector without interacting with its sensitive material. Light generated by scintillation escapes from the back end of the detector's container through a transparent crystal to a photomultiplier tube, see section 4.2.3. The inside of the container is coated with a layer of magnesium oxide, which acts as a mirror reflecting light within the container.

Detector Properties. There is a wide variety of other inorganic scintillators, as shown in Table 4.3. The Table lists some of the properties of these scintillators, including the amount of scintillation (light photons) produced per unit energy of incident radiation energy deposited in the scintillator (light yield), and the electron yield in a PM tube as a result of the incident scintillation light (photoelectron yield). The latter quantity depends on the nature of radiation used, and the convention is to report it with respect to that of NaI(Tl), which produces the highest yield per amount of radiation absorbed in all known scintillators. Note that the absolute efficiency for NaI(Tl) is about 12%; i.e. 12% of the energy deposited in the scintillator by the incident radiation is converted into light. Obviously, the larger these yield values, the more effective is the detector. Table 4.3 also reports the decay-time of the primary light emitted in the scintillator, as well as whether the scintillator is hygroscopic (sensitive to water moisture) or not, and the percentage of light emission that remains following the initial pulse (afterglow). A brief description of the main features of each of the scintillators described in Table 4.3 is given below in alphabetical order.

BaF_2. The scintillator with the fastest response is barium fluoride (BaF_2), which has less than a nanosecond decay-time. Therefore, it is the detector of choice for systems requiring very fast-timing measurements, such as time-of-flight and fast-coincide measurements, see section 4.5. The fast scintillation light is emitted in the frequency range of ultraviolet light, requiring the use of a photomultiplier tube, or photodiodes, sensitive to this range of light frequency. A photomultiplier tube with a quartz window is necessary for use with this scintillator, to extend the sensitivity of the photocathode to the higher frequency of ultraviolet light [103, 15]. Along with the fast component, a more efficient slow component exists, but at a higher frequency (shorter wavelength) than that of the fast component. The relative ratio between the intensity of the fast and slow components depends on the type of the incident particle; for instance alpha-particles hardly produce a fast component. Therefore, pulse-shape discrimination, see section 4.5, can be used to discriminate against charged-particles that may coexist with photons. Although BaF_2 is not hygroscopic and is resistant to radiation damage, its energy resolution is considerably poor, due to its low light yield.

Table 4.3. Some common inorganic scintillators used in photon detection [103, 105].

Scinitllator (Hygro-scopic)	Density (kg/m³)	PE* yield (%)	Light yield (per keV)	Decay time (ns)	Afterglow after (ms)	(%)
BaF_2 (No)	4880					
	Fast →	5	2.5	0.6 – 0.8		
	Slow →	16	6.5	630		
BGO (No)	7130	15 – 20	8 – 10	300	3	0.005
CaF_2(Eu) (No)	3180	50	19	940	6	< 0.3
$CdWO_4$ (No)	7900	25 – 30	12 – 15	20×10^3, 5×10^3	3	0.1
CsF (very)	4640	5 – 7	2	3 – 5	6	0.003 – 0.06
CsI(pure) (Slightly)	4510	4 – 6	2	16		
CsI(Na) (Yes)	4510	85	38 – 44	630	6	0.5 – 5.0
CsI(Tl) (Slightly)	4510	45	52 – 56	1,000	6	0.5 – 5.0
GSO (Ce) (NO)	6710	20 – 25	0.2**	30 –60		
NaI(Tl) (Yes)	3670	100	38	230	6	0.3 – 0.5
YAP (Ce) (No)	5550	40	18	27		

* Photoelectron yield relative to that of to NaI(Tl) at room temperature for gamma rays, when coupled to a PM tube with a bialkali photocathode.
** Estimated from reference [106].

BGO. Bismuth Germanate ($Bi_4Ge_3O_{12}$), known as BGO, provides the largest probability, among common scintillators, of photon-absorption per unit volume, with a very small rate of afterglow. This is due to its high density and the presence of bismuth (which has a high atomic-number: 83). Both these factors increase the probability of photon interaction in the detector, via photoelectric-absorption and Compton scattering, see section 3.4. Unfortunately, however, the light yield of BGO is not very high, which lowers its energy resolution and makes it difficult to use in conjunction with a photodiode (which does not provide light multiplication as a photomultiplier tube). Although, BGO has a long decay-time, its afterglow is quite low due to the absence of the long decaying components that exist in other scintillators. BGO has some characteristics that accommodate harsh industrial conditions: it is rugged, non-hygroscopic and can be formed in various configurations. Like other scintillators, the light output of BGO decreases with temperature [15]. It is also susceptible to radiation damage at radiation doses from 1 to 10 Gy, but such damage is largely reversible and a good quality crystal can recover rapidly to its original performance [103].

CaF_2(Eu). The main attraction of CaF_2(Eu) is its physical, chemical and mechanical properties. It is not hygroscopic and has very low vapor pressure. In addition, it is relatively chemically inert, and is insoluble in water and most organic solvents. It has a high resistance to thermal and mechanical shocks, and can be manufactured in various shapes. Therefore, it is useful for use in harsh environmental conditions. However, because of its low photon yield, due to its low density and the relatively low atomic-number of its elements, it has a low energy resolution, particularly at high photon energy.

$CdWO_4$. Cadmium tungstate ($CdWO_4$) offers both a relatively high light-yield and low afterglow. Although, it is a high density, high atomic-number scintillator, its light yield is lower than that of NaI(Tl). It has two main light components, one with a decay-time of 5 μs and the other is slower with 20 μs, which is the highest among all detectors. Therefore, its energy resolution is poor. However, its low afterglow makes it attractive for use in applications involving sequential measurements, such as in the case of imaging. Another attractive feature of this scintillator is that its response stays quite stable over a temperature range from about -20°C up to about 60°C [103], a handy feature in some applications. It is also radiation resistant. However, it is a fragile detector susceptible to mechanical and thermal shocks. Moreover, because the crystal can absorb some of its own scintillation light, it is not fabricated in large sizes.

CsF. Cesium fluoride (CsF) is another fast-response detector that is attractive for use in applications involving timing measurements or high count rates. Its light decay-time is comparable to that of organic scintillators. Since its photoelectron yield is only 5 to 7% of that of a NaI(Tl) detector, the energy resolution of CsF detectors is poor, from 18 to 20% for 662 keV gamma-rays for small crystals, vs. about the 10% or so offered by NaI(Tl) detectors. CsF is also susceptible to irreversible radiation damage at doses above 100 Gy [103]. Being extremely hygroscopic, it is provided in hermetically sealed detector assemblies.

CsI. Cesium iodide detectors are either doped, CsI(Tl) and CsI(Na), or undoped CsI. These detectors are quite rugged, as they resist well thermal and mechanical shocks, and as such are well-suited for harsh industrial conditions. The material is soft and plastic and can be shaped easily into different geometries. The pure (undoped) CsI has the attraction of a low decay-time, making it useful for use in fast-timing applications. Because of its low light-yield, it is suited only for medium to high energy photons, where a larger amount of energy can be deposited to take advantage of its relatively high density and high mass-number. However, its energy resolution is 18% at 662 keV photon energy, in comparison to the 10% or so obtainable with NaI(Cl) detectors. This detector is quite sensitive to temperature changes, as its scintillation intensity decreases significantly with temperature [103]. It is, however, quite resistant to radiation damage.

CsI(Na) is the closest detector in photoelectron yield to NaI(Tl). It also operates well at high temperature, giving its maximum yield at 80°C [103].

CsI(Tl) has the largest light yield of all scintillators. But the main attraction of CsI(Tl) is that most of its light emission is at a wavelength that is well-suited for photodiodes; though this comes at the expense of a lower energy-resolution due to electronic noise. For best performance, the size of the photodiode should be such that it can detect the maximum amount of scintillation light. When a PM tube, with bialkali photocathode is used, its photoelectron yield is less than half that of NaI(Tl), due to the mismatch between the frequency of its light and that to which the PM tube is sensitive. CsI is also a relatively slow scintillator, which puts a limit on the count rate the detector can accommodate. High radiation fields, above 10 Gy, can damage the crystal, but some of the damage is reversible [103]. In addition, the light yield of the detector tends to increase with temperature, up to about 40°C where it begins to decline.

GSO:Ce. Gadolinium silicate doped with cerium is a fast scintillator with relatively good efficiency. Its main attraction is in its extremely high radiation resistance, up to 10^7Gy exposure [106], which makes it suitable for space and nuclear applications involving high radiation fields.

NaI(Tl). This is the most widely used photon detector. It produces the highest yield per amount of radiation absorbed of all known scintillators. Also, it hardly self-absorbs the scintillation light. Since NaI is hygroscopic, the detector is sealed in an airtight metallic container, equipped with an entrance aluminum foil, about 25 μm thick, through which the photons pass. NaI(Tl) detectors have a decay-time of 0.23 μs, and a pulse rise-time of about 0.5 μs. Theses are relatively long times for timing measurements, such as those involving coincidence measurements. Nevertheless, this decay-time can accommodate count rates in the order of 10^6 counts per second. The light output from a NaI(Tl) detector increases with temperature up to about 30°C, after which it gradually drops [103]. The crystal can also be damaged by prolonged exposure to radiation above a level of about 1 Gy. NaI(Tl) is usually fabricated in a single-crystal structure, which makes the detectors vulnerable to thermal and mechanical shocks. However, more rugged polycrystalline crystals are also available [103]. This ruggedness allows the use of such detector in the harsh environment of some industries, and permits the manufacturing of various detector geometries.

YAP:Ce. Yttrium Aluminum Perovskite doped with Cs, $YAlO_3(Ce)$, known as YAP:Ce, is a fast scintillator with a good light output. It is, therefore, well-suited for high count rates and timing applications. Its non-hygroscopic nature, good mechanical and thermal properties (its light output changes is only 0.01%/°C), makes it attractive for use in industrial applications [103]. It has, however, a threshold detection window for photons of about 5 keV. This can be advantageous in eliminating low-energy photons, but prevents the use of the detector below this photon energy.

Phoswich Detectors. If two scintillators of considerably different decay-times are combined together and coupled to the same PM tube, they form the so-called phosphor sandwich (or a *phoswich detector*). The difference in the decay-time, hence pulse shape, between the two scintillators can be used to distinguish between events that occur in one scintillator but not in the other, by pulse-shape discrimination, see section 4.5.5.4. For example, the first layer of the sandwich can be made thick enough to stop low-energy photons, or for that matter charged-particles, but not high-energy radiation, which will reach the second scintillator. Phoswich detectors made of combinations of NaI(Tl) and

CsI(Tl), CsF_2(Eu) and NaI(Tl), or BGO and CsI(Tl), etc. have been utilized [15, 92].

4.3.3. Semiconductor Detectors

The principle by which semiconductor materials detect radiation is described in section 4.2.4 for charged-particles. Since photons can produce electrons in matter via the photoelectric effect and Compton scattering, see section 3.4, they are readily detectable in the relatively high-atomic number of semiconductors. However, as discussed in section 4.2.4, the detection process takes places within a small area defined by the depletion zone between the donor (n) and acceptor (p) regions of a semiconductor junction. Although the depth of the depletion zone is quite small, even if increased by applying a high reverse-voltage, it does not exceed two or three millimeters in common semiconductors. This is a small interaction length for photons, which unlike charged-particles do not continuously interact with matter and can travel some distance without interacting at all, as indicated in section 3.4. Since, as explained in section 4.2.4, the depth of the detection (depletion) area is inversely proportional to the concentration of impurity in a semiconductor, larger depletion zones can be established if a material with high purity is utilized. However, the purity level, required to provide a depletion layer sufficiently wide to detect photons, is too high to be practically attainable in most materials. Germanium is, however, an exception, where ultrapurity can be achieved, creating depletion zones as large a 10 mm in depth in the so-called high-purity germanium (HPGe, or simply Ge) detectors. The gap between the valence and the conduction bands in Ge is quite small (only 32 eV). Thus, at room temperature or higher, the thermal mobility of atoms is sufficient to create charge-carriers, resulting in a signal noise called the reverse leakage-current, that is sufficiently large to compete with the charge-carriers produced by the incident radiation. Therefore, Ge detectors are operated at a very low temperature to suppress this thermal agitation. However, there is no need to cool the detector during storage.

Si(Li). Another way to expand the depletion zone is counter doping, e.g. doping a p-type crystal with n-type impurity (such as lithium) to increase the depth of the depletion zone in Si or Ge semiconductors. The Ge(Li) and Si(Li) detectors are based on this concept. Although Ge(Li) detectors are no longer produced, due to the availability of HPGe detectors, Si(Li) detectors are still being produced, with a depletion depth of up to 5 mm. This depletion depth is too small for use in detecting high-energy photons, but is sufficient for the detection of the lower-energy x-rays. Detector-doping is achieved in a process known as lithium-ion drifting, which introduces n-type lithium atoms into a p-type silicon. After accomplishing this process, the detector needs to be kept at a very low temperature, since at room temperature the lithium atoms are mobile and can diffuse away from the junction. Therefore, Si(Li) detectors are kept at a very low temperature during operation and extended storage. Operating at this low temperature reduces also thermal charge-carrier generation to an acceptably low level.

Operating Temperature. The low temperature required for the operation of Ge detectors, and operation and storage of Si(Li) detectors, is achieved by cooling the detector to less than 100 K. The detector is housing in a clean vacuum chamber, called a cryostat, which is attached to or inserted into a dewar (vacuum container) filled with liquid nitrogen, to maintain a temperature of 77 K. Electromechanical refrigeration systems are also available [91]. These systems can cool a detector to as low as 50 or 60 K, but attention must be given to the associated mechanical vibrations (microphonics, see section 4.5.7.1) that affect the resolution of the detector. At such low operating temperatures, the reverse leakage current can be reduced to the insignificant range of nanoamp, or even the picoamp, range. Operating below 40 K can, however, deteriorate the performance of the detector due to the decrease in mobility of charge-carriers that can lead to their recombination or trapping by the non-doping impurities that may be left in the material. In the presence of a high field of fast-neutrons, the detector should operate as close to 77 K as possible to reduce neutron radiation damage [91]. The recoiled nuclei generated by fast-neutrons leave behind, in the semiconductor junction, negatively charged traps that capture holes and can cause deterioration in the performance of the junction as a photon detector. Damaged detectors can, however, be repaired by heating the detector in an annealing process, [91]. Another source of damage is high magnetic fields, which can create an electric discharge within the detector, and if the field is sufficiently large (several hundred militesla) it can make the detector inoperable, unless special provisions are made to prevent the occurrence of this discharge [91].

Detector Types. The remaining low-level of impurity in an HPGe crystal defines the type of semiconductor, whether an acceptor (p-type) or a donor (n-type). Heavily doped electrical contacts of opposite types are deposited on opposite sides of the material to facilitate the application of the reverse voltage. A voltage gradient sufficiently high (about 100 V/mm) is usually applied so that the entire crystal constitutes the depletion zone; the junction is then called *a fully depleted detector.* For small detectors sizes (a few tens of mm in diameter and up to about 20 mm in thickness), the junction is usually formed as a small disk, with the electrical contacts embedded on its flat surface forming a *planar* detector. For larger active volumes a long cylinder is used, with its core removed to place an electrical contact over the inner cylindrical surface, with the other contact embedded on the outer surface of the cylinder. This so-called *co-axial* detector offers a much larger active volume, and hence better efficiency and resolution, while also minimizing the capacitance of the detector for reasons discussed later in this section. A *well*-detector is a special case of a co-axial detector, in which the inner core does not penetrate the entire length of the cylinder. This creates a hole (or well), in which a small sample of a radioactive material can be placed. The electrical contacts represent in effect a dead layer, since they are not part of the depletion zone. For low-energy photons, absorption within this region can significantly reduce the detector's efficiency and energy resolution. Therefore, to minimize absorption of low-energy photons, the contacts on a low-energy planar detector are formed by implanting a low atomic-number material such as boron, while the flat surface of a co-axial detector is equipped with a thin window made of beryllium, carbon fiber, or aluminum (very thin).

Energy Resolution. The most attractive feature of semiconductor detectors in photon detection is their superb energy resolution. This, as discussed in section 4.2.4, is due to the large number of charge-carriers produced per unit energy, since only a small amount of energy (about 3 eV in Ge) is required to produce a charge-carrier. Table 4.4 compares the typical energy resolution obtained by various types of detectors in full-width-at-half-maximum (FWHM) at various photon energies, see section 4.5.6.2 for definition of energy resolution. The Table reports the energy resolution at three representative energies very low (5.9 keV, K line of Mn x-rays from a ^{55}Fe source), medium (122 keV from ^{57}Co), and high (the 1.332 MeV peak of a ^{60}Co source), see sections 2.2.2 and 2.2.3 for more information on these sources. The superb energy resolution of Ge detectors at all energies is obvious. Note that the absent resolution values in Table 4.4 indicate that the detector is not suitable for providing energy information at these photon energies, either because full photon energy-deposition within the detector is not possible, or because the photon energy is so low for the photons to be able to penetrate the detector's thin metallic (usually aluminum) casing or the dead (nonsensitive) layer on the periphery of the depletion zone. The expected value of $\frac{\text{FWHM}}{E}$, where E is the photon-energy is from 0.08% to 0.1% for E between 1.5

to 10 MeV [91]; however, the actual measured value depends on the quality of the crystal. For energies up to 1.5 MeV, the energy resolution is controlled by the noise associated with the detector and electronics and can be empirically expressed as FWHM $= \sqrt{n^2 + kE}$, where n is the noise level and k is an empirical constant equal to 2 eV [91]. The value of n depends on the capacitance (and consequently size) of the detector, and varies from 0.570 eV up to about 2 keV for Ge detectors, depending on the detector's capacitance [92]. Recalling that Ge detectors are operated at low-temperate (77 to 100 K) to reduce thermal noise, it is also important to maintain a constant temperature to preserve the level of energy resolution during measurements. The capacitance of the detector affects its energy resolution as it determines the performance of the preamplifier used with the detector, the lower the better. Therefore, large size detectors, which have a larger capacitance, provide a higher FWHM (lower resolution) than a small-size detector, at the same photon energy. Reference [107] provides more details on photon spectrometry with semiconductor detectors.

For Ge detectors, the peak-to-Compton ratio, evaluated using a ^{60}Co source by dividing the height of the 1.332 MeV peak by the average height of the Compton continuum between 1.040 and 1.096 MeV, can be as high as 90:1 for a large-size detector, and as low as 37:1 for a small size detector [91]. The lower-value for small detectors is due to the increase in the photon escape probability, which reduces the intensity of the photopeak, making the Compton continuum appear stronger in smaller detectors.

Efficiency. The efficiency of a Ge detector is measured, in accordance to the IEEE standard test procedures for germanium detectors [108], using a ^{60}Co source with known intensity positioned 250 mm from the face of the detector, and taking a fixed-time count for the 1.332 MeV peak of the source. The convention is to report this efficiency with respect to that of a standard 76 mm diameter, 76 mm long NaI(Tl) scintillator, which has an absolute photopeak efficiency of 12% at 1.332 MeV for a source 250 mm away from the detector [15]. The ratio between these two absolute efficiencies is used to define the conventionally-reported relative efficiency of Ge detectors. The relative efficiency of a Ge detector is above 10%, but can be greater than 100% for a co-axial Ge detector 75 mm in diameter (about two kg of germanium is required for the latter detector [104]). Detectors of relative efficiencies up to 150% have been produced [91]. At high energies, the total active detector volume is the dominant factor in determining the detector's efficiency. At low-photon energies, where photon absorption occurs in a small volume near the

Table 4.4. Typical energy resolution for various types of photon detectors [104].

Detector	Photon Energy, E (keV)		
	5.9	122	1332
		FWHM (keV) ($\frac{\text{FWHM}}{E}$)	
Proportional Counter	1.2 (20%)	—	—
NaI(Tl) (for x-rays)	3.0 (51%)	12.0 (10%)	—
NaI(Tl), 76 mm × 76 mm	—	12.0 (10%)	60 (4.5%)
Si(Li)	0.16 (2.7%)	—	—
Planar Ge	0.18 (3%)	0.5 (0.4%)	—
Coaxial Ge	—	0.8 (0.65%)	1.8 (0.135%)

detector's surface, the detector efficiency depends on the cross-sectional area of the detector and the thickness of its window. Therefore, semiconductor detectors, such as Si(Li), are equipped with a beryllium window to increase their efficiency and improve their intrinsic energy response, by lowering absorption outside the detector's active area.

Pulse Rise-Time. The pulse formation process in a semiconductor detector, as discussed in section 4.2.4, is affected by the motion of both the electrons and holes created by the incident radiation, since the mobility of both charge-carriers are about the same in magnitude, approximately 10^5 m/s. For photon detectors with depletion layer 10 mm depth or more, the charge collection-time, and hence the pulse's rise-time, would be in order of 100 ns. This rise-time is relatively long, say in comparison to the faster organic detectors, and hence puts some limitations on the use of these detectors in fast-timing applications. The relatively large volume of the signal also makes the rise-time vary from pulse to pulse, depending on the location where the charge-carriers are created. The farther away a charge-carrier is from its attracting electric contact, the longer its distance of travel, and the slower is the rise time, and vice versa. In a large detector, this creates a wide variation in the rise-time of the pulses emitted from the detector, making it difficult to obtain precise pulse-timing information from the detector. Once the entire charge of the incident particle is collected, the detector is ready to receive a new signal. Therefore, the time-constant of pulse-shaping electronics, see section 4.5, is the determinant factor in both count rate and energy-spectrum measurement applications. Energy spectrum measurements require a reasonably long time-constant to give the integrated pulse sufficient time to reach its full value. For high count-rates, the time constant of the pulse-shaping amplifier should be shortened so that the detector can be quickly ready to receive a new signal. The ANSI/IEEE Standard 325-1996 [108] recommends the use of amplifier time-constants that are a compromise between optimum energy resolution and reasonable count rates; typically between 6 and 10 μs.

Other Types of Detectors. The disadvantage of Si(Li) and Ge detectors is the need for cooling at a very low temperature, to reduce the effect of thermal generation of charge-carriers. However, other semiconductors with a wider band width between the valence and the conduction band, to enable operation at room temperature, are available. Unfortunately, most high atomic-number semiconductors, needed for good photon detection, tend to have low charge-carrier mobility. Nevertheless, a number of room-temperature semiconductors have been emerging as suitable photon detectors. Perhaps the most promising of these detectors is the Cadmium-Zinc-Telluride (CZT) detector. The high-atomic number of the elements of this material increases its efficiency and makes it attractive for use in imaging applications, where small-size detectors, arranged in an array, are required. The detector also offers good energy discrimination ability and can handle high count-rates [109]. The $Cd_{1-x}Zn_xTe$ detectors have a high mass-density 5800 kg/m^3, and an effective atomic-number of 49.6 (Cd0.9:48,Zn0.1:30,Te:52) which allows for a decreased detector thickness, and consequently improved spatial resolution (resulting from the fine mircochip collector size and high gain), high detection-efficiency, good linearity, high stability, high sensitivity, and a wide dynamic range [95].

CdTe is another room-temperature detector. With the use of proper contacts, a much higher bias voltage can be applied to this detector, creating a wider depletion layer [110]. This in turn provides higher detection efficiency and resolution than CdZnTe detectors. Mercuric iodide (HgI_2) is another interesting semiconductor material, because of its high-atomic number. It has also a low dielectric-constant, hence reduced electrical noise [111].

4.3.4. Radiographic Films

Radiograhic films provide a convenient way for qualitatively detecting photons, and are widely used in medical and industrial radiography. This detection process was described in section 4.2.1 for charged-particles, explaining that the detection process is essentially via a chemical reaction that "sensitizes" the emulsion of silver halide grains in the film. For this process to be useful for photon detection, photons passing through the film must interact to produce electrons, that can in turn trigger the chemical reaction. The amount of material in an ordinary photographic film is too small to cause considerable photon interactions within a film. Therefore, radiographic films are made considerably thicker than photographic films, with coatings of emulsion on both sides of the film. For additional enhancement of electron production, films are enclosed between thin foils (20 to 150 μm) made of a high-atomic number material, such as lead, which act as *converter* (photon-to-electron) screens. Another approach is to use a light-emitting screen, made of a phosphorus material such as calcium tungstate, called intensifier screens. In such material, electron excitation by photon absorption produces fluoroscopic visible light. These intensifying screens, also called fluorescent screens, can considerably increase the amount of film exposure to light. The intensifier can be used to produce an image on a television (or cathode-ray tube) screen. The light emerging from the fluorescent screen ejects electrons from the photocathode screen. The electrons are then electrically accelerated to produce a small, but very intense, visible image on another fluorescent screen. The image can then be viewed through a television monitor or recorded by a camera.

4.3.5. Electrostatic Plates

An electrostatic process is also employed for photon detection. This technique uses a photo-conducting selenium plate (same material used in paper photocopying), which is initially electrostatically charged. Since selenium is a good insulator, it maintains its charge until it is exposed to photons, which makes it partially conducting. A charged plate is enclosed in a light-tight cassette before being exposed to radiation. Incident radiation neutralizes the initial discharge, leaving a track, or image, of the intensity of incident photons. The exposed plate is then transferred to a light-tight box where it is developed by exposure to a cloud of charged (toner) particles, just as in normal photocopying, hence the process is called Xeroradiography. The charged toner particles are deposited on the electrostatic image, making it visible.

4.4. Neutrons Detectors

The neutron is a neutral particle that does not directly produce an electric charge. Unlike photons, neutrons are not affected by the electrons of the atom, and interact directly with the nucleus. Therefore, a nuclear modifying reaction that produces a charged-particle is essential for the detection of neutrons. Once the charged-particle is produced, it can be detected by one of the methods discussed in section 4.2. The neutrons can interact with a gas (to enable detection by direct ionization), a scintillator, semiconductor material or some emulsion material. Since the production of a charged-particle by a neutron-nucleus interaction usually requires neutron absorption, as indicated in section 3.5, the detection material should have a high absorption cross-section for neutrons. The probability of neutron absorption is highest at low neutron energies. Therefore, slow-neutron detectors are readily available. However, the response of thermal-neutron detectors is affected by the temperature of the medium in which the neutrons are thermalized. The number of thermal neutrons per unit energy changes with temperature, in accordance with the Maxwell-Boltzmann distribution, Eq. (3.89). In addition, the reaction cross-section changes with neutron energy (proportional to $\frac{1}{\sqrt{E}}$), see section 3.5.6. Therefore, at high-temperature, compensation must be made to the reading of a thermal-neutron detector, see reference [112].

The detection of fast-neutrons is a more challenging task, since they are not readily absorbed. However, the fact that the neutron and proton have about the same mass, makes hydrogen an attractive material for the detection of fast-neutrons; as the neutron loses, on average, most of its energy when interacting with a hydrogen nucleus (a proton), see section 3.5.1. The recoiled proton becomes then the charge-carrier that conveys information about the intensity and energy of the neutron, resulting in the so-called proton-recoil detectors, which are widely used in fast-neutron detection. The recoiled protons are either detected by gas ionization or scintillation, as discussed below.

4.4.1. Gas Detectors

Three types of gases are widely used for converting neutrons to charged-particles, for their subsequent detection by ionization: boron-trifluoride (BF_3), helium-3 (^3He) and common organic gases. The first two gases are used for detecting slow-neutrons due to their high absorption cross-section, while organic gases are used to detect fast-neutrons by the recoil of protons (hydrogen nuclei). In addition, gas detectors lined with a material that produces charged-particles are also employed.

The main characteristics of each of these detector types are discussed below.

4.4.1.1 Boron Trifluoride (BF_3) Detectors

Boron is an attractive element for detecting neutrons, because of the high cross-section of the (n,α) reaction with one of its isotopes (^{10}B). The cross-section for this reaction is 3840 b at thermal-neutron energy (0.025 eV), but drops rapidly with energy, in proportion to $\frac{1}{\sqrt{E}}$, where E is the neutron energy. In order to utilize boron in a proportional counter, it has to be in the form of a gas. The abundance of ^{10}B in natural boron is only 19.9%; the rest is ^{11}B. Therefore, highly ^{10}B-enriched (to about 96%) boron is utilized, in the form of a boron-trifluoride, BF_3 gas. This gas has good properties for use in proportional counters, see section 4.2.2.2.

In a BF_3 counter, the following two neutron reactions take place [113]:

$$\begin{aligned} \mathrm{n} + {}^{10}\mathrm{B} &\overset{(93\%)}{\longrightarrow} {}^{7}\mathrm{Li}\ (836\ \mathrm{keV}) + {}^{4}\mathrm{He}\ (1464\ \mathrm{keV}) + \gamma\ (480\ \mathrm{keV}) \\ \mathrm{n} + {}^{10}\mathrm{B} &\overset{(7\%)}{\longrightarrow} {}^{7}\mathrm{Li}\ (1010\ \mathrm{keV}) + {}^{4}\mathrm{He}\ (1770\ \mathrm{keV}) \end{aligned} \tag{4.5}$$

The alpha-particle and the ^{7}Li nucleus released in the reaction cause ionization of the BF_3 gas operating as a proportional counter, see section 4.2.2.2. The 480 keV gamma-ray produced in the first reaction of (4.5) is the result of the decay of the excited state of ^{7}Li (half-life of about 10^{-3} s). Both reactions are exoergeic reactions, in which a reaction energy, Q, is released. The value of Q is either 2.30 MeV, when ^{7}Li is left in an excited state, or 2.78 MeV, when the ^{7}Li nucleus is in the ground state. In either case, the Q-value is much larger than the energy of slow-neutrons, which is in the eV range. The energy carried by the charged-particles is, therefore, essentially equal to the Q-value. This energy is shared between the alpha-particle and the ^{7}Li nucleus in inverse proportion to their mass. However, since both are positively charged-particles that are eventually attracted to the cathode in a proportional counter, they produce a pulse carrying an energy equal to the Q-value. For slow-neutrons, 93% of the time, ^{7}Li is left in an excited state. Therefore, a somewhat idealized pulse-height response of a BF_3 detector would appear as shown in Figure 4.6, where a stronger peak corresponding to the 2.30 MeV for the excited state of ^{7}Li appears, along with a weaker peak corresponding to the 2.78 MeV ground state of ^{7}Li. Since some of the produced charged-particles lose their energy by interacting with the walls (cathode) of the detector, instead of ionizing the gas, a tail of a smaller pulse usually appears along with the peaks corresponding to the Q-values. This tail may show two shoulders, corresponding to

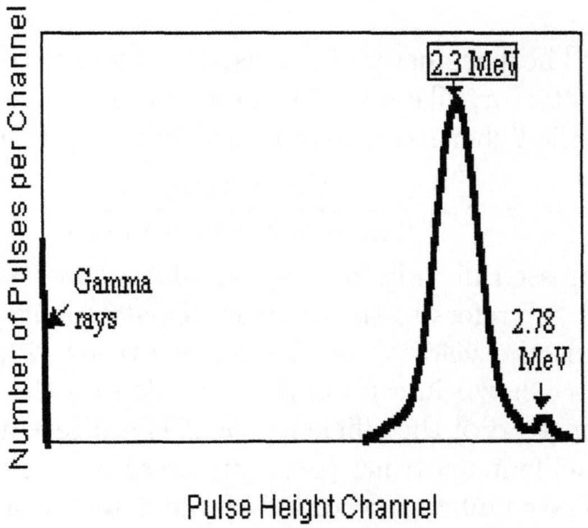

Figure 4.6. Response of a large BF_3 detector to slow neutrons.

the alpha and ^{7}Li particles depositing their full energy, of $\frac{7}{11}Q$ and $\frac{4}{11}Q$, respectively, in the detector walls, and nothing in the gas. These wall-induced effects can be suppressed by increasing the detector size, so that charged-particles have a higher probability of depositing their energy in the gas before reaching the detector.

Efficiency. The efficiency, ϵ, of a BF_3 detector can be expressed as:

$$\epsilon = 1 - \exp\left[-1.7 \times 10^{-2} \frac{p(\text{kPa})L(\text{m})}{\sqrt{E(\text{eV})}}\right] \tag{4.6}$$

where L is the distance traversed by a neutron within the gas, p is the gas pressure, and E is the neutron energy. Eq. (4.6) shows that the detector's efficiency deteriorates rapidly with increasing energy, making the detector mainly a slow-neutron detector. For a typical cylindrical detector, 0.3 m long with 80 kPa gas pressure, with neutrons incident parallel to its axis, ϵ is equal to 92% at a neutron energy of 0.025 eV (thermal neutrons), 12% at 10 eV and only 4% at 100 eV. Therefore, a neutron of energy, E, should add a pulse of a height corresponding to an energy of $Q + E$ in Figure 4.6, but the amplitude of the pulse would be so weak that it is of no practical value. Therefore, no pulse-height information are normally recorded for a BF_3 detector, and the detector is considered as a counter, with no energy information. The

BF_3 detector is mainly a slow (thermal) neutron detector, due to its feeble response to faster neutrons. However, if the detector is surrounded by a neutron moderating material, fast-neutrons are slowed-down and become detectable.

Sensitivity. The parameter often used to specify the efficiency of a slow-neutron detector, like the BF_3 detector, is not its efficiency, but rather the so-called thermal-neutron sensitivity, S, defined as:

$$S = \frac{\text{Count rate}}{\text{Thermal Neutron Flux}} \tag{4.7}$$

where S is expressed in units of cps/nv, where "cps" stands for counts per second and "n" refers to the neutron density (neutrons per unit volume) and "v" is the velocity of thermal neutrons (2200 m/s at room temperature); as shown in section 3.6 "nv" defines the flux. The sensitivity is used, instead of the efficiency, as it is a direct indication of how many of the incident neutrons (per unit area) are actually counted in the detector. The count-rate recorded by the detector obviously depends on the number of neutrons incident on the detector (flux), as well as on the volume and density of the detector, and the reaction's cross-section which is defined by the detector's type. Therefore, the detector sensitivity is a parameters that reflects both the detectors density and its volume; thus sensitivity is a better measure of the detector's efficiency, which depends on the path length of neutrons through the detector, the parameter L in Eq.(4.6). The sensitivity of BF_3 detectors varies from 0.07 to 25 cps/nv, depending on their size and gas pressure [93, 114], with "nv" in neutrons per cm^2 per second.

Gas Pressure and Operating Voltage. As Eq.(4.6) indicates, increasing the gas pressure can increase the detector's efficiency. However, there are practical limits to how much the detector can be pressurized. Obviously, a very high pressure will require a thick metallic wall, which increases neutron absorption within the wall and prevents neutrons from reaching the gas. Though aluminum is preferred as a wall material because of its low absorption cross-section for thermal neutrons, it does not provide the same resistance to stress as steel and the absorbed neutrons produce alpha-particles that cause undesirable ionization within the gas. Moreover, at very high gas pressure, ion recombination and negative ion formation occur, due to electron sticking to the atoms, which deteriorate the performance of the gas as a proportional counter; see section 4.2.2.2. Therefore, a pressure from 20 to 180 kPa is used in these detectors [93]. The detectors are usually cylindrical in shape, but square blocks (quadrilateral) and spherical detectors are also available. Cylindrical detectors can be as long as about 3 m and as small as 0.3 m, while their diameter varies from 12 mm to 76 mm. The operating voltage in these proportional counters, see section 4.2.2.2, varies from 500 to 2000 V depending on the detector's size and gas pressure. Boron trifluoride detectors can usually be operated up to a temperature of 100°C, above which the detector's performance deteriorates.

Although BF_3 has some good characteristics as a gas for a proportional counter, it is not the best gas for this purpose. As section 4.2.2.2 indicates, inert gases are more suited. Therefore, an argon/BF_3 gas mixture is often used to improve the performance of the detector as a proportional counter, by providing a more stable plateau, see section 4.2.2.2. However, this comes at the expense of reduced counting efficiency; but this can be accommodated by increasing the detector's length.

Lined Detectors. The interior wall of an otherwise conventional proportional counter can be coated with boron enriched in ^{10}B. However, in such a detector, only one of the recoil particles, ^{7}Li or an alpha-particle, will be directed towards the gas, while the other is directed away from it, to maintain momentum balance. The end result is that the net energy deposited per a neutron interaction is lower than that registered in a BF_3 counter, making it more difficult to discriminate against the lower amplitude events produced by gamma-rays and to operate the detector as a stable proportional counter [15]. Nevertheless, ^{10}B-lined neutron proportional counters are available, filled with argon, typically to a gas pressure of 27 kPa [93], providing a thermal sensitivity of a few cps/nv, with "nv" referring to neutrons per cm^2 per second. Boron-10 lined detectors are also available as ionization chambers, filled with hydrogen to a gas pressure of typically 33 kPa. These chambers are useful in high neutron fields, such as in a nuclear reactor, as they can be operated in the current mode, see section 4.5.3. Such high neutron-fields are also accompanied by high gamma-fields, that also cause ionization of the gas. However, gamma-compensated ^{10}B-lined neutron ionization chambers can be used to separate neutron-induced events from gamma-induced ionization. This is accomplished by employing two concentric cylinders, one lined with ^{10}B and the other with no lining, with the latter providing a gamma-only pulse that can be subtracted from the former. The chambers are filled with hydrogen, nitrogen or xenon, to a gas pressure of typically 100 kPa.

Gamma-ray Sensitivity. The 0.480 MeV gamma-rays produced in the first reaction in (4.5) release electrons by interacting with the detector metallic walls. These electrons deposit some of their energy in the gas as they travel through the detector, resulting in an increase in the detector's counts at small pulse heights. External gamma-rays that almost always accompany neutron fields also affect the detector in a similar manner. Although gamma-rays hardly affect the gas of the detector due to its low density, their interaction with the detector's metallic walls, and the subsequent liberated electors, produce a detectable electronic signal. Thanks to the high Q-value of neutron interactions with ^{10}B, the amplitude of the signal produced by photons is much smaller than that of the neutron signal. These low-amplitude signals can be easily discriminated against using a single-channel-analyzer, as explained in section 4.5. As noted by reference [113] there is a "neutron event free-energy window" below the pulse-height corresponding to a deposition energy in the detector less than 836 keV; the smallest energy deposited by a charged-particle in the detector according to the reactions of (4.5). This energy-window may be used for detecting gamma-rays without interfering with neutron counting. In high gamma fields, pulse pile-up may produce a cumulative signal that can interfere with neutron pulses. In addition, if the gamma field is quite high, gamma-rays can themselves interact with the gas, breaking its chemical bonding and deteriorating the performance of the detector as a proportional counter. Nevertheless, common BF_3 detectors have been successfully operated in gamma fields as high as 0.1 Gy/hour [15], and are generally regarded as insensitive to gamma radiation.

4.4.1.2 Helium-3 Detectors

As a gas, helium-3 is more suited than BF_3 for use in proportional counters. Helium-3 is an inert monatomic gas and thus does not suffer from the electron attachment and compound disintegration problems associated with BF_3. Moreover, ^{3}He has a reaction cross-section of 5330 barns for thermal-neutrons (0.025 eV), in comparison to the 3840 b of the ^{10}B(n,α) reaction, but both cross-sections have a $\frac{1}{\sqrt{E}}$ energy dependence, where E is the neutron energy. However, ^{3}He is a more expensive gas than BF_3.

Pulse-Height Distribution.

Full-Energy Peak. Helium-3 produces the charged-particles needed for neutron detection via the reaction [113]:

$$\mathrm{n} + {}^3\mathrm{He} \longrightarrow {}^3\mathrm{H}\ (191\ \mathrm{keV}) + \mathrm{p}\ (573\ \mathrm{MeV}) \tag{4.8}$$

This is an exoergeic reaction that releases an energy of 764 keV (Q-value). The energy, in addition to the neutron energy, E, is shared by the proton and the triton in inverse proportion to their mass, that is the proton carries 75% of the Q-value and the rest goes to the triton. Unlike the case with the $^{10}B(n,\alpha)$ reaction in the BF_3 detector, the Q value of the $^3He(n,p)^3H$ reaction is sufficiently low and the (n,p) cross-section for fast-neutrons is reasonably high, to allow the energy of fast-neutrons to be observable in the pulse-height spectrum of the detector. Therefore, a full-energy peak corresponding to an energy of $Q + E$ appears in the pulse-height spectrum for neutrons of energy E, as schematically shown in Figure 4.7. If charged-particles do not deposit all their energy in the gas, and instead bombard the detector walls, two shoulders bellow the peak would appear in the pulse-height spectrum at energies of $0.75(Q + E)$, and $0.25(Q + E)$ corresponding, respectively, to complete loss of proton and triton energy.

Epithermal Peak. In addition, slow-neutrons that almost always exist with fast-neutrons, as neutrons bouncing off surrounding shielding walls produce a peak, called the epithermal-peak, at a pulse height corresponding to Q, since their energy is considerably low compared to the Q-value. The wall effect in a small detector also produces shoulders below the epithermal peak at energies of 0.75 Q and 0.25 Q, for the same reasons discussed above.

Elastic-scattering Continuum. Figure 4.7 also shows a continuum between the full-energy and the epithermal peaks, that starts at an energy of 0.75 E, and extends to zero energy. This continuum is attributed to the elastic scattering between incident neutrons at energy, E, and the ^{3}He nuclei. The cross-section for this reaction is reasonably high to compete with the (n,p) reaction. As discussed in section 3.5, when a neutron is elastically scattered by a nucleus of mass A, it can lose energy, depending on the angle of scattering, from nothing to a maximum of $(1 - \alpha)E$, where $\alpha = \frac{A-1}{A+1}$. For a ^{3}He nucleus, the maximum neutron energy loss would by 0.75 E. This neutron energy loss is an energy gain for the ^{3}He nucleus, which in turn can cause ionization in the detector. Therefore, the energy spectrum for the recoil ^{3}He nuclei extends, as shown in Figure 4.7, from 0.75 E when the energy gained by the nucleus is maximum, to zero when the neutron loses no energy. Note, that for $E \leq 1$ MeV, the recoil spectrum will be below the epithermal peak, and can be discriminated against. However, at $E > 1.02$ MeV, the recoil spectrum will appear beyond the thermal peak, and will overlap with full-energy peaks produced by neutrons of energies greater than Q.

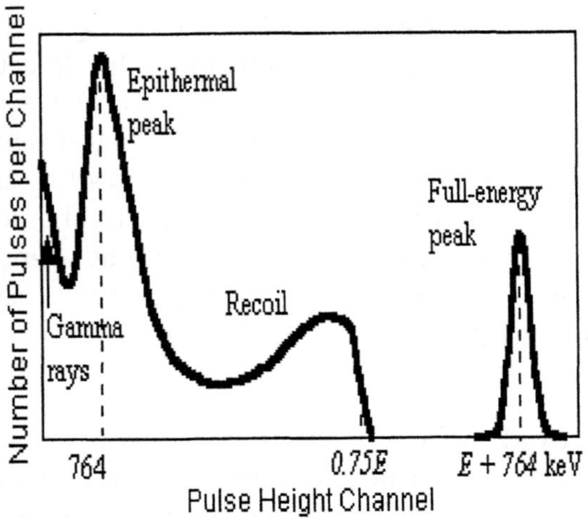

Figure 4.7. Response of a ^{3}He-3 detector to neutrons of energy E and to thermal-neutrons.

Gas Pressure. Helium-3 detectors can be operated at high pressure, from 200 kPa up to 2 MPa, due to the higher saturation (condensation) pressure of the gas. This not only provides a higher detection efficiency, but also makes its easier to minimize the wall effect. However, because of its lower atomic-number, the range of charged-particles in ^{3}He is long and the wall effect is more significant than that in BF_3 detectors. A heavier gas, such as krypton, is usually added to shorten the range of charged-particles and stop them from reaching the wall.

Efficiency. The efficiency of a ^{3}He counter can be expressed approximately as:

$$\epsilon = 1 - \exp\left[-2.4 \times 10^{-2}\, \frac{p(\text{kPa})L(\text{m})}{\sqrt{E(\text{eV})}}\right] \tag{4.9}$$

where E is the neutron energy, L is the detector length and p is the gas pressure. For a typical detector 0.3 m long, with a gas-pressure of 400 kPa, the detection efficiency would be 100% at 0.025 eV, 25% at 100 eV and 9 % at 1 MeV. These efficiencies are much larger, particularly at higher neutron-energies, than those of a typical BF_3 detector, see discussion following Eq. (4.6). This makes ^{3}He detectors suitable for counting fast-neutrons, and for investigating the neutron spectrum in the 20 keV to 2 MeV range. Below 20 keV the energy is too low to be distinguishable from the epithermal peak (see Figure 4.7). Above 2 MeV, the detection efficiency is too low and the ^{3}He recoil effect, discussed above, adds a very broad spectrum (from 0 to $0.75E$) that smears out the pulse-height spectrum, making it difficult to extract neutron energy information in the MeV range. However, recoil pulses tend to have a shorter rise-time, as they are heavier and thus have a shorter range than those of the proton and the triton. Therefore, pulse-rise discrimination, using one of the time pick-off techniques discussed in section 4.5.5.1, can be used to discriminate against these recoil particles [15]. The thermal-neutron sensitivity of ^{3}He detectors, as defined by Eq.(4.7), varies from 0.5 to as high as 880 cps/nv, with "nv" in neutrons per cm^2 per second, depending on the detector's size, shape and gas pressure [93, 114]; values that are an order of magnitude higher than those obtained for BF_3 detectors.

Geometry and Operating Voltage. Helium-3 detectors are available as cylindrical detectors that vary in size form 0.09 to 1.3 m in length, and from 12.5 to 50 mm in diameter. They are also available in block shapes (quadrilateral) and spherical shapes [93]. The operating voltage for a ^{3}He counter is between 600 and 5000 V, depending on the detector's shape, size, and gas pressure. It can also tolerate operation at temperatures as high as 150°C.

Position-Sensitivity. Position-sensitive ^{3}He detectors are also available. Neutron-position information, as discussed in section 4.2.2.2, can be obtained by using high resistance anode wire, and collecting the pulses at both sides of the anode. The pulse rise-time, and the amount of charge collected, at either side of the anode, depend on the position at which the charge is deposited in a detector, and in turn on the neutron's position. Position information can also be obtained using a detector with a multiwire chamber [115], as discussed in section 4.2.2.2.

Ionization Chamber. A ^{3}He detector can be operated as an ionization chamber, usually in the gridded configuration described in section 4.2.2.1. The advantage of operating in this mode is the superior quality of pulse-height, and consequently energy resolution, arrived at by avoiding the fluctuations introduced by the avalanche process in proportional counters, see section 4.2.2.2. A mixture of ^{3}He, argon and methane, is usually employed in such chambers, to reduce the wall effect. However, the pulse amplitude obtained is quite low and the charge collection time is long, which makes them susceptible to noise and limits their use to low count rates.

Sensitivity to Gamma-rays. Gamma-rays, as in the case of BF_3 detectors, produce a detectable signal by releasing electrons via interactions with the metallic walls of detectors. However, in the case of ^{3}He detectors, the signals produced by electrons arising from gamma-rays are more difficult to discriminate against by their amplitude due to the relatively low Q-value of the neutron interaction with ^{3}He. The fact that ^{3}He detectors operate at higher gas pressure than BF_3 detectors also contributes to this problem by the increased energy deposition by electrons as they traverse a detector, which in turn increases the amplitude of the signal associated with gamma-rays. Therefore, it commonly stated that ^{3}He detectors are more sensitive to gamma-rays than BF_3 detectors, due to the difficulty of discriminating against gamma-rays in ^{3}He detectors. Reference [113] noted that there is a "neutron event free-energy window" below the pulse-height corresponding to a deposition energy in the detector of less than 191 keV; the smallest energy deposited by a charged-particle in the detector, according to the reactions of (4.8). This energy window may be used for detecting gamma-rays without interfering with neutron counting. Being a monatomic gas, helium does not decompose under intense gamma exposure, and as such ^{3}He detector can sustain very high gamma fields.

4.4.1.3 Fission Chambers

The neutron fission process is also employed in neutron detection, as fission produces heavy charged-particles in the form of fission fragments; see section 3.5. The large mass of these fission-fragments and their high energy (about 160 MeV per fission) results in the production, in gas detectors, of pulses of amplitudes that are an order of magnitude higher than those produced by other methods. Therefore, these detectors are operated as ionization chambers, since there is no need for charge multiplication. The chambers are lined with a fissile material such as ^{235}U. Although, fissile materials also emit alpha-particles that can cause gas ionization, the pulses produced are small in amplitude, in comparison to those of the fission fragments, and can be easily discriminated against. In order to allow fission fragments to be able to leave the detector walls without being absorbed within, a very thin layer (in the micrometer range) of fission material is used. Uranium, highly enriched ($> 93\%$) in ^{235}U, is utilized in these detectors to maximize the fission probability for thermal-neutrons, but ^{238}U or ^{232}Th can be used for the detection of faster neutrons. However, due to the small thickness of coating that must be used, the efficiency of fission chambers is quite low; about 0.5% for thermal-neutrons (0.025 eV), decreasing with $\frac{1}{\sqrt{E}}$, where E is the neutron energy, to only 0.1% at 0.5 eV.

Gas Pressure and Operating Voltage. Fission chambers are usually filled with argon or P-10 (10% methane, 90% argon), with the latter used to shorten the pulse width, as in the case of proportional counters, see section 4.2.2.2. The gas pressures is typically 100 kPa, and the applied voltage varies from 50 to 800 V. These chambers vary in length from about 40 to 300 mm, with diameters from about 6.5 to 51 mm. A fission chamber can also be used in the current mode, see section 4.5.3, since it does not need to rely on gas multiplication due to its heavy charged-particles and the large current produced. Since these chambers rely on the fission process, they are not affected much by photons, except for the small and low-amplitude pulses produced by direct ionization of the gas. Therefore, they are well-suited for use in high gamma fields, such as in the core of a nuclear reactor.

Sensitivity. The sensitivity of a fission chamber (count rate per unit flux) is expressed as [116]:

$$S = (\pi d L t) N \sigma_f \left(1 - \frac{t}{2R_f}\right) \tag{4.10}$$

where d is the diameter of a cylindrical fission chamber, L is its length, N is the atomic-density of ^{235}U, t is the thickness of the uranium coating, σ_f is the fission microscopic cross-section, and R_f is the range of the fission fragments in the coating. For a typical chamber, enriched to 93% in ^{235}U, with $d = 25$ mm, $L = 0.2$ m, coated with 1 μm ($= t$) uranium dioxide with $N = 2.23 \times 10^{23}$ atoms of uranium per m^3, $R_f = 15$ μm, $\sigma_f = 536$ barns (536×10^{-24} cm^2) for thermal-neutrons (0.025 eV), the thermal-neutron sensitivity, see Eq. (4.7), is equal to about 0.18 cps/nv, with "nv" in neutrons per cm^2 per second. Now assuming that 10^3 thermal-neutrons per second are incident on the coated surface of the detector, which has an area of πdL, for the above typical values these neutrons would produce a flux, "nv" of about 6.5 neutrons per cm^2 per second. With a sensitivity of 0.18, this flux would result in about one count per second. This is comparable with the values reported by one manufacturer of such chambers, which quotes a typical sensitivity of one count per 10^3 neutrons [93]. As Eq. (4.10) indicates, the sensitivity can be changed by varying the coating thickness, t and/or density, N.

4.4.1.4 Cadmium-Based Detectors

The high absorption cross-section of cadmium makes it an attractive candidate for neutron detection. The isotope with the largest thermal-neutron absorption cross-section is ^{113}Cd (12.22% abundance in natural Cd), with a cross-section of 20.6 kb [13] at 0.025 eV. However, neutron absorption by cadmium results in the emission of a 558.46 keV gamma-ray, the detection of which is not as efficient as the detection of the charged-particles resulting from interactions with ^{10}B, ^{3}He, or ^{6}Li. Nevertheless, the detection of gamma-rays resulting from the ^{113}Cd(n,γ)^{114}Cd reaction using a Geiger-Müller tube, covered with a cadmium foil, was reported [117]. Although the operation of such detector is simple, it is not a very efficient detector and produces a high background count rate due to the sensitivity of the Geiger-Müller tube to background radiation.

4.4.1.5 Proton-Recoil Detectors

The neutron detectors discussed above produce the charged-particles required for neutron detection using a material that absorbs neutrons. Such absorption process tends to favor slow-neutrons, as they have a higher probability of interaction than faster neutrons, with the reaction cross-section decreasing with $\frac{1}{\sqrt{E}}$, where E is the neutron energy. This makes such detectors quite inefficient for detecting fast-neutrons. However, a reaction that has a high cross-section for fast-neutrons is the elastic-scattering of neutrons by light nuclei. If these light nuclei are in the form of a gas, they would be free to move and would recoil. The recoiled ions would carry with them some of the energy imparted by the bombarding neutrons, and in the process the gas becomes ionized. This process is utilized in fast-neutron detection. The most common light element employed in this detection process is hydrogen, which upon recoiling becomes a proton, hence the name proton-recoil detectors. Hydrogen is used, not only due to its wide availability as a gas, but also because it receives more energy from collisions with neutrons than any other element, as the hydrogen nucleus is about equal in mass to a neutron. As explained in section 3.5.1, a neutron can lose energy anywhere from nothing to all of its energy upon colliding with a hydrogen nucleus, depending on the angle of scattering, losing on average half of its energy. This process also tends to be isotropic, in the center-of-mass frame of reference, up to a neutron energy of 10 MeV. This makes it equally probable for a neutron to lose any amount of energy, from zero to full energy.

Response Function. For monoenergetic neutrons of energy, $E_n <$ 10 MeV, recoil-protons have a uniform energy distribution. That is, the probability that a recoil-proton will have an energy between E_p and

$E_p + dE_P$ is:

$$p(E_p)dE_p = \begin{cases} \frac{dE_p}{E_n} & \text{for } 0 \leq E_p \leq E_n \\ 0 & \text{for } E_p > E_n \end{cases} \qquad (4.11)$$

Therefore, the response of a proton-recoil detector is as schematically sketched in Figure 4.8. Such a simple response-function will subsequently produce, by ionization, a pulse-height distribution of the same shape. This simple shape makes it relatively easy to determine the neutron-energy using the slope of the distribution, which increases indefinitely at a pulse height corresponding to the neutron energy. As Eq. (4.11) indicates, the neutron-energy is equally distributed among the recoil protons. Therefore, $\frac{dE_p}{E_n}$ is lower at high values of E_n. Moreover, at high neutron-energies, the recoil-protons also possess high energy and can reach the detector walls without depositing their entire energy, which further lowers the pulse-height. Above about 2 MeV neutron energy, the recoil protons have a sufficiently high energy that they do not deposit their entire energy in the wall, making the wall effect so severe that the detector ceases to be useful for energy measurement. The response function of Figure 4.8 is based on a single-collision of neutrons within the counter. Multiple scattering will distort such a simple response-function, as lower-energy scattered neutrons can add their own proton-recoil response. However, in gas-filled detectors, the probability of occurrence of such multiple-scattering events is quite low, given that the probability of occurrence of the first scatter itself is low, as explained below. Ionization by electrons induced by gamma-rays, mainly on the metallic walls of the detector, results in pulses that are smaller in magnitude, than those produced by recoil-protons. Moreover, electrons, being lighter in mass than protons, deposit their energy over a longer range. Electrons produce, therefore, pulses with longer rise-time than that associated with recoil protons. This difference in rise-time can also be used to differentiate against gamma-rays, with pulse-shape discrimination electronics as discussed in section 4.5.

Filling Gas. Gas proton-recoil detectors usually operate as proportional counters, see section 4.2.2.2, and employ hydrogen, or one of its compounds, such as methane (CH_4). Although hydrogen provides the simple response function depicted in Figure 4.8, its ability to fully absorb the energy of recoil-protons is limited because of its low density. Detectors filled with hydrogen are, therefore, limited to use at low neutron-energies, below about 100 keV, where the proton-recoil energy is relatively low so that it can be absorbed within the detector. At higher neutron-energies, methane is preferred over hydrogen, due to

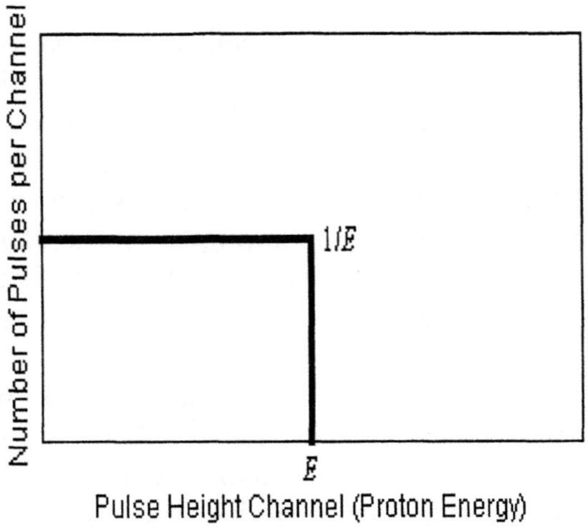

Figure 4.8. Ideal response of a proton-recoil detector to monoenergetic neutrons of energy E.

the former's higher ability to absorb the energy of recoiled-particles. As in conventional proportional counters, see section 4.2.2.2, argon is added as a quench gas to improve the performance of the detector. In a methane-filled detector, neutrons also interact with the carbon of the methane. However, as the discussion in section 3.5.1 indicates, a neutron can lose no more that 28.4% of its energy upon collision with carbon nuclei. Therefore, a 2 MeV neutron will produce carbon recoils with energy that varies from zero to 0.568 MeV. This energy is added to the proton-recoil spectrum that covers the range from 2 MeV to zero. This slightly complicates the analysis of the pulse-height distribution, but it can be accommodated by taking the presence of carbon recoils into account. Below 100 keV neutron energy, the effect of carbon recoils becomes quite significant, as it is imposed on top of a narrower proton-recoil spectrum. Therefore, methane-filled detectors are not recommend for use below 100 keV. They are also not very attractive for use at energies above 2 MeV, as the probability of neutron interactions becomes quite low, and the wall effect becomes quite significant, as indicated earlier. However, the interaction probability can be increased by lining the walls of the counter with a thin layer of polyethylene (a hydrocarbon plastic), to enhance the production of recoil particles.

Efficiency. The detection efficiency for a hydrogen-filled proton-recoil detector can be approximated in a fashion similar to that of other gas detectors by:

$$\epsilon = 1 - \exp\left[-\left(\frac{1.8 \times 10^{-2}}{\sqrt{E_n(\text{keV})}} - 6.9 \times 10^{-5}\right) p(\text{kPa})L(\text{m})\right] \qquad (4.12)$$

where E_n is the neutron energy, p is the gas pressure, and L is the path-length of the neutron in the detector. For a typical detector operating at a gas-pressure of 100 kPa, with a neutron passing through a length of 0.12 m, the detection efficiency within the recommended energy range for this detector would be equal to only 0.6% at 1 MeV, 2% at 100 keV energy and about 19% at 1 keV. Spherical counters are used for hydrogen-filled detectors to obtain the same neutron path length [93], and subsequently, detection efficiency, for all neutrons having the same energy regardless of their direction of incidence. This further simplifies the process of obtaining the neutron energy spectrum from the recorded pulse-height distribution of the detector. For a CH_4-filled detector, the counting efficiency, for the same gas pressure and detector dimensions, is slightly lower due to scattering by carbon, but Eq. (4.12) can be used to provide a rough estimate of the proton-recoil efficiency of a methane-filled detector.

Energy-Resolution and Calibration. The efficiency of a detector affects its energy resolution, since a lower efficiency means that a smaller number of ion-pairs is generated, making the signal quite susceptible to statistical fluctuations. The energy resolution of a gas-filled proton-recoil proportional counter is in the order of 4% at 615 keV, and 6% at 1 keV [92]. Knowing the value of neutron energy requires its calibration against a source of neutrons of known energy. Since monoenergetic neutron sources are not readily available, an indirect calibration process is required. Some detectors include a small amount of helium or nitrogen to produce protons via the reactions: ^{3}He(n,p)^{3}H and ^{14}N(n,p)^{14}C, respectively. At thermal-neutron energy, the reactions release an energy, Q, of 764 keV in the case of ^{3}He and 626 keV for ^{14}N, which are carried by the proton. Either of these proton energies can be used to calibrate the pulse-height spectrum versus proton energy, assuming a linear relationship between the pulse-height and proton energy. This assumption is reasonable, since the amount of ionization produced is essentially proportional to the proton's energy (above 10 keV) [15, 92]. Obviously this calibration process requires a thermalized neutron source, to produce a distinct and identifiable peak in the pulse-height distribution. However, materials that have strong absorption resonances, see section 3.5, can be used to produce dips in the spectrum at higher (than thermal) neutron energies. These dips can be used as markers on the pulse-height spectrum of different neutron energies, and as such can be employed in the calibration process. Among useful filters are ^{27}Al (34.7, 88.5 keV), ^{56}Fe (27.7 keV), NaCl (2.85 keV) and ^{19}F in Teflon (27.0, 49.1, 97.0 keV), where the value in brackets refer to the neutron resonance energies [118].

Geometry and Specifications. Proton-recoil detectors are available, typically in the from of cylinders, 25 or 50 mm in diameter, with length varying from 0.1 to 0.5 m [93]. Hydrogen-filled detectors are also available as spherical detectors. Gas pressure varies from 60 kPa to 430 kPa, but smaller-size detectors with gas pressure as high 1.2 MPa had been manufactured [118]. The operating voltage varies from 1.1 to 3.2 V, depending on the size of the detector and the filling gas. These detectors can operate typically to a temperature up to 50°C. Proton-recoil hydrogen-filled detectors are usually operated as proportional counters, to take advantage of the ion multiplication effect, see section 4.2.2.2. They are also available as ionization chambers, which can operate in the current mode, see section 4.5.3, and can accommodate very high count rates, such as those encountered in a nuclear reactor. In order to increase detection efficiency, large ionization chambers, almost 0.2 m in diameter, and a few m long, are used. In addition, a high gas-pressure, close to 2 MPa, is required to attain good neutron sensitivity [93]. Even then, the overall neutron sensitivity is low, in the order of 10^{-14} amp/nv, where "nv" is per cm^2/s and the amp is used here, instead of the count rate, see Eq.(4.7), since these chambers are usually operated in the current mode. This very low sensitivity limits the use of such chambers to the cores of nuclear reactors, where the neutron flux is quite high.

4.4.2. Scintillation Detectors

Scintillator detectors offer a higher detection-efficiency than gas-filled detectors, due to the increased density of the scintillation material, whether liquid or solid. Scintillators loaded with a neutron absorbing material, such as lithium, boron or gadolinium, are used for detecting slow-neutrons, while proton-recoil scintillators are employed for fast-neutrons. As indicated in section 4.2.3, photomultiplier tubes, or photodiodes, are needed to detect and amplify the collected light and convert it into an electronic signal.

4.4.2.1 Lithium-6 Loaded Scintillators

Lithium is both a good neutron-absorbing and a good scintillating material. As a lithium iodide, it is chemically similar to common NaI scintillators used in photon detection, see section 4.3.2. As with NaI(Tl) detectors, a small amount of impurity, in this case europium, is added to the lithium crystal to enable light emission in the visible part of the spectrum. The ^{6}Li isotope, which constitutes 7.40% of natural lithium, has a large microscope cross-section, 940 barns, at thermal-neutron energy for the (n,α) reaction. Like other neutron-absorbing elements, the cross-section for this reaction decreases with $\frac{1}{\sqrt{E}}$, where E is the neutron energy, up to about 100 keV. Above this energy the cross-section exhibits some broad resonances, the strongest of which is at about 250 keV. These resonances increase the value of the reaction cross-section by an order of magnitude or so beyond its $\frac{1}{\sqrt{E}}$ nominal value [15]. However, this cross-section still remains below the corresponding cross-sections of ^{10}B and ^{3}He, used in gas-filled detectors, see section 4.4.1. The reaction, however, releases considerable amount of energy, Q, via the interaction:

$$^6\text{Li} + \text{n} \longrightarrow {}^3\text{H} + \alpha + 4.78 \text{ MeV} \tag{4.13}$$

Characteristics. As as result of the high Q-value of this reaction, the energy added by the slow-neutrons is negligible, and thus the detector shows very little sensitivity to neutron energy, for slow neutrons. Moreover, the low-amplitude pulses produced by gamma-rays are much easier to discriminate against, as they deposit a small amount of energy in comparison to that produced by neutrons. While slow-neutrons will produce a pulse with a height corresponding to 4.78 MeV, the Q-value of the reaction, a photon would produce a pulse not exceeding its energy. This makes discrimination against photons of relatively low energy reasonably easy. Attention should, however, be given to discriminating against gamma-rays, since the detector, being made of a solid material, has a higher detection efficiency for photons than gas-filled detectors. Note also that the sensitivity of a ^{6}Li-based detector to fast-neutrons can be increased by surrounding it with a moderating material, to slow neutrons down before reaching the detector.

Specifications. The solid nature of Li scintillators makes it possible to design small-size detectors. For example, a 100 mm thick ^{6}LiI (Eu) crystal of a highly enriched ^{6}Li (to 95%) have a detection efficiency of almost 100% for neutrons up to 0.5 eV in energy, with the efficiency dropping to 60% or so at 10 eV. The high density also makes it easier for the produced charged-particles to deposit all their energy before leaving a detector, thus eliminating the wall effect observed in gas-filled detectors. Solid, ^{6}Li-loaded scintillators require protective enclosures, since LiI(Eu) is very hygroscopic. Glass scintillators are also available, formed by the melting of ^6LiO, AL_2O_3, Ce_2O_3 and SiO_2. Though not hygroscopic, the lithium content of glass scintillators is lower than that of LiI(Eu), and thus their detection efficiency for thermal-neutrons is limited only to about 80%. Glass scintillators are also less sensitive to photons than LiI(Eu), as they contain elements with lower atomic-number. Detectors can also be manufactured using glass fibers (Ce-^{6}Li) to construct detectors ranging from a single 150 μmm fiber to tens of thousands of fibers with areas as large as 5 m^2 in length [119]. Lithium-6 is also embedded in a ZnS(Ag) matrix to form scintillators, with efficiency of 60% for thermal-neutrons, with very low sensitivity to photons, see for example reference [103] for the BC-702 scintillator. These scintillators are quite small in size, 6.35 mm thick, and are usually employed in thermal-neutron survey instruments. On the other hand, large scintillators are available as liquid scintillators with a relatively low ^{6}Li concentration [103]. The scintillation decay-time for LiI(Eu) is about 300 ns, while for Li-glass it is about 100 ns, which makes the former more suitable for timing applications, and/or higher count rates.

Energy Resolution. As indicated above, ^{6}Li-loaded detectors produce a single peak corresponding to 4.78 MeV, the Q-value of the (n,α) reaction. Since this Q-value is quite high, deducing the neutron energy from the pulse-height distribution can be difficult. However, for high energies (above several hundred keV), the total energy deposited into the detector would be equal to $Q + E$, and can in principle be distinguishable from the Q-value. A LiI(Eu) detector exhibits a nonlinear response to the alpha and the triton particles produced in the reaction. This broadens the energy distribution of the pulses, and makes it difficult to obtain a good energy resolution. Cooling a detector to liquid-nitrogen temperature reduces this nonlinear response, and helps improve the resolution, but provides an energy resolution of only about 20% [15]. The energy-resolution of glass scintillators is even worse, due to their poor light output, in addition to the nonlinearity of the response. However, their short decay-time (100 ns or less), and their small size, makes them suitable for use in time-of-flight measurements to determine the velocity, and hence energy, of traveling neutrons, see section 4.5.

4.4.2.2 Proton-Recoil Scintillators

The most widely used method of detecting fast-neutrons is through the use of hydrogen-containing organic scintillators, via the proton-recoil process; the same process described for gases in section 4.4.1.5. Unlike gases, scintillators, being liquid or solid (plastics), are more dense, thus provide a small stopping-distance (range) for protons, allowing recoil-protons to deposit their full energy in a small-volume of detector material. Therefore, the ideal response function of a proton-recoil detector, shown in Figure 4.8, is easier to attain in a proton-recoil scintillator detector than a gaseous one. The high scattering cross-section for hydrogen (about 2.5 barns at 2 MeV) allows the use of a relatively small detector (about 20 mm in thickness), while obtaining a reasonably high detection efficiency (about 10% at 2.5 MeV neutron energy). Other advantages of organic scintillators include their low cost, ease of formulating into detectors of various sizes and shapes, and the directional-dependence of their scintillation response.

Characteristics. One can state that the most important characteristics of a neutron detector are all met in organic scintillator detectors [120]. They have a high efficiency, good timing resolution, ability to discriminate against gamma-rays and to distinguish between neutron energies, and a good pulse-height resolution. The decay-time of organic scintillators is in the order of a few nanoseconds, see Table 4.5, making it possible to accommodate high count-rates and to perform signal-timing analysis, when needed. The low-density of organic scintillators makes them inefficient for detecting photons; however, large-size scintillators may have a photon-detection efficiency as large as their efficiency for neutrons. In scintillators, the rise-time of the pulse, τ' in Eq. (4.2), depends on the time it takes to populate the optical levels in a scintillator, see section 4.2.3. Electrons, as those arising from photon interactions, have the fastest rise-time, followed by recoil-protons, induced by neutrons, and heavier charged-particles, if present at all. Therefore, the pulse rise-time of a scintillation pulse can be used to distinguish various types of particles, in particular to discriminate against gamma-rays that almost always accompany neutron fields. The pulse-shape discrimination methods, described in section 4.5, can be employed for this purpose.

Pulse-Height Distribution. The energy, and hence pulse-height, behavior of a proton-recoil scintillation detector can be best described by first considering a very large detector. When a scintillator is sufficiently large, so that all incident neutrons interact within the detector, each neutron will lose, on average, one-half of its energy to the recoil-proton in its initial collision in the detector, as discussed in section 4.4.1.5. The neutron will then undergo further collisions, producing more proton-recoils, and depositing most of its energy in the detector before it escapes or is absorbed. The light generated from all these recoil protons, for a single neutron, produces a number of scintillation photons, and subsequently photoelectrons in the detector's photomultiplier tube, resulting in a single pulse for each incident neutron. The magnitude of the pulse is slightly less than that produced by a proton, whose energy equals that of the incident neutron, due to the presence of non-recoil neutron interactions (such as neutron absorption). Moreover, the light-output of an organic scintillator is not linearly proportional to the proton energy. Therefore, the total light output for a given neutron energy depends on the energy distribution of the recoiling protons. For example, four 1-MeV protons produce less than half the light output of one 4-MeV proton [120]. The pulse-height distribution of a small scintillator will differ from the ideal distribution, depicted in Figure 4.8, due to the escape of the generated recoil-protons, before they deposit their full energy in the detector.

Deviation from Ideal Response. The response of a proton-recoil scintillation detector differs from the ideal response function of Figure 4.8 for the following reasons:

1 For neutrons above 10 MeV in energy, neutron scattering is not isotropic in the center-of-mass system, which skews the ideal rectangular distribution.

2 Protons can escape from a detector before depositing their full-energy (the wall, or edge effect).

3 A neutron may scatter more than once before leaving its detector, producing a second proton recoil.

4 The light output is nonlinearly proportional to the proton energy, as discussed above for a large scintillator.

5 Neutrons interact with carbon, producing carbon recoils with different energy distribution, see section 4.4.1.5. Elastic and inelastic scattering by carbon become particularly significant above 10 MeV neutron energy.

6 The light collection (to the cathode of the photomultiplier) is not necessarily uniform, as it can vary at different parts of the scintillator.

7 There is a statistical variation in the number of photoelectrons produced by a given light intensity at the photocathode.

8 The photocathode itself may not be uniform, resulting in variation in the electron output, depending on the point of light incidence.

9 The multiplication (gain) of the photomultiplier tube can also vary with the intensity of the light output.

In spite of these factors, measuring neutron energy with proton-recoil detectors remains one of the most viable options, as these problems can be overcome to a great extent, as explained below after introducing various types of organic scintillators.

Detector Types. Many hydrogen-containing scintillation materials are available, some of which are listed in Table 4.5. Organic crystals, such as Anthracene and Stilbene, offer a relatively high density, and hence high detection efficiency. Anthracene produces the highest light output per unit energy, and thus is used as a reference; (much as NaI(Tl) for gamma detectors; see section 4.3.2, with respect to which Anthracene provides 34.4% light output). However, the long time-decay constant of its signal makes it unsuitable for use in fast-counting applications, in addition to making it difficult to distinguish, by pulse-shape discrimination, between the slower photon-induced electrons and the faster particles produced by neutrons. Stilbene has a shorter decay-time, but a lower scintillation efficiency (light output). Both scintillators, being crystalline, are expensive and fragile, i.e. subject to damage by mechanical and thermal shocks. Another disadvantage of these crystals stems from the directional variation of their light output, as protons have preference to channel themselves along different crystallographic directions. In plastic scintillators, an organic scintillant is incorporated in a bulk matrix of polymerized hydrocarbon, while in liquid scintillators the organic scintillant is dissolved in hydrogen-containing organic solvents. Obviously, liquid scintillators need to be encapsulated, leaving a vacant volume to accommodate their expansion. Each of the scintillators listed in Table 4.5 is designed to favor a particular application, as discussed below.

Pilot F (BC-408). This is a plastic scintillator with a higher-density and larger light-output than most other scintillators, and has a short decay-time. Thus, it is useful in time-of-flight measurements (see section 4.5), where a small and fast detector is needed.

Liquid Basic (BC-501, BC-501A and BC-505). BC-501, also known as NE213, is a favorable scintillator because it has good pulse-shape discrimination properties, which makes it possible to discriminate against gamma-rays. It is also has a high light output. BC-501A is quite similar to BC-501. BC-505, also known as NE224, has the highest light output among liquid scintillators. It is also quite transparent to light, making it useful for use in large volume detectors, where the light has to travel a long distance to the photocathode.

Table 4.5. Some organic scintillators for neutron detection [15, 103].

Scintillator	Form	Density (kg/m^3)	Light Output (Relative)	Decay Time (ns)	H/C Atomic Ratio
Anthracene	Crystal	1250	100	30	0.715
Stilbene	Crystal	1160	50	4.5	0.858
BC-408 (Pilot F)	Plastic	1032	64	2.1	1.104
BC-436 Deuterated	Plastic	1130	52	2.2	1.001 (D/H = 20.4)
BC-454 Natural Boron loaded	Plastic	1026		2.2	
		(1%)	60		1.19
		(5%)	48		1.170
		(10%)	38		1.240
BC-501 (NE213)	Liquid	874	78	3.7	1.213
BC-501A	Liquid	874	78	3.2	1.212
BC-505 (NE224)	Liquid	877	80	2.5	1.330
PSD capability BC-509	Liquid	1610	20	3.1	3.5×10^{-3}
BC-521 (NE323) Gd Loaded (0.5%)	Liquid	890	68	3.6	1.314
BC-523 (NE311) Natural Boron loaded (0.96%)	Liquid	916	65	3.7	1.738
BC-523A Enriched (90%) Boron loaded (4.41%)	Liquid	916	65	3.7	1.741
BC-525 Gd Loaded (0.5%)	Liquid	880	55	3.8	1.557
BC-537 (NE230)	Deuterated Liquid	954	61	2.8	0.991 (D/H=114)
BC-720	ZnS(Ag) in Plastic	–	–	200	

Hydrogen-Free (BC-509). This scintillator contains only 0.18% hydrogen, but about an equivalent amount of carbon and fluorine. It is, therefore, useful when slowing-down by hydrogen in other scintillators is to be avoided.

Deuteron-Based (B-436 and BC-537). B-436 is a deuteron (rather than a proton) recoil detector, manufactured as thin plastic discs. Deuterium is used in this detector to avoid the slowing-down of neutrons by hydrogen, but this comes at the expense of reduced detection efficiency. The detector, therefore, is principally used for studying the (n,d) reaction. BC-537 is another deuteron-recoil detector, but is a liquid scintillator and offers good pulse-shape discrimination properties, for separating photon from neutron signals.

Plastic Boron-Loaded (BC-454). This is a plastic loaded with natural boron (19.7% ^{10}B). Boron, produces scintillation via the heavy charged-particles in the reaction ^{10}B(n,α)^{7}Li, which releases 2.310 MeV in energy; the same reaction used in the BF_3 proportional counters discussed in section 4.4.1.1. Fast neutrons are likely to encounter elastic collisions with the hydrogen in the detector, producing hydrogen and carbon recoils, and consequently pulse-heights similar to those generated by other proton-recoil scintillators. However, neutrons are likely to undergo many collisions with the hydrogen of the detector before being captured by ^{10}B. Therefore, the signal produced in a detector due to ^{10}B absorption is delayed by the so-called capture time. This capture time depends on the probability of neutron capture, which decreases exponentially with time (as less neutrons become available for absorption) with a time constant, τ, expressed as [121]:

$$\tau = \frac{1}{N_B \sigma v} \tag{4.14}$$

where N_B is the atomic density of ^{10}B, σ is the cross-section for the ^{10}B(n,α)^{7}Li reaction and v is the neutron speed [121]. Since σ is proportional to $\frac{1}{v}$ for slow-neutrons (less than 200 keV), then the capture-time constant, τ, is independent of the neutron energy and decreases with increasing ^{10}B concentration, N_B. For 5% concentration of ^{10}B in this detector type, the capture time-constant is 2.74 μs [121]. The light output introduced by neutron capture by ^{10}B is due to the energy liberated in the ^{10}B(n,α)^{7}Li reaction, see section 4.4.1.1. The energy carried by the resulting ^{7}Li and alpha-particles produces a pulse (delayed by the capture-time). The presence of both the proton-recoil pulse (from the scintillator material) and the ^{10}B capture pulse is then a definite indication that the detected pulse is due to neutrons. This fact is useful for discriminating against pulses produced by gamma-rays, or charged-particles, which may introduce pulses that coincide with either the proton-recoil or ^{10}B capture pulses produced by neutrons, but not with both. Lithium-7, as indicated in section 4.4.1.1, can be in an excited state, emitting 408 keV photons as it decays to a stable state, see reaction (4.5). These photons also cause scintillation in the detector, but can be discriminated against by their shorter rise-time, as described in the beginning of this section. This detector is manufactured in sheets, films, or fibers, and the concentration of boron can be increased up to 15% per weight boron.

Reference [122] presents an interesting method of using BC-454 as a fast-neutron detector that rejects gamma-rays and low energy neutrons. Recoil-protons produced by the interaction of fast-neutrons with the hydrogen of the scintillator, generate a detectable (first) light pulse for energy depositions above about 0.5 MeV. Neutrons that are slowed-down in the detector interact with ^{10}B in the scintillator, giving rise to a second pulse, delayed by at least 350 ns. Coincidence-gate measurements that accepts only the occurrence of the two signals within the gate width (25.6 μs in this case), ensures that only fast-neutrons are detected. This scatter-and-capture coincidence gating ensures that neutrons above 0.5 MeV in energy are detected, since lower-energy neutrons will not produce the scatter (first) pulse. In addition, light produced by the electron of Compton-scattered gamma-rays is also discriminated against, since a photon cannot produce two pulses separated by the 350 ns minimum time difference, but may produce a low background measurement due to random coincidence.

Liquid Boron-Loaded (BC-523 and BC-523A). BC-523, also known as NE311, is a liquid scintillator loaded with natural boron (19.7% ^{10}B). Similar to the plastic scintillator, BC-454, this scintillator provides sensitivity to both slow and fast neutrons (by proton recoils). The capture-time by boron for energy below 200 keV is about 1.4 ms, for the 0.96% boron concentration in this detector. As with the BC-454 detector, this capture-time can be used to distinguish pulses introduced by the alpha-particles, emerging from neutron interactions with ^{10}B, from those induced by recoil-protons. BC-523 is similar to BC-523, but contains boron enriched in ^{10}B, which enhances its detection efficiency for slow-neutrons. The neutron capture time is about 0.5 ms, for neutrons below 200 eV. This detector has excellent pulse-shape characteristics, useful for discriminating against photons.

Gadolinium-Loaded (BC-521 and BC-525). Gadolinium has the highest cross-section value for capturing thermal-neutrons (46 kb). It produces beta-particles and several photons, carrying an energy of 8 MeV in total. Beta-particles, and the electrons produced by photon interactions, being lighter in weight, deposit their energy over a longer distance, thus generating a delayed pulse, in comparison to that produced by recoil-protons. This facilitates discrimination between pulses induced by fast and slow neutrons, by timing analysis or pulse-shape discrimination, see section 4.5. BC-521 is a liquid scintillator with a high light output and is usually used in large tanks (hundreds of liters in volume). BC-525 is another gadolinium loaded scintillator, (see BC-521, above). However, being mineral-oil based, it is a better light transmitter, and has a higher flashing point (66°C) than other gadolinium-loaded scintillators. It is, therefore, more suited for use in large tanks.

ZnS(Ag)-Incorporated (BC-720). This detector is designed for ease of discrimination against photons, in which ZnS(Ag), a scintillator, is incorporated into a transparent plastic material. The gamma pulses produced in this detector have a low amplitude and can be discarded by simple pulse-height rejection (rather than the more sophisticated pulse-shape discrimination method used in other proton-recoil scintillators). Therefore, a plastic-scintillator with a longer decay time can be used.

Parametric Characterization. For a proton-recoil to be used for measuring the energy spectrum of neutrons, a number of parameters need to be determined. These parameters are examined here for, the most commonly employed, and basic, organic scintillators for neutron spectroscopy (BC-501/501A, NE213), using the approximate method of reference [82]. The number of neutrons that interact within a detector depends on its efficiency, which can be approximated as:

$$\epsilon(E) = \frac{N_h \sigma_h}{a L_d} \left[1 - \exp(-a L_d)\right] \tag{4.15}$$

where E is the neutron energy, L_d is the detection length, and $a = N_h\sigma_h + N_c\sigma_c$, with N_h and N_c being the number of hydrogen and carbon atoms per unit volume, while σ_h and σ_c are the hydrogen and carbon microscopic cross-sections respectively, evaluated at E. Eq. (4.15) considers single-scatter events of neutrons by hydrogen. This is not a bad assumption for a small detector in the MeV neutron-energy range, since for a detector 125 mm in diameter and 125 mm in length, the total efficiency (including all reactions and scattering event) at 1 MeV is 14.67%, while that assuming single-scattering with hydrogen is 14.60%. For 14 MeV neutrons, the respective efficiencies are 2.14% and 2.95% [123]. Nevertheless, the second scatters of neutrons and the loss of recoil-protons through the ends of a detector can be accounted for by multiplying the efficiency by the empirical correction (shape) factor:

$$B = 1 - 0.780\frac{R_m}{L_d} + 0.9N_h\sigma_h L_d + 0.077N_h\sigma_h' r_d \tag{4.16}$$

where r_d is the detector radius, σ_h' is the hydrogen cross-section at 0.068 E, and R_m is the range of a proton that receives the full energy of the neutron, and is given by:

$$R_m = 0.01738\ (E + 0.1504)^{1.8194}\ \text{mg/mm}^2 \tag{4.17}$$

The scattered neutrons produce recoil-protons with a uniform spectrum that extends from the neutron energy, E to zero energy, as depicted in Figure 4.8. The magnitude of this uniform spectrum for one neutron would be $\frac{1}{E}$. For $\phi(E)$ neutrons per unit area per unit energy, incident on the detector, only $\phi(E)A\epsilon(E)BdE$ will be detected, where A is the cross-sectional area of the detector, and ϵ and B are as defined by Eqs. (4.15) and (4.16), respectively. Then, the number of protons of energy E_p per unit energy, called here $Q(E_p)$, can be expressed as:

$$Q(E_p)dE_p = \phi(E)A\epsilon(E)B(-dE)\ \frac{1}{E} \tag{4.18}$$

where the negative sign reflects the fact that the neutron energy-loss ($-dE$) results in a proton energy gain (dE_p). Ideally the slope of the distribution of $Q(E_p)$ for monoenergetic neutrons of energy E would be zero everywhere, except at energy E where it goes to infinity, or in practice it becomes very large. This is the principle behind the method of unfolding the energy distribution by differentiation, proposed in reference [82]. However, a detector does not measure the proton-recoil distribution, but rather it registers the photoelectrons produced by its scintillation light. Therefore one needs to convert the proton-energy distribution to the detector's pulse-height distribution. The scintillators response to protons is nonlinear with proton energy, and can be expressed by the Birck's formulae [15]:

$$\frac{dL}{dr} = S\ \frac{dE_p}{dr}\ \frac{1}{1 + k\frac{dE_p}{dr}} \tag{4.19}$$

where L the scintillation response (light output), r is the proton range (distance traveled before full absorption), S is the scintillation efficiency and k is a constant that relates the proton-energy loss to the amount of ionization or excitations of electrons in the scintillator. For electrons of energy E_e, $\frac{dE_e}{dr}$ is sufficiently small that $\frac{dL}{dr} = S(\frac{dE_e}{dr})$, which is in agreement with the observed linear response of scintillators to electrons. The response to protons, and other charged-particles, is therefore expressed in terms of electron equivalence. A unit called MeVee (MeV electron-equivalent) is used to express this equivalence, where one MeVee is the kinetic energy of an electron stopped in a scintillator that produces the same amount of light as a certain heavy-particle stopped in the same scintillator. For protons, the electron-equivalence relationship can be expresses as:

$$P = c_1 E_p^{c_2} \tag{4.20}$$

where P is the electron-equivalence in MeVee, E_p is the proton energy and c_1 and c_2 are calibration constants. Reported values vary depending on the system used, but typically $c_1 = 0.150$ and $c_2 = 1.5414$, with E_p expressed in MeV. That is, a 1 MeV proton produces the light output equivalent to that of 0.15 MeV electrons. Relationship (4.20) also facilitates calibration using the electrons produced by a monoenergetic source of gamma-rays. To facilitate this calibration, the light output (pulse-height) is defined in terms of light-units, where one light-unit is equal to the light produced by 1.25 MeV electrons. For example, a ^{22}Na gamma-source, emits 1.275 MeV photons, see Table 2.11, that would deposit electrons of maximum energy of 1.062 MeV, using Eq. (4.3). Therefore, the upper edge of the pulse-height distribution emerging from a ^{22}Na source corresponds to 0.85 light-units ($\frac{1.062}{1.25}$). A ^{60}Co source can also be used in this calibration process, as it emits two photons at 1.17 and 1.33 MeV, producing an average energy of 1.125 MeV. Using either source, the pulse-height (or the channel number of a multichannel analyzer, see section 4.5.4), can be expressed in light-units; making use of the scintillator's linear response to electrons and the fact that at zero energy the light output should be zero. Given that one light-unit corresponds to 1.25 MeVee, the pulse-height can be converted to MeVee, hence to proton energy using Eq. (4.20). Note that the gamma-ray calibration source can also be used to adjust against changes in the gain of the electronic system.

To obtain the neutron energy from a given pulse-height spectrum, the above processes can be followed in reverse order; that is, one converts the pulse-height to light units, the light units to electron-energy-equivalence (MeVee), the electron-energy to proton-energy, and finally differentiate the proton spectrum to obtain the neutron energy. The number of neutrons at that energy is then obtained from the distribution of recoil-protons using Eq. (4.18). There are more sophisticated methods for inversion using pre-determined (measured [124] or calculated [125]) detector response functions. A number of computer codes for unfolding the pulse-height distribution of neutrons measured with organic scintillators are available, see reference [126].

4.4.3. Other Detection Methods

4.4.3.1 Semiconductors

Semiconductors offer the best energy-resolution ability of a radiation measuring material, due to the large number of charge-carriers produced in each reaction, see section 4.2.4. Unfortunately, however, most semiconductor materials, are not very good neutron absorbers. Although lithium is incorporated as a doping material in semiconductor detectors, such as Si(Li), it is included as an impurity at a very small concentration; increasing the concentration hinders the material's ability to function as a semiconducting material. However, a number of semiconductor-based neutron detectors exist. These detectors mainly depend on using a thin layer of neutron-sensitive material to produce the charged-particles that can be detected by the semiconductor junction. Lithium and gadolinium are used for this purpose.

Sandwiched Detectors. The nonlinear response of ^{6}Li detectors to neutrons, due to the distribution of energy between the alpha-particles and the tritons, emerging in the neutron interaction with ^{6}Li , see detection 4.4.2.1, can be avoided by separately detecting these charged-particles. This is made possible by the fact that, due to momentum conservation, the two particles are emitted at two opposite directions. Therefore, if a thin film of a ^{6}Li-containing material, typically lithium fluoride, is sandwiched between two surface-barrier silicon semiconductor diodes (see section 4.2.4), each of the diodes will sense simultaneously one of the particles. Coincide-measurements, see section 4.5, can be used to identify these events. The energy deposited, and consequently, the pulse-height measured by each detector, should be equivalent to the energy of the incoming neutrons plus the Q-value of the reaction (4.78 MeV), enabling direct deduction of the neutron energy. However, the performance of this type of detector is limited by a number of factors. First, the sandwiched film has to be quite thin, otherwise the emerging charged-particles will lose a considerable amount of their energy before reaching the diodes. The use of a thin layer obviously lowers the detector's detection efficiency (about 10^{-6} for 2 MeV neutrons). Another limitation is that the charged-particles are emitted at opposite directions, only if the momentum of the incoming neutrons can be neglected, i.e. at low neutron-energy. At high neutron-energy, emerging charged-particles will have some momentum in the direction of the incoming neutron and will not be emitted exactly in the opposite direction. It then becomes possible for some neutron reactions in the detector to give rise to two particles that strike one detector, but not the other, and thus will not register in coincide measurements. If a large number of slow-neutrons is present, most the signal obtained will be due to these neutrons, as the detector is most efficient at low neutron energy. This makes it difficult to analyze signals produced by fast-neutrons. In addition, because of the use of semiconductors, the detector is susceptible to radiation damage.

4.4.3.2 Plates and Screens

In principle, photographic films can detect neutrons, as they are made of material rich in hydrogen cellulose $(C_6H_{10}O_5)_n$. The energy carried by the recoil of the hydrogen's protons in the film in turn affects the chemistry of its emulsion material, as described in section 4.2.1. However, this is a very ineffective detection process due to the very small thickness of the film. A sheet of a hydrogenous material (such as polyethylene), so-called hydrogen-radiator, can be placed in front of the photographic film, or on both sides of the film, to produce more proton-recoils and enhance the detection efficiency. However, the sheet has to be quite thin to allow recoil-protons to escape to the photographic film. This makes this process quite inefficient (detection efficiency in the order of 10^{-5}). The recoil-protons emitted from such a plate can also be measured using any of the charged-particle detectors presented in section 4.2. Hydrogen-radiators are perhaps the only means of detecting fast-neutrons on a film, or a plate, a process usually desirable in radiographic imaging. However, the efficiency of detection of slow-neutrons by films can be greatly enhanced by using screen converters containing a neutron-absorbing material such as boron, gadolinium or lithium. Lithium-6, mixed with zinc sulfide (ZnS) to form a screen, or embedded in cerium-activated glass which acts as scintillation materials, leads to the production of light upon neutron bombardment, see section 4.4.2.1. The emitted light can then be detected by photographic films, or captured by a video camera, typically for use in radiography, see section 13.2. A layer of gadolinium oxysulphide (Gd_2O_2S) deposited on a screen can also be used to convert neutrons to light, as neutron absorption leads to the generation of beta- and gamma-radiation that causes scintillation. However, these screens tend to be sensitive to photons, as well as neutrons. Thin foils (a few microns in thickness) containing natural gadolinium have been used for neutron detection [127], in conjunction with silicon diodes, providing an efficiency for thermal-neutrons (0.025 eV) of about 25%, doubled if the detectors is surrounded from both sides by silicon diodes.

4.4.3.3 Detection by Activation

Neutrons can also be detected indirectly by inducing radioactivity in a material and measuring in turn the intensity of the secondary radiation emitted from these materials. This indirect method has the advantage of being sensitive only to neutrons, and thus is useful when the gamma-ray background is quite intense. The material has to be highly absorbing of neutrons, made of thin solid foils or wires, to minimize perturbation of incident neutrons by the foil material and to minimize absorption of the secondary radiation within the material itself. The amount of induced activity can be measured by placing a suitable radiation detector directly around the activation foil, forming the so-called activation counters. More commonly, however, the induced activity is measured in a neutron-free laboratory, where both the intensity and energy of the emitted radiation can be measured, away from the background effect of the original radiation field. In this case, the activation products have to have a reasonable half-life to allow time for the material to be transmitted to a separate laboratory. Activation producing gamma-rays is preferred over that generating charged-particles, since photons are less likely to be absorbed within the foil material, and measuring the intensity and energy of gamma-rays is relatively easier to perform, see section 4.3.

The choice of the foil material determines the detected neutron-energy range. For slow-neutrons, targets made of cobalt, copper, dysprosium, gold, indium, manganese and silver are used [15], all in their natural abundance. Fast-neutrons can also be detected by foils made of materials that are only activated with neutrons exceeding a certain energy; called the threshold energy, and is in the MeV range. A number of these foils can be arranged together to provide detection windows at various neutron energies. Computer codes are available to unfold the measured activities from various foils into a neutron-energy distribution, see reference [128]. These types of detectors are mainly used in reactors, or in conjunction with neutron beams extracted from reactors or produced by accelerators, where the neutron flux is quite high. References [15] and [92] list a number of materials and reactions that exhibit this threshold behavior.

4.5. Signal Processing and Analysis

The discussion above in this chapter indicated that radiation detectors produce electronic pulses that can carry information on the nature, intensity and energy of radiation. The number of these pulses is also an indication of the intensity of detected radiation. A measured signal has no value unless it is converted into some useful *indication*. This is done by electronic processing, which aims, not only at counting the number of pulses, but also at extracting some other useful information, such as pulse-height distribution, and pulse rise and decay times. This section summarizes the essential aspects of this process. The basic electronic components used in signal processing are first described.

4.5.1. Basic Components

A radiation detector represents a capacitance, C, along which a charge pulse, Q, is introduced by each detected radiation quantum. This charge burst produces a voltage, $V = \frac{Q}{C}$, across the capacitance, as well as a current , i ($= \frac{dQ}{dt}$, t refers to time), that can be drawn from the charge. The detector signal is usually transmitted by a cable to an area outside the radiation field for further electronic processing. Such a cable presents an added capacitance to the electronic circuit, which further reduces the voltage derived from the charge. Since the electric-charge produced in most detectors, except GM counters and some scintillators, is quite small, the resulting voltage is low (typically in the mV range). The signal produced by the detector can be easily lost in transmission lines and/or overwhelmed by electronic noise. Therefore, it is often vital to equip the detector with a preamplifier, placed as close as possible to the detector.

4.5.1.1 Preamplifiers

A preamplifier acts, in effect, as a capacitance terminator that prevents deterioration of the detector signal. In addition, the preamplifier matches the high electric impedance of the detector with the low impedance of the coaxial cable connected to subsequent signal processing components. Thus, signal amplification provided by an preamplifier, if any, is of secondary importance to its role as an impedance matcher between the detector and the rest of the signal processing system. The preamplifier also acts as a convenient bridge to supply a detector with the required high voltage, through a resistance, called the load or bias resistance.

Types. Preamplifiers are designed so that their output is directly proportional to the detector's voltage, charge, or current, thus are classified, respectively, as voltage-sensitive, charge-sensitive, or current sensitive. The current-sensitive amplifier is the only type of preamplifier that is relied upon for pulse amplification, and is used with photomultipliers to transform its current pulses into amplified voltage pulses. Voltage-sensitive amplifiers are commonly employed with various detectors. If a detector's capacitance, C, stays constant during operation, the output voltage, V, of a voltage-sensitive detector, would also be proportional to the charge, Q, provided by the detector, since $Q = VC$. The output of a voltage-sensitive amplifier is also proportional to its input pulse, supplied by the detector. However, a detector's capacitance can change with the applied voltage, which may vary due to instabilities in power supply. This can affect the proportionality between the voltage of the signal and its charge, and makes it difficult to extract the charge-related information required in energy-spectrum measurements. In semiconductor detectors, the detector's capacitance can change with operating conditions, see section 4.2.4. Charge-sensitive preamplifiers are designed to deal with this problem by providing a voltage-output that is proportional to the charge supplied by the detector, provided that the pulse duration is shorter than the time-constant of the preamplifier, see references [15, 92] for design details. Most preamplifiers are equipped with a test input that allows the introduction of an external pulse, via a pulse generator, to assess the performance of the preamplifier by allowing direct comparison of its output to its input. Preamplifiers do not change the shape of the input pulse, but can affect its rise-time and decay-time. However, the rise-time of a pulse is kept as short as the charge collection time of the detector (typically less than 60 ns), and its decay-time is made quit long, a few tens of microseconds, to allow full-collection of the charge produced in the detector. Commercially available preamplifiers are designed to accommodate pulses produced by different detector types.

4.5.1.2 Nuclear Instrument Modules (NIMs)

Aside from the preamplifier, which must be kept in the proximity of the detector, radiation signal processing components conform to the nuclear instruments module standards, which require each module, called a NIM, to be 34.5 mm wide and 222 mm height; though modules occupying integral multiple widths are permitted. A crate that accommodates 12 of these NIMs is called a NIM bin, and is equipped with direct-current (DC) supply voltages of ±6, ±12, and ±24 V. Signal processing and test units, such as pulse generators, amplifiers, high-voltage power suppliers, scalers (counters), timers, etc. are manufactured according to the NIM standards, so that they occupy one or more of the slots of the NIM bin. The function of such modules is described in the following sections. However, since the use of an amplifier is a basic need for almost all pulse processing systems, its use is addressed first.

4.5.1.3 Amplifiers

The function of an amplifier as its name implies is to amplify pulses provided by a detector, typically via a preamplifier. The amplification can be as high as 5000 times, but the produced pulse does not exceed 10 V in magnitude, no matter how high the amplification is. The amplifier also has an other important function, that is to shape the pulse to make it usable for further processing. Therefore, the presence of an amplifier is a must in almost all pulse processing systems. There is a variety of pulse shaping methods, see reference [15], that accommodate different goals of pulse processing. For a simple low-rate counting, i.e. when the count-rate is sufficiently low so that pulses do not pile up on top of each other, allowing sufficient time for processing the pulse, the type of pulse-shaping used in not important. Also at low count rates, the exact value of amplification and drifting in the amplification level (due to temperature changes) does not affect the counts, since each signal of an amplitude above the noise level will introduce a count. On the other hand, particular attention needs to be given to the pulse-shaping method when the count-rate is high, or when performing pulse-height analysis (for energy-spectrum measurement), pulse-timing analysis, or pulse-shape discrimination.

Pile-up. At high count-rates, pulses can interfere with each other, with the front end of an incoming pulse superimposing itself on the tail end of a preceding pulse. The degree of overlapping (called pileup) depends on the time of arrival of the new pulse, as well as on the decay time of the preceding pulse. Pulse-shaping can reduce this decay-time by employing a pulse-shaping circuitry with a small time-constant (shaping time). Note, however, a too-short shaping-time can increase the so-called ballistic deficit of the amplifier, i.e. the prevention of the pulse from reaching its full amplitude, thus losing some of the radiation energy indication carried by the pulse height. At high count-rates, the use of pulse-shaping methods that produce bipolar pulse (i.e., a pulse with positive and negative voltage components) is preferred over those that produce unipolar (positive only pulses). This is because shaping-circuits employ a capacitance, which is an alternating current (AC) device that cannot pass direct current (DC). A unipolar-pulse resembles a DC signal, as it has no negative components, thus for such a pulse to pass through a capacitor, a negative component (known as an "undershoot") must be produced. This shifts the baseline (zero-voltage reference line) of the signal, and results in amplified-pulses with reduced apparent pulse amplitude. The problem is more severe at high count-rates, due to pulse pileup. Amplifiers are also equipped with the so-called pole-zero cancellation circuit that prevents pulse undershoot, by modifying the shape of the signal to prevent it from ever having a voltage less than zero. The term pole-zero emerges from the automatic-control theory, where the transfer (input-to-output) function is constructed such that a zero-pole is not possible. This process is achieved by strategically adding a variable resistance in the amplifier that can be adjusted, in accordance with the decay-time of the amplifier used until the signal undershoot is eliminated; this is usually done by observing the signal on an oscilloscope.

Delay Lines. Baseline-shifting can be prevented, or at least reduced, by the use of bipolar pulses, since a bipolar pulse is an AC signal in which the positive and negative portions of the pulse tend to cancel each other out, thus eliminating (completely if the pulse is symmetrical) the DC component that causes the baseline-shift. There are also other methods that employ electronic circuits which actively restore baseline-shift, see reference [15]. Double delay-line shaping is also useful at high count-rates, as it reduces the pulse width and introduces no baseline-shift as it uses symmetrical bipolar pulses. A single delay-line is simply an electric transmission line that is shortened at one end, as a result it reflects back a pulse passing through it and inverts its polarity, while delaying the pulse by the propagation-time through the line. The overlapping, of the original pulse with the reflected pulse has the effect of eliminating the tail end of the original pulse, thus in effect reducing the width of the pulse to a time-interval equal to the travel time (back and forth) of the pulse through the line (which can be in the order of microseconds). If a second delay-line is further added, a bipolar pulse will be produced, since the pulse reflected in the second line will have a negative polarity and will overlap with the zero baseline of the pulse shortened by the first delay-line. Therefore, with two identical delay-lines, a symmetric bipolar pulse is produced, which eliminates the baseline-shift problem. The total width of the resulting pulse is equal to the propagation time in both lines, and can be controlled to produce an overall short pulse-duration, which is helpful in remedying the pulse-pileup problem at high count-rates. Some other active pileup restoration/rejection circuits are available, see reference [15].

Spectroscopy Amplifiers. In energy-spectrum measurements (spectroscopy) the pulse-height information must be maintained, as it is a direct indication of the energy giving rise to the pulse in the detector. Then the demands on pulse processing are in effect opposite to those imposed at high count-rates. Spectroscopy requires a unipolar-pulse with a long width, so that there is sufficient time to capture and analyze the full pulse-height. Bipolar pulses are not preferred in spectroscopy, since in effect the latter half of the equivalent unipolar pulse changes polarity, which reduces the duration of the portion of the pulse having the largest amplitude. If spectroscopy has to be performed at a high count-rate, a compromise must be made between the opposing demands of these two operations. It is also important in spectroscopy to employ high-quality amplifiers that have precise amplification and are immune to drift by temperature changes. Noise reduction is also crucial in spectroscopy involving low-energy signals. Then, pulse-shaping circuits that produce a good signal-to-noise ratio need to be used. This tends to exclude pulse-shaping methods employing delay-lines, due to the noise produced as the signal is reflected in the lines, and favors circuits producing Gaussian or triangularly shaped pulses, due to their superior signal-to-noise characteristics, see reference [15]. However, if the pileup problem is present, so that two pulses may arrive at about the same time (causing the so-called peak pileup), the use of a delay-line in a pulse-rejection circuit may be necessary in spectroscopy amplifiers, see section 4.5.7.

Fast Amplifiers. When pulse timing, i.e. time of arrival, information is needed, a detector that provides a pulse with short duration is required. The shaping-time of the amplifier has then to be set to a low value (smaller than the charge collection-time in the detector), so that the voltage signal produced by the amplifier has a shape almost identical to the current (charge per unit time) produced in the detector. This is done to preserve the fast-timing information of the detector's pulse. On the other hand, amplifiers are often designed to increase the pulse-height to a level useful for use in a time pick-off electronic system, see section 4.5.5.1. Common amplifiers are, therefore, not suited for use in fast-timing, as they are designed to filter out the high-frequency components of the signal to preserve pulse-height information. The frequency, f_i, means here the frequency of the $i$the $\sin(2\pi f_i t)$ components, where t is time, to which a signal can be decomposed using a Fourier series. Therefore, fast-timing analysis requires filtering out the low-frequency components, to remove the long tail-end of the pulse, thus preserving only its leading (rise-time) edge. However, removing the tail-end of the pulse makes it more vulnerable to baseline-shifting (see discussion earlier in this section), as the pulse duration, along which the baseline is situated, is shorter. It is, therefore, desirable to preserve the low-frequency components. Therefore, wide-band amplifiers are used in fast-timing applications. These amplifiers, also known as fast amplifiers, neither filter nor shape the signal, but merely replicate the incoming pulse, with an amplification. Such amplifiers can also reverse the polarity of the pulse, if needed.

Applications requiring information on the shape of the pulse rely on the change of the pulse-height with amplitude (time profile). In the fast-timing application discussed in the above paragraph, the pulse's time-profile is preserved, since no pulse shaping needs to be performed. In the case of slow-timing, pulse-shaping distorts the pulse's time profile. However, the rise-time of the shaped-pulse can be used as indication of whether the pulse is slow (small slope) or fast (large slope). Moreover, if the detector's pulse is shaped to a bipolar pulse, the time at which the pulse crosses the baseline (from positive to negative) depends on the pulse-shape and its rise-time. This approach is used in pulse-shape discriminating systems, see section 4.5.5.4, thus requires amplifiers with shaping circuits that produce bipolar pulses, or the use of double delay-line shapers (discussed earlier in this section).

4.5.2. Pulse-Mode Counting

Pulse-mode operation is the modality most commonly used in radiation measurements, where individual voltage pulses emitted by the detector are analyzed, with each pulse corresponding to a particle sensed by the detector. The simplest form of analysis is pulse counting, where the number of pulses within a given counting period is measured. Since pulses emitted from a detector, and magnified by an amplifier, are analog (voltage) pulses, the first step in counting pulses is to convert the analog signal to a digital signal that can be tallied. For the purpose of counting, all is needed is a binary indication (one or zero) that determines whether a pulse has arrived from the detector or not. This can be provided by a single-channel analyzer.

4.5.2.1 Single-Channel Analyzer (SCA)

When a pulse is detected, its presence can be manifested as a logic signal by a SCA, which issues a rectangular (0.5 to 10 μs wide) pulse of a fixed height (usually 10 V), regardless of the height of the incoming pulse. This is done in a logic electronic circuit, which issues the rectangular pulse every time the height of the incoming pulse exceeds a certain value, called the lower-level discrimination (LLD). No pulses are issued at all, if the pulse height is below the LLD value. The value of LLD is selected by the user to eliminate electronic noise, and any other background signal of low amplitude. This logic-circuit, known as integral discriminator, is incorporated in the the Single Channel Analyzer (SCA) NIM. As the name indicates, a SCA defines a channel, with a LLD and an upper level discriminator (ULD). The maximum value of ULD is usually 10 V, but can be reduced to a lower value, so that pulses with heights exceeding the ULD value are rejected, producing no logical pulse. The use of LLD and ULD together provides differential discriminator, which is useful, for instance, in counting pulses within a selected range of magnitudes (particle energy).

4.5.2.2 Scaler/ Timer

Once digital rectangular pulses are generator by a SCA, they can be counted by a scaler or a *counter.* These are simply electronic digital registers that can accumulate counts within a preset time period or to a preset count level. In either case, a Timer is needed. This can be a separate NIM, or can be combined with a Scaler. In either case, the Timer can issue a gate signal to the Scaler to start and stop counting, or can be simply record the time period within which a preset count is accumulated in the Scaler. Some Scalers are also equipped with an interface port that can be connected to a printer, a computer or some other data acquisition system, to facilitate automatic gathering of data.

4.5.2.3 Rate Meter

In situations where a digital count is not necessary, and a visual or an analog indication is desirable, a counting rate-meter can be sued. A rate-meter, in effect, converts the digital signals obtained from the SCA to analog signals, that can be useful for example for control of other devices. The rate-meter consists essentially of a capacitance, C, which accumulates the charge of the incoming digital pulses, and a resistance, R, across which a voltage, V, can be measured. If the pulses are incoming at a rate of $\dot{c}$ counts per second, each carrying a charge q, then C will be charged at a rate of:

$$\frac{dQ}{dt} = \dot{c}q - \frac{V}{R} \tag{4.21}$$

This is simply a current-balance equation, where Q is the charge in the capacitance, $\frac{dQ}{dt}$ is in turn the current in C, $\dot{c}q$ is the incoming current, and $\frac{V}{R}$ is the current in R. Substituting $V = \frac{Q}{C}$ in Eq. (4.21), since R and C are in parallel, Eq. (4.21) can then be solved, with the initial condition $Q = 0$ at $t = 0$, and with $V = \frac{Q}{C}$, one obtains:

$$V(t) = \dot{c}qR\left[1 - \exp\left(-\frac{t}{RC}\right)\right] \tag{4.22}$$

when the time t exceeds considerably the time-constant of the rate-meter, RC, the measured voltage reaches a saturation value of $\bar{V} = \dot{c}qR$, a value that is independent of C and is directly proportional to the count rate, $\dot{c}$. However, for a fast response, the time-constant of the rate meter, RC, should be small so that V reaches its saturation value quickly. Since pulses arrive at various irregular times at the counter, even at steady state of counting, the rate-meter will display a voltage that fluctuates along the saturation value, $\bar{V}$. The variance, Var(V), of the measured voltage signal is equal to $\frac{q^2 R\dot{c}}{2C}$, which can be derived from Eq. (G.23) using the fact that $\text{Var}(V) = \frac{\text{Var}(Q)}{c^2}$. The variance in the count-rate is $\frac{\dot{c}}{2RC}$, using Eq. (G.24) with a long time-constant. This indicates that lowering the statistical-variance requires reducing the time-constant for the rate-meter, RC, which comes at the expense of reducing the time response of the meter, as Eq. (4.22) indicates.

4.5.3. Current-Mode Operation

In some applications, and with some detectors, such as ionization chambers, fission chambers and GM tubes, the current-mode, in which the collected-charge is integrated over time, is preferred to pulse-mode operation. The current-mode of operation is best suited for high count-rates, where the time between successive pulses is too short to allow pulse-by-pulse processing. This mode of operation is quite simple, as it only requires the use of a current measuring device. Standard current meters, galvanometers, rely on detecting the magnetic-field induced by the current, through the movement of a magnetic-needle or a coil. However, the current generated by most detectors is quite small (in the picoamp range), to be detectable by a galvanometer. Therefore, a device known as an electrometer is used, which measures the voltage drop across a resistance, connected to the detector, through which the current flows. By making the resistance across which the current moves very large, 1 giga to 1 tera ohms, the signal is amplified from a small current value to a measurable voltage drop. The presence of the resistance creates, in effect a resistance-capacitance (R–C) circuit, in which R and C are in parallel, since the detector and the connecting cable are electrically equivalent to a capacitance. Then the current, I, produced by the detector can be related to the voltage, V, across the resistance, R, by the relationship (known in electrical engineering as Kirchhoff's first law):

$$I = \frac{dQ}{dt} + \frac{V}{R} = C\frac{dV}{dt} + \frac{V}{R} \tag{4.23}$$

where Q is the charge developed over the detector's capacitance, ($Q = CV$) and t refers to time. The solution of Eq. (4.23) is:

$$V(t) = IR + R(I_0 - I)\exp\left(-\frac{t}{RC}\right) \tag{4.24}$$

where I_0 is a constant, equal to the current at time zero. Eq. (4.24) shows that the response of the measured voltage responds exponentially to the current in the detector, at a rate determined by the time-constant, RC. For a fast-response, the value of RC should be as small as possible. Since the value of R needs to be large, for amplification purposes as indicated earlier, a fast-response with an electrometer is attained only when the detector has a very small capacitance. On the other hand, if the resistance R is made as large as possible, the detector charge will simply accumulate over the capacitance, C, providing a time-integrated current. Then, by monitoring the change in the measured voltage, the total current over the measuring period can be estimated.

Detectors emit pulses continually, while a current-measuring device, regardless of the method of measuring the current, provides a time-averaged indication. The time-period upon which this averaging occurs depends on the response-time of the ammeter or the electrometer, and is typically in the order of a second. This time-averaging process provides the average of individual current-pulses produced by the detector. At constant interaction rate in a detector, the recorded current will consist of a constant-current baseline, with a fluctuating component, caused by statistical variation in the detector, imposed on top of the baseline. The deviation of these fluctuations from the baseline value, over the measuring time, provides an estimate of the statistical variance of measurements.

4.5.4. Energy Spectroscopy

Spectroscopy requires measuring the energy distribution of radiation, i.e. determining the intensity of radiation incident on a detector as a function of energy. The radiation energy is determined by the height (amplitude) of the pulse, in detectors capable of providing radiation-energy information. Therefore, radiation-energy spectroscopy is in effect a pulse-height analysis, which can be attained with the aid of a *multichannel analyzer* (MCA).

4.5.4.1 Multichannel Analyzer

A multichannel analyzer receives analog (voltage) pulses from an amplifier, converts them to digital signals of equivalent digital values. The pulse-height information is then stored into a digital-memory, and sorted into different channels in an ascending order in accordance to their height. Since this is a digital operation, the number of channels is a multiple of 2^n, that is $n = 10$ provides 1024 channels (1 k of digital memory). Channel memory varies from as low as 0.25 k (256 channels) to 64 k (65 536 channels). A MCA displays the number of counts in each channel, thus proving the pulse-height distribution. Stand-alone MCA's are equipped with interface cards that allow down-loading the data into a computer for further analysis. MCA cards are also available, with supporting software, for direct insertion into personal computers. A number of computer codes for spectral analysis are also commercially available, the performance of which is compared in reference [129]. The software is usually equipped with data smoothing, channel calibration and other manipulation features that facilitate analysis of the pulse-height spectrum. Energy calibration of the channel is usually achieved using calibration sources with known distinct energies, and relating the source energy to the corresponding channel number. It should be kept in mind, however, that is some detectors the pulse-height distribution is not directly related to the particle energy, as in proton recoil detectors, requiring application of a numerical unfolding process. Even in the cases where peaks in the pulse height correspond directly to particle energy, as in detecting photons, peaks may overlap over pulse-heights caused by incomplete energy deposition in the detector, due to Compton scattering and pair production interactions, see section 4.3.

4.5.5. Timing Measurements

4.5.5.1 Time Pick-Off

Timing the arrival of a pulse to a detector can be useful in some applications. For example, measuring the time-of-flight of fast neutrons over a fixed distance can be used to calculate their velocity, hence energy. The arrival of a pulse is best measured with a detector of fast response. Therefore, fast collection of charge-carriers is needed, for timing information to be confined within the shortest possible time interval. Marking the time-of-arrival of a pulse can be performed by monitoring its first appearance, indicated by its leading (rise-time) edge. Any voltage momentarily exceeding the baseline-voltage can be taken as an indication of the arrival of a new signal. This simple approach would, however, be susceptible to signals generated by electronic noise. Therefore, pulse-timing methods, rely on the so-called time pick-off methods that aim, among other things, at minimizing the effect of electronic noise. The simplest approach is to consider a new signal to have arrived when a pulse-height exceeds a certain threshold-value, preset above the noise level. A single-channel analyzer (see section 4.5.2) is used to issue a logic (rectangular) pulse each time an incoming signal exceeds the threshold value. This leading-edge triggering method, though simple, produces time-information that depends on the slope of the signal, which determines the intersection of the signal and the threshold-amplitude level. This slope can change as the amplitude of the pulse changes (an effect known as time-walk), and by random fluctuations in the shape and size of the signal, an effect known as time-jitter. A method that is less susceptible to the time-walk effect is the zero-crossover method, which triggers (using a specially designed NIM) a logic-pulse when a bipolar-pulse crosses the baseline (zero-line) from negative to positive. Such a bipolar pulse can be produced by pulse-shaping in the amplifier, see section 4.5.1.3, and the zero-crossover time is independent, at least theoretically, of the pulse-height. This method is, however, more susceptible to time-jitter due to the additional noise introduced in the bipolar shaping of the pulse, and because of the reliance on the zero-level as a timing indicator. Dependence on the amplitude, which causes the time-walk, can also be eliminated, without increasing time-jitter, by triggering a timing logic-signal when the pulse reaches a pre-specified fraction, typically 10%, of its full amplitude. This is done by a specially designed NIM known as the Constant-Fraction Discriminator. This method is more electronically sophisticated than the other two methods, since it is more effective in reducing time-walk (amplitude-dependence) effects, and is less susceptible to random fluctuations (time-jitter).

4.5.5.2 Time-Amplitude Converter

While a time-pick off method will give the time of arrival of a particle to a detector, it does not quantify its time of arrival. Since measurement of time is a relative quantity, time is quantified by the time difference between two pulses. For electronic processing, it is useful to have this time expressed as a voltage. This is done in a NIM known as Time-Amplitude Converter (TAC), which converts the difference in the timing of two pulses into a voltage pulse, between zero and 10 V. The two pulses for which the time difference is found are designated as "start" and "stop" pulses, and are fed to two separate connectors in the TAC NIM. The time-difference information, now expressed in voltages, can be fed into a multichannel analyzer, see section 4.5.4, to provide the so-called time spectrum; which is analogous to the pulse-height spectrum, except for the fact that the height now corresponds to time rather than pulse amplitude. This time-spectrum is useful in a number of applications. It can be used, for instance, to distinguish a fast-rising pulse from a slowly-rising pulse for the purpose of discrimination against one of them.

Timing information can also be used to determine whether two pulses are arriving at the same time or not, in coincidence measurements. Pick-up timing information is also useful in time-of-flight measurements, to determine the time it takes a particle to fly from a source to a detector.

4.5.5.3 Multichannel Scaler

A Multichannel Scaler (MCS) is a digital device that records counts as a function of time, by storing the recorded counts in separate locations in memory every time interval, Δt, called the dwell-time. This dwell-time can vary from nanoseconds to hours [130], depending on the demand of an application. At the end of the scanning process, i.e. when the digital memory of the device is completely filled, the distribution of the count rate versus time can be displayed and analyzed. This device is useful in applications that require time-analysis, such as time-correlated counting, event lifetime measurement, neutron time-of-flight and Mössbauer experiments.

4.5.5.4 Pulse-Shape Discrimination

The shape of pulses often carries information on the nature of the incident radiation, or in some cases the authenticity of the pulse. Although the shape of the pulse can be viewed on an oscilloscope, electronic processing of shape information requires quantifying the signal, by the pulse rise-time and decay-time. There are a variety of pulse-shape discrimination methods, see references [15, 92]. The most basic method relies on shaping the input pulse into a bipolar-pulse and using leading-edge triggering to issue a "start "pulse and the zero-crossover method to issue a "stop" pulse to a TAC NIM, see section 4.5.5.1 for a description of these methods. The "start" signal indicates the beginning of the pulse, while the "stop" signal, which corresponds to the zero-crossover point, is dependent of the original pulse's rise-time and shape, as indicated in section 4.5.5.1. Therefore, the voltage produced by the TAC will depend on the rise-time and shape of the original pulse, before being shaped into a bipolar pulse. In a mixture of a variety of pulse types, the time-distribution of a TAC, as displayed on a multichannel analyzer, can be used to set discrimination windows for detecting a particular type of pulses.

4.5.5.5 Coincidence Measurements

Coincidence measurements are useful in detecting particles generated, or reaching a detector, at the same time. This can be used, for example, to identify the two (511 keV) photons instantaneously emitted as a result of the annihilation of a positron by an electron, see section 3.4, or when detecting particles/photons, emitted simultaneously from a source. Since it is not possible to achieve exact coincidence measurements, pulses arriving within a small time interval, known as the resolving-time are considered to be in coincidence for all practical purposes. The time-spectrum produced from a TAC and displayed on a MCA, can be used to define a "time window" within which the incoming particles/photons are in coincidence. Alternatively, a NIM known as the Coincidence Unit can be used. This unit is analogous to the single-channel-analyzer, see section 4.5.2, in the sense that it provides a single-channel discrimination window (unlike the multichannel nature of the TAC/MCA combination). Therefore, a Coincidence Unit will accept a pair of incoming pulses only if they arrive with a pre-set resolving-time, and issue accordingly a logic-pulse; otherwise the pulses are rejected and no logic pulse is issued.

4.5.5.6 Time-of-Flight Measurements

Time-of-flight measurements are used for measuring the energy, E, of a particle, with a resolution better than any other method, by determining the particle velocity, v, from the time, t, it takes to travel between two fixed stations in air at a distance d apart, from the relationship $v = \frac{d}{t} = \sqrt{\frac{2E}{M}}$ where M is the particle's mass. Therefore, the time-of-flight, t, in nanosecond over a distance d is given by:

$$t\ (\mathrm{ns}) = 10^9\ d\ (\mathrm{m})\sqrt{\frac{M}{2E}} = 10^9 \left(\frac{10^9 d}{c}\right)\left(\frac{Mc^2}{2E}\right) \tag{4.25}$$

The energy of the particle, E, can be calculated from t, given that $c = 2.997924 \times 10^8$ m/s, as:

$$E = \frac{1}{2}\left(\frac{10d\ (\mathrm{m})}{2.997924\ t\ (\mathrm{ns})}\right)^2 (Mc^2) \tag{4.26}$$

This allows the expression of E and the rest-mass energy, Mc^2 in the same units, say MeV.

For neutrons, $Mc^2 = 9.395656 \times 10^2$ MeV, and the energy E_n in MeV is determined from the time-of-flight as $E_n = 0.5\left(\frac{10d}{2.997924t}\right)^2 (9.395656) = 52.2704\left(\frac{10d}{t}\right)^2$. At high neutron-speeds ($E > 6.5$ MeV), the relativistic effects (see appendix C) should be taken into account. Using Eq. (C.8), one obtains the following relationship for the kinetic-energy of a neutron:

$$E_n = 52.2704\left(\frac{10d}{t}\right)^2 + 4.3619\left(\frac{10d}{t}\right)^4 + 0.4044\left(\frac{10d}{t}\right)^6$$

$$+0.0394\left(\frac{d}{t}\right)^8 + 0.0039\left(\frac{10d}{t}\right)^{10} \qquad (4.27)$$

where E_n is in MeV, t is in nanoseconds and d in meters.

Since photons travel at the speed of light, their velocity is constant and time-of-flight measurements are not suited for measuring their energy, but can be used for neutrons and charged-particles, particularly heavy ions. If the energy of the particle is known, determining the velocity of a particle enables the calculation of its mass. Time-of-flight measurements require two stations along which the particle passes and is time-tagged. This can be done either by using a pulse (timed) source and a detector, or by employing two detectors intersecting the path of radiation. Since the detection of neutrons, see section 4.4, necessarily invokes the loss of some neutron energy, if not all, the detector-to-detector method is not suited for neutrons, but can be used with energetic charged-particles that can be detected in a thin detector without losing much energy. The timing of a pulse reaching a detector involves the use of one of the time pick-off methods described in section 4.5.5.1.

4.5.6. Statistics

4.5.6.1 Counting Statistics

Radiation is emitted from a radioisotopic source in a random fashion. Therefore, any measurement based on such sources will contain statistical fluctuations that are attributed to this randomness. The same argument also applies to radiation produced by nuclear interactions, say by an accelerated particle bombarding a target (see section 2.3.1.3), since not every incident particle is likely to interact with the target nuclei, and the interaction will occur randomly, though according to a probability governed by the cross-section of the reaction (see chapter 3). However, both radioactive decay and reaction-induced processes have the following statistical characteristics:

1 There is only two statistical states for a nucleus, it decays or not; or in the case of nuclear interactions, interacts or not.

2 Both involve discrete (not continuous events), since each nucleus represents an entity independent from other surrounding nuclei.

3 Each event (disintegration or interaction) is independent of any other; that is the decay of a nucleus has no effect on the decay of another nucleus, and the interaction of an incident particle with a target nucleus has no effect on the interaction of another particle with another nucleus.

4 The population of nuclei subject to decay or interaction is very large, as the number of nuclei in any material is very high; one mole contains 6.022×10^{23} atoms, (Avogadro's number), e.g. 12 g of ^{12}C contain 6.022×10^{23} atoms.

5 Only a very small fraction of the nuclei are likely to decay (or interact with a bombarding particle) within the measurement time.

The first three characteristics (1, 2, 3) dictate that the statistical distribution governing these processes should be a *binomial* distribution (see any classical textbook on statistics). A binomial distribution that has a large population (characteristics 4) and a small probability of event occurrence (characteristic 5) becomes, on the limit, a *Poisson* distribution. Therefore, the statistics of radiation counting is governed by the Poisson distribution. The variance of a Poisson distribution, σ^2, is equal to its mean, m, that is $\sigma = \sqrt{m}$. However, the third and forth characteristics (3 and 4) lead to a normal (Gaussian) distribution, which with characteristic 5 approaches a Poisson distribution when the average of the measured events (counts) is 20 or more. Based on these facts, statistical analysis is greatly simplified as one can evaluate the variance using Poisson statistics ($\sigma^2 = \sqrt{m}$), and use the confidence intervals of the normal distribution (e.g. 68% of the measurements will lie within $m \pm \sigma$). Based on these concepts, Appendix G provides some guidelines and procedures for accumulating counts with adequate statistics, and for estimating the statistical variability associated with obtained measurements. The appendix also addresses some practical questions, such as: how long should a count rate be, how large a signal should be compared to the background, how statistically good are the data, and how could the statistical error be reduced, etc.

4.5.6.2 Energy Resolution

The energy-resolution of a detector is governed, not only by its intrinsic nature, but also by the statistical fluctuations in the number of charge-carriers created in a detector by incident radiation. These charge-carriers are the ion-pairs in an ionization detector, the photoelectrons in a scintillation detector, and electron-hole pairs in a semiconductor detector. Although a specific amount of energy deposited in a detector generates on average the same number of charge-carriers, the number generated by different particle varies statistically around this average value. The statistics of this charge-carrier generation have the same statistical characteristics as radiation counting, discussed earlier in this section 4.5.6.1. Therefore, if N charge-carriers are recorded, their variance would be equal to $\sigma_N^2 = N$, according to the Poisson statistics of counting. The relative standard-deviation $\left(\frac{\sqrt{\text{variance}}}{\text{mean}}\right)$ is equal to $\frac{\sqrt{N}}{N} = \frac{1}{\sqrt{N}}$. Therefore, detectors that provide a large number of charge-carriers will have a lower relative variability, thus have a good resolution. That is why semiconductor detectors provide good energy-resolution, as they generate a large number of charge-carriers. The pulse-height produced by most detectors is linearly proportional to the number of charge-carriers, and therefore will have the same statistical distribution as that of the charge-carriers. According to the same statistical features, the distribution of the pulse-height generated by a stream of radiation particles of a certain (fixed) energy will also follow a normal (Gaussian) distribution with a mean pulse-height corresponding to a variance equal to σ_N^2. The width of that normal distribution defines the energy resolution of a detector. For ease of measurement, the width is measured by the width of the Gaussian peak corresponding to half its maximum value, the so-called full-width at half-maximum (FWHM). It can be analytically shown that the area under the FWHM is equal to 2.35σ, where σ^2 is the variance of the pulse-height distribution. Therefore, the confidence interval associated with the FWHM is 76%, that is the area under the peak defined by the FWHM occupies 76% of the total area of the distribution.

Fano Factor. Experimental observations have shown that some ionization detectors and semiconductor detectors provide better energy-resolution than estimated by the statistical arguments made above. This can be attributed to the correlated nature of the generated charge-carriers, that is the formation of one charge-carrier is not totally independent from others. This violates one of the statistical feature (characteristic 3) discussed in section 4.5.6.1. Fortunately, however, this correlated nature brings regularity into the charge-carrier formation process, reducing its randomness. Consequently, the observed variance for these detectors is lower than that based on statistical arguments presented above. The amount of reduction in the variance, hence improvement in energy-resolution, is know as the *Fano factor*; that is a Fano factor of 0.5 implies that the variance of pulse height of a certain detector is 50% of the statistically estimated value. For semiconductor detectors, the Fano factor can be as low as 0.06; a typical value for germanium detectors is 0.16 [92]. For a proportional gas detectors, the Fano factor varies from 0.05 to 0.2 [15].

4.5.7. Problems in Pulse Analysis

4.5.7.1 Electronic Noise

Electronic noise of most importance in pulse processing is that generated in detectors and the preamplifiers, since it is subsequently magnified and propagated across the rest of the subsequent electronic modules. Such noise can be produced by unsteady current-leakage from a detector, or as a result of thermal-motion of the charge-carries in detectors, resistors, cables, etc., as well as by the noise picked up from power lines, communication lines, etc. The frequency range of noise can be quite wide, and is generally represented by a "white noise", i.e. uniformly distributed in frequency over a broad range. Mechanical vibrations, as those produced by an operating pump, also introduce a low-frequency component, known as "microphonics". The superposition of noise on detector signal adds a small amplitude component at different times to the signal, which has the effect of broadening its peaks, and thus distorting the pulse-height distribution.

Filtering. Most detectors produce signals with a frequency somewhere in the middle range, thus low-pass filtering is used to eliminate the contribution of the high-frequency component of noise, while high-pass filtering is utilized to get rid of the low-frequency component. This filtering process is carried out at the amplifier stage, see section 4.5.1.3, in which a frequency band is defined beyond which incoming signals are attributed to noise and are subsequently rejected. The extent of noise filtering in an amplifier defines to a great extent its quality. Fortunately, however, noise that passes the filtering process are low in amplitude, and can be discriminated against using an integral discriminator, or a single channel analyzer, see section 4.5.2.

Pulse-Height Cut-off. The noise level can also be observed on a multichannel analyzer (see section 4.5.4), which displays the pulse-height distribution of incoming signals. Since noise is usually low in amplitude it appears at the first few channels on the MCA. A user can then determine the channel number below which noise signals appear. Since the incoming signal has usually a maximum amplitude of 10 V, and the MCA divides this 10 V by the total number of channels, one can convert the channel number to a corresponding voltage. For example, a MCA set to resolve 1024 channels, provides $\frac{10}{1024}$ V per channel, so that channel number n corresponds to $n \times \frac{10}{1024}$. The voltage corresponding to the channel below which the noise is to be eliminated can then be used as the lower-level discrimination window of a SCA. A user can test this procedure using a pulser emitting pulses at a level corresponding to the cut-off voltage, and verify that these pulses do not pass the SCA (i.e. no counts are produced) but appear at the proper channel number on the MCA. Increasing the amplitude of the pulser's signal should also allow the SCA to issue logical signals that trigger a scaler.

Gain Stabilization. When making measurements over a long period of time, a system's gain can drift. This is typically the case with detectors employing photomultiplier tubes, but can occur also with other systems. Such drift deteriorates the energy resolution of the system, as it affects directly the relationship between the radiation energy and the signal's amplitude (channel number in a MCA). To correct for such drift, a reference source of known pulse-height is used. Electronic circuitry can then adjust the system's gain to keep the reference peak at the same channel number. Some MCA's have an automatic gain stabilizer that monitors periodically the position of an energy peak. Light sources (light-emitting diodes) can be used with scintillation detectors, and the 60 keV photons of an ^{241}Am source can be utilized in gamma and x-ray detection.

4.5.7.2 Radiation Background

Radiation background can be a problem in radiation measurements, if it constitutes a significant portion of the gross counts. Although there are statistical means to determine the counting period needed to statistically accommodate a certain level of background measurement, as shown in appendix G, it is always desirable to physically reduce background radiation to improve the signal-to-background ratio. Background-radiation has a number of origins. It can be reaching a detector directly from a source, without passing through the material or object it is intended to interact with (source component). Radiation scattered off surrounding shielding walls, floors, source and detector collimators, etc. also contributes to the background-radiation (surroundings component). Background-radiation also emerges due to natural radiation, from cosmic-rays and natural emissions from concrete walls, the Earth's surface, shielding materials etc. (natural component). These components are not necessarily independent from each other, as discussed below. Judicious system design and selection of its materials can help reduce the background signal, see section 17.3.1. However, active methods can be used to reduce the background-radiation, such as pulse processing or coincidence (anticoinidence) measurements, see reference [15].

Source Background. Source background can emerge directly from a source or indirectly from surrounding material. The direct component emerges when one is not interested in directly detecting the source radiation, but rather in detecting its interactions (or lack of) with some test material. Obviously shielding the detector from direct exposure to the source can help eliminate the direct component. However, shielding has its complicating effects, as it can scatter source radiation towards the detector and/or emit some secondary radiation that can be sensed by the detector. For example, a lead shielding for gamma-rays produces 170 keV and 77 keV fluorescent x-rays by photoabsorption of gamma-rays, see section 3.30, while a neutron shielding material can produce gamma-rays by neutron activation (see section 3.5.3) that may be counted by some detectors.

Background from Surroundings. Surroundings include radiation shielding, source and detector collimator walls, as well as building walls and floors and any other objects that happen to be around. Surroundings not only scatter unwanted radiation towards detectors, but can emit secondary radiation (by activation in the case of neutrons, and fluorescent emission in the case of photons and charged-particles). In addition, surroundings can also produce their own natural activity as discussed below. Adding and removing a particular surrounding component, when possible, and monitoring the corresponding background-count rate, can help identify the component(s) most responsible for elevating the background signal. Such components should then be rearranged, repositioned, or even replaced, to reduce the background count rate.

Natural Background. Natural radioactivity is present in many common materials and can increase the background signal in some applications, see section 2.4. Potassium-40 is a common radioactive material (1.26×10^9 year half-life) that decays by electron capture, emitting 1.460 MeV gamma-rays (11% yield), and producing ^{40}Ar which in turn emits characteristic x-rays. In addition, ^{40}K produces beta-particles with a maximum energy of 1.314 MeV (89% yield). Natural potassium contains ^{40}K at a concentration of 0.012% per weight, and in turn can be found in many construction materials, aluminum, steel magnesium, copper, rubber, cement, epoxy, quartz, etc. The same materials also contain traces of thorium, and uranium. These two materials are part of long decay sequences that produce a variety of alpha, beta and gamma rays that contribute to background measurements, see reference [15]. For very low-background measurements, high-purity materials free of such contaminants should be sought.

Natural activity can also be airborne, such as the radon gas that arises from the decay of uranium (^{222}Rn) and thorium (^{220}Rn). In addition cosmic radiation (highly energetic charged-particles and heavy-ions) bombarding the Earth's atmosphere produce a variety of radioactive particles, that are themselves highly energetic and usually pass most detectors without interacting with them. However, they can induce some radioactivity in the surrounding materials; e.g. fast-neutrons from cosmic-rays can be slowed-down by a hydrogen-rich material, and in turn be absorbed by hydrogen producing 2.223 MeV gamma-rays. Such natural radioactivity is usually difficult to control, but fortunately represents a very small component of the background radiation in many applications.

4.5.7.3 Pile-up Problems

Pile-up of pulses, as the name indicates, occurs when two consecutive pulses are superimposed on each other. The result is an apparent increase in the pulse amplitude. This typically occurs at the preamplifier stage, since the decay-constant of an preamplifier is normally kept relatively long (50 to 100 μs) to permit complete collection of detector charge, see section 4.5.1.1. Usually pulse-shaping by a subsequent amplifier shortens the duration of signals, and to a great extent prevents this pile-up problem. The shorter the shaping-time of an amplifier, the shorter the duration of the signal, the less likely is pulse pile-up. However at very high count-rates, preamplifiers can be overloaded, or saturated, resulting in a severe distortion of the produced signals. This is particularly a serious problem in charge-sensitive amplifiers, where the output is directly proportional to the charge and is independent of the detector's capacitance, see section 4.5.1.1. Reducing the electrical-resistance of a charge-sensitive preamplifier shortens the decay of the pulse and help eliminate this effect, albeit at the expense of increased noise level.

Capacitance-based Preamplifiers. The approach of reducing the amplifier resistance can be taken to the limit by removing it, and simply accumulating the detector's charge on an preamplifier's capacitance until it reaches a value close to the saturation level where it is actively reset with a light-emitting diode or a transistor reset [15]. The sudden drop from a maximum charge to zero is masked by an "inhibit" output-time period (called the reset period), to avoid processing of such abrupt a signal by a subsequent amplifier. This reset period is essentially a dead-time that should be minimized. The transistor-reset provides a shorter period (a few μs) than the optical diode method, and as such is preferred in high-count rate applications; while the light-emitting diode is favored in applications requiring low noise.

Peak Pileup. When pulse pile-up occurs, it distorts the pulse-height, by increasing it. Therefore, pulse-height data required for energy-spectrum measurements (discussed in section 4.5.4) are severely affected by the pile-up problem. In pulse-height analysis, this problem is dealt with at the amplifier level, where pulse-shaping takes place. Pulse-pile can worsen the problem of baseline (zero-voltage reference line) shifting observed in unipolar pulses, as discussed in section 4.5.1.3. However, baseline-restoration methods and the pole-zero cancellation approach, discussed also in section 4.5.1.3, can help remedy this problem. The most severe case of pulse-pileup occurs when two pulses are combined into a single pulse, as they arrive at the amplifier at about the same time, and cannot be discriminated against by electronic processing. This problem is known as peak pileup, and is not as easy to deal with as the pileup that occurs at the tail-end of a preceding pulse that can be overcome by baseline-restoration or pole-zero cancellation. Peak-pileup, not only distorts the shape of the pulse-height distribution, by showing pulses at higher pulse heights, but it also affects the area under a peak. Most spectroscopy amplifiers are equipped with a pileup-rejection circuitry that eliminates pile-up, typically by passing the incoming signal simultaneously through both fast and delayed lines, and allowing, through a gating process, a pulse to be issued from the amplifier only if one pulse passes through the fast-line during the delay process. This obviously produces some dead-time losses that need to be taken into account, see section 4.5.7.4. One way of quantitatively estimating the amount of peak pile-up is to use a random pulse generator (a commercially available NIM) and connect its output at a given pulse-height to the same preamplifier used for the detector, and monitor its pulse height, on a multichannel analyzer, in a region away from the detector's pulse height. Since the pulse-rate of the pulser is precisely known, and the number of counts in the MCA corresponding to the pulser's signal, can be measured, the ratio between the count rate and the pulse-rate provides an indication of the amount of pulse-pileup. Note also that the same procedure can be applied using a regular-period pulser, provided that its emission-rate is a small fraction (less than 10%) of the pulse-rate produced by the used detector, to provide a realistic amount of pileup, otherwise an excessive amount of pile up will be produced.

4.5.7.4 Dead-Time losses

Dead-time losses occur when a detector or a pulse-processing component become inactive or "dead" after receiving a pulse for a certain period of time, called the dead-time. This differs from the pulse pile-up problem, discussed in section 4.5.7.3, in the sense that while in pile-up two consecutive pulses are detected at the same time, thus may be counted as one pulse, dead-time is a period during which no pulses are counted at all. As discussed in section 4.5.7.3, pulse pile-up occurs at the preamplifier and amplifier stages of the pulse processing train. Dead-time losses, on the other hand, tend to occur in the detectors and electronic components that require some processing time, such as the case with multichannel analyzers, or can occur when pulsed sources are utilized.

Detector Dead-Time. Since radiation is emitted randomly, it also reaches a detector at random times. There is, therefore, a finite probability that more than one radiation particle/photon can reach a detector at the same time, while the detector is still occupied with generating a pulse from a previous radiation particle. The detector, in effect, becomes "blind" for a while, unable to produce pulses for newly incoming radiation. Obviously, the higher the count-rate, the more likely the loss of pulses due to the detector's dead-time. If the dead-time associated with each recorded count is assumed to be constant and equal to τ, and n particles cause the generation of charge-carries in the detector within a time period t, while only c counts are observed, then the detector will be dead for a period of $c\tau$, or equivalently "alive" for a period of $t - c\tau$. Since during the live-time period the detector records c counts, then the observed count rate is equal to $\frac{c}{t-c\tau}$. This count-rate should be equal to the actual count-rate expected of the detector, if it had no dead-time losses, $\frac{n}{t}$, since the "alive" count-rate should be the same in each case. That is:

$$\frac{c}{t - c\tau} = \frac{n}{t}$$

leading to,

$$ct = n(t - c\tau)$$

and upon dividing by t^2,

$$\dot{c} = \dot{n} - \dot{c}\,\dot{n}\tau$$

consequently,

$$\dot{n} = \frac{c}{1 - c\tau} \tag{4.28}$$

where $\dot{c} = \frac{c}{t}$ and $\dot{n} = \frac{n}{t}$. Relationship (4.28) enables the determination of the "true" count-rate from the measured one, provided that the dead-time, τ, is known and constant. However, the assumption of a constant τ can be replaced by an equally plausible assumption of using an average value for the dead-time. Some detectors are known to have a large value of τ, particularly the GM tubes, see section 4.2.2.3. In such tubes, the produced charge-carriers are so strong that they diminish the strength of the applied electric-field, to the extent that it can no longer produce the required avalanche in the detector. In GM tubes, the dead-time can be as high as 50 to 100 μs.

Dead-Time in Counting System. In order to measure the dead-time of a detector, a counting system must be used. Such a system can, however, introduce its own dead-time losses. Therefore, the actual dead-time is a combination of the dead-times of a detector and the associated measurement system. In most cases, dead-time losses are caused by limitations in the counting system, rather than the detector itself. It is also often difficult to distinguish between dead-time losses in the detector and those in the counting system, since the two have to be used in conjunction with each other. The single channel analyzer (SCA), see section 4.5.2, is usually the component most responsible for dead-time losses, as it takes some finite time to produce a logic signal from an incoming analog signal. The dead-time of a SCA is typically a microsecond, or two, larger than the width of the pulse it processes [15]. Therefore, it is desirable to shape the pulse in an amplifier using a short shaping-time to minimize dead-time losses in the SCA; this also helps reduce the pile-up problem discussed earlier in this section.

Although the dead-time of a counting system and the associated detector can be measured, using two sources of different intensities, see reference [15] and [92], the process is tedious, and is at best approximate. Therefore, effort should be made to minimize dead-time losses, either by changing the physical layout of the experimental setup or by proper selection of the detector and associated electronics. The amount of dead-time losses can be estimated by comparing the measured count-rate to the expected count rate, with the latter theoretically estimated or determined by scaling up the count rate obtained with a less intense source. In general, one should aim at a dead-time losses of no more than 10 to 20%, as higher losses make the counting system quite vulnerable to small changes in the "true" count rate.

Dead-Time in Multichannel Analyzer. The most notable and quantifiable dead-time losses are observed in the multichannel analyzer

(MCA). As explained in section 4.5.4, a MCA utilizes some processing time to convert incoming analog pulses into logical pulses and sort them in accordance with their height into corresponding channels. This processing-time is predictable and most MCA's are equipped with an electronic gate that blocks incoming signals during the dead-time period. This gate also drives a meter that determines the fraction of the live-time during which the gate was closed, i.e. the dead-time fraction. As in counting, one should aim at reducing dead-time losses, which should not exceeds 30 or 40% to avoid severe distortions of the recorded pulse-height. It is obvious that MCA dead-time losses are dependent on the MCA circuitry, and should not be confused with other types of dead-time losses.

Pulsed-Source Dead-Time. Dead-time losses can occur in an electronically triggered pulsed source, see section 2.3.1.3, which emits many radiation particles within a short burst-period, and remain inert for some time, depending on the frequency of pulsing. Then a detector will be exposed to a large number of particles within a short interval, say T. If the dead-time of the counting system, τ, is larger than T, but less than the time spacing between pulses, then only one particle per source pulse can be detected. The loss in counts does not depend on the dead-time of the detector or the associated counting system, but on the number of particles emitted per source-pulse, or the probability that more than one radiation particle is produced per pulse. Obviously, if τ is much smaller than T, the source would appear to the detector as a continuous source, at least during the duration of the pulse, and dead-time losses will follow the relationship given by Eq. (4.28). However, the problem becomes more complex, if τ is just slightly smaller than T. Then the counting losses will depend on both the number of source particles emitted per pulse and the dead-time of the detector.

Chapter 5

RADIATION SAFETY

5.1. Introduction

Individuals working with nuclear and atomic radiation must deal with it with the utmost attention and care. The ionizing effect of radiation, discussed in chapter 3, can also take place in the body, destroying the chemical bonds in molecules that control the functioning of the human body. Body mechanisms are often capable of restoring the damage, but the same repair process can cause the production of faulty cells that can multiply rapidly. Being faulty cells, the immune system will attack them, but if these cells grow out of control they can cause cancer. There is some evidence that exposure to low levels of radiation has some beneficial effect (so called hormesis effect); "hormesis" is the Greek word for "hormaein" which means to "excite". Low-level radiation is, therefore, thought to stimulate biological anticancer defense mechanisms to deal with both spontaneous and externally-induced DNA damage. Even if the premise of the hormesis effect is acceptable, there is no reason to unnecessarily expose oneself to radiation. Therefore, regulating authorities in different countries have designated radiation exposure limits beyond which the risk to the human body is not acceptable. The level of acceptability depends on whether the risk of exposure to radiation is countered by some benefit to the subjected individual. Therefore, the dose limit for a casual member of the public is much lower than that of a professional working with radiation (often called atomic-radiation, or nuclear-energy worker). Nevertheless, users must also do their best to minimize the risk of exposure to radiation.

Everyone, before undertaking any work involving radiation, must have a good understanding of the principles of radiation safety, be quite familiar and comfortable with the regulations in their jurisdiction, and know the details of the work they are involved in. Institutions involved with radiation work have their own radiation safety manuals, and often their radiation safety (also called health physics) staff that can train and guide radiation workers. In installations where such infrastructure is not available, potential radiation workers should identify and contact the authority responsible for regulating radiation safety in their jurisdiction. Radiation users should also be very familiar with their institution's security and emergency procedures and policies. This chapter provides a summary of the main points radiation workers should be familiar with.

5.2. Principles and Definitions

The International Basic Safety Standards (BSS) of radiation protection states the following [131] principles:

Justification of Practice: There should a benefit to the use of radiation that outweighs the risk of radiation exposure.

Optimization of Protection: Exposure to radiation should be kept as low as reasonably achievable within economic and social considerations, and constrained to prevent an unfair distribution of exposure or risk.

Dose Limits: Individual exposure should not exceed pre-assigned dose limits, above which the dose or risk is deemed unacceptable.

The objective of these standards is to protect individuals against short-term effects of high-doses and to restrict the likelihood of occurrence of long-term effects. To this end, BSS promotes: (1) the development of a safety culture that encourages a questioning and learning attitude and critical thinking, and discourages complacency; (2) the establishment of quality assurance programs related to protection and safety; (3) controlling human factors by providing appropriate training and education to limit the contribution of human error; and (4) the availability of experts to advice and observe. Sound technical planning is also advocated by BSS through: (a) securing radiation sources to prevent theft or damage; (b) applying a defense in depth approach to radiation safety that ensures that the failure of one layer of safety is compensated for by subsequent layers; (c) adopting good engineering practices that abide by codes and standards and are supported by reliable management and organization; (d) verification of safety by regular monitoring for compliance; and (4) transporting radioactive material in safe and reliable manner in accordance to national protocols and international conventions.

The ALARA Principle. Although regulating authorities specify the maximum allowable limits for different categories of individuals, one should always aim at keeping radiation exposure to *as low as reasonably achievable.* This so-called ALARA principle is the cornerstone of radiation safety and clearly dictates that every reasonable effort should be made to minimize exposure of individuals to radiation. That is, the designated limits are upper bounds one should attempt to stay below, as much as practically possible.

Exposure and Absorbed Dose. The terms "exposure" and "dose" are used, often interchangeably, in radiation protection when dealing with photons, for reasons that will become apparent shortly. Exposure refers to the intensity of radiation, e.g. the number of neutrons, photons or charged-particles incident on a material, integrated over time. The dose, D, refers to the amount of radiation imparted to matter per unit mass, that is:

$$D = \frac{d\bar{\varepsilon}}{dm} \quad \text{J/kg or Gy} \tag{5.1}$$

where $\bar{\varepsilon}$ is the mean-energy absorbed by a matter of mass m. The special name for the unit of absorbed dose is gray (Gy). An older unit called *rad* (radiation-absorbed dose) is still often used, (1 rad = 10^{-2} J/kg = 10^{-2}Gy); it is convenient to recall that 1 rad = 1 cGy (centigray). The definition of D in Eq. (5.1) is independent of the type of radiation, but is difficult to measure, particularly in body tissue. However, measurements can be readily made in air, as it is easier to measure the electric-charge rather than the energy deposited. Therefore, the roentgen (R) unit emerged earlier in radiation-protection measurements to designate the amount of photons (x- or gamma-rays) that produce in 1 cm^3 of dry air, at 0°C at atmospheric pressure, ions carrying one electrostatic unit (or 2.58×10^{-4} C/kg) of electric (positive or negative) charge. This quantity is also called *exposure* as it is related to the intensity of ionizing radiation; note that it does not apply to directly to neutrons. This exposure can be measured with an ionization chamber containing air (or an equivalent material), see section 4.3, and then converted to Gy, making use of the following facts:

1 esu is produced by 2.082×10^9 ion pairs.
Each ion pair requires an energy deposition of 34 eV ($34 \times 1.603 \times 10^{-19}$J) in air.
1 cm^3 of dry air at 0°C and 1 atom weighs 1.293×10^{-6} kg.
Therefore,

$$1 \text{ Gy (in air)} = \frac{1 \text{ J}}{1 \text{ kg}} = \frac{1}{1.603\times10^{-19}} \frac{\text{eV}}{\text{kg}} \equiv \frac{1.293\times10^{-6}}{34\times1.603\times10^{-19}} \frac{\text{ion pairs}}{\text{cm}^3}$$

$$1 \text{ Gy} \equiv \frac{1.293\times10^{-6}}{(34\times1.603\times10^{-19})(2.082\times10^{9})} \frac{\text{esu}}{\text{cm}^3} \equiv 114 \text{ R}$$

The amount of energy absorbed in a material can be related to the amount of absorption in air using the ratio of the so-called energy absorption coefficient, a quantity analogous to the total attenuation coefficient, see section 4.3, but is defined on the basis of the average-energy loss per photon interaction. An exposure of 1 R results in the absorption of about 0.01 Gy in body soft-tissue (though a larger amount of energy is deposited in hard tissue, such as bone). That is, 1 Gy in soft tissue is equivalent to 100 R, which is very close to the 114 R in air, and the R and Rad (1 cGy) are often used interchangeably when dealing with photons, though one quantity refers to exposure and the other to energy deposition in soft tissue. The term *kerma* (kinetic energy released in a material) is typically used to describe the amount of energy deposited in non-biological materials.

Equivalent Dose. From a biological point of view, although it is useful to know how much radiation-energy is deposited in tissue, it is as equally important to know how the absorbed-radiation affects the tissue. For example, one would anticipate that, for the same amount of energy deposited, alpha-particles will cause much more damage to the body than electrons (directly induced by a source or through gamma interactions), since alpha-particles deposit all their energy within a much shorter distance, due to the lighter weight of electrons. Therefore, a quantity known as the dose-equivalent, H, is used, and is defined as:

$$H = Dw \tag{5.2}$$

when D is in Gy, H is expressed in a unit known as sievert (Sv), and w is a dimensionless quantity known as the *radiation weighting factor*, or *quality factor* [132]. The value of w determines the biological effect for a given type of radiation. For photons and electrons, of all energies, $w = 1$. For neutrons, of energy, E less than 10 keV, $w = 5$; 10 keV $< E <$ 100 keV, $w = 10$; 100 keV $< E <$ 2 MeV, $w = 20$; 2 MeV $< E <$ 20 MeV, $w = 10$; E $>$ 20 MeV, $w = 5$. For protons, (other than recoil protons) with $E >$ 2 MeV, $w = 5$, while for alpha-particles, fission-fragments, heavy-nuclei, $w = 20$. It should be emphasized here that, while the values of H for different radiation types can be added to each other to obtained the total dose-equivalent, the values of the dose D in Gy should not be added (unless w is equal to unity). The old unit of H, which is still used by some, is called rem (roentgen equivalent man), where 1 rem = 0.01 Sv = 1 cSv.

Dose Rate. The radiation dose, in Gy, and the corresponding dose-equivalent, in Sv, do not depend on the time during which the dose is delivered to the body. However, the absorbed dose-rate, $\dot{D}$ in Gy per unit time (usually Gy per hour) is often a more practical quantity to calculate or measure. Then D is estimated as:

$$D = \int_0^T \dot{D} dt \tag{5.3}$$

where T is the exposure time. The dose limits are normally expressed in Sv per year, e.g. 10 mSv per year (whole body exposure) for an atomic-radiation (nuclear energy) worker. These limits can then be used to budget exposure time during a specific period of working time, keeping in mind that the common practice is to impose a quarterly limit (i.e. per quarter of a year) to avoid large exposures within a short period.

5.3. Principles of Radiation Protection

For the external sources, four basic principles are employed to minimize the radiation dose:

Source Strength: The strength or activity should as small as practically as possible.

Exposure Time: The period of exposure to radiation should be minimized.

Distance from Source: Workers should stay away as much as possible from radiation sources.

Shielding: A barrier should be provided to avoid direct exposure of personnel to the source of radiation, using the proper type and amount of radiation-absorbing material.

Exposure. The amount of absorbed dose is directly and linearly dependent on the source strength and the time of exposure. Therefore, in designing an instrument, or in acquiring a device, one should aim at using the smallest amount of source strength that meets the detection and indication requirements. Once the source strength is determined, every effort should be made to minimize exposure to radiation, since as Eq. (5.3) indicates, the accumulated dose is directly dependent on the exposure time. Minimizing exposure-time can be done by planning ahead a strategy that minimizes the working-time around the source. If an extensive amount of work is to be done, it is often useful to exercise ahead with a dummy (not radioactive) arrangement to identify in advance any possible delaying obstacles and find ways to overcome them.

Distance from Source. Distance is a passive, inexpensive, yet a very effective way of reducing radiation exposure. As the divergence law discussed in section 3.6.2 indicates, the radiation flux, and consequently the intensity and exposure, are inversely proportional to R^2, for a point source, where R is the distance from the source to the receiver. Although this law applies strictly to a point source, at a large distance the source appears to the receiver as a point. The inverse-square nature of this dependence makes distance a very strong ally to radiation workers.

Shielding. As a common practice, all radiation sources are provided with an appropriate shielding, both during use and storage. Section 16.3 discusses the methods used for shielding design. Different shielding materials are utilized for different types of radiation. For alpha-particles, shielding against external exposure to the body is not necessary, since the dead layer of the skin stops alpha-particles emitted from isotopic sources. For beta-particles, safety glasses should be worn to protect the eyes, and plastic or aluminum foils can be used for shielding, (but the use of lead or steel should be avoided as they produce secondary x-rays). Photons (x- or gamma-rays) are effectively shielded using electron-rich materials (heavy metals), such as lead, steel and concrete, since photons interact mainly with the atomic-electrons as discussed in section 4.3. Neutrons, on the other hand, lose most of their energy when interacting with the hydrogen nuclei (see section 3.5). Therefore, hydrogen-rich materials, such as water, paraffin wax or polyethylene, are used for neutron shielding. Hydrogen slows-down neutrons, which can then be captured by a neutron-absorbing material such as boron, lithium and cadmium. Neutrons-emission is almost always accompanied with gamma-radiation, and neutron-absorption often results in the release of gamma-rays by activation, see section 3.5.3. Therefore, gamma shielding is often required along with neutron shielding.

5.4. Monitoring and Dosimetry

Monitoring radiation exposure is not only mandated by regulating authorities, but is a sensible way of ensuring that the practices discussed above are all well-planned, and that they remain so. Individual-radiation workers are to be provided with radiation badges. These dosimeters are *personal* and not to be shared with any other worker. Monitors should be worn at the waist or lapel position, and should be periodically read to provide an estimate of the person's dose exposure. Personal monitors are either film badges or thermoluminescent dosimeters TLD's, the principle of the operation of each is described below.

Film Badges. Film badges utilize the principle of photographic emulsion described in section 4.2.1. Photographic films are mounted on a small holder, covered with a light-tight wrapping. The film is also covered with a set of small pieces of filters made of different materials to discriminate against different types of radiation, e.g. to stop beta-rays, low-energy photons, or convert thermal-neutrons (using gadolinium or cadmium) to detectable radiation (beta-particles and gamma-rays). These filters enable the determination of the dose contribution of different types of radiation. Although fast-neutrons are not absorbed by the film, they scatter in the emulsion material creating proton-recoil tracks that can be counted under a microscope and used to estimate the fast-neutron dose.

Thermoluminescent Dosimeters. TLD's are made of inorganic crystals that have thermoluminescence properties. As explained in section 4.2.4, the close-packing of atoms in solids causes the formation of valence and conduction bands of electrons. Electron movement is only allowed between these two bands, and is not permitted in the gap in between, the forbidden band. However, due to the presence of impurities and lattice defects, energy levels can be created in the forbidden band, allowing the formation of electron traps, that capture electrons in this band. Radiation-deposition can move electrons from the valence band to these electron traps. Electrons remain trapped until the material is heated, and the trapped electrons gain sufficient energy to free the electrons and return them to their original state, and in the process release their excitation energy in the form of light; hence the name thermoluminescence. This electron trapping process allows accumulation of the effect of radiation by generation more trapped electrons. The heating process and subsequent light emission allows the release of the accumulated trapped electrons, and hence gives an indication of the amount of radiation absorbed. One of the common materials used in TLD's is $CaSO_4$(Mn), where magnesium is added as an impurity to create the electron traps. However, LiF is the most widely used material. It is an intrinsic thermoluminescent material, i.e. it does not require the addition of an impurity, as the electron traps are created by crystal defects. Since it has a low-atomic number, LiF is close to tissue in its response to photon absorption (effective atomic-number for photoelectric absorption of 8.4, compared to 7.42 for tissue), and has a linear response over a wide exposure range and its response is quite stable at room temperature. Moreover, natural lithium contains 7.4% ^{6}Li, a good thermal-neutron absorber. Therefore, TLD's are somewhat sensitive to thermal neutrons, via the same (n,α) reaction used in ^{6}Li neutron detectors, see section 4.4.2.1. However, TLD's highly enriched in ^{6}Li are available for providing enhanced sensitivity to thermal-neutrons. Also, TLD's highly depleted in ^{6}Li, i.e. enriched in ^{7}Li, are available and are used to provide a neutron-free dose, so that when jointly used with a TLD containing natural lithium (or ^{6}Li enriched), the difference between the readings of the two would provide an estimate of the neutron dose. TLD's made of Al_2O_3, covered with polyethylene, has been also used for fast-neutron dosimetry, with the hydrogen in the polyethylene producing recoil-protons that in turn cause thermoluminescence.

Survey Meters. In addition to personal dosimeters, radiation laboratories should be equipped with hand-held survey instruments. These instruments are essential for immediate determination of the radiation dose, and should be used frequently to monitor radiation at different locations surrounding radiation sources. These are radiation detectors specially calibrated and equipped with audio and visual indicators and alarms. The detectors are mostly gas-filled detectors, see chapter 4, operating as ionization chambers, proportional counters or GM tubes, as appropriate. For example, a pocket chamber that provides a direct visual indication of gamma exposure is simply an air-filled ionization chamber, while a GM tube can be used to provide an audible warning of high photon and beta radiation fields. The instrument widely known as the "cutie-pie" is another air-filled ionization chamber used to measure the dose rate for photons, up to 2 MeV of energy, and is also sensitive to high-energy beta radiation. When such a device is equipped with a thin window, it allows the passage of both beta-particles and photons, while a thicker window can be used to filter out beta-radiation. Proportional counters are also widely used to measure gamma and alpha radiation. Although gas-filled detectors are the dominant type of survey instruments utilized in detecting photons and charged-particles, solid-state digital devices based on silicon-diodes, see section 4.2.4, are currently used as electronic personal dosimeters for detecting photons and high-energy beta-particles. Other forms of miniature digitized personal dosimeters are also becoming increasingly available.

The neutron dose-rate is measured with a thermal-neutron detector, one of those discussed in section 4.4, positioned at the middle of a polyethylene sphere (named the Bonner sphere after one of its first developers). The polyethylene sphere acts as a neutron moderator that slows-down fast-neutrons so that they can be detected by the enclosed thermal-neutron detector. Visible indications of the fast-neutrons (above 200 keV) can be measured by a bubble detector, see section 4.2.2.4. This is a glass-tube filled with superheated liquid droplets which vaporizes upon exposure to neutrons leaving a visible mark on an elastic polymer; the number of bubbles formed is proportional to the neutron dose-equivalent, H.

Leak Testing. When sealed sources are used, radiating contamination can only occur if the seal of a source is ruptured. The common practice is to perform a leak-test every six months on sealed sources to ensure their integrity. Alpha and beta-sources are particularly vulnerable to developing leaks in the coverings, which must be thin enough to allow penetration of radiation. Leak-testing is done by wiping the source, or the outside surface of its container, with a filter paper or other absorbent material, which is then counted for radioactivity. Accelerator-based sources, see chapter 2, require special safety precautions, due to the possible leak of radioactivity from the accelerator's chamber. Users of such sources should become familiar with radiation, high-voltage, and other safety precautions associated with theses devices, see section 17.2.4.

PART II: TECHNIQUES

Part I of this book presented three of the basic elements of NDE: source, modifying physics, and detection. Part II addresses the other fundamental element of NDE: the technique, which aims at providing indications that are amenable to a meaningful physical interpretation. The technique is the process of combining the source(s), modifying physics and detector(s) together to form a device, or an examination system, and displaying the measurements in a from understandable by the user. Since there is a wide variety of techniques and applications, they need to be classified in some structured fashion. Such a classification process naturally depends on the reader's point-of-view. One may want to classify techniques according to the utilized physical processes, the nature of the indication provided (probing, gauging, or imaging), the type of radiation sources used, the measured physical property, or even the industrial field of application. All these methods of classification are accommodated in this book in the following manner:

Indication Technique: each method is addressed in a separate chapter in this Part of the book.

Nature of Indication: probing, inspection, bulk gauging, detailed imaging, or elemental or composition analysis; each category is covered in one of the chapters of Part III.

Applications/Industrial Field: example applications are given in Part III, stating the industrial field in which they are applied, and a special application-area index is given in page **??**.

Source/Detected Radiation: are identified for each application discussed in Part III.

Measured Physical Parameter: the nature of the measured physical property is identified when discussing techniques in this Part of the book, and when presenting applications in Part III.

A technique is "a way of achieving a purpose". The purpose of a technique is to obtain an *indication* of some physical parameter. The most important aspect of NDE is arranging the source and detector, along with the inspected object, in a technique that can produce the desired indication. Therefore, each chapter in this Part of the book is devoted to a different technique. Since the source/detector type and arrangement determine the nature of the indication obtained, the discussed methods are classified in each chapter in accordance to the type of indication provided.

The process of indication is dictated by the way the source/detector are arranged, to monitor certain physical processes, as schematically shown in the accompanying Figure.

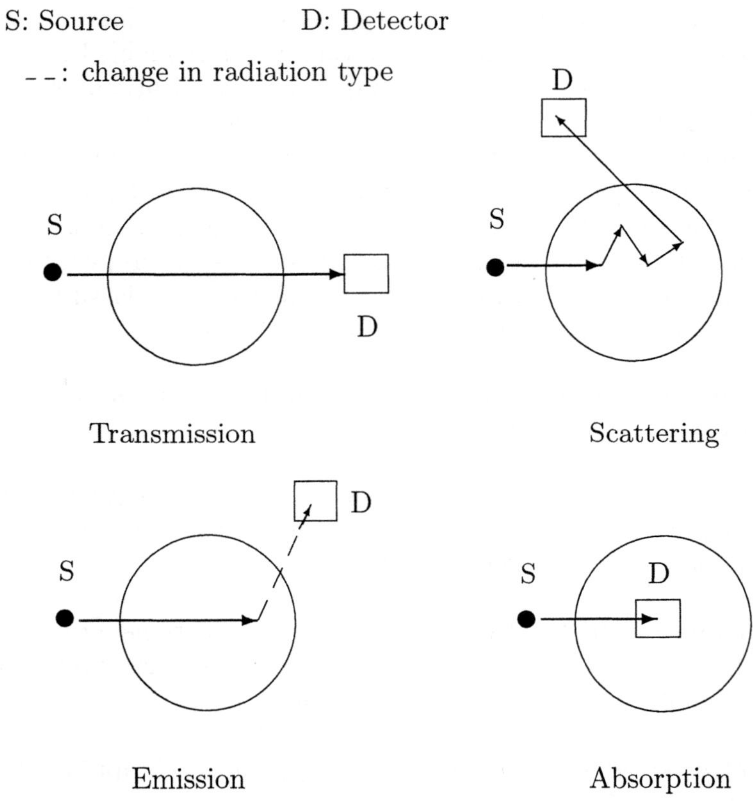

Schematic of techniques and indications.

Transmission: source radiation that succeeds in penetrating the object. Such radiation has to survive absorption or dispersion (scattering) by the object. In essence, the intensity of transmitted radiation provides an indication of the ability of the object's material to "remove" radiation from its original path.

Scattering: radiation that re-emerges from the object after being deflected by its material. This indication can be readily monitored, provided that provisions are made to avoid detection of radiation emerging directly from the source or scattered by the surroundings.

Secondary Emission: the release of radiation that is different in type from the incident radiation. The indications provided by such emissions are strongly dependent on the elemental composition of the material, making it useful for material characterization.

Absorption: radiation that does not emerge from the interrogated object. Since radiation disappears in this process, there is no direct way of providing an indication of radiation absorption. Therefore, the effect of radiation absorption is deduced from indirect indications, such as a reduction in the intensity of the incident radiation after it traverses the interrogated object, i.e. transmitted radiation, or as a depression in the radiation-flux within the object itself.

One should keep in mind that measuring the effect of one of those modalities does not necessary eliminate the effect of the others, though no detector is set to measure them. For instance, measured transmitted radiation can include a component of scattered radiation that succeeds in reaching the transmission detector, while secondary emissions can affect transmission and scattering detectors, if such detectors are sensitive to the secondary radiation. Such interferences can introduce some difficulties in interpreting the measured indications.

Interpretation of acquired measurements requires good understanding of the nature of the examined object, as well as the underlying physics of the applied NDE technique. A measurement model can facilitate the interpretation process, by mathematically relating the measurements to the nature of the object's material and geometry. Even, if such model is not used directly in the interpretation process, it will be helpful in providing a basic understanding of the problem at hand. Therefore, a measurement model is given for each of the techniques discussed in this Part of the book.

Chapter 6

TRANSMISSION METHODS

Owing to the special nature of charged-particles, their transmission process is discussed separately at the end of this chapter, in section 6.6.

6.1. Measurement Model

Radiation transmission is by far the most widely used technique in radiation applications, because of its simplicity and ease of interpretation. Common radiography and tomography techniques are based on the concept of radiation transmission, as shown in sections 6.3 and 6.4, respectively. The intensity of radiation transmitted through an object is indicative of the *total attenuation* (removal) capability of the encountered material. The italic emphasis has significant physical implications that must not be forgotten, or ignored. Other subtle aspects of the transmission process will be become apparent, and will be indicated in italics, as the widely-used exponential relationship of radiation attenuation by transmission is derived.

Let us consider the simple case shown schematically in Figure 6.1, where a *well-collimated narrow* beam of radiation is made incident on a slab of thickness t, and the intensity of the transmitted radiation is monitored by a suitable detector. For neutral particles, photons and neutrons, the measurement model of the transmission process can be described by the law of attenuation of radiation, Eq. (3.147). If a beam of intensity $I_0(E)$ at energy E is made incident on one side of the slab, it will be attenuated to intensity I as it reaches the other side of the slab, such that:

$$I(E) = I_0(E) \exp[-\Sigma_{total}(E)\ t] \tag{6.1}$$

where $\Sigma_{total}(E)$ is the total cross-section at energy E of the encountered material for the given radiation type. Note that while Eq. (3.147) uses radiation flux density, ϕ, the radiation *intensity*, I, is utilized in Eq. (6.1), where I is simply the flux intensity integrated over some area. That is, $I = \int \phi \, dA$, where A refers to area. Therefore, I_0 and I must be defined over the *same* size of area. In practice, this is overcome by measuring I_0 with the same detector, and from the same source, with air being the intervening material between the source and the detector. This implies that radiation *attenuation in air is neglected*, which is a reasonable assumption in many applications; nevertheless it is an assumption that one must be aware of. In addition, the assumptions associated with the simple law of attenuation, discussed in section 3.6.3, also apply to relationship (6.1). In addition to the requirement of a narrow well-collimated beam, the assumptions addressed in section 3.6.3 necessitate the system be at *steady state* (i.e. a stable source and non-varying interrogated medium), and that *no radiation enters the beam from outside its domain, nor be generated within.* Some of these assumptions can be satisfied in many applications. However, the derivation of the attenuation-law from the particle-transport equation requires the absence of an external source, the Q term in Eq. (3.118). Obviously, as Figure 6.1 shows, an external source is involved. However this restriction is overcome by keeping in mind that the attenuation-law does not apply to radiation transport within the volume occupied by the source. Recall also that the attenuation-law does not apply to charged-particles (though empirically used with beta-particles, as they interact continuously with matter, see section 3.6.5).

Limitations. Even with the many assumptions inherent in arriving at the measurement-model of Eq. (6.1), this simple exponential equation has other inherent limitations of practical importance. First, it assumes that the medium, within which the distance t exists, is occupied by the same material. This assumption can be relaxed by expressing relationship (6.1) as:

$$I(E) = I_0(E) \exp\left[-\int_0^t \Sigma_{total}(x, E) \, dx\right] \tag{6.2}$$

where $\Sigma_{total}(x, E)$ is the total cross-section at distance x within the object, for source energy E. Direct indication of the material-dependent quantity, $\int_0^t \Sigma_{total}(x, E) \, dx$, is indicated by the ratio, $T = -\ln I(E)/I_0(E)$, referred to as transmittance. The practical implications of Eq. (6.2) is that, with all the above mentioned assumptions being satisfied, one would be measuring an *integrated* quantity, representing

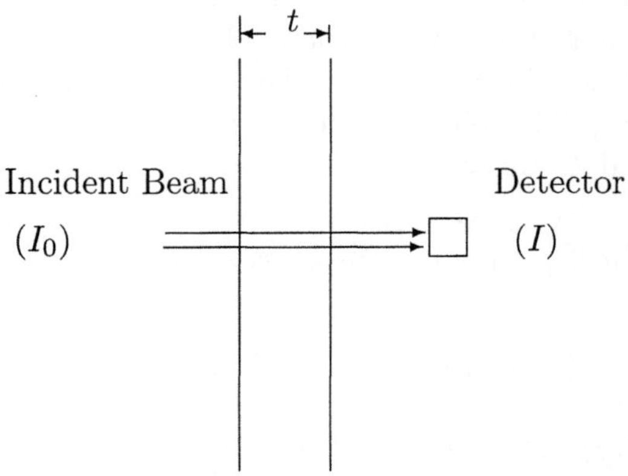

Figure 6.1. Idealized radiation transmission system.

the radiation-attenuation over the entire beam path. In other words, the material information along the beam-path is *compressed* (projected) into a single indication point, provided by the transmission measurement. When a narrow-beam is used, the radiation-path along the beam traverses a material chord, and therefore, the indication obtained is often described as a *chordal* indication. The physical implication of the nature of this indication is that one cannot obtain localized information along the beam path. Moreover, if the material thickness is not known, or is a varying quantity, the indication will not only reflect the radiation-attenuation property of the material, given by Σ_{total}, but it also will be geometry-dependent. Consequently, for a homogeneous material, the transmission-indication can be used for thickness measurement.

Effective Energy. For a multienergetic beam of radiation of a distribution described by the function $\chi(E)$, with $\int_0^{E_m} \chi(E)\, dE = 1$, where E_m is the maximum radiation energy, and if radiation-transmission is measured over all energies, relationship (6.2) takes a more complex from as:

$$I = I_0 \int_0^{E_m} \chi(E) \exp\left[-\int_0^t \Sigma_{total}(x, E)\, dx\right] dE \qquad (6.3)$$

where I_0 is the total beam intensity, subsequently $I_0\chi(E)$ is the fraction of the beam undergoing attenuation with a total cross-section $\Sigma_{total}(x, E)$. Therefore, when transmission is recorded for a multienergetic beam, no direct indication of $\int_0^t \Sigma_{total}(x, E)\, dx$ at any energy is easily obtainable from Eq. (6.3). The transmittance, $T = -\ln \frac{I}{I_0}$, would then be an indication that incorporates two integrals, over E and x. Therefore, the indication becomes dependent on many parameters, and is more difficult to quantitatively interpret. Therefore, dependence on energy is often excluded by employing the so-called *effective energy*, E_{eff} (see appendix F). This effective-energy can be chosen so that the integration over x and E are separated.:

$$\left(\int_0^{E_m} \chi(E)dE\right) \exp\left[-\int_0^t \Sigma_{total}(x, E_{eff})\, dx\right] =$$

$$\int_0^{E_m} \chi(E) \exp\left[-\int_0^t \Sigma_{total}(x, E)\, dx\right]\, dE$$

Then, the value of E_{eff} can determined by finding the energy that produces a total cross-section equal in value to $\Sigma_{total}(x, E_{eff})$ evaluated using Eq. (6.3). Since $\int_0^{E_m} \chi(E)dE = 1$, then the use of E_{eff} makes Eq. (6.3) equivalent to Eq. (6.2), with E_{eff} replacing E. The difficulty, of course, is in obtaining an acceptable value for E_{eff}, given the fact that $\Sigma(x, E)$ is material dependent and thus the right-hand-side of Eq. (6.3) can only be evaluated if the material distribution with the object is known . This makes it impossible to obtain a unique value for E_{eff} that satisfies all possible conditions. Since in NDE and imaging, the nature and distribution of the inspected material may not be known in advance, the value of E_{eff} cannot be pre-determined. This problem is overcome in practice by evaluating E_{eff} for a single uniform material (called the base material), thus eliminating the dependence of $\Sigma(x, E)$ on x, and using the same effective-energy for materials of similar radiation attenuation properties. For example, water can be used as a representative of hydrogen-rich and hydrocarbon materials for the attenuation of neutrons. One then has to be aware that an approximate value of E_{eff} is used and consequently the indication obtained from transmission may not exactly reflect the attenuation characteristics of materials that are significantly different from the base material. Appendix F provides other methods for dealing with multienergetic beams.

Equivalent Attenuation-Coefficient. Another approach to overcoming the energy-dependence of $\Sigma(x, E)$ in Eq. (6.3) is to measure in advance an *equivalent attenuation-coefficient* for the materials one expects to encounter, using the relationship derived from Eq. (6.3) for a single material:

$$\Sigma_{equiv} t = -\ln \frac{I}{I_0} = -\ln \int_0^{E_m} \chi(E) \exp[-\Sigma_{total}(E)]\, dE \tag{6.4}$$

The obtained values for different materials can then be used if quantitative information is to be obtained from transmission indications. This approach would be useful in many industrial applications in which a uniform material, with some minor intrusions, is inspected.

Beam Hardening. The problem of energy-dependence of radiation attenuation is further complicated if radiation absorption by the material is such that it removes lower-energy radiation from the incident beam. This in effect reduces the value of $\chi(E)$ to zero at low energies, as the beam progresses through the material. Then the lower-limit of the integration over x in Eq. (6.3) becomes itself a function of energy. As lower-energy radiation is removed from the beam, it becomes *harder*, i.e. its effective energy increases. This phenomenon is known as *beam hardening* and has the effect that the transmittance, $T = -\ln(\frac{I}{I_0})$, is no longer a linear-function of the thickness; as the total cross-section in Eq. (6.1) becomes thickness-dependent. Therefore, beam-hardening can cause error in calculating proper values for Σ_{total} from transmission indications.

Buildup. Another factor that further complicates the exponential-attenuation process is the *buildup* effect. As its name implies, buildup is an increase in the value of the measured transmission intensity, caused by radiation scattering. That is, some of the radiation removed from the beam by scattering can reach the transmission detector. This apparent increase in the radiation transmission is not in accordance with the law of attenuation of radiation, which only considers radiation removal from the beam, see section 3.6.3. Therefore, the exponential relationships of Eqs. (6.1), (6.2) and (6.3) should be corrected by a buildup factor B. For instance, Eq. (6.3) should be rewritten as:

$$I = I_0 \int_0^{E_m} B(\Sigma_s(E), t)\chi(E) \exp\left[-\int_0^t \Sigma_{total}(x, E)\, dx\right] dE \qquad (6.5)$$

where the buildup factor is shown as a function of the material's thickness and its macroscopic cross-section, Σ_s, with the latter being a function of the radiation type and energy, and is dependent on the material density, see section 3.2.3. The effect of buildup can be reduced by proper choice of the source type and energy for a given material and thickness so that the value of B approaches unity. Buildup can also be minimized by proper collimation of the detector so that its field-of-view is limited as much as possible to minimize detection of scattered radiation. Reference [133] provides a review of methods for calculating the buildup factors for gamma-rays.

Wide Beam. One of the basic limitations of the attenuation-law, upon which Eqs. (6.1), (6.2) and (6.3), are based is the assumption of a narrow incident beam. This assumption, as explained in section 3.6.3, is dictated by the fact that the attenuation-law represents a one-dimensional form of the particle-transport equation, under certain conditions. However, one of the most common applications of radiation transmission is radiography, which usually involves exposing the interrogated object to a wide beam of radiation, see section 6.3. The attenuation law is implicitly used in interpreting the indications obtained in radiography, usually by the darkness of a film that detects the transmitted radiation. Although this may appear to be in violation of the attenuation-law of radiation, if certain precautions are taken the use of the attenuation law can be justified. If one views such a wide-beam as consisting of a set of narrow adjacent elementary beams, each one of them projecting a point on the radiographic film, then the attenuation-law can be applied to each of these elementary-beams, provided there is no cross-talk between the beams, i.e. the effect of one elementary beam is independent of others. This independence can be achieved, for example, by placing a set of pin-hole collimators, so that each hole is only exposed to one of each of these elementary beams.

Divergence. Another basic assumption imposed by the attenuation-law is that it is only applicable to an ideal (non-diverging) pencil-beam. This assumption is included since the presence of radiation source was not considered in the derivation in section 3.6.3 of the attenuation law. The exclusion of the source is only possible if the source is very far-away from the target material. In practice, the source is usually brought close to the object to make effective use of the emitted radiation by maximizing the density of the radiation flux incident on the target. Therefore, source divergence must be taken into account, in accordance with the law of divergence discussed in section 3.6.2. Divergence also means that the incident beam is never a narrow (pencil-beam), but rather a cone (wide) beam. Beam divergence implies that the attenuation-law in its simple exponential form is never applicable in reality. However, if one considers the incident fan-beam as a set of elementary independent beams, as discussed above for a wide-beam, then the attenuation-law can be applied to each elementary beam, though each elementary beam will have its own direction, i.e. x in Eqs. (6.1), (6.2) and (6.3) will take different directions in space. The divergence of the beam can then be taken into account by applying the $\frac{1}{R^2}$ law of divergence to each elementary beam, where R is the distance from the source to the detector. Alternatively, by measuring I_0 in air (with no object present), the divergence terms in I and I_0 become identical, and are eliminated when evaluating the transmittance, $T = -\ln \frac{I}{I_0}$.

The above discussion elucidates how the simple attenuation law commonly used to express radiation transmission is laden with many simplifying assumptions. Therefore, in interpreting information obtained from transmission measurements, one should be aware of the implications of such assumptions. Given that, let us now examine more carefully the nature of the information provided by transmission.

Nature of Indication. Aside from the thickness of the object along which transmission measurements are performed, the basic quantity that affects radiation transmission, as the attenuation-low indicates, is the total cross-section, Σ_{total}. As discussed in section 3.2.3, the macroscopic cross-section can be expressed as a function of the microscopic cross-sections, σ_{total}, as:

$$\Sigma_{total}(E) = N\ \sigma_{total}(E) = \frac{\rho}{A\mathrm{u}}\ \sigma_{total}(E) \tag{6.6}$$

where N is the number of atoms, or molecules, of the material, per unit volume, ρ is its mass density, A is the atomic or molecular number of the material, and u is the atomic mass unit. Eq. (6.6) indicates that the value of Σ_{total} is not only dependent on the radiation energy, E, but also on the nature of radiation, the material type and density. Therefore, if E is fixed, and t is constant, transmission provides an indication of the material's density, ρ, and its atomic/nuclear properties, σ_{total} and A. In addition, if density is constant, transmission will provide directly a measure of the nature of the material via σ_{total} and A. As discussed in details in chapter 3, the information provided by $\sigma_{total}(E)$ depends on the nature of the radiation used and on its energy. Therefore, one should be quite familiar with the nature of the radiation source and the inspected material, in order to extract atomic/nuclear information from transmission measurements. The reader should consult chapter 3 for the underlying physics. These physical characteristics will also be emphasized when discussing specific techniques in the ensuing sections. Different arrangements of radiation transmission are possible in the view of the nature of the source used, whether it is a single narrow (pencil) beam, a wide beam, or a set of separate beams (narrow or wide). These arrangements result in pencil-beam probing, radiography or tomography, as discussed in the following sections.

6.2. Pencil-Beam Probing

Pencil-beam transmission enables the inspection of a single line, or a chord of the object, at a time. The indication is then a one-measurement (or a point indication) that reflects both the length of the chord and the radiation-attenuation ability of the material within the chord. Pencil-beam transmission is the simplest form of radiation transmission and the easiest one to interpret, as its indications directly abide by the inherent assumptions behind the simple exponential-law of the measurement model of Eqs. (6.1), (6.2) and (6.3). This attenuation law is, therefore, used here to explain the nature of the indications provided by different types of radiation. However, the practicality of obtaining a pencil-beam is first examined.

Source Collimation. A pencil-beam as the name implies is a narrow mono-directional beam of radiation. Such a beam is described as a "pencil" beam in analogy with the well-collimated light beam easily obtainable in optics with the aid of a lens. In most of the energy range of interest in NDE applications, radiation behaves like a particle, not like a wave, see chapter 3. Therefore, there is no such a thing as a lens to collimate radiation emerging from x-ray, gamma-ray and neutron sources. The only possible method for collimating such radiation sources is to use particle-elimination by absorption, so that only radiation passing through the collimator's opening is allowed to pass through; as schematically shown in Figure 6.2. The opening of the collimator can be shaped into a hollow-long cylinder to provide a pencil beam. However, due to radiation divergence, the physical size of the source, radiation scattering from the collimator walls, etc., the process of radiation collimation is not that simple, and becomes a design problem with some trade-offs, as discussed in section 15.1.

A pencil-beam can also be used in radiography, see section 6.3, when it is necessary to image only a small portion of an object or when a series of radiographs of certain locations are needed. The pencil-beam can be defined either at the source side, using a source collimator with a narrow slit, or at the film side by placing a collimator slit in front of the film. The latter arrangement is particularly useful when imaging a radioactive material with an external source, to reduce the "fogging" on the film introduced by the radiation emitted from the radioactivity of the material [134]. In either case, the collimator and the source can be moved together to probe various regions in the object.

Detector Collimation. It is also important in transmission measurements to collimate the detector, or in other words to confine its field-of-view to that of a "pencil"beam. In effect, a detector's collimator is what defines the inspected volume. Naturally, the portion of an object that is not within the field-of-view of a detector does not contribute

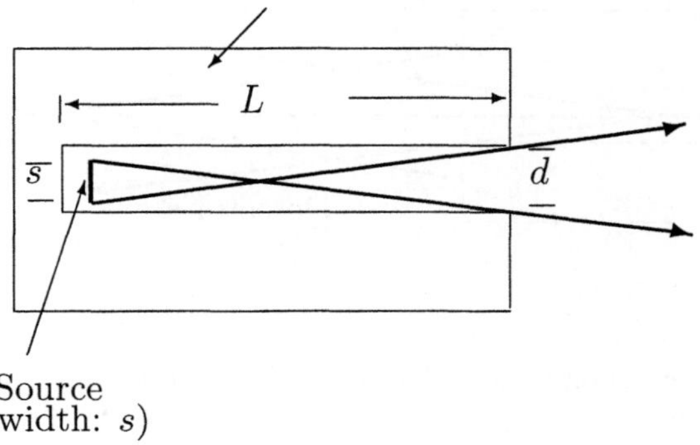

Figure 6.2. Typical collimator geometry.

to the transmission measurement. Also, if a detector's field-of-view is not covered by the incident radiation, no transmission indication will be recorded. It is, therefore, important to align the source and detector collimators so that they both coincide with each other. This can be done in practice by insuring that the detector records the maximum radiation intensity in air, since if the detector's collimator is not properly aligned with that of the source it will miss a portion of the incident radiation and produce a lower count. Like source collimation, detector collimation is also a relatively complex process that involves many design parameters, see section 14.6.3. Both the source and detector fields-of-view diverge with distance, as schematically shown in Figure 6.3 for a source and detector each collimated with a long hollow cylinder to define as much as possible pencil-like fields-of-view. Therefore, even in this idelaistic transmission arrangement, the pencil-beam arrangement of Figure 6.1 is not strictly achievable. Note also that only a fraction of the source radiation is recorded by the detector as transmitted radiation. Therefore, the intensity I_0 in any of the measurement models of Eqs. (6.1), (6.2) and (6.3), does not correspond to the source strength, even after accounting for the effect of the source collimator. Therefore, the detector's response recorded in the absence of the object should be used for the value of I_0, assuming that attenuation in air is negligible. In the arrangement of Figure 6.3, the divergence of the field-of-view of the detector increases the probability of recording scattered radiation. Scattering can be produced within the volume defined by the union (intersection) of the fields-of-view of the source and detector collimators. However, a smaller amount

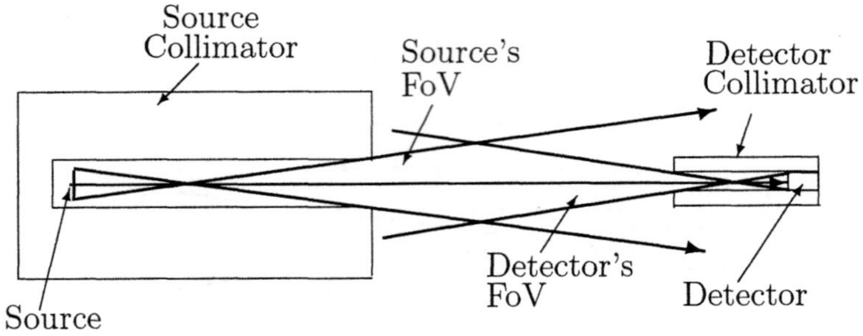

Figure 6.3. Intersection of field-of-views (FoV's) source and detector in a transmission measurement.

of radiation can also reach the detector by scattering outside this union. Note that buildup in the detector response can also arise from scattering outside the inspected material.

Detection Energy. One other added complication to transmission measurement is the effect of a detector's intrinsic characteristics on the measurement, i.e. its efficiency and its variation with energy. For a monoenergetic source, the detector can be made to measure only radiation at the same energy as that of the source. Then the detector measures in effect the *uncollided* radiation, i.e. the radiation that did not collide or interact within the material subjected to radiation. The measurement of I_0 in void with the same detector will then automatically incorporate the detection efficiency. Indeed, one added advantage of using a monoenergetic source is that any scattered radiating, causing buildup, would occur at an energy lower than the source's energy, since scattering lowers the radiation energy. Therefore, energy discrimination in the detector can be used to eliminate the effect of buildup by discarding radiation that has an energy below that of the source energy. For multienergetic sources, such elimination of buildup by energy discrimination is not that straightforward, since higher energy radiation can scatter to an energy in the domain of a lower-energy component of the source. Moreover, the energy-spectrum of the transmitted flux can also change with the thickness and nature of the material, as the total cross-section Σ_{total} in Eqs. (6.1), (6.2) and (6.3) can vary with energy. Since the efficiency of most detectors varies with energy, another complicating effect is introduced into the problem. However, these effects can be included in the process of accommodating changes in the energy spectrum of the source, as discussed in section 6.1, by incorporating the detection-efficiency in calculating the effective energy, and in measuring the equivalent total cross-sections for various materials using the same detector. Since the thickness of the object alters the energy-spectrum of the detected radiation, the values of the obtained effective-energy, or equivalent cross-section, should be used within a certain thickness range for a given set of materials, to minimize the errors introduced in lumping those effects into single effective, or equivalent, numerical values.

6.3. Radiography

Radiography, as the name indicates, is the process of obtaining a "picture" of object using radiation. The indication picture, like a photographic picture, is a two-dimensional representation of the object. Like photography, radiography requires a wide-source to cover the part of the object being interrogated. Therefore, a wide-beam radiation source is required. Radiation sources lend themselves naturally to creating wide beams of radiation, as the emitted radiation spreads in all directions. However, as indicated in section 6.1, the use of a wide-beam does not enable direct application of the exponential-relationship of radiation attenuation, Eqs. (6.1), (6.2) or (6.3), unless effort is made to create the impression that the wide-beam is constituted of a set of adjacent narrow beams, each of which independently contributes to the detector. This is achieved as indicated in section 6.1 by reducing the buildup effect in the detector via the use of detector collimators. However, in order to obtain a picture in radiography, a wide detector (typically an emulsion film) is used. It becomes then difficult to equip such a large receptor of radiation with a proper collimator. Grids of radiation-absorbing materials can be placed on top of the film to absorb scattered photons, so that the holes in the grid mainly allow the passage of radiation directly emitted from the source. While metallic grids are used in medical radiography, the higher density material usually encountered in industrial applications, makes the use of such grids not very effective. Therefore, one must resort to other means to reduce the occurrence of scattering within the inspected object. One of the main design choices in industrial radiography is the source energy, which needs to be sufficiently low so that radiation is removed from the object to provide an indication, but high enough so that the amount of radiation scattering (buildup) at the radiographic film is minimized. Moving the film away from the proximity of the inspected object can also help reduce the buildup effect, as scattered radiation would be given a chance to diverge away from the film. This approach tends to reduce the sharpness of the image due to the divergence of the source radiation, unless the source is very small in size as discussed below. Aside from the scattering effects, the quality of a radiographic image is affected by the film characteristics and the extent of its exposure to radiation, as well as the layout of the source, object and film with respect to each other. These factors are briefly explained in section 6.3.1, for film radiography. Some variations to film radiography are subsequently discussed. X-ray machines, as well as radioisotopic sources (mainly ^{60}Co, ^{137}Cs, ^{192}Ir, ^{169}Yb, ^{170}Tm and^{75}Se, see Table 2.10) are used in industrial radiography. References [4], [7], and [135] provide more details on radiographic techniques, while section 13.1 presents some applications of radiography along with specialized forms of radiography.

6.3.1. Film Radiography

6.3.1.1 Film Characteristics

Photographic emulsion, discussed in section 4.2.1, is the basic component of film radiography. Like in photography, a film needs to be exposed to a sufficiently intense radiation and for a long enough time to produce a good quality image. Therefore, *film exposure* is quantified by the quantity, $\mathcal{E}$, defined by:

$$\mathcal{E} = IT \tag{6.7}$$

where I is the intensity of the radiation incident on the film, and T is the length of the time during which the film is exposed to radiation. The log of *relative exposure*, $\mathcal{E}_R$, is usually used instead of the absolute value of $\mathcal{E}$, where:

$$\log(\mathcal{E}_R) = \log \frac{\mathcal{E}}{\mathcal{E}_0} \tag{6.8}$$

with $\mathcal{E}_0$ being some reference exposure, conveniently taken as unity, and log is the logarithm to base 10. The use of the log-to-base-ten is for historical reasons.

After developing the film, light is used to view the image. The *light transmission density*, D, also called *optical density*, or *darkness*, is expressed as:

$$D = \log \frac{J_o}{J} \tag{6.9}$$

where J_0 is the intensity of light incident on the film and J is intensity of the light transmitted through the film. Note that $D = 0$ means 100% transmission of light through the film. Each commercial film has its own characteristic logarithmic exposure curve (D v.s. $\log \mathcal{E}$) that defines its response rate, see for example reference [4]. The *film contrast* at a density D is defined by the slope of the film characteristic curve:

$$G_D = \frac{dD}{d\log\mathcal{E}} = 2.3\frac{dD}{d\ln\mathcal{E}} = 2.3\mathcal{E}\frac{dD}{d\mathcal{E}} = 2.3I\frac{dD}{dI} \quad (6.10)$$

where the factor 2.3 appears due to the conversion to natural log, and use is made of the definition of exposure, Eq. (6.7). The higher the value of G_D, the better the film contrast. Faster films have a larger value of D for a given $\mathcal{E}$, as the grains in film are chemically affected sooner, but higher-speed films tend also to have larger grains and may not produce the required detail.

The edge of an object can appear blurred in an image, due to the lack of film definition caused by poor response characteristics. This image *unsharpness* is quantified by the blurring distance observed on a developed film that corresponds to a change in darkness, D, in response to a step change (a sharp edge) in object geometry. This blurring distance, designated as U_f, is then called *film unsharpness*.

6.3.1.2 Geometric Effects

The finite size of the source and radiation divergence introduce the *penumbra* effect, as schematically shown on Figure 6.4. This unsharpness effect describes the system's inability to reproduce faithfully the boundary of a given object. In x-ray sources, the source size is known as the focal spot size, F, and is equal to the effective width of the source (or width of the target area subjected to electron bombardment). For a radioisotopic source, F is equal to the physical width of the source. The geometric unsharpness, U_g, is then defined by:

$$U_g = \frac{FL}{L_0} \quad (6.11)$$

where L is the film-to-object distance, and L_0 is the source-to-object distance, as shown in Figure 6.4.

Total image unsharpness, U_t, is then a combination of the film unsharpness, U_f, and the geometric unsharpness, U_g, and is given by:

$$U_t = \sqrt{U_f^2 + U_g^2} \quad (6.12)$$

Note that U_f and U_g have units of distance and are combined using the Euclidean distance (minimum distance between two points in space) to provide the most probable value, in accordance to the procedure for combining errors, Eq. (G.4).

While a high-quality film can help improve the sharpness of an image, geometric effects should also be optimized to improve image sharpness.

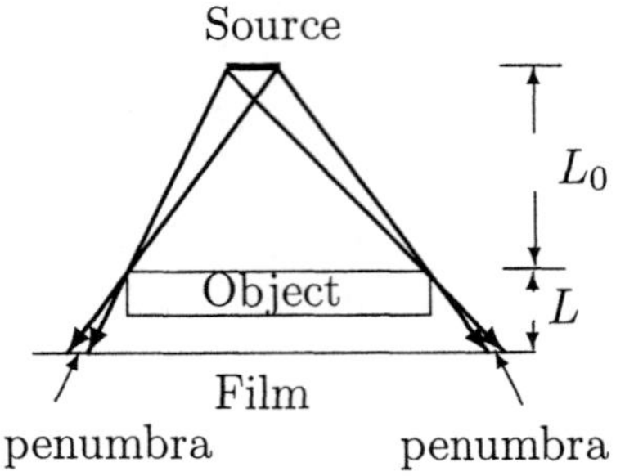

Figure 6.4. The penumbra effect in radiography.

This can be done for example, as indicated by Eq. (6.11), by increasing the value of L_0, i.e. having the source as far as possible from the object, which in effect makes the source appear to the object as a point source. Increasing L_0 comes, however, at the expense of reduced film exposure, $\mathcal{E}$, as I in Eq. (6.7) is decreased by beam divergence (see section 3.6.2). Eq. (6.11) also indicates that the image sharpness can be improved by reducing the value of L, i.e. bringing the film as close as possible to the object. Although this is easy to do in practice, and is the norm in most applications, it can be problematic if radiation scattering (buildup) is considerable. As mentioned earlier, the buildup effect can be reduced by moving the film away from the object. The contradicting requirements of increased sharpness and decreased buildup can be accommodated by the use of a source of a small physical size, such as a microfocus x-ray tubes. This enables bringing the source closer to the object to increase intensity of exposure, while moving the film away from the object (to reduce radiation buildup), without severely reducing the sharpness of the image. Bringing the source closer to the object has also the added advantage of enlarging the obtained image, which can be useful in the inspection of flaws in small objects.

One can view radiography as a geometric projection of an object on the film, with the projection process performed along the direction of the incident radiation, as Figure 6.4 schematically shows. Therefore, failure to align the edge of the object with the direction of the beam, or any motion of the object, the beam or the film, can also cause image unsharpness, not accounted for in Eq. (6.12).

6.3.1.3 Axial Effects

The quality of an image, i.e. the ability to produce a detailed image of an object, is affected by many factors. Therefore, a number of parameters are used to quantify the quality of an image. These parameters can be classified in radiography as those dealing with *axial* changes, along the direction of the beam, and *lateral* changes, in a direction normal to the beam. The lateral-changes are caused by the unsharpness effects discussed above. The axial-changes are affected by the change in the intensity of the transmitted beam, and can be quantified by considering the measurement model of the transmission process. Let us consider the measurement model of Eq. (6.1), dropping energy dependency for simplification. Then a change ΔI in the beam intensity can be obtained by differentiating Eq. (6.1) to obtain:

$$\frac{|\Delta I|}{I} = \Sigma_{total}\ \Delta t + t\ \Delta\Sigma_{total} \tag{6.13}$$

Therefore, for a certain material (constant Σ_{total}), $\frac{\Delta I}{I} = \Sigma_{total}\ \Delta t$ reflects the sensitivity of the transmission signal to changes in the thickness. Obviously good sensitivity, i.e. small value of Δt, requires a material that has a large total cross-section for the applied radiation, or the use of radiation type and energy that provides a large value of Σ_{total} for the material. Figure 6.5(a) shows schematically this effect, where a large value of Σ_{total} gives a steeper response curve, and consequently more sensitivity to changes in t, for thin objects. For thick objects, the sensitivity (slope or curve) decreases, until it becomes almost zero as the thickness increases considerably. For a given thickness of an object, Eq. (6.13) shows that $\frac{\Delta I}{I} = \Delta\Sigma_{total}\ t$, which reflects the sensitivity of I to material changes. It is obvious that changes in Σ_{total} are easier to see in thick objects, as schematically shown in Figure 6.5(b), for small cross-sections. However, if Σ_{total} is so large that most of the beam is attenuated within the material, it will be difficult to resolve the measurements, and the sensitivity will be lost. The loss in sensitivity at large thickness and total cross-sections is further illustrated by examining the change of I with respect to $t\Sigma_{total}$, that is:

$$\left|\frac{\partial I}{\partial(t\Sigma_{total})}\right| = I \tag{6.14}$$

which indicates that the sensitivity of I to a change in $t\Sigma_{total}$ is proportional to the value of I. If I is small due to a high value of $t\Sigma_{total}$, as Eq. (6.1) indicates, then $\frac{\Delta I}{I}$ would be small and it would be difficult to observe changes within the object. Plotting I versus $t\Sigma_{total}$ also shows this effect, as this exponential function Eq. (6.1), would tend to become flat (almost constant) at large values of $\Sigma_{total}t$, making the value of I insensitive to changes in either Σ_{total} or t. Figure 6.5(c) shows schematically this effect.

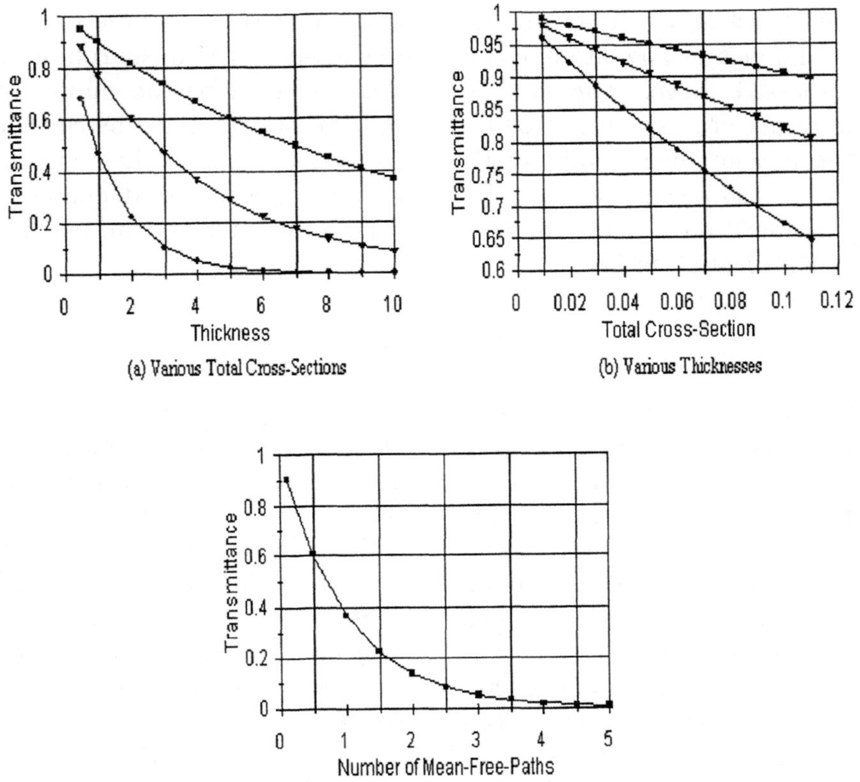

(a) Various Total Cross-Sections

(b) Various Thicknesses

(c) Number of Mean-Free-Paths

Figure 6.5. Change in transmittance with: (a) thickness, t, for same material, top curve has lowest Σ_{total}; (b) Σ_{total} for a fixed t, top curve has smallest thickness; and (c) $\Sigma_{total}t$.

6.3.1.4 Image Quality

The quality of radiographic imaging can be quantified using a number of parameters as discussed below.

Sharpness. This is the ability to clearly see abrupt changes in an object. Unsharpness, as defined in section 6.3.1.2, is the parameter used to quantify the lack of sharpness. Both the quality of the film and the geometric setup of the system affect the sharpness of the image, as discussed in section 6.3.1.2. This is obviously a parameter that indicates the lateral quality of an image.

Spatial Resolution. Another lateral image-quality parameter that defines the ability to distinguish in an image the presence of two separate objects placed close to each other, in a direction normal to the direction of the incident radiation, is spatial resolution. Image unsharpness directly affects the spatial resolution of the image, and therefore, U_t of Eq. (6.12) can be used as a measure of the spatial-resolution obtainable with a radiographic system. Spatial resolution is often quoted in line-pairs per mm (ℓp/mm). The terminology of line-pairs comes from the definition of resolution in television monitors. This means simply that in a system having a spatial resolution of 10 ℓp/mm, opaque objects 0.1 mm apart will be just resolved, i.e. become visually observable. Typical resolutions are from 30 ℓp/mm for high detail x-ray films to one ℓp/mm in television-display systems.

Material Contrast. The ability to distinguish in the image the presence of two different materials is called contrast. This is an axial quality parameter that depends directly on the ability of the employed radiation to cause a detectable change in the intensity of the transmitted signal. That is, if the radiation intensity in some reference material a is I_a and in another material b is I_b, the contrast, $C_{a,b}$ between materials a and b is defined as:

$$C_{a,b} = \left| \frac{I_b - I_a}{I_a} \right| \tag{6.15}$$

Either Eq. (6.1), Eq. (6.2), or Eq. (6.3) can be applied as appropriate, to materials a and b, to determine the value of the contrast. For simplification, let us apply Eq. (6.1), then:

$$C_{a,b} = \left|1 - \exp\left(\Sigma^a_{total}(E)t_a - \Sigma^b_{total}(E)t_b\right)\right| \tag{6.16}$$

where the superscripts a and b are added to identify the two materials. From Eq. (6.16), it is obvious that, for the same thickness of material, $t_a = t_b$, the contrast depends on the difference between the total cross-sections of the two materials. This difference can be enhanced by proper choice of the radiation type and radiation energy, since Σ_{total} depends on both. It is also interesting to note that in Eq. (6.16), the intensity of the incident radiation, I_0 in Eq.(6.1), does not appear and thus has no effect on the contrast, in other words the contrast cannot be simply improved by increasing the intensity of the incident radiation. The intensity alters the blackness of the film, hence affects the film contrast.

Film Contrast. Defined by Eq.(6.10), film contrast is the ability to distinguish two close values of film exposure by visual observation of changes in the film darkness. This is important for preserving the material contrast of the system, otherwise changes in the radiation intensity due to changes in material may not be noticeable on the film. It also affects the lateral quality of the image, as the size of the grains of the film's emulsion material, after development, can result in an uneven distribution of grains, producing the so-called "graininess" in the image.

Thickness Sensitivity. Another axial image-quality parameter is sensitivity, S, which is measured by the detectable change in material thickness, along the radiation beam path, Δt in a thickness t, i.e. $S = \frac{\Delta t}{t}$. Sensitivity is a measure of the ability to detect small changes in thickness in the same material. Note that a large value of S indicates poor sensitivity, and vice versa. The smallest detectable thickness change in thickness is that produces an observable change in darkness ΔD. This Δt can be related to ΔD, at a constant value of Σ_{total}, as follows:

$$|\Delta t| = \frac{\Delta I}{I_c \Sigma_{total}} = \frac{2.3(\Delta D) I_m}{I_c \Sigma_{total} G_D} = \frac{2.3(\Delta D) B}{\Sigma_{total} G_D} \tag{6.17}$$

where use is made of Eq. (6.14) with the subscript c associated with I to indicate that it is a calculated value, while when using Eqs. (6.10) the subscript m is used to indicate a measured intensity (related to exposure and film darkness). The factor B in Eq. (6.17) is the buildup factor, accounting for radiation scattering. Therefore, the sensitivity, S, can be expressed using Eq. (6.17) as:

$$S = \left| \frac{\Delta t}{t} \right|_{\Sigma_{total}} = 2.3 \frac{(|\Delta D|) B}{(t \Sigma_{total}) G_D} \tag{6.18}$$

which is the same expression reported in the literature [4, 7], and the absolute value is used to accommodate both decreases and increases in thickness. Eq. (6.18) shows that poor sensitivity (large value of S) results if a large change in darkness is needed to see the change in thickness, if the amount of radiation scattering (buildup) is high, if the film is of poor quality (small value of G_D), or if the material has a low total cross-section. As discussed in Chapter 3, $t\Sigma_{total}$ is equal to the number of mean-free-paths. Therefore, Eq. (6.18) indicates that better sensitivity (small value of S) would be attained if a large number of interactions occur. This implies that for a given material, increasing the thickness would enhance the sensitivity of a radiographic system. However, this effect is countered by a corresponding decrease in radiation intensity recorded by the film, necessitating a large value of ΔD to observe the change in thickness, with a subsequent increase in the value of S. Note that the expression for S of Eq. (6.18) could have been used to describe material sensitivity, $\frac{\Delta\mu}{\mu}$, that is, the factors that affect material contrast are the same factors that influence thickness sensitivity.

It is interesting to note that, except for S, all quality indicators discussed above refer to lateral changes in the image (i.e. perpendicular to the direction of the incident radiation). This is due to the fact that the formed image is an imprint of the projection of the incident radiation on the film, which tends to be in a direction perpendicular to the incident radiation. Note that the above parameters may be defined differently by other authors, so readers should be careful in interpreting quoted values for image resolution contrast and sensitivity.

Pentrameters. Although the above parameters are useful in providing quantitative indicators of the quality of an image, it is not always practical to calculate or measure them. Therefore, another practical image-quality indicator is employed. This involves the use of so-called *sensitivity indicators* or *pentrameters*. These are small thin pieces made of the same material as the inspected object, and placed on the side of the object close to the source. If the thickness of the pentrameter is Δt and the thickness of the inspected object is t, then when the presence of the pentrameter is clearly observed on the film (by decreased darkness) one can state, according to Eq. (6.18), that the radiographic system has a thickness-sensitivity of $\frac{\Delta t}{t}$. The smallest observable thickness of the pentrameter determines the best sensitivity of the radiographic setup. If a desired level of sensitivity is not attainable, one should alter the layout of the setup, guided by Eq. (6.18), until the required level of sensitivity is achieved. For such process, a step-type pentrameter with increasing thickness can be useful. The lateral dimension of the pentrameter can also be used to provide indications of the spatial resolution of the system. For instance, pentrameters in the form of wires of varying diameters can be used to determine the smallest detectable width. In plate-type pentrameters, drilled circular holes of varying diameters, are used to determine the smallest detectable flaw size. A number of standard pentrameters are available for photon radiography, see for example reference [7]. The same concept can be used to determine whether a radiographic system has the desired material contrast, by using pentrameters made of two different materials.

The non-uniform thickness of most objects makes it difficult to obtain the same quality of image at various thickness. This problem can be overcome in a variety of ways. Multiple exposures at different energies, each optimized to a different energy range, may be used. Also, a cut-off filtering, see section 6.46, can be applied to increase the effective energy of an incident radiation, so that radiation can penetrate the thicker sections. Another approach is to apply *compensators*, for the "missing" material in thin sections. Lead in the form of a sheet, a salt solution, a clay, small spheres (shots), and tungsten powder can be used as compensators in this masking process [7].

6.3.1.5 Quantification

To quantify information obtained from radiography, a recorded image must be converted to numerical values at each pixel in the image. In film radiography, the film needs to be exposed to a light source, and the darkness, D, as defined by Eq. (6.9), be measured at each pixel, and subsequently digitized. Then a relationship between D and the radiation-exposure, $\mathcal{E}$, must be established. This can be achieved with the aid of a standard calibration (reference) block of known thickness, t_r, and total cross-section, Σ_r. Let us assume that the intensity of the radiation incident on the film, I, and the film darkness, D, are linearly related so that:

$$D = HI + J_b \tag{6.19}$$

where H is a response-coefficient that depends on the film's quality and the exposure time, and J_b is the light's background darkness that may result from light leakage to the film or electronic noise in the digitization of the signal. Now with the aid of the measurement model, Eq. (6.1), one can write the following relationships for an object of thickness, x, and macroscopic cross-section, Σ,

$$\begin{aligned} D &= H[I_0 \exp(-\Sigma x) + I_{ss} + I_{so}] + J_b && (6.20a)\\ &= H[I_0 \exp(-\Sigma x) + I_{so}] + D_b && (6.20b)\\ &\approx H[I_0 \exp(-\Sigma x)] + D_b && (6.20c)\end{aligned}$$

where I_0 is the intensity of the radiation reaching the film in the absence of the object, I_{ss} is the intensity of the radiation scattered by the surrounding towards the film, I_{so} is the intensity the radiation scattered within the object towards the film, and $D_b = HI_{ss} + J_b$ is the total background component of the imaging system. In Eq. (6.20c), it is assumed that the radiography system is well-designed so that the contribution of scattering within the object is much less than the contribution of the transmitted radiation, i.e. $I_{s0} << I_0 \exp(-\Sigma x)$. With the same assumption, Eq. (6.20c) can be rewritten for a reference block of known thickness, x_r, and macroscopic cross-section, Σ_r, as:

$$D_r = H[I_0 \exp(-\Sigma_r x_r)] + D_b \tag{6.21}$$

In the absence of any object at all, Eq. (6.20b) becomes:

$$D_0 = HI_0 + D_b \tag{6.22}$$

Combining Eqs. (6.21) and (6.22) leads to:

$$D_b = \frac{D_r - D_0 \exp(-\Sigma_r x_r)}{1 - \exp(-\Sigma_r x_r]} \qquad (6.23a)$$

$$HI_0 = \frac{D_0 - D_r}{1 - \exp(-\Sigma_r x_r)} \qquad (6.23b)$$

That is, from the background measurement in the absence of any object, D_0, and a measurement with a reference block of known thickness and macroscopic cross-section, the background film-darkness component, D_b, and the exposure caused by the radiation in the absence of the object, HI_0, can be calculated. This also enables the calculation of the value of Σx, using Eq. (6.20b), and hence quantifying the image, by applying the above correction procedure at every pixel of the image.

In materials that are not strong radiation attenuators, such as porous media and some fluids, $\Sigma x << 1$, Eqs. (6.20c) and (6.21) can be reduced to:

$$D = H[I_0(1 - \Sigma x)] + D_b \qquad (6.24a)$$

$$D_r = H[I_0(1 - \Sigma_r x_r)] + D_b \qquad (6.24b)$$

Eqs. (6.24a), (6.24b) and (6.22) lead to the expression:

$$\frac{\Sigma t}{\Sigma_r t_r} = \frac{D_0 - D}{D_0 - D_r} \qquad (6.25)$$

which quantifies the value of Σt for the inspected object with respect to that of the reference object at each pixel of the image, without having to know any cross-section, or thickness values. Reference [136] used this method for calculating the void-fraction of various multiphase flow materials with thermal-neutron radiography, employing the liquid phase of light water (0.832 mm), heavy water (6.39 mm), liquid potassium (64.1 mm) or liquid sodium (26.1 mm), as the reference materials in a pipe of known diameter; where the number in brackets is the object's thickness for $\Sigma t = 0.278$ which leads to 5% error ($e^{-0.287} = 0.7505 \approx 1 - 0.278$) due to the approximation of relationships (6.24a) and (6.24b).

6.3.2. Variations of Film Radiography

Although, film radiography, discussed in section 6.3.1, is the most commonly used radiographic method, some variations on film radiography have emerged either to improve the quality of the image or to provide some other means of recording images. Supplementary and alternative methods to film radiography are discussed in this section.

Intensifying Screens. In order to enhance the absorption of photons on a film, intensifying screens are often applied. These are usually made of thin metal foils pressed as closely as possible to the film emulsion. The photons absorbed in the screen produce electrons that are in turn absorbed in the film emulsion. Lead screens are typically used for this purpose, but gold, copper and steel are also employed as screen material. In addition, screens made of cadmium, indium and dysprosium were also considered for high-energy radiography [137]. Fluorescent screens act as intensifiers by the emission of light upon exposure to photons. They are widely used in medical radiography, but not as much in industrial applications, because of their poor resolution [7]. Fluorescent screens made of calcium tungstate, or of salts based on rare earth elements, gadolinium, terbium or lanthanum, are utilized as intensifying screens [7].

Instant Films. Instant films, widely known as Polaroid photography, were initially introduced in 1951 for x-ray imaging [138]. However, such films are sensitive to visible light, and therefore are used in radiography in conjunction with a fluorescent screen, i.e. screens that emit light upon bombardment by radiation, see section 13.1.2. The Polaroid film is butted against the screen and protected from the ambient visible light by a cartridge. A positive (in contrast to the film's negative) print of the image becomes available momentarily, after the exposed Polaroid film is removed from the cartridge [139].

Printed Images. A printed image of a radiograph can also be obtained using electrostatic imaging (Xeroradiography). In this technique the film receptor is replaced by a photo-conducting selenium plate, initially charged with high voltage (1500 V or so). Selenium is a good insulator, and thus maintains its charge until it is exposed to either light or x-rays, which renders it partially conducting. The charged selenium plate is loaded into a light-tight cassette and exposed to x-rays, resulting in photo-conduction that discharges the exposed surface leaving an image in the form of an invisible charge-pattern. The exposed plate is then transferred to a light-tight box where it is developed by exposure to a cloud of charged (toner) particles, which are deposited on the electrostatic image, making it visible. The toner particles are then transferred from the selenium plate onto a plastic-coated paper and fused into it. The sheet provides an image in the form embodied by the powder on it. Alternatively, the powdered plate can be photographed to produce the image [140]. The permanent and positive images produced by Xerography makes it particularly useful in radiographing art work [141] and artifacts [142].

Stereoradiography. This technique is used in medical imaging to help visualize x-ray images in three dimensions [140]. It takes advantage of the naturally stereoscopic ability of human vision that mentally combines two images received by the eyes to provide a three-dimensional image. Similarly, stereography involves the recording of two x-rays images (films) of the same object, at two source positions separated by the normal interpupillary distance [134]. The films are then placed in front of two viewing boxes and arranged so that the right-eye sees the film taken with the source in the right-hand-side position while the left-eye sees the other. The films are observed through mirrors adjusted so that the images are superimposed, and the observer sees the image in three dimensions.

Parallax Radiography. In industrial applications, a process similar to stereoradiography is used to determine the depth of a flaw. A radiograph represents a geometric projection of all the points encountered along the path of radiation. Therefore, the position of flaw in the plane normal to the film can be determined from its projection on the film. However, the depth of a flaw, that is, its location along the path of the incident radiation, cannot be determined from a single exposure. Nevertheless, the position of a flaw can be determined from two exposures at different source positions, as schematically shown in Figure 6.6, by the so-called parallax principle. The position of the flaw will be displaced by a distance, x, as the source position is horizontally shifted by a distance X. Knowing the source-to-film distance, Y, the vertical distance, y, from the position of the flaw to the film can be determined from the geometry of the problem as shown in Figure 6.6. The principle of parallax is further extended by applying many intersecting exposures in the tomographic process discussed in section 6.4.

Digital Radiography. As the name implies, in digital radiography an image is obtained in a numerical form that can stored in computer memory for subsequent processing and display [143]. Image enhancement techniques and pattern-recognition algorithms can then be used to improve the quality of the image and heighten the capability of detecting defects [144, 145, 146, 147, 148]. In order to obtain a digitized image, measurements corresponding to each pixel on the image plane must be obtained. Transmission measurements can be collected individually for a number of narrow-beams spanning the object, by moving the x-ray source and detectors linearly along the entire length of the object, or by moving the object past a stationary source-detector assembly. X-ray images can also be digitized after being recorded on a

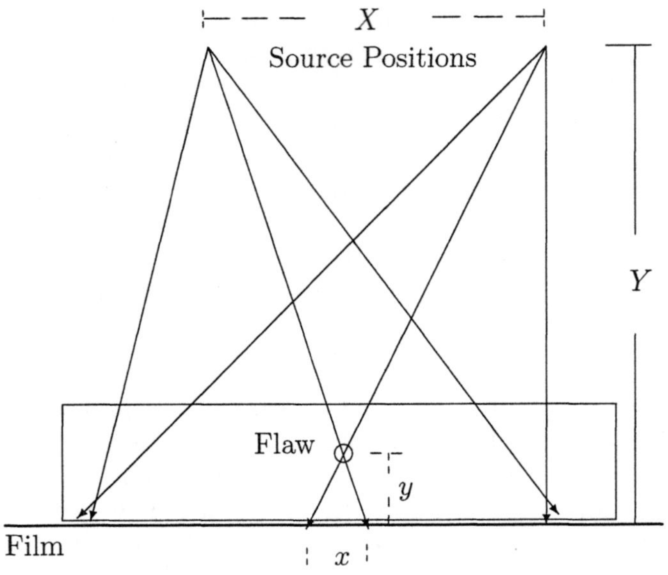

Figure 6.6. Flaw-depth determination with parallax radiography: $y = \frac{Yx}{X+x}$.

fluoroscopic screen (section 13.1.2), or using a fiber-optic scintillator in conjunction with a charge-coupled device (CCD), with the scintillator acting as the x-ray detector [149, 150]. Flat-panel imaging devices [149] are also used for digital radiography. These devices produce electrons directly by the interaction of photons with a flat panel of material such as amorphous silicon [151] or amorphous selenium [152], and the electrons are directly collected and processed. Note also that a digital image, suitable for printing or further digital processing, can be produced from a radiographic silver-halide film using a laser imager, or a Polaroid Helios imager [153]. To replace film radiography, large-area solid-state x-ray imaging arrays, configured to resemble a screened film cassette in size and shape, were developed [154]. The analog signal provided by the current supplied by such arrays can be digitized using an analog-to-digital converter.

6.4. Tomography

6.4.1. Problem Formulation

Radiography techniques have the disadvantage that they provide compressed information, integrated over the chord of radiation transmission,

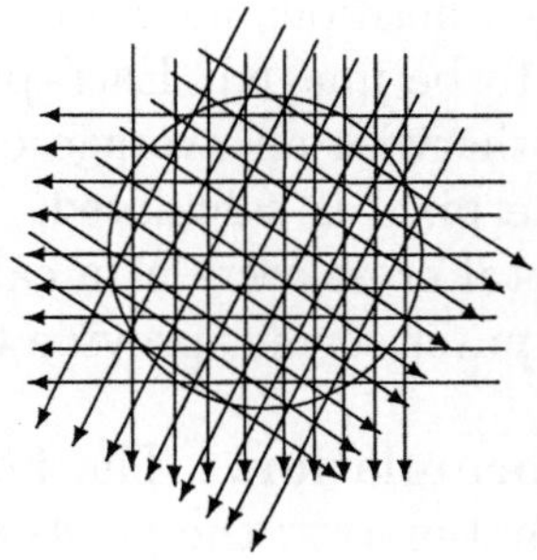

Figure 6.7. Schematic of a basic tomographic system.

as explained in section 6.3. Thus the attributes of objects along the path of the incident radiation are added together. This can lead to masking and smearing out of information. Computed *tomography* (CT) solves this problem by employing many intersecting beams of radiation along a section of the object at different directions, as schematically shown in Figure 6.7, and recording the intensity of each beam. A mathematical algorithm is then applied to unfold the measurements, i.e. determine the total cross-section of radiation at different points (or more accurately small area elements, called pixels) in the section, thus providing a detailed two-dimensional image. The Greek word, "tomos" means a slice, hence the term "tomography". Thus three-dimensional views of an object can only be obtained by reconstructing images of a series of adjacent slices.

Basic Setup. To illustrate the mathematical basis of computed tomography, let us consider the case of a set of parallel pencil beams, each coupled to a radiation detector. To obtain a tomograph, the beams and associated detectors move together so that a section of the object is scanned by a series of intersecting rays, as schematically shown in Figure 6.7. At each direction of exposure, the set of beams provides a *projection* of the object. For each ray, the intensity of transmitted radiation, i.e. I, is measured and stored on a computer for further processing. Since a pencil beam is used, in the case considered here, any of Eqs. (6.1), (6.2) or (6.3) can be employed. For simplification, Eq. (6.1) will be used, with the reference to energy dropped, i.e. assuming that a monoenergetic source is utilized, or that the concept of equivalent energy is employed (see section 6.1). One should keep in mind that all the inherent assumptions associated with the use of the exponential-law of attenuation, discussed in section 6.1, apply to transmission-tomography. Let us superimpose a fictitious mesh of small area elements (called pixels) on the section to be imaged. Each pixel is assumed to contain one type of material, otherwise an average of the cross-section of the material present in the pixel is calculated. The mathematical problem is to determine the total cross-section in each element. The results, when displayed on a computer screen, provide the required image.

Mathematical Formulation. The integration in Eq. (6.1) can be replaced by a summation over the pixels encountered along a particular beam. Replacing Σ_{total} in Eq. (6.1) by μ to avoid confusion with the summation sign, the fraction of the radiation transmitted along that beam, i, can be expressed as:

$$-\frac{I}{I_0} = \exp\left[-\sum \mu(x,y)\Delta t(x,y)\right] \tag{6.26}$$

where $\mu(x, y)$ is the total cross-section of the material in a pixel located at position $(x,\ y)$ in the tomographed section, and Δt is the path-length of the radiation across that pixel, with the summation carried over all pixels along the path of the radiation beam. The value of Δt is predetermined by the geometry of the problem. The transmittance $-\ln\frac{I}{I_0}$ is referred to in tomography as the *ray-sum*, p, with the ray being the incident radiation beam (which in reality has some physical width) being represented by a single ray or line. Note that the *ray-sum* is the value provided by Eq. (6.26) for a single beam, while a *projection* is the set of ray-sums given by a set of parallel beams at a given direction. The value of I_0 can be obtained from a calibration measurement in air. Taking the natural logarithm of Eq. (6.26), and replacing $-\ln\frac{I}{I_0}$ by p, one obtains:

$$p_i = \sum_{j=1}^{N} \Delta t_{ij}\ \mu_j, \quad j = 1 \cdots M \tag{6.27}$$

where the subscript i defines ray-sum number i, and the subscript j designates a pixel in the imaged section, M is the total number of ray-sums (measurements) from all projections and N is the total number of pixels in the section. Note that if the beam corresponding to ray-sum i does not pass through pixel j, then the distance traversed by this beam is equal to zero. The system of linear equations represented by Eq. (6.27) can be expressed in a matrix form as:

$$\{p\} = [H]\{\mu\} \tag{6.28}$$

where $\{p\}$ is a one-column matrix (vector) of length $M$of ray-sums from all projections, $\{\mu\}$ is a vector of length N of the total cross-section of the pixels of the section, and $[H]$ is a matrix in which an element H_{ij} is equal to the length intersected by beam i as it traverses pixel j, with $H_{ij} = 0$ if beam i does not intersect pixel j.

Although the formulation of Eq. (6.28) was derived for parallel pencil beams, an incident wide fan beam can also be accommodated. This is done by considering the beam to consist of a set of narrow elementary beams emerging from the same source point. The distance of intersection of each elementary beam with a voxel then defines an element in the matrix $[H]$ of Eq. (6.28). These elementary beams are defined in practice with the aid of detector collimators, designed so that the field-of-view of each detector coincides with a portion of the fan beam, froming an elementary beam.

The matrix H is usually a large-size matrix, since an image has to consist of a sufficiently large number of pixels, hence a large value of N, to be able to discern any image details. Inversion of the matrix $[H]$ to obtain the unknown vector $\{\mu\}$ using standard matrix-inversion methods is numerically demanding, in terms of computer storage requirements, and therefore in not very practical. Moreover, the number of recorded measurements, M, is usually chosen to be larger than the number of pixels in the section, N, by a factor of two to three, to reduce the effect of measurement error and other uncertainties. The system of linear equations represented by Eq. (6.28) becomes overdetermined, i.e. with more known measurements than unknowns (μ_j's), $M > N$. The matrix H would be an $M \times N$ non-square matrix that is not possible to mathematically inverse.

Least-Squares Solution. In principle the common method of least-squares inversion can be used to reconstruct the image represented by the vector $\{\mu\}$, by minimizing the quantity:

$$\chi^2 = \sum_{i=1}^{M} w_i^2 \left(p_i - \sum_{j=1}^{N} H_{ij}\mu_j \right)^2 \tag{6.29}$$

where w_i^2 is a weight-factor usually included to favor measurements with low statistical variability over measurements with poor statistics, by equating it to the inverse of variance of the measurement. At the minimum value of χ^2, $\frac{\partial \chi^2}{\partial \mu_k} = 0$, yields:

$$\sum_{i=1}^{M} w_i^2 H_{ik} \left(p_i - \sum_{j=1}^{N} H_{ij}\mu_j \right) = 0, \quad k = 1 \text{ to } N \tag{6.30}$$

which in a matrix form gives:

$$\{\mu\} = \left[H^t W H\right]^{-1} [H^t]\,[W]\{p\} \tag{6.31}$$

where the subscript $[H^t]$ designates the transpose of the matrix $[H]$ and $[W]$ is a diagonal matrix whose elements are the weight factors, w_i^2. The inversion of Eq. (6.31) is numerically feasible, since the inverted matrix $[H^t W H]$ is always a square matrix. However, this matrix can be a large matrix that is computationally demanding, in terms of computer storage requirements, to be directly invertible. Even if the computer memory is sufficient to accommodate such a process, serious round-off error can arise that make the solution subject to severe numerical oscillation problems. Nevertheless, a generalized matrix inversion algorithm is available for use in CT image reconstruction [155]. However, other common methods are employed to avoid direct matrix inversion, as discussed below.

6.4.2. Back-Projection

The oldest and simplest of the image reconstruction techniques is the back-projection method [156, 157]. It involves distributing the contribution of a particular ray-sum equally among all the pixels through which the incident beam has passed, i.e. in effect projecting a measurement back into its contributing pixels. That is, if the value of Eq. (6.27) represents the contribution of n_i pixels along beam i, then each pixel, j, in the beam-path will receive a contribution of:

$$\delta_{ij} = \frac{p_i}{\sum_{j=1}^{n_i} \Delta t_{ij}} \tag{6.32}$$

The summation term in Eq. (6.32) represents the total thickness of the pixels along beam j, and thus δ_{ij} represents the average total cross-section along beam i.

The process is repeated for all beams passing through pixel i to estimate its total cross-section as:

$$\hat{\mu}_j = \sum_{i=1}^{m_j} \delta_{ij} = \sum_{i=1}^{m_j} \frac{p_i}{\sum_{j=1}^{n_i} \Delta t_{ij}} \tag{6.33}$$

where $\hat{\mu}$ indicates that an estimated value of μ_j is obtained and m_j is the number of beams that crosses pixel j. By repeating this process for each pixel, the entire image is reconstructed. A pixel that contributes strongly to more than one measurement, i.e. has a large value of μ, will gather a large number of contributions from such back-projections. Thus the aggregate effect of these back-projections would provide at the

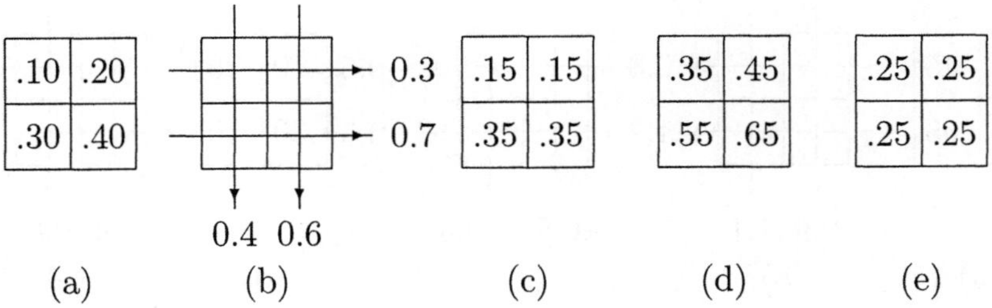

Figure 6.8. Example for reconstructing an image using back-projection.
(a) Actual image (total cross-section values) in square pixels of unit width.
(b) Expected error-free (ideal) ray-sums, in accordance to Eq. (6.27).
(c) Image after back-projection of beams pointing downward.
(d) Image after back-projection of beams pointing sideways, added to previous back-projections.
(e) Background (difference between reconstructed image and actual image).

end an image reflecting the value of μ in each pixel; though the image will be superimposed on a background (fogging) resulting from the back-projection process, since even a vacant pixel will receive a minimal amount of contribution from the back-projection process. Although this process is not currently used, its simplicity makes it attractive for obtaining quick and rough indications.

To illustrate this method, let us consider the simple case of imaging a square-section, divided into 2×2 square pixels, each of unit width. At least four measurements are required to enable reconstructing the image of this section. Figure 6.8 demonstrates the image-reconstruction process, using the simplest form of the back-projection process. Although the resulting image maintains the basic feature of the actual image, in terms of relative order, the obtained values are in effect added to a constant background value. In this example, since the actual values of the image parameters were known, the intensity of the background can be calculated. However, in practice, one cannot determine in advance the intensity of the "fogging" of the background.

6.4.3. Successive Approximation

The intensity of the background associated with the basic back-projection method of section 6.4.2 can be reduced by processing the image via a series of successive approximations. After reconstructing an image, the ray-sums that would be produced by the *estimated* image are *calculated.* The *difference* between the calculated ray-sums and the *measured* ones is indicative of how the reconstructed image is different from the actual one. The difference between the measured and calculated

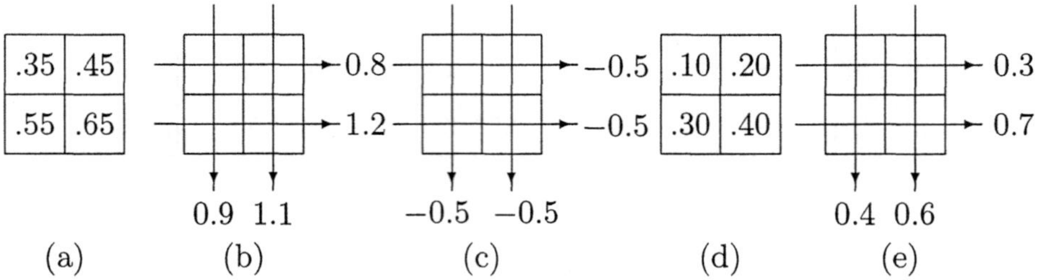

Figure 6.9. Corrected back-projection image reconstruction for example problem of Figure 6.8.
(a) Image reconstructed by back-projection (see Figure 6.8).
(b) Calculated ray-sums for reconstructed image, in accordance to Eq. (6.27).
(c) Difference between given ray-sums (Figure 6.8a) and calculated ones.
(d) Image after back-projection of difference between ray-sums (identical to actual image of Figure 6.8a.)
(e) Calculated ray-sums for reconstructed image, which match the given values of Figure 6.8b.

ray-sums is then used to *correct* the reconstructed image, in an attempt to produce ray-sums that better match the measured ones. The process can be repeated in a series of approximations, until an acceptable image is produced.

Back-Projection of Difference. A simple way to correct the estimated image is to simply back-project the difference between the calculated and measured ray-sums. Figure 6.9 demonstrates the application of this process to the example of Figure 6.8. Although in this example one back-projection of the difference was sufficient to reproduce the actual image, shown in Figure 6.9, this hardly happens in practice. Measurement uncertainties and errors associated with the use of the exponential law for the measurement model (see section 6.1), voxel size, etc. require the application of a number of successive steps of corrections (or iterations). Even then, there is no guarantee that the problem will converge to an adequate solution, or that it will converge at all.

General Formulation. The correction process described above is just one of many ways that can be used to reconstruct an image iteratively, i.e. through a series of successive approximations of the solution. In general, the iterative process can be represented by:

$$\mu_j^{(\ell)} = \mu_j^{(\ell-1)} + \sum_{i=1}^{m_j} \Delta\mu_{ji}^{(\ell)} \tag{6.34}$$

where ℓ refers to iteration number and $\Delta\mu_{ij}^{(\ell)}$ is the correction applied to pixel j from beam i at iteration ℓ. The first guess (μ_j^0) can be taken as the image reconstructed by uncorrected back-projection, according to Eq. (6.32), or any other reasonable guess. To apply such a correction process one must decide on a number of things:

- How to determine the value of $\Delta\mu_{ji}^{(\ell)}$.
- Which pixel or beam sequence to follow in applying the correction process.
- How to control, or guide, the iterative process, so that it converges to an adequate solution.
- How to determine when to stop the iterative process.
- How to determine whether the problem has converged to an acceptable solution.

These aspects are described in the following subsections.

6.4.3.1 Correction Value

Additive Correction. The correction value, $\Delta\mu_{ij}^{(\ell)}$ in Eq. (6.34), has to depend in some way on the difference between the measured ray-sums, p_i's, and the ray-sums calculated based on the estimated values of $\hat{\mu_j}$'s. If the calculated ray-sum for beam i is $p_i^{(\ell)}$ at iteration ℓ, and the corresponding measured value is p_i, then there is a difference of $\Delta p_i = p_i - p_i^{(\ell)}$ between the two values. Obviously, if Δp_i is negative, then the μ values in the pixels contributing to this ray-sum are overestimated and need to be reduced and vice versa. The question now is how to distribute Δp_i among the contributing pixels. This can be done, as was shown in the example of Figure 6.9, by back-projecting the value of Δp_i in proportion to the distance traveled by beam i in the cell, which can be mathematically expressed, with the help of Eq.(6.32), as follows:

$$\Delta\mu_{ij}^{(\ell)} = \frac{\Delta p_i \Delta t_{ij}}{\sum_{j=1}^{n_i} (\Delta t_{ij})^2} \tag{6.35}$$

This is known as the additive correction method. In the simple example shown in Figure (6.8), this correction method was applied very effectively, leading to the correct solution after correcting only once for the first set of back-projection values.

Multiplicative Correction. The reconstructed image can also be modified via a multiplicative correction process in the form:

$$\Delta\mu_{ij}^{(\ell)} = \frac{\Delta p_i \mu_j^r}{p_i^{(\ell)}} \tag{6.36}$$

where the superscript r refers to the most recently used value of μ_j, which could be the value obtained in the previous iteration, or within an intermediate step in the current iteration. This multiplicative correction process of Eq. (6.36) distributes Δp_i in proportion to the contribution of a pixel to the calculated values; note that $\sum_i^{n_i} \mu_j^r = p_i^{(\ell)}$. Since $\Delta p_i = p_i - p_i^{(\ell)}$, one can show easily that:

$$\mu_j^r + \Delta\mu_{ij}^{(\ell)} = \mu_j^r \frac{p_i}{p_i^{(\ell)}} \tag{6.37}$$

which indicates that the ratio between the measured and calculated values of the ray-sums can be used to calculate the corrected value of μ_j. There is no obvious preference between the use of the additive and multiplicative correction schemes.

Minimum Difference. Another approach is to chose the correction value such that the difference between the measured ray-sums and the calculated ones is minimized. This in effect replicates the least-squares process of Eq.(6.31), applied on a pixel-by-pixel basis, leading to the correction factor [158]:

$$\Delta\mu_j^{(\ell)} = \frac{w_i^2 H_{ij}\left(p_i - p_i^{(\ell-1)}\right)}{\sum_{i=1}^{M}(w_i H_{ji})^2} \tag{6.38}$$

Note that the zero values of H_{ij} ensure that only the beams passing across pixel j contribute to its correction factor. This least-squares correction process enables one to reduce the effect of poor quality measurements through the weight factors, w_j's, which are normally taken as the inverse of the measurement variance.

6.4.3.2 Correction Scheme

The sequence of applying one of the correction processes mentioned above can have an impact on the propagation of error in the iterative process. It, therefore, affects the overall performance of the iterative process. There are three main correction strategies, as discussed below.

Using All Beams. In this scheme, all values of p_i^ℓ are calculated and subsequently all values of $\mu_j^{\ell-1}$ are corrected to μ_j^ℓ. This correction scheme usually employs the least-squares correction process of Eq.(6.38), and thus the technique is known as the Iterative Least-Squares Technique (ILST). No intermediate corrections are performed, rather correction is applied to all pixels at the end of the iteration. Therefore, in this process each pixel will receive a correction $\Delta\mu_{ij}^{(\ell)}$ from all the m_j beams passing through it. This will lead to an excessive amount of correction that can easily lead to solution divergence. To avoid this over-correction, some dampening of the correction factors needs to be introduced. The most common approach is to restrain the correction process using dampening factors. A dampening factor that provides the best least-squares fit to the measured projections is given by [158]:

$$\delta = \frac{\sum_{i=1}^{M}(w_i^2 c_i)\left(p_j - p_j^{(\ell-1)}\right)}{\sum_{i=1}^{M}(w_i^2 c_i^2)} \quad \text{with } c_i = \sum_{j=1}^{N} H_{ij} \sum_{i=1}^{m_j} \Delta\mu_{ij}^{(\ell-1)} \tag{6.39}$$

The calculated $\Delta\mu_{ij}$ values are multiplied by the dampening factor, δ, before being used to correct the image.

Using Pixel-Specific Beams. In this scheme only the m_j beams passing through a pixel j are considered at a time. The projections for these m_j beams are calculated, and the obtained values are used to correct the $\mu_j^{(\ell-1)}$ value for this pixel only, leading to an intermediate corrected value μ_j^r. Then next pixel $j+1$ is considered, and the ray-sums for all m_{j+1} beams passing through it are calculated, using the intermediate values for the previous pixels. The correction process is then repeated. This method is known as the *Simultaneous Reconstruction Technique* (SIRT), as it involves simultaneous correction of all beams passing through a pixel. The advantage of this method is that it allows gradual changes in the image, which prevent the over-correction of the all-beams method.

Using One-Projection Beams. One set of beams in a given direction (projection angle) are considered at a time, and the correspondence corrections, using Eq.(6.35) or Eq.(6.36), are applied to all pixels, producing an intermediate set of values in the iteration. The values of the ray-sums are then updated and the correction process is applied using another projection at a different angle (usually orthogonal to that of the previous set of beams to provide the domineering influence of two sets of beams at adjacent angles, which will tend to have similar influence). The process is then repeated using another set of beams at a different angle, and so on. This is the process applied in the example of Figure 6.9, and is commonly known as the *Algebraic Reconstruction Technique* (ART).

This approach protects the problem from over-correction, since intermediate corrections are introduced within each iteration (those produced by each set of beams at a given direction). The method is, however, more susceptible to error propagation than the other two methods, since the image values at the end of the iterative process reflect more the effect of the last set of beams than those used early on in the iteration. Thus, the method in essence follows projections from one direction to another, without averaging the effect of beams emerging from various directions as done in the two other methods. Therefore, if measurements are inconsistent, due to statistical variability, electronic instabilities, etc., iterations can diverge as the method attempts to accommodate the conflicting measurements. Nevertheless, for good quality data, this method tends to converge rapidly to a solution, This is demonstrated in the example of Figure 6.9 for exact data, where converge to the actual solution occurred after utilizing only one projection and without even completing the first iteration. ART has also been considered for reconstruction of three-dimensional objects from cone-beam projections [159].

Other Schemes. A combination of the schemes discussed above is also possible. For instance, one can use ART for the first couple of iterations, as it tends to converge rapidly, then apply ILST (which is faster than SIRT) to provide some averaging out of the uncertainties. More sophisticated algorithms proceed from one iteration to another using steps in the direction in which the χ^2 function of Eq. (6.31) is minimized, which happens to be in the direction opposite to the gradient of the function [155]. The method of *steepest descent* is based on this premise. Note that the gradient of χ^2 is itself a vector, and thus the method is in effect applied to all pixels at the same time. The *conjugate gradient* method is a variant of this method, in which convergence is accelerated by making the change from one iteration to another in a direction that is orthogonal (vector-wise) to that taken by all previous iterations. Reference [155] provides algorithms for these two powerful methods.

6.4.3.3 Solution Guidance

Measurement uncertainties and noise can affect the iterative process in a number of ways:

- cause an oscillatory behavior (high values followed by low values),
- produce unrealistic values of the total cross-section (negative or too large),

- converge the solution to values completely different from the actual ones (as more than one solution set can produce calculated projections that match the measured values within the margin or error), or even
- drive the problem to divergence, i.e. leading to no solution whatsoever.

Therefore, in such iterative solutions, some guidance is given to the iterative process, based on some *a priori* physical knowledge of the nature of the problem at hand, or simply by experience from trial and error. The following guiding criteria are typically used [156]:

- If a measured ray-sum is zero, or is sufficiently small in value to be practically equal to zero, the μ_j values in the pixels along the beam corresponding to this zero ray-sum are set to zero and are not allowed to change within the iterative process.
- Knowing that the total cross-section can never be negative in value, any value of μ_j less that zero is set to zero at any stage during the iterative process.
- The value of μ_j is constrained so that it does not exceed some preset maximum value, determined from prior knowledge of the object being imaged.

The above processes are *solution bounding* measures that prevent further propagation of the effect of such unrealistic values to the next stages of the iterative process.

6.4.3.4 Stopping Criteria

Knowing that measurements are noisy, one cannot expect a perfect match between the reconstructed image and the actual one, or between the measured and the calculated ray-sums. Therefore, some tolerance level must be accepted beforehand. This tolerance level can be controlled by a stopping criteria to terminate the iterative process, considering the problem to have reached a solution within error bounds. The Euclidean norm (measure of distance) is usually used as an indication of deviation between values, as it provides the most probable error (inherently assuming that the fluctuations have a normal (Gaussian) statistical distribution). For instance the difference between the μ_j values obtained between two successive iterations can be defined by the Euclidean norm:

$$||\epsilon_\mu|| = \frac{1}{N}\sqrt{\sum_{j=1}^{N}\left(\mu_j^{(\ell)} - \mu_j^{(\ell-1)}\right)^2} \tag{6.40}$$

The iterative process can then be stopped if $||\epsilon_\mu||$ is less than some predetermined value, at such point one can consider the problem to have converged to within an acceptable error-range (also known as the error-ball, with $||\epsilon_\mu||$ representing the radius of an N dimensional ball).

Alternatively, one can terminate the iteration by observing the difference between the calculated ray-sums and the measured ones:

$$||\epsilon_p|| = \frac{1}{M}\sqrt{\sum_{i=1}^{M}\left(p_i - p_i^{(\hat{\ell})}\right)^2} \tag{6.41}$$

where $\hat{\ell}$ means that this norm is measured either at an intermediate step with iteration ℓ or at the end of the iteration. The iteration process can then be terminated when $||\epsilon_p||$ is less than a preset value. One can also use an "either/or" stopping criteria, that is, stop the iterative process when either $||\epsilon_\mu||$ or $||\epsilon_p||$ reach their preset values. Note also that one can simply terminate the iterative process after a preset number of iterations, predetermined from trial-and-error for similar type of images. Generally five to ten iterations are sufficient for the corrected back-projection methods described earlier in this section. It is also wise to always have a preset maximum number of iterations, so that if the problem never converges, the iterative process will not continue unnecessarily for a long time. One can then examine the results and attempt to deal accordingly with the problem, for instance by switching to a different iterative scheme.

6.4.3.5 Convergence

In a well-behaved problem, one would expect the solution norm, $||\epsilon_\mu||$, given by Eq. (6.40), to continuously decrease (or contract) as it converges to the solution. Therefore, one should monitor the behavior of this norm and ensure that it is continually contracting. The rate of change of contraction of this error (difference in value between two successive iterations) can also be used as a stopping criteria. However, a problem will not converge, if the measurements are not consistent (due to measurement uncertainties), or if the iterative scheme is susceptible to error propagation. One should be careful, not to assign too low values for the termination criteria. On one hand, it is not realistic to expect an image of a quality (error) better than the quality (uncertainty) of measurements. On the other hand, even if measurements are exact (which of course is impossible), computer truncation error (due to the finite precision of computer processors) can also propagate through the iterations. Attempting to reach a solution beyond that is tolerated by truncation error is not only unrealistic, but can also lead to an erratic behavior of the solution, and the error norms.

6.4.4. Modal Approximation

The successive approximation methods discussed in section 6.4.3 are relatively computationally time-consuming, and there is no guarantee that the solution will converge to a true solution of the problem. In addition, as indicated in section 6.4.1, direct solution by matrix inversion, say using Eq. (6.31), is not also computationally efficient in practical problems, where a large number of pixels are involved, due to computer storage requirements. Therefore, reconstruction methods based on modal expansion of the unknowns are often used. The unknown total cross-sections are assumed to obey some known series of modes (analytical functions), weighted with appropriate coefficients. Although the expansion functions are pre-assigned, their weighting coefficients are not known. Image reconstruction then becomes the problem of determining the coefficients of these expansions. This class of solution methods is often referred to as analytical methods, due to the use of analytical functions. Modal methods are widely used in numerical analysis to decompose a solution into a set of fundamental functions. A modal method that uses orthogonal polynomial expansions is described in reference [155]. However, the most common method in tomography relies on the use of harmonic modes, similar to those encountered in simple harmonics motion. The frequency of these harmonics can then be controlled to eliminate noise in measurements, since noise and uncertainties in measurements are not governed by the same physical process and thus produce different harmonics.

Fourier Expansion. As in the simple harmonics motion of a pendulum, the frequency of modes used in CT reconstruction are spatial frequencies, not time frequencies. The wave number, k, is used to express the spatial frequency (having units of inverse distance). The expansion of the desired solution, μ, is usually expressed in terms of a Fourier series (sum of sine and cosine functions of different wave-numbers), as it provides a physically meaningful and numerically efficient means of series expansion (Fast Fourier transform algorithms are now widely available). The superposition of sine and cosine waves can be represented by a complex exponential function, with its real part being the cosine function and its imaginary part being the sine function. Therefore, the total cross-section, μ, at an image point (x, y) on the imaged section can be expanded to different modes (also called partial waves) using the Fourier integral:

$$\mu(x,y) = \int_{-\infty}^{\infty}\int_{-\infty}^{\infty} F(k_x,k_y)\exp[i(k_x + k_y y]dk_x\, dk_y \tag{6.42}$$

where k_x and k_y are wave-numbers in the x and y directions, respectively, and $F(k_x,k_y)$ is the coefficient of expansion (or the amplitude of the partial wave) at wave-numbers k_x and k_y, with i being the imaginary value $\sqrt{-1}$. The infinite limits in the integral of Eq. (6.42) assume that all frequencies are possible, which is not the case as explained later on this section.

Inverse Transformation. Now, to determine μ, one needs to find the values of F at different wave numbers, keeping in mind that k_x and k_y are both less than some maximum value, k_m. The expansion coefficients can be determined by the inverse transform of the Fourier integral:

$$F(k_x,k_y) = \int_{-\infty}^{\infty}\int_{-\infty}^{\infty} \mu(x,y)\exp[-i(k_x x + k_y y)]dx\, dy \tag{6.43}$$

where $F(k_x,k_y)$ is the two-dimensional Fourier transform of $\mu(x,y)$. Note that outside the image domain, the μ values are zero, and the portion of the integral in Eq. (6.43) outside the image vanishes, but the infinite limits of the integral are preserved here for mathematical rigor.

Fundamental Equation. Now if the x and y axes are rotated by an angle θ, with $\tan\theta = \frac{k_y}{k_x}$, to define a polar set of coordinates (r,θ), with $r = x\cos\theta + y\sin\theta$ being in the radial direction, $\vec{r}$, and $\vec{s}$ in the tangential direction (normal to $\vec{r}$), then Eq. (6.43) can be rewritten as:

$$\begin{aligned} F(k_x,k_y) &= \textstyle\int_{-\infty}^{\infty}\int_{-\infty}^{\infty} \mu(x,y)\exp(-ikr)ds\, dr \\ &= \textstyle\int_{-\infty}^{\infty}\left\{\int_{-\infty}^{\infty} \mu(x,y)ds\right\}\exp(-ikr)dr \\ &= \textstyle\int_{-\infty}^{\infty} p(r,\theta)\exp(-ikr)dr \end{aligned} \tag{6.44}$$

with $k^2 = k_x^2 + k_y^2$, $p(r,\theta)$ being the ray-sum at some distance $r = x\cos\theta + y\sin\theta$ in a direction $\vec{s}$ perpendicular to $\vec{r}$, and the exchange of order of integral in the second line of the equation is made possible since k is not a function of x or y. The last line of Eq. (6.44), when compared to the from of Eq. (6.43), indicates that the Fourier coefficient of $F(k_x,k_y)$ is equal to the Fourier coefficient of the ray-sum $p(r,\theta)$ with respect to r. Let us call this transformation $P(r,\theta)$, then one has the following fundamental and important set of equations defined by the relationship:

$$F(k_x,F_y) = P(r,\theta) \tag{6.45}$$

Central-Slice Theorem. Eq. (6.45) physically states that the coefficient of the two-dimensional Fourier transform of $\mu(x, y)$ at k_x and k_y is equal to the corresponding one-dimensional Fourier coefficient of the ray-sum taken at the same angle, $\theta = \frac{k_y}{k_x}$. This is known as the *central-slice theorem* [158], and is the essence of reconstructing CT images with Fourier Transforms. The method involves taking one-dimensional Fourier transforms of all the ray-sums, generating the two-dimensional array of Fourier coefficients, and inverting to produce the image. Fast Fourier-Transform algorithms are readily available to perform such processes. However, the difficulty is in generating the two-dimensional set of coefficients from the one-dimensional ones. This difficulty arises from the fact that the number of ray-sums is always limited, which leaves a gap in the frequency space, and causes the absence of some arbitrary frequencies that may be required to generate a two-dimensional set of coefficients. However, interpolating between the one-dimensional coefficients can be utilized to estimate the absent values.

Sampling. Interpolation requires that the ray-sums be closely spaced, otherwise erroneous values will be produced. According to the sampling theorem, it is possible to exactly reconstruct a function from its values at discrete set of sampling points, if the function's Fourier transform vanishes outside a finite frequency region; such a function is called *band-limited* [158]. A band-limited function would then have non-zero Fourier transforms only when $|k| \leq k_{max}$, where k_{max} is the maximum wave-number (frequency). The band-width of the function is then equal to $2k_{max}$, as it covers the frequency range from k_{max} to $-k_{max}$. If one limits the Fourier frequencies to some value k_{max}, it is then reasonable to limit the spatial sampling frequency of the data, accordingly. If the spacing between the beams in one direction in a CT similar to that of Figure (6.7) is W, which would also be the width of a pixel, then the sampling frequency would be equal to $\frac{1}{W}$. This sampling frequency must be equal to or greater than the band width, i.e. $W^{-1} \geq 2k_{max}$ [158]. Therefore, band-limiting, usually by taking $W = \frac{1}{2k_{max}}$, is the only approximation required to utilize Fourier transforms for the nodal solution of the image reconstruction problem.

Artifacts. Restricting k_{max} to a value of $\frac{1}{2W}$ has the effect of cutting-off terms in the integral of Eq.(6.42) for which $|k| > k_{max}$. Sharp boundaries within an image, due to abrupt changes in material, can produce frequencies greater than k_{max}. The eliminated high-frequencies may reappear disguised as lower frequencies, an artifact[1] known as *aliasing*. If these high frequencies are cut-off altogether, an artifact known as the *Gibbs phenomenon* is produced. While aliasing is usually a small effect, the Gibbs phenomenon can produce an overshoot, or ringing, at sharp boundaries, where for example, a low-density ring can appear artificially inside a high-density boundary [156].

6.4.5. Filtered Back-Projection

Both the successive-approximation and modal methods of solution, discussed in sections 6.4.3 and 6.4.4, respectively, eliminate the background associated with the basic back-projection method given by Eq. (6.32). While successive-approximation deals with the background problem by iterative corrections, the modal method provides a direct solution that does not involve the computationally demanding matrix-inversion process. Another method that takes advantage of the simplicity and speed of the back-projection process, while eliminating the background problem, involves artificially altering the measured projections with a *filtering* function that when back-projected will eliminate this background.

Recall that the background artifact is the result of giving equal contribution to all pixels in the direction of the back-projected rays, regardless of their relative contribution to the measurement. The result is that a pixel which contributes nothing to a ray-sum receives the same amount of contribution as all other pixels. The result is an over-contribution that creates the fogging of the background, as the example of Figure 6.8 demonstrated. Filtering aims at eliminating this over-contribution by spreading the influence of each ray-sum beyond its original direction, so that its back-projection produces negative contributions to the pixels in the present path of the neighboring beams. When these neighboring beams are back-projected, their over-contribution is eliminated, or at least minimized. Eventually the last beam in the scanning process will also reduce the background effect of the first beam.

Convolution Filter. A number of filtering methods are available [156, 158]. The choice of a filtering technique does not greatly affect the performance of the image reconstruction process. One method filters p_i

[1]An artifact is defined as any image feature which does not correspond to an object feature in the image plane [160].

to p_i^* as follows:

$$p_i^* = (p_i k_{max})W - \frac{1}{\pi^2}\sum_{k=1}^{M} \frac{p_k}{(i-k)^2} \quad \text{for odd} \quad i-k \tag{6.46}$$

with the summation taken over only odd values of $i - k$, and all other notations are as used earlier in this section. This filter is called a *convolution filter* as it involves a convolution kernel that relates one point in the system (beam i) to another point (beam k) by the separation distance between them (represented by $i-k$). The effect of the negative term in Eq.(6.46) is in effect to let the neighbors reduce the effect of the back-projection of the ray-sum of beam i by an amount that decreases with the inverse square of the distance from the beam; note that $i - k$ is a measure of proximity of beam i to beam k. The even values of $i-k$ are excluded from Eq. (6.46) as in effect the filter presented by this equation represents a squared-sine wave originating at beam i and peaks at the next neighbor [156]. In the first term in Eq. (6.46), $p_i k_{max}$ is the estimated highest spatial rate of change of p_i along a distance perpendicular to the beam. The first-term of the equation dampens the measured ray-sum, to further reduce its over-contribution in the back-projection process. Eq. (6.46) is an approximation of a continuous integral, which is replaced by a summation in the equation, on the assumptions that $W^{-1} \geq 2k_{max}$ as required by the sampling theorem, see section 6.4.4. Taking the case of $W = \frac{1}{2k_{max}}$, the first term in Eq. (6.46) becomes simply equal to $\frac{p_i}{2}$.

Filtered back-projection is applied by filtering the ray-sums of a projection, say according to Eq. (6.46), and back-projecting with Eq. (6.32) using the filtered ray-sums instead of the measured one. The process is then repeated to all other projections. Note that due to the limitation of the band-limiting assumptions inherent in this method, the reconstructed images are also susceptible to the aliasing and Gibbs phenomenon artifacts, discussed at the end of section 6.4.4. Reference [155] provides a number of algorithms for a variety of filtering and convolution functions.

6.4.6. Image Quality

Image quality naturally depends on the quality of the measurements. This depends, in turn, on its level of uncertainty, the number of measurements available and on the nature of the system setup. These factors are discussed below.

Measurement Uncertainty. If the statistical uncertainty in a measured ray-sum, p, is given by the standard deviation, $\sigma(p)$, then the standard deviation in the reconstructed image, σ_μ, can be approximately estimated to be [156]:

$$\frac{\sigma_\mu}{\mu} \cong \sqrt{n}\,\frac{\sigma(p)}{p} \tag{6.47}$$

where n is the number of beams per projection in a setup similar to that schematically shown in Figure 6.7. Although Eq. (6.47) provides an approximate relationship for uncertainty in the total cross-section of the reconstructed image, it shows that the relative uncertainty in the measurements, $\frac{\sigma(p)}{p}$, is magnified by a factor of $\sqrt{n}$ when propagated to relative error in μ of $\frac{\sigma_\mu}{\mu}$. Though increasing the value of n increases the number of measurements in a projection, it does also increase the number of pixels in the image. Therefore, Eq. (6.47) reflects the fact that for a finer image (more pixels), the uncertainties in the measurements have to be quite low to achieve low uncertainties in the reconstructed image. However, uncertainties in the reconstructed image can be reduced by filtering the measurements, smoothing the reconstructed image, or overdetermining the problem.

Number of Measurements. Overdetermining the problem by having more measured ray-sums, M, than the total number of pixels, N, provides redundant measurements that help reduce the effect of measurement uncertainties, as well as errors introduced by the reconstruction algorithm. On the other hand, if the problem is underdetermined, $M < N$, some assumptions must be made explicitly or implicitly within the reconstruction algorithm to compensate for the missing information, such as assuming the image to be smooth or circularly symmetric. In an underdetermined problem, an iterative scheme (section 6.4.3), will tend to converge to the smoothest possible solution [156], while the modal method (section 6.4.4) and the filter back-projection method (section 6.4.5) make the implicit assumption that the missing ray-sums are similar to the measured ones, thus produce symmetric images.

An algorithm that produces images from an underdetermined problem with no bias towards any particular orderly structure relies on maximizing an entropy function defined as [155]:

$$S(\{\mu\}) \;=\; -\sum_{j=1}^{N} \frac{\mu_j}{T} \ln \frac{\mu_j}{T} \tag{6.48}$$

where $T = \sum_{j=1}^{N} \mu_j$. Maximizing this entropy function, while satisfying the measurement model of Eq. (6.28) and ensuring that $\mu_j \geq 0$, leads to an image that has less structured features than any other solution that matches the measurements.

Inherent Assumptions. The reconstruction process in CT is based on the exponential relationship of Eq. (6.26), which models transmission measurements. As discussed in section 6.1, there are many inherent assumptions behind this simple exponential relationship, as it is only applicable to a beam that is narrow (no radiation scattering of radiation to detector), well-collimated (non-diverging), and monoenergetic. In practice, these assumptions are not all fulfilled. Therefore, in addition to the sources of errors discussed above, there are inherent errors that are the result of using a mathematical formulation (measurement model) that does not faithfully represent reality. The impact of these assumptions are discussed below.

Collimation. Source and detector collimator geometry (see Figure 6.2) defines the narrow beam required for the use of Eq. (6.28). The physical size of the source beam (collimator opening), no matter how narrow, is not zero, but is typically a few mm wide. The height of the collimator opening defines the width of the slice (section) of the image. Although it is desirable to reduce the beam height as much as possible, to approach the narrow beam assumption, narrowing both the beam width and height has the effect of eliminating most of the source radiation, allowing only a very small portion of the source radiation to reach the object. This is not only an inefficient use of the source, but it also produces a low number of counts in the detector, necessitating a longer counting period. Given the fact that many measurements are required in tomography, a long time-period is not often practical. Beam heights are, therefore, typically three to four times larger than the beam width. The wider beam height produces in effect a volume element (called a voxel) that is relatively large in volume, and the reconstructed μ values are in turn volume-averaged over this voxel. For example, if a volume is partially filled, the reconstructed μ value in this voxel would be underestimated (since μ for air is almost zero), producing the so-called *partial-volume effect.*

Beam Divergence. The other factor that challenges the assumption of exponential attenuation is beam divergence with distance. Radiation intensity decreases with the inverse square of the distance of propagation, see section 3.6.2. The result is that the voxel volume increases with distance from the source, as the beam broadens both along its direction of propagation and in the direction perpendicular to it.

Moreover, the intensity of the beam at the center of the voxel would be higher than that at its edge, due to the increase in distance of travel of the source beam. Therefore, a material at the center of the beam would contribute more to measurements than a material at its edge, producing an artifact that can affect the sharpness of the image. The increase in the size of the beam due to divergence and the use of a collimator of finite width can produce radiation buildup within the detector, as radiation can scatter within the beam volume and reach the detector. Detector collimation can help reduce the buildup effect.

Beam Hardening. One last fundamental assumption behind the measurement model of Eq. (6.26) is that it is expressed for a monoenergetic source. As discussed in section 6.1, an effective-energy, or equivalent total cross-section, can be used to overcome this assumption. However, the possible absorption of low-energy radiation in the early stages of the transmission process (beam hardening), increases the value of the effective energy. Beam-hardening occurs in the pixels down-stream of the beam, but those same pixels will be upstream of an opposite beam during the rotation process of the projections (see Figure 6.7). This will tend to cancel out, or at least reduce, the effect of beam-hardening for pixels near the edges of an image. However, pixels at the center of the image will tend to be subjected to the same amount of beam-hardening, and thus are more vulnerable to beam-hardening effects, producing the so-called *cupping effect* [156].

6.5. Special Methods

6.5.1. Combined with Scattering

A number of efforts were made to take advantage of radiation scattering in conjunction with transmission tomography. Reference [161] presented an interesting limited-view transmission imaging system that relies on producing the interrogating radiation by the Compton scattering of two pencil beams of gamma-rays (produced by two collimated ^{137}Cs sources) incident axially on each side of an aluminum rod. Single scattered photons produced by the rod have an energy that depends on their angle of scattering, according to Eq. (3.37). Therefore, the energy of the photons, measured by a high resolution gamma-detector, determines their angle of scattering. Photons scattered, in the form of fan beam, from each of the two axial halves of the rod, are measured with separate detectors. The interrogated object is placed in the field-of-view of the scattered photons. Detectors then provide two fan-beam transmission projections of the object. More projections can be obtained by rotating the object, or the source-scatter-detector assembly.

Use was also made of radiation scattering to supplement transmission measurements. Reference [162] proposed the use of backscattered radiation that accompanies a parallel-beam computed-tomography system. A similar approach was presented for a fan-beam system [163, 164]. While transmission-tomography reconstructs the attenuation coefficient (total cross-section), scatter imaging reconstructs the electron-density, in photon imaging. At photon energies where the photoelectric effect is reasonably prevalent, the ratio of the attenuation-coefficient and the electron-density becomes dependent on the atomic-number of the material, and thus can provide composition indications.

6.5.2. Region-of-Interest Imaging

Region-of-interest (RoI) is the imaging of a particular area of interest within a larger structure. Obviously to reach such region, radiation has to travel through the entire object. However, one can only expose the pixels within the RoI to a sufficient number of projections to reconstruct an image of these pixels. Naturally, the obtained image can contain significant artifacts due to the influence of the pixels outside the RoI. One way to correct for these artifacts is to construct a low-resolution (large pixel) image of the region outside the RoI (or the entire object), and utilize its information to correct the image of the RoI for the effect of the surroundings. This method is called the *dual-sampling* method [165].

6.5.3. Dual Transmission

Two independent transmission measurements, along the same path length, can be useful, not only in eliminating the dependence of measurements on the thickness of the object, but also in providing some information on the nature of the encountered material [166]. Independence of one set of transmission measurements from another can be attained by using either the same radiation type at two different energies, or by employing two different types of radiation. Both options are discussed below after introducing the theoretical framework of the concept.

Dual Energy. Consider two radiation beams A and B for which the examined object has total cross-sections Σ_A and Σ_B, respectively. If two beams transverse the same distance t in the object, then one would obtain two measurement, I_A and I_B, which can be modeled using Eq. (6.1) (assuming each source has a single energy or that the total cross-sections are evaluated at some equivalent energy) as:

$$I_A = I_{A,0} \exp[-\Sigma_A t] \tag{6.49a}$$
$$I_B = I_{B,0} \exp[-\Sigma_B t] \tag{6.49b}$$
(6.50)

where the zero subscript indicates a measurement along the same chord in the absence of the object (i.e. in air). The ratio of the transmittance of the two measurements, r, eliminates the thickness, t, as follows:

$$r = \frac{-\ln \frac{I_B}{I_{B,0}}}{-\ln \frac{I_A}{I_{A,0}}} = \frac{\Sigma_B}{\Sigma_A} = \frac{\sigma_B}{\sigma_A} \tag{6.51}$$

where the microscopic cross-section, σ, replaces the macroscopic cross-section Σ, as they correspond to the same material in the object, hence the same atomic density, see Eq. (3.6). The elimination of the atomic density does not, however, remove the dependence of the ratio r of Eq. (6.51) on the nature of the material of the object, unless of course the cross-sections σ_A and σ_B have the same level of dependence on the material. For example, if a photon source is used for both beams A and B in the range where Compton scattering is dominant, according to Eq. (3.40), the microscopic cross-sections become dependent on the atomic-number and the photon energy, and the ratio r of Eq. (6.51) becomes dependent only on the energy of the two beams, rather than on the material itself. On the other hand, if one of the beams, say A, is at an energy where Compton scattering is dominant, while the other beam, B, is at a lower energy at which the photoelectric effect is dominant, then using Eqs. (3.40) and (3.30), the ratio r of Eq. (6.51) becomes strongly dependent on the atomic number, Z, of the material, since:

$$r = \frac{\sigma_A}{\sigma_B} = \text{Constant } Z^{n-1} \tag{6.52}$$

where n varies from 3 to 5, depending on Z and the photon energy used for beam B. Since most materials are composed of compounds rather than single elements, the ratio r is dependent on the effective atomic-number Z_{eff}, see appendix E. This dependence on the atomic-number enables classification of materials into organic (low Z) and inorganic (high Z) materials, which is the essence of the *dual-energy radiography* method used in some medical applications [167, 168, 169] and at airports for luggage inspection, see section 13.5.

Dual Radiation-Type. Independent transmission measurements can also be obtained by combining neutron and gamma-transmission. Consider a binary mixture consisting of two materials, 1, and 2, with mass-fractions w and $1 - w$, respectively, evaluated along the same transmission chord, then the macroscopic cross-sections of the neutron beam, say, A and a gamma-ray beam B, can be expressed in terms of the macroscopic cross-section as:

$$\Sigma_A = [Nw\sigma_{A,1} + N(1-w)\sigma_{A,2}] \tag{6.53a}$$

$$\Sigma_B = [Nw\sigma_{B,2} + N(1-w)\sigma_{B,2}] \tag{6.53b}$$

where N is the atomic-density of the mixture, and the subscripts define the radiation beam and the material. The transmission ratio r of Eq. (6.51) can be expressed as:

$$r = \frac{\Sigma_A}{\Sigma_B} = \frac{w(\sigma_{A,1} - \sigma_{A,2}) + \sigma_{A,2}}{w(\sigma_{B,1} - \sigma_{B,2}) + \sigma_{B,2}} \tag{6.54}$$

Therefore, if the nature of the two constituents of the material is known, and their microscopic cross-sections, for the two types of radiation, are predetermined, the ratio, r, of Eq. (6.54) can be used to determine the mass-fraction, w, of component 1; with its complement being the mass-fraction of the other material. This is a useful means for analyzing the relative content of a material consisting of two main constituents [170, 171]. Binary mixtures in which this method is useful include water in soil and wood, and fat in boneless meat, cream/skim milk mixtures and wool, etc, see section 12.4 for more details.

6.5.4. Resonance Mapping

As indicated in chapter 3, the cross-sections of some materials exhibit resonances at certain radiation energies. Monitoring radiation transmission at these resonance energies can be used to detect the presence and concentration of such elements. With a number of intersecting projections, a tomographic image of each element can be reconstructed using one of the methods discussed in section 6.4. This *elemental mapping* process can be used either with neutrons [172, 173], or photons [174, 175]. In order to detect such resonances, the radiation source has to have an energy that spans the resonances of interest. For example, neutrons exhibit strong cross-sections for nitrogen and oxygen in the energy range from 1 to 3 MeV [176], thus requiring a source within that range. Neutrons produced by linear accelerators, say the Be(d,n) reaction, or isotopic sources [177], such as Am/Be or ^{252}Cf, provide neutrons in this resonance range. However, nuclear resonances in the nuclear cross-sections of photons are encountered at higher energies, e.g. nitrogen at 9.17 MeV and chlorine at 8.21 MeV. Therefore, for photon nuclear-resonance measurements, powerful accelerators are required. On the other hand, atomic resonances occur at lower energies, when the photon energy matches the binding energy of the electrons in the element(s) of interest. This methods is called critical-edge absorption and is discussed in section 12.2.2.1. Resonances at lower photon energies are also possible by the Mössbauer effect, see section 6.5.5.

To simplify the analysis of measurements, it is preferable to have a uniform-intensity source, the so-called "white" source, offered by the Be(d,n) reaction for neutrons. Such a uniform source provides a constant value for $I_0(E)$ in Eq. (6.1). It is also possible to correct for variation in the intensity of the incident source, by prior measurement of the source intensity in air, $I_0(E)$.

Resonance measurements require the use of a detector capable of measuring the radiation energy-spectrum. This can be achieved by employing an energy-sensitive detector, see section 4.5.4. For neutrons, it is also possible to use the time-of-flight measurement method to measure the neutron velocity, see section 4.5. This requires, however, a timing signal to determine the time of emission of the source, which can be achieved by a pulsed source. Note, however, that this resonance method can only detect certain elements in the object, those that have resonances in the range of the applied source of radiation.

6.5.5. Mössbauer Spectrometry

If the nuclei of a material are frozen in place, as in the case of a solid matrix, they cannot significantly recoil (gain momentum) when bombarded by an external source of radiation, and the entire solid recoils as a single rigid mass. If that source is a gamma-ray source with an energy equal to the difference between two nuclear states in the target nucleus, radiation will be strongly (resonantly) absorbed (the Mössbauer effect) bringing the target nucleus to an excited state. To have a gamma-source with the same energy as one of the excited states of the target nucleus, the source may consist of the excited nuclei of the same isotope as the absorber. A number of such sources are available, see section 2.2.2.3.

Most of Mössbauer spectroscopy analysis is performed using ^{57}Fe as a source, because it is the strongest recoilless absorption of gamma-rays at room temperature. Therefore, the technique is most useful in the analysis of iron, or iron containing samples, since natural iron contains 2.1% ^{57}Fe. As indicated in section 2.2.2.3, ^{57}Fe is produced by the decay of ^{57}Co, emitting a 14.4 keV gamma-ray. The low penetrability of photons in solids at this relatively low photon-energy limits the technique to the analysis of thin samples (a few μmm thick).

Doppler Effect. If the source, or the target, are moved at some speed relative to each other, a Doppler effect takes place by virtue of the change in the frequency (i.e. energy) of the incident photons as seen by the target. Since photons move at the speed of light, c, moving the source at a velocity v will change the energy of photons. This shift in energy will result in a mismatch between the energy of the photons (as received by the nucleus) and the difference in energy between its resonant states. Then the target will cease to be a resonant absorber of the incident photons, and the photons will be removed by the usual photon interaction process discussed in chapter 3. Since the nuclear levels are characteristic of each type of nucleus, the energy of the Mössbauer effect, in combination with the Doppler effect, can be used to detect the presence of certain elements within an object and its concentration.

The Mössbauer effect can be observed in the transmission modality by positioning the target between a Mössbauer source and a photon detector, and moving the source or the target to introduce the Doppler effect. Both the nuclei of the source and the target have to be incorporated into a solid matrix to meet the recoilless conditions of the Mössbauer effect. By varying the relative velocity between the source and target, in both the approaching and the receding direction, the amount of radiation transmitted through (i.e. not absorbed within) the material, varies from a minimum when the relative velocity is zero to a maximum as the target moves away from the source, provided of course an element with a nuclear energy level that matches the source energy is present. The variation of the count rate with source velocity is known as the Mössbauer spectrum.

The Mössbauer spectrum provides some useful indications. Photons absorbed by the target nucleus bring it to an excited state (an *isomer* of the parent nucleus). This affects the electrostatic attraction between the nucleus and its atomic electrons, as in effect the charge distribution of the nucleus is altered by its excitation. This effect shifts the Mössbauer spectrum. The *isomer shift* depends on the electron density near the target nucleus, which in turn depends on the *chemical composition* of the sample. The isomer shift can, therefore, provide information on the chemical composition of the sample, such as its oxidation state, or the metallurgic (phase) nature of the solid. Analytical information that can be deduced from the Mössbauer spectrum are discussed in a number of monographs; a detailed summary can be found in reference [178].

6.6. Charged-Particle Transmission

The continuous energy loss suffered by charged-particles, discussed in section 3.3, makes their transmission through materials different in nature from that of photons and neutrons. Moreover, while photons and neutrons can be transmitted through an object without encountering any collisions, hence maintaining their energy, transmitted charged-particles always lose some of their energy. However, as discussed in section 3.3.2, the broad energy distribution of beta-particles makes it possible to describe their transmission (attenuation) by an exponential relationship similar to that of photons and neutrons. Another noticeable difference between charged-particles and photons and neutrons, is that air has a noticeable effect on the attenuation of charged-particles, but a negligible effect on neutral radiation. Therefore, provisions should be made to account for air in transmission measurements with charged-particles. This can be done by performing the measurements in vacuum, or by reducing the source-to-detector distance to minimize losses in air.

6.6.1. Alpha Particles

The main removal mechanism of alpha-particles, as indicated in section 3.3.1, is inelastic scattering with the bound electrons of the atom. Although, this is a scattering process, the scattered alpha-particles do not significantly change their direction, because the mass of an alpha particle is much larger than that of the electrons it bombards. Therefore, the intensity and/or energy of the transmitted alpha-particles can be used as material property indicators. In addition, since the energy gained by the electrons can excite or ionize the atom, the energy emitted by the de-excitation of the atom can be used as an atom identifier (see section 8.7.3). If ionization occurs, the current resulting from it can be used as property indicator. This ionization process is utilized in smoke detectors based in the ionization principle, see section 10.5.2.

Energy-based Transmission. The energy carried by alpha-particles transmitted through a medium is determined by the value of their energy decrement per unit length (the stopping power), which is a function of the material density, as discussed in section 3.3. Therefore, for a fixed material density, the energy of transmitted particles will be indicative of the distance traveled by alpha-particles, hence the material's thickness. On the other hand, for a fixed distance, but for a given material composition, the energy of alpha-particles will be indicative of the material's density. The material composition has to be fixed, since as Eq. (3.14) indicates, the range of an alpha-particle is proportional to the mass-number of the material. Another obvious restriction, is that the thickness of the inspected material has to be less than the range of the alpha-particle for the material at the given source energy, otherwise no particles will be transmitted at all. This limits the applicability of alpha-particle transmission to material density of no more than 0.02 mg/mm^2 [179].

6.6.2. Beta Particles

The wide energy distribution of beta-particles permits the empirical description of their attenuation using an exponential relationship, Eq. (3.23), for modeling the transmission of beta-particles. Although this expression is similar to those used to model the transmission of photons and neutrons in section 6.1, the relationship for beta-particles applies at a distance less than the range of the beta-particles, since these particles cannot penetrate a distance greater than their range. Within that distance, the transmission of beta-particles provides indications similar to those of photons and neutrons, but are obviously limited to thin materials of areal density (total cross-section times distance) less than 3 mg/mm^2. Therefore, thickness, density and even radiography measurements are possible with beta-particles, as shown in chapters 11 and 13.

6.6.3. Electron Radiography

Electrons induced by x-rays bombarding a thin metallic foil and transmitted through thin sheets can be used to image thin low-atomic materials. The advantage of using x-rays to produce electrons, rather than a beta source, is the high intensity that can be produced by the former. High-energy (several hundred kV) x-rays are usually used to produce electrons by the photoelectric effect and Compton scattering, typically in a lead foil about 0.1 mm thick [134]. The image is then recorded on a radiographic film placed directly beneath the object. Electrons emitted from the foil, and consequently from the object, are diffused in many directions, necessitating the use of a fine-grain (slow) film to improve the quality of the image. The film should also be placed as close as possible to the object, preferably in a vacuum-cassette, to minimize the dispersion of the transmitted electrons. Low-energy photons passing through the object, and subsequently the film, would leave a weak uniform image, due to the low absorption of such photons in the object and in the film. This background image reduces the contrast of the image produced by the electrons. It is, therefore, desirable to filter-out the low-energy tail of the x-ray spectrum by equipping the x-ray machine with a several millimeters of copper to cut-off low-energy photons [134].

Chapter 7

SCATTERING METHODS

7.1. Introduction

Radiation scattering offers some advantages for probing, gauging and imaging over the more widely used transmission-through techniques, discussed in chapter 6. Scattering is a *localized* event, that is, it emerges from a point within an object, unlike the line-integrated signal of transmission. This is why the natural process of visual imaging, seeing by the naked eye, and the artificial process of photography, are both based on light reflection, or scattering. In both cases, a light source is required, the light is reflected off the surface of the viewed object and is recorded by a detector, the naked eye (assisted by spectacles, if needed), or a photographic film (behind a lens). Since visible light is reflected off the surface and does not penetrate deep within an object, a direct image of the *surface* of the object is obtained. However, radiation penetrates matter, and thus the scattered radiation signal is more complex than that of light scattering. This complicates the interpretation of recorded indications. However, unlike transmission, scattering does not require access to two opposite sides of the interrogated object, as is the case with transmission. This provides flexibility in arranging the measurement setup, and enables the investigation of structures where accessibility to two opposite sides of the object is not possible, or when the object is too thick, or too thin. Note that meaningful transmission indications are difficult to obtain in very thick (or very dense) and very thin (or very light) objects, where a very weak signal is obtained in the former and a hardly altered signal is recorded in the latter. While in transmission only one measurement is obtainable for every narrow incident beam of radiation, many detectors can be used to monitor scattering from the same radiation beam simultaneously, since radiation scatters in various directions. The ability to use many scattering detectors is helpful in improving the level of confidence in the measurements, or to provide multiple indications, with the same source. Such multiple indications can be used to obtain additional information, e.g. on object's homogeneity or even its composition. While scattering from a point (or a small point) in an object is useful in probing, overall scattering from a larger region of the object can be used in radiation gauging for bulk measurement. Imaging can also be accomplished in scattering by combining indications arising from many points of the object. Since the nature of scattering of charged-particles differs from that of photons and neutrons, the scattering of charged-particles is discussed separately in section 7.10.

7.2. Measurement Model

In spite of the many advantages of scattering discussed in section 7.1, its use is hindered by the complexity of its interpretation process. This complexity arises from the fact that the scattering signal from a point within an object is attenuated by the surrounding material, as radiation travels from the source to the scattering point and from that point towards the detector. This is reflected in the measurement models discussing in this section. Like all measurement models, the scattering measurement model has some inherent physical assumptions that are emphasized here.

To model the scattering process, let us focus first on single (once scattered) events, i.e. the first scattering event a radiation source encounters within an object. At that point of scattering, let us create a fictitious infinitesimal volume, ΔV, as schematically shown in Figure 7.1. Let us consider a *point source* that emits radiation *isotropically* with an intensity (particles per unit time) of $I_0(E)$ per unit energy at energy E.

From Source. The number of particles directed towards ΔV per unit area, with an energy E within a small energy interval E to $E + \Delta E$, i.e. the radiation flux, $\psi_i(E)$, *directed towards* ΔV will then be equal to:

$$\psi_i(E) = \frac{1}{4\pi R_{in}^2} \int_E^{E+\Delta E} I_0(E) dE \tag{7.1}$$

where R_{in} is the distance from the source point to the scattering point.

Pre-Scattering Attenuation. As the source radiation approaches the scattering volume, the incident flux will be attenuated by the material present between the source and the scattering point, by the *attenu-*

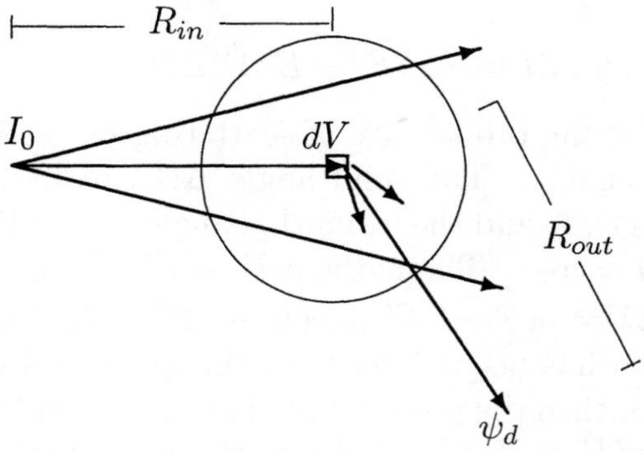

Figure 7.1. A schematic showing scattering from an infinitesimal volume, ΔV, within an object.

ation factor, $f(E)$:

$$f(E) = \exp\left(-\int_0^{R_{in}} \int_E^{E+\Delta E} \Sigma(E,x) dE dx\right) \tag{7.2}$$

where $\Sigma(E,x)$ is the total (absorption plus scattering) cross-section at the source energy E of the material present at point x along the path of the incident radiation. The exponential term in Eq. (7.2) accounts for the attenuation of radiation along the path of travel from the source to the point of scattering. Then, the radiation flux *incident on* ΔV will be equal to :

$$\psi(E) = \psi_i(E) f \tag{7.3}$$

Scattering within Volume. Recalling that $\Sigma(E)\psi(E)$ is equal to the interaction rate per unit volume at energy E, see section 3.6, the number of radiation collisions, or scatters, N_s, that take place within ΔV is equal to:

$$N_s = \Sigma_s(E)\psi_V(E)\Delta V \approx \Sigma_s(E)\psi(E)\Delta V \tag{7.4}$$

where $\Sigma_s(E)$ is the scattering cross-section (probability per unit length) at energy E and $\psi_V(E)$ is the radiation flux within ΔV at energy E, which is assumed to be equal to the flux, $\psi(E)$, incident on the surface of ΔV, with ΔV being infinitesimally small. If this is not the case, the attenuation of the incident radiation within ΔV should be included in Eq. (7.3).

Towards Detector. Radiation reaching ΔV can scatter to any direction. The number of scatters directed towards a specific detector located along the direction $\vec{\Omega}$, covering a small cone define by $\Delta\Omega$ can be expressed as:

$$N_{\Omega}\Delta\Omega = N_s p(E \rightarrow E'; \vec{\Omega})\Delta\Omega \tag{7.5}$$

where $p(E \rightarrow E'; \Omega)$ is the probability of scattering per solid angle and E' is the scattering energy. The solid angle $\Delta\Omega$ can be expressed in terms of the polar angle, θ, and the azimuthal angle, ϕ, see Figure 3.1, as $\Delta\Omega = \Delta\mu\Delta\phi$, with $\mu = \cos\theta$. Therefore, $p(E \rightarrow E'; \vec{\Omega})$ can be expressed as: $p(E \rightarrow E'; \vec{\Omega})\Delta\Omega = p(E \rightarrow E'; \mu)p(E \rightarrow E'; \phi)\Delta\mu\Delta\phi$. Since the scattering of radiation has no preference in the azimuthal direction, as discussed in chapter 3, then the probability of scattering in the azimuthal direction is simply, $p(E \rightarrow E'; \phi) = \frac{1}{2\pi}$, as the azimuthal angle varies from 1 to 2π. The probability of scattering, $p(E \rightarrow E'; \mu)$ is more complicated and depends on the type and energy of radiation and the scattering nucleus/atom (see chapter 3). Thus,

$$p(E \rightarrow E'; \vec{\Omega}) = \frac{p(E \rightarrow E'; \mu)}{2\pi} \tag{7.6}$$

Post-Scattering Attenuation. Radiation traveling, from the scattering point, a distance R_{out} to reach a detector will be attenuated by a factor, $g(E')$:

$$g(E') = \exp\left(-\int_0^{R_{out}} \int_{E'}^{E'+\Delta E'} \Sigma(E', y)dE'dy\right) \tag{7.7}$$

where $\Sigma(E', y)$ is the total cross-section of the material present at point y along the path of the scattered radiation, E' is the energy of the scattered radiation determined by the kinematics of scattering (see chapter 3). Then, the flux, ψ_d, of radiation reaching a detector of a view area ΔA_d can be expressed as:

$$\psi_d(E') = \frac{N_{\Omega}\Delta\Omega}{\Delta A_d} g(E') \tag{7.8}$$

With $\Delta\Omega = \frac{\Delta A_d}{R^2}$, and using Eqs. (7.5) and (7.6), Eq. (7.8) becomes:

$$\psi_d(E') = N_s \frac{p(E \rightarrow E'; \mu)}{2\pi R_{out}^2} g(E') \tag{7.9}$$

Overall Model. From, Eqs. (7.4) and (7.3), Eq. (7.9) is expanded to:

$$\psi_d(E') = \Sigma_s(E)\psi_i(E)f(E)\Delta V \frac{p(E \to E'; \mu)}{2\pi R_{out}^2} g(E') \tag{7.10}$$

The scattering volume ΔV is equal to $\Delta A_s \Delta x$, where ΔA_s is the area of the surface of ΔV facing the incident radiation beam and Δx is the width of ΔV along the direction of incident radiation. Using Eq. (7.1), for an isotropic source, Eq. (7.10) can be rewritten as:

$$\psi_d(E') = \left[I_0(E)\frac{A_s}{4\pi R_i^2}\right][f(E)]\,[\Sigma_s(E)\Delta x]\left[\frac{p(E \to E'; \mu)}{2\pi R_{out}^2}\right][g(E')] \tag{7.11}$$

Eq. (7.11) is rearranged so that each term within the square brackets relates to a different step in the interaction process: the source, attenuation before scattering, within the scattering volume, scattering towards the detector, and attenuation following scattering, respectively.

Inherent Assumptions. In arriving at the measurement model of Eqs. (7.10) or (7.11), in addition to the fundamental assumption of *single scattering*, it should be kept in mind that a number of *lumping* processes were introduced: in energy, volume and solid angles. That is, source particles within a certain energy interval, E to $E + \Delta E$, were lumped into a single energy, E, the scattering energy interval was represented by a single energy, E', and particles scattered in the direction $\vec{\Omega}$ within the solid angle Ω and $\Omega + \Delta\Omega$ were assumed to have the same probability of scattering. Moreover, all scattering events within the volume ΔV were lumped into a single point, enabling the definition of a single angle of scattering. Of course, such lumping is adequate as long as the intervals used for energy, solid angle and volume are sufficiently small. For larger intervals, this lumping process will introduce some error due to the averaging effect it inherently introduces. It should also be noted that the measurement model of Eqs. (7.10) or (7.11) *does not account for the detector's efficiency or response function.*

Point-wise Indication. Probing with a confined or collimated source beam, enables monitoring a single point around ΔV. However, the attenuation factors, exponential terms in Eqs. (7.10) and (7.11), must be known in advance. This is possible if one is examining a uniform structure, of known properties, for defects. Then the amount of radiation scattering will reflect changes in $\Sigma_s(E)p(E \to E'; \mu)$, since all the other parameters are determined by the source-detector setup. Scatter probing can then be used to detect the presence of inclusions or deterioration in an otherwise uniform material. If on the other hand, scattering is used to probe for void flaws, then $\Sigma_s(E)p(E \rightarrow E'; \mu)$ would be equal to zero, and no scattering signal, aside from scattering from the surroundings, would occur. Scattering can then be used not only to enable the detection of a flaw, but it also to determine its location, since the position of the scattering volume is determined by the source detector arrangement. The range of Δx in Eq. (7.10) or Eq. (7.11) within which no scattering occurs will also be indicative of the width of the flaw in the direction of the incident radiation.

Bulk Indication. For bulk gauging, one would need to cover a large volume of the object. Then the models of Eqs. (7.10) and (7.11) will not be directly useful, since one will be monitoring many scattering events. However, the overall indication will still be reflective of the scattering cross-section of the medium, Σ_s, and thus can be used to provide some useful volume-averaged physical properties of the medium as shown later in section 7.5.

Indications for Imaging. The models of Eqs. (7.10) or (7.11) reflect the complexity of imaging with scattered radiation. Although scattering lends itself naturally to point-by-point imaging, since scattering from individual points can be monitored, scattering from each point as the models of Eqs. (7.10) or (7.11) indicate, depends on the nature of the adjacent points, through the attenuation (exponential) factors in the model. This complicates the imaging process and makes image reconstruction more challenging than that for transmission, discussed in section 6.4.

Eqs. (7.10) or (7.11) also reflect the fact that the nature of the information obtained from scattering will depend on the type of radiation and the nature of the interaction involved. Since photons and neutrons differ in their interactions with matter, as discussed in chapter 3, scatter techniques from these radiation types are discussed separately in the ensuing sections.

7.2.1. Model for Compton Scattering

In the energy range of 0.3 to 2 MeV, Compton scattering is the dominant mode of photon interaction in most materials. In Compton scattering, as discussed in section 3.4.2, photons interact with the (relatively) free electrons of the matter. Scattering measurements can, therefore, be utilized to determine the electron density, and consequently the mass density, of the scattering material. This is because, the scattering cross section, Σ_s in the measurement models of Eqs. (7.10) or (7.11), is related to the electron density and mass density, in accordance with Eqs. (4.5) and (3.51), respectively. This facilitates writing the measurement model of Eq. (7.10) in terms of a measured detector response $S(E)$, instead of the particle flux, $\psi_s(E)$, as [180]:

$$S(E) = Kf\Sigma_{0,s}(E)\rho_e p(E' \to E; \mu)g \tag{7.12}$$

where now $f(E)$ and $g(E')$ of Eqs. (7.2) and (7.7), respectively, can be expressed as:

$$f = \exp\left(-\int_0^{R_{in}} \rho_e(x)dx \int_E^{E+\Delta E} \Sigma_{0,s}(E)dE\right) \tag{7.13}$$

and

$$g = \exp\left(-\int_0^{R_{out}} \rho_e(y)dy \int_{E'}^{E'+\Delta E'} \Sigma_{o,s}s(E')dE'\right) \tag{7.14}$$

Eq. (7.12) is simplified by incorporating all the geometry-related parameters in Eq. (7.10) into a system constant K. Once the source and detector location and collimators are fixed, the volume of the scattering voxel, ΔV, and subsequently the distances from the source to the detector, and the detector's sensitive area, become constant. The constant K also incorporates other non-varying parameters, such as the source strength, as well as the detector's efficiency, and thus enables the use of the measured detector's response, S, instead of the flux at the detector site, ϕ_d. The direct proportionality between the scattering cross-section and the electron density enables the replacement of the scattering cross-section, Σ_s in Eq. (7.10), with that of some reference material, say water. The relative density of material with respect to the reference material is denoted by ρ_e in Eq. (7.12). Also, the dominance of Compton scattering over all other types of interactions enables the replacement of the total cross-section, $\Sigma's$ in Eq. (7.8), by the scattering cross-section, hence $\Sigma_{0,s}\rho_e$ in Eq. (7.12). This also facilitates the separation of the double integrals in Eqs. (7.2) and (7.7), in energy and space, into two separate integrals in Eqs. (7.13) and (7.14). The probability of Compton scattering of photons of energy E to energy E' by an angle $\cos^{-1}\mu$, $p(E' \to E; \mu)$ in Eq. (7.12), is material independent, since it is related to the interaction of a single photon and a single electron. Hence, the Klein-Nishina formula of Eq. (3.41) can be used to evaluate this probability. However, a convenient and a good approximation for $p(E' \to E; \mu)$ in Eq. (7.12) is given by Eq. (3.45).

Another interesting fact that can be utilized in Compton scatter probing and imaging is that the kinematics of scattering dictates a unique relationship between the energy of incident photons and that of the scattered ones, E and E' in Eqs. (7.10) and (7.11), according to Eq. (3.37). This is useful as it enables the determination of the direction of scattering by measuring the energy of the scattered photons with an energy-sensitive detector. Therefore, the energy spectrum of singly-scattered photons can be directly related to their direction of scattering, if a narrow source beam is employed.

7.2.2. Model for Neutron-Elastic Scattering

As discussed in section 3.5, neutrons can scatter elastically or inelastically. The latter reaction is accompanied by gamma-ray emission, as part of the neutron energy is converted into excitation energy of the target nucleus. Therefore, inelastic-scattering is mainly employed in emission techniques, discussed in chapter 8. Neutrons at low energy (cold neutrons: 10^{-4} to 10^{-3} eV) scatter as waves (i.e. diffract), see section 3.5.5.1. In the literature "neutron scattering" is often associated with neutron diffraction techniques used in studying the lattice structure of materials. The probing methods discussed here are those relying on elastic interactions.

Neutron elastic-scattering is particularly attractive for NDE applications, since neutrons interact with the nucleus, thus unlike photons, they provide element-specific information. The kinematics of neutron scattering, manifested by Eq. (3.80), indicate that neutrons lose more energy when scattered by low mass-number elements than by heavy ones. Therefore, neutron elastic-scattering is particularly suited for diagnosing hydrogen-rich materials. This becomes obvious if Eq. (3.80) is rewritten for hydrogen, $A = 1$, leading to:

$$\frac{E'}{E} = \cos^2\theta \quad \text{for hydrogen,} \quad A = 1 \tag{7.15}$$

where E' is the outgoing neutron energy, E is the energy of the incoming neutron, θ is the angle of scattering. Eq. (7.15) indicates that the same amount of fractional energy-loss are attained at two angle cosines: $\pm\cos\theta$. This is not, however, physically possible, since a neutron and a hydrogen nucleus (a proton) are about equal in mass, and the nucleus is considered to be at rest. This situation is similar to that observed in a billiard (pool) game, where a collision between an incoming ball and identical ones at rest leads only to movement of the incoming ball in the forward direction. This *forward-scattering* phenomenon should be kept in mind when applying the single-scatter model of Eq. (7.10) to neutron collisions with hydrogen. Another interesting observation is

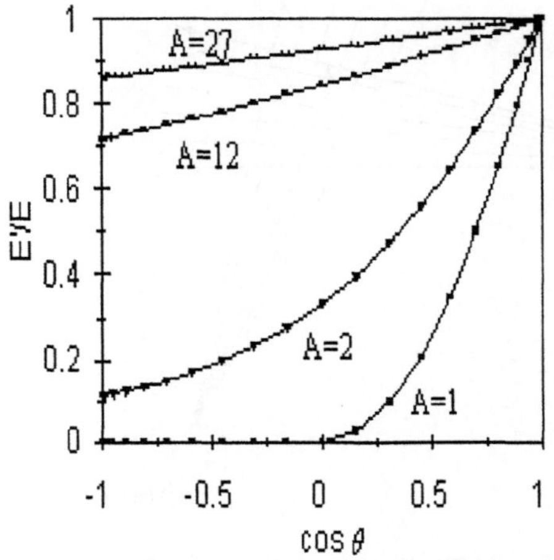

Figure 7.2. Energy-angle relationship for single neutron elastic-scattering with various nuclei: A = mass number, $\frac{E}{E'}$ = incident/scattering energy, θ = scattering angle.

that, according to Eq. (7.15), the neutron can lose all of its energy if it scatters by hydrogen at angle of 90°.

Figure 7.2 shows a plot of the angle of scattering with the relative energy-change, using Eq. (3.80), for some light elements. The Figure indicates that neutrons singly scattered by an element cannot have an energy below a certain minimum energy, given by Eq. (3.81), and this energy depends on the mass-number of the scattering nucleus. This fact can be used to distinguish between various elements, say by monitoring the energy of single-scattered neutrons from a monoenergetic source, say at 90°.

Another point that should be kept in mind in applying the scattering model of Eq. (7.10) or Eq. (7.11) to neutrons: unlike photons, the probability of scattering $p(E' \longrightarrow E, \mu)$ is element-dependent. The total cross-sections, $\Sigma(E)$ and $\Sigma(E')$, are also material-dependent, and may not always be approximately equal to the scattering cross-section, Σ_s. In addition, $p(E' \longrightarrow E, \mu)$ is not easy to analytically express as a mathematical function. This makes direct application of the scattering measurement model often difficult.

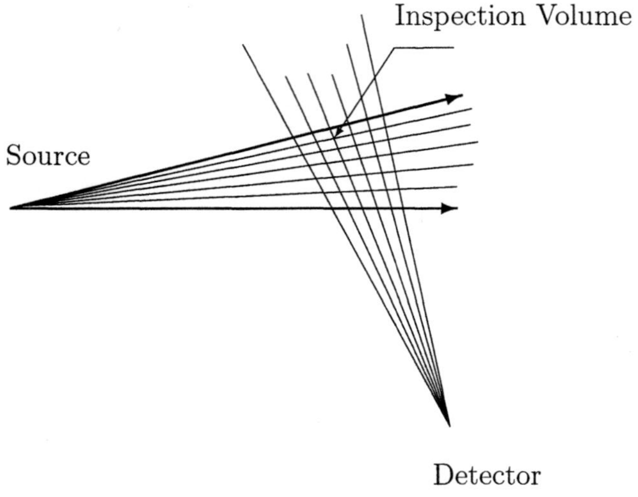

Figure 7.3. A schematic of probing with radiation scattering.

7.3. Point Probing

Scattering is uniquely suited for probing within a small volume inside an object. Scatter probing can be accomplished with the aid of a source and a detector with confined fields-of-view. The intersection of the two fields-of-view defines the *inspection-volume*, as schematically shown in Figure 7.3. If the inspection-volume is small, one can speak of an inspection point. Even if the inspection-volume is not sufficiently small, probing will provide an average estimate of the measured property within the volume, i.e. probing lumps the inspection-volume into a point. One must, however, design the system to eliminate or reduce the contribution of multiple scattering to the detector, to limit the indication to single-scatter events occurring within the inspection-volume. Multiple scattering can be reduced by proper detector collimation and by electronically discriminating against multi-scattered radiation by its lower energy. Note that because of radiation divergence, the size of the inspection-volume depends on the distance between the inspection-volume and the source and detectors; the larger the distance the larger the volume.

Attenuation Effects. The measured property in scatter probing is the scattering cross-section, Σ_s in Eq. (7.10). However, in order to determine this property, one must account for the attenuation of the incident and the scattered radiation beams, the $f(E)$ and $g(E')$ factors in Eq. (7.10); unless they are known in advance as in the case of inspecting an object of known properties for unexpected discontinuities (flaws) or (inclusions). In general, however, these attenuation factors may not be known in advance, and need to be determined along with the unknown scattering cross-section. The probing problem would then have three unknowns: the scattering cross-section and the two attenuation factors, before and after scattering. Obviously, one scattering measurement would not be sufficient to solve a problem of three unknowns. One simple method for dealing with this problem in NDE, when inspecting for small discontinuities (flaws or inclusions), is to utilize predetermined values for f and g calculated from *a priori* knowledge of the constituents of a reference (flawless) object [181]. However, a number of other methods have emerged for dealing with this *attenuation problem*, as discussed below. Since photons are generally easier to collimate than neutrons, most of the developed methods are based on photon scattering, particularly Compton Scattering. However, probing with Rayleigh scattering and neutron scattering is also used, as discussed in sections 7.3.10 and 7.3.11, respectively.

7.3.1. Neglected Attenuation

The attenuation factors can be considered to be approximately equal to unity, if the argument of their exponential term, see Eqs. (7.2) and (7.7), is close to zero. This requires satisfying one of the following conditions:

1 $\Sigma(E, x)$ and $\Sigma(E', y)$ are small in value, for all values of x and y, i.e. at all points surrounding the inspection-volume; provided that R_{in} and R_{out} are not too large.

2 If R_{in} and R_{out} are short distances provided that $\Sigma(E, x)$ and $\Sigma(E', x)$ are not too high in value.

The macroscopic cross-section, as discussed in section 3.2.3, is a multiplication of the atomic density, N, which is related to mass-density by Eq. (3.2), and the microscopic cross-section, σ. Therefore, condition 1 is satisfied if the material surrounding the point of interest has a low mass-density, and in turn low atomic-density. This can be the case, for example, if one is examining a high-density inclusion within a low density medium. Alternatively, condition 1 is satisfied, if the microscopic cross-section of the surrounding material for the employed radiation is low. The latter condition is usually achieved at high radiation energy, since the cross-section of all radiation types tends to decrease with energy. This so-called *high-energy approximation* was utilized [182] with an X-ray beam derived from a 5.6 MeV linear accelerator, to measure the density of internal body tissue. The same approach can be used also in industrial applications, when a high-energy source is available.

Condition 2 requires in effect that the distance radiation travels within the inspected object be small. This is because, as schematically shown in Figure 7.1, portions of R_{in} and R_{out} are in the air outside the object, where the total cross-sections is practically zero, due to the very low density of air. Therefore, Condition 2 is satisfied when examining small-size objects. Then, one can directly measure the scattering cross-section, Σ_s, of the object. This approximation was used in examining breast tissue extracted from female patients undergoing mastectomy [183]. The same approach can also be applied in the inspection of small industrial samples, say to examine consistency in their material content.

It should be kept in mind that satisfaction of condition 1 requires that condition 2 is not grossly violated, and the opposite is true; otherwise the argument of the exponential term in Eqs. (7.2) or (7.7) can become quite large. It should also be noted that by setting the attenuation-factors to unity, an approximate value of the third unknown, Σ_s, is obtained. This can be helpful in detecting large changes in the value of Σ_s, even when the attenuation factors are not close to unity.

7.3.2. Signal Modulation

Another approach to correct for radiation attenuation is to adjust the scattering signal by an amount that depends on the location of the scattering point along the path of the incident beam. This can be done by multiplying the response of the scattering detector by a factor that increases in value with the depth of inspection-volume along the incident beam. A similar approach can be used to correct for attenuation along the path of scattered radiation, the farther away the inspection-volume from the detector, the higher the correction factor. In a system that scans more than one inspection point, signal adjustment can be done on-line by controlling the amplification of the recorded scattering detector response, by an amount that depends on the relative position of the inspection-position with respect to the source and the detector [184, 185]. The amount of correction can be determined in advance using a reference object of properties similar to those of the inspected object [186, 187, 188].

7.3.3. Attenuation Averaging

Another way to approximately eliminate the effect of attenuation of the incident beam is to expose the object to radiation from two opposite sides. The average of the two measurements should then be nearly independent of the position of the point within the object [184, 189]. If the thickness of the object traversed by the incident beam is d, and scattering is monitored at a distance x along d from the direction of the first beam, see Figure 7.4, then the average of the two scattering measurement is:

$$\bar{S}(x) = K\ \bar{I}(x)\ \Sigma_s(x)\ g \tag{7.16}$$

where K is a system constant, g is the attenuation factor of the scattered beam (same in both exposures) and $\bar{I}(x)$ is the average intensity of two opposite beams at the point of scattering,

$$\begin{aligned} \bar{I}(x) &= \frac{I_0}{2}\left\{\exp(-\Sigma x) + \exp[-\Sigma(d-x)]\right\} \\ &= \frac{I_0}{2}\exp(-\Sigma x)\left\{1 + \exp[-\Sigma(d-2x)]\right\} \end{aligned} \tag{7.17}$$

where Σ is the total cross-section of the incident radiation. As Eq. (7.17) indicates, the overall reduction of the beam intensity at x, $\exp(-\Sigma x)$, is dampened by the term in the curly brackets, making $\bar{S}(x)$ of Eq. (7.16) less dependent on the attenuation of the incident beam. When Σx is not very large, then $\bar{I}(x)$ becomes approximately equal to $\frac{I_0[1+\exp(-\Sigma d)]}{2}$. Then a transmission measurement recorded for one of the beams can enable the determination of $I_0 \exp(-\Sigma d)$, hence providing an estimate of $\bar{I}(x)$. Note, however, that this method of *two opposite beams* only addresses the attenuation of the incident, not the scattered, radiation. However, the average of the response of two symmetrical scattering detectors can also be used to reduce the attenuation effect [184]. The method has the advantage of averaging also the divergence effect of the incident radiation, since if one of the beams diverges quite significantly at one side of the object, the opposite beam would not have diverged as much; as one can see by drawing two opposite cones of the same width on the same axis.

7.3.4. Constant-Transmission

This method attempts to minimize the effect of radiation attenuation on the scattering signal by placing an amount of water in the way of the incident radiation to make the transmission measurement about constant and equal to some pre-determined value [190]. Now the transmission signal becomes equal to:

$$I = I_0 \exp[-\Sigma_w x_w - \Sigma(x_i + x_o)] \tag{7.18}$$

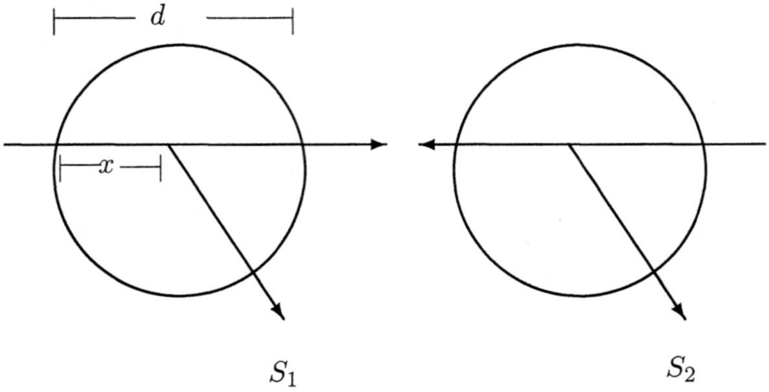

Figure 7.4. Schematic for scatter probing with two opposite exposures.

where I_0 is the intensity of the incident beam and w refers to water, x to a distance along the incident beam, i to the path of the incident radiation before it reaches the scattering point, and o to the path of radiation from the scattering point to the position of the transmission detector. Now, the constant I can be made to approach a constant value by varying x_w, so that the attenuation of the beam before reaching the scattering point, $\exp[-\Sigma_w x_w - \Sigma x_i]$, becomes essentially constant because of its dominance over the attenuation of the incident beam over x_o, i.e. $\exp[\Sigma x_o]$. Consequently,

$$I \approx I_0 \exp[-\Sigma_w x_w - \Sigma x_i] \tag{7.19}$$

Now with the scattering measurement modeled as:

$$S = K \exp[-\Sigma_w x_w - \Sigma x_i] \rho_e \exp[-\Sigma' y] \tag{7.20}$$

where y is the distance of travel along the scattering beam, Σ' is the cross-section of the scattered radiation, ρ_e is the electron-density at the point of scattering and K is a system constant. Using Eq. (7.18) and (7.20), the ratio between the scattering and the transmission measurements becomes:

$$\frac{S}{I} = \left[\frac{K}{I_0}\right] \rho_e \exp[-\Sigma' y] \tag{7.21}$$

Then $\frac{S}{I}$ is dependent on the electron-density of the scattering point and the attenuation of the scattered radiation. By combining the response

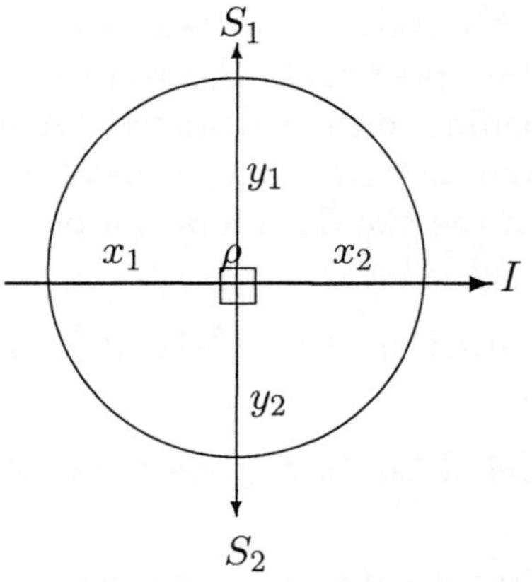

Figure 7.5. Schematic of scatter/transmission method for normalized response.

of symmetrically positioned scattering detectors, the effect of the attenuation of the scattered radiation is diminished, and $\frac{S}{I}$ becomes strongly dependent on ρ_e. This method would work reasonably well, for not-too-thick objects, where the addition of water has an effect on dampening the attenuation of the incident radiation in the object. If the object is thick, the addition of water would not affect much the attenuation of radiation within the object, and the use of this method can introduce a significant amount of error.

7.3.5. Normalized Scattering and Transmission

Reference [191] presented a method that approximately removes the effect of attenuation by normalizing scattering and transmission measurements with respect to a reference material of density, ρ_r. The concept of this method is illustrated using Figure 7.5. A collimated monoenergetic gamma-source (^{137}Cs, 662 keV) is directed towards the point (voxel) of interest and the scattering off the voxel is measured at two identical angles at opposite directions of the object (90°), either at the energy corresponding to single scattering, using Eq. (3.37), (288 keV in the case of ^{137}Cs), or using collimated detectors. This enables the application of the single-scatter model of Eq. (7.12).

The ratio between the scattering measurement S_1 at some unknown electron density, ρ, in the voxel, and the measurement $S_{1,r}$ at some reference density, ρ_r, recorded for some known material (uniformly occupying the entire volume of the object) can be expressed as:

$$\frac{S_1}{S_{1,r}} = \frac{\rho}{\rho_r} \exp[-\mu(\rho_{x1} - \rho_r)x_1 - \mu'(\rho_{y1} - \rho_r)y_1] \tag{7.22}$$

where μ and μ' are the mass attenuation-coefficients at the incident and scattering energies, respectively, the distances x_1 and y_1 are as shown in Figure 7.5, and the subscripts refer to the average density of the material in the chords corresponding to these distances. A similar expression can be written for the second detector, S_2, so that the geometric average of the relative response of the two detectors can be expressed as:

$$\sqrt{\frac{S_1 S_2}{S_{1,r} S_{2,r}}} = \frac{\rho}{\rho_r} \exp[-\mu(\rho_{x1}-\rho_r)x_1 - 0.5\mu'(\rho_{y1}-\rho_r)y_1 - 0.5\mu'(\rho_{y2}-\rho_r)y_2] \tag{7.23}$$

The transmission model of Eq. (6.2) gives the relative response:

$$\frac{I}{I_r} = \exp[-\mu(\rho_{x1} - \rho_r)x_1 - \mu(\rho_{x2} - \rho_r)x_2)] \tag{7.24}$$

Assuming that the densities along the chords of Figure 7.5 are all approximately equal to some value, say ρ_c, then from Eqs. (7.23) and (7.24) one can show that:

$$\sqrt{\frac{S_1 S_2}{S_{1,r} S_{2,r}}} \left[\frac{I_r}{I}\right]^{\frac{\mu+\mu'}{2\mu}} = \frac{\rho}{\rho_r} \times \exp\left[-(\rho_c - \rho_r)\{0.5\mu(x_2 - x_1) + 0.5\mu'(x_1 + x_2 - y_1 - y_2\}\right] \tag{7.25}$$

The exponential term in Eq. (7.25) can be assumed to be approximately close to unity, if the difference in distances in its arguments is small. With this approximation, the local density, ρ, in the voxel can be estimated as:

$$\rho \approx \rho_r \sqrt{\frac{S_1 S_2}{S_{1,r} S_{2,r}}} \left[\frac{I_r}{I}\right]^{\frac{\mu+\mu'}{2\mu}} \tag{7.26}$$

In spite of its assumptions, this method can be useful in NDE applications where the object is otherwise uniform, and one is interested in detecting local anomalies.

7.3.6. Single Low-Energy Source Transmission-Assisted

This approach applies to photon scattering, when the source energy, E, is chosen so that it is much smaller than the rest-mass of the electron. Then, as Eq. (3.37) shows, the energy of the scattered photons, E', becomes independent of the scattering angle, and approximately equal to the energy of the incident photons. Consequently, the dependence on energy of the total cross-sections, $\Sigma_(E, x)$ and $\Sigma_(E', y)$, in Eqs. (7.2) and (7.7), respectively, can be eliminated. That is, if the path of the

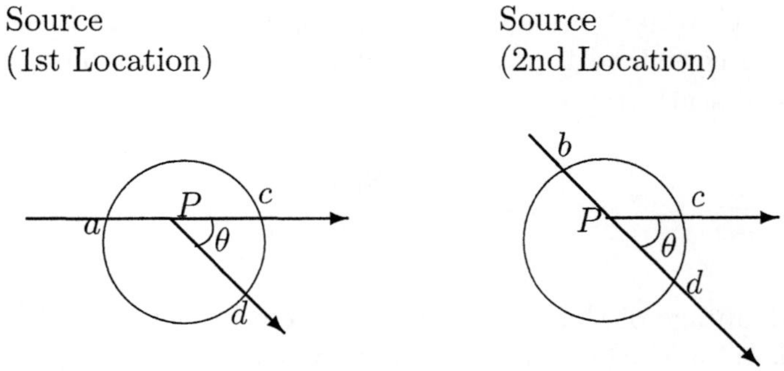

Figure 7.6. Schematic for scatter probing with a single low-energy source, transmission-assisted

incident radiation (to the scattering pointy) is exchanged with the path of the scattered radiation (from the same scattering point), one would get the same attenuation factors. This fact was utilized to correct for attenuation by moving the source, after collecting the first scattering measurement, to a location along the path of the scattered radiation in the first measurement [192], as shown in Figure 7.6. Advantage was also taken of the transmission measurements to provide supplementary measurements.

For the this system, with the aid of the schematic of Figure 7.6 and the scattering and transmission models of Eqs. (7.12) and (6.2), respectively, one can write the following equations:

$$S_1 = Kf(a \to P)\rho_e g(P \to d) \tag{7.27a}$$

$$S_2 = Kf(P \to d)\rho_e g(P \to c) \tag{7.27b}$$

$$I_1 = I_0 f(a \to c) = I_0 f(a \to P) f(P \to c) \tag{7.27c}$$

$$I_2 = I_0 f(b \to d) = I_0 f(b \to P) f(P \to d) \tag{7.27d}$$

where K is a constant incorporating all the unchanged parameters in the model of Eq. (7.12), and $f(x \to y)$ and $g(x \to y)$ designate, respectively, the attenuation (exponential) factors of the primary and the secondary beams as they path from point x to point y, and ρ_e is the electron-density at the scattering point, P. Note that the exponential nature of the attenuation factors enables the separation of the overall attenuation factors for transmission measurements into two separate factors, covering two connected segments of the path of the transmitted radiation. Since the incident and scattered energies are assumed to be independent of energy, the attenuation factors $f(P \rightarrow c)$ and $g(P \rightarrow c)$ can be taken to be equal to each other. Then, $f(P \rightarrow c) = g(P \rightarrow d)$, and Eqs. (7.27a) to (7.27d) can combined to give an estimate of the electron-density, ρ_e at the point of scattering as:

$$\rho_e = \frac{1}{K}\sqrt{\frac{S_1 S_2}{I_1 I_2}} \tag{7.28}$$

The disadvantage of this approach is the requirement that the source energy be less than 511 keV ($= m_0c^2$), which limits the penetration depth of the probing method. The method was utilized for lung and bone (body extremities) densitometry, [193, 194], owing to the low density of the former and the relatively small size of the latter. The sources used in these two medical applications had energies in the 100 KeV range, but other low-energy sources are also available, see Table 2.7.

7.3.7. Two-Source Transmission-Assisted

The method of 7.3.6 attempted to equate the attenuation factors of the incident and scattered radiation by approximately employing a low-energy photon source. The same approach can be applied without having to introduce the low-energy approximation by employing two photon sources, with energies, E and E'. The energy of the second source, E', must correspond the scattering angle used with the first source, in accordance with the Compton kinematics of Eq. (3.37). Equivalently, the scattering angle has to satisfy Eq. (3.37) for a given E and E' values. For example, let us choose ^{60}Co for the first source, which has an energy $E = 1250$ keV (average of its two close dual peaks, see Table 2.10). Let ^{137}Cs be the second source, then $E' = 662$ keV, as shown in Table 2.10. Then, according to Eq. (3.37), the scattering measurements have to be taken at a scattering angle, θ, equal to 50.44°.

In this method [192], transmission measurements, I_1 and I_2, are first recorded for the two sources, positioned as shown in Figure 7.7.a, so that their incident paths intersect at an angle, θ determined by Eq. (3.37). The second, lower-energy source, is then removed, and a scattering measurement, S_1, is recorded for the first source, Figure 7.7.a$'$. The test object (or the source-detector assembly) is subsequently rotated by 180°, to give a second scattering measurement, S_2, Figure 7.7.b. Using the models of Eqs. (7.12) and (6.2), the following expressions can be written for each of the four recorded measurements:

$$I_1 = I_0(E) f(a \rightarrow P, E) f(P \rightarrow c, E) \tag{7.29a}$$

$$\begin{aligned} I_2 &= I_0(E')f(b \rightarrow P, E')f(P \rightarrow d, E') && (7.29b) \\ S1 &= K_1 f(a \rightarrow P, E)\rho_e g(P \rightarrow d, E') && (7.29c) \\ S2 &= K_2 F(c \rightarrow P, E)\rho_e g(P \rightarrow b, E') && (7.29d) \end{aligned}$$

where the parameters are analogous to those used in section 7.3.6, with the subscripts 1 and 2 referring to measurements taken with the first source (of energy E) and the second source (of energy E'), respectively. Since the scattered beam and the second source possess the same energy, the attenuation factors $g(P \rightarrow d, E')$ and $f(P \rightarrow d, E')$ are equal, and the same applies to $g(P \rightarrow b, E')$ and $f(b \rightarrow P, E')$. With $g(P \rightarrow d, E') = f(P \rightarrow d, E')$ and $g(P \rightarrow b, E') = f(b \rightarrow P, E')$, the relationships of Eqs. (7.29a) to (7.29d) can be combined to yield the electron density at point P:

$$\rho_e = \sqrt{\frac{I_0(E)I_0(E')}{K_1K_2}\frac{S_1S_2}{I_1I_2}} \qquad (7.30)$$

This method was utilized in bone [195, 196] and lung density measurements [197].

7.3.8. Dual-Energy: Special Case

In a system consisting of two materials, one enclosing the other, as schematically shown in Figure 7.8, the need to move the source, or rotate the object, as required in the method discussed above, can be eliminated. This can be done by employing a dual-energy photon source, or two sources each with a different energy [198]. If a source emits photons with two energies, E_1 and E_2, then for the same angle of scattering, there would be two scattering energies, E_1' and E_2', according to Eq. (3.37). For these four energies, one can determine in advance the total cross-sections for the two materials in the inspected structure, say materials A and B. Then, if $\Sigma^A(E), \Sigma^A(E'), \Sigma^B(E)$ and $\Sigma^B(E')$ are all known (or measured) in advance, one can determine the electron-density at the point of scattering from the corresponding scattering measurements recorded at E_1' and E_2'. The corresponding system of equations, in accordance with the model of Eq. (7.12), are:

$$\begin{aligned} S_1 &= K_1 \exp[-\Sigma^A(E_1)x_A - \Sigma^B(E_1)x_B]\rho_e \\ &\quad \exp[-\Sigma^B(E_1')y_B - \Sigma^A(E_1')y_A] && (7.31a) \\ S_2 &= K_2 \exp[-\Sigma^A_(E_2)x_A - \Sigma^B(E_2)x_B]\rho_e \\ &\quad \exp[-\Sigma^B(E_2')y_B - \Sigma^A(E_2')y_A] && (7.31b) \end{aligned}$$

where K_1 and K_2 are systems constants and the other parameters are analogous to those used in section 7.3.6. In addition to the two equations

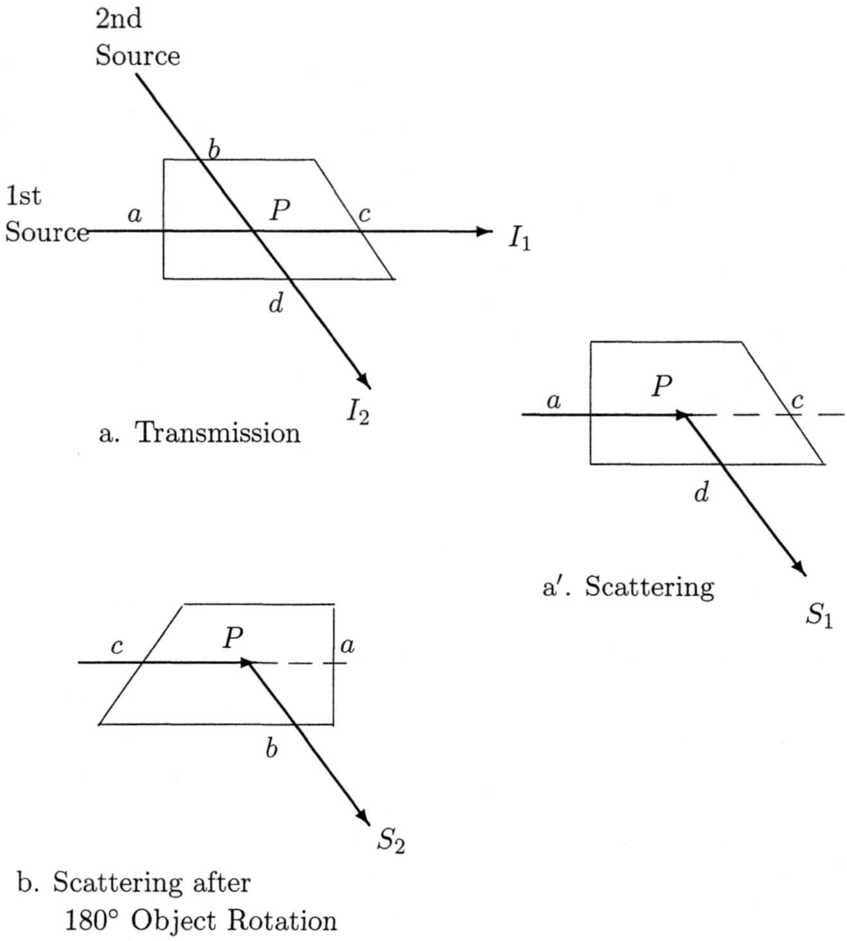

Figure 7.7. Schematic for the two-source transmission-assisted probing method.

(7.31a) and (7.31b), one can write the following three equations from the geometry of the problem, by referring to Figure 7.8:

$$x_A + x_B = x \tag{7.32a}$$

$$y_A + y_B = y \tag{7.32b}$$

$$x + y = d \tag{7.32c}$$

where x and y are distances determined by the positioning of the source, detector and scattering angle, and can be calculated in advance, and d is a constant distance dictated by the system geometry. The two equations

(7.31a) and (7.31b), and the three equations (7.32a), (7.32b) and (7.32c), can now be solved simultaneously, to obtain the main problem unknown, ρ_e, and the other four unknowns: x_A, x_B, y_B and y_A.

A version of this method was used for measuring the density of bone surrounded by muscle layers [199]. However, the solution of the problem was simplified from the simultaneous solution of five algebraic equations to the solution of only three equations, by taking advantage of the fact that $\Sigma(E)$ and $\Sigma(E')$ are related to each other, for the same material, via the photon energy, see section 3.4. This relationship can be determined in advance, and an equivalent cross-section can be calculated by the weighted average of the cross-sections at the incident and scattering photon energies for each material, see appendix F. These equivalent (also called effective) cross-sections can be measured experimentally, by scattering measurements in known configurations. Then one can designate a single effective cross-section, Σ_{eff}, for each source energies and each material. Therefore, Eqs. (7.31a) and (7.31b) can be expressed as:

$$S_1 = K_1\rho_e \exp[-\Sigma^A_{eff}(E_1)(x_A + y_A) - \Sigma^B_{eff}(E_1)(x_B + y_B)] \tag{7.33a}$$

$$S_2 = K_2\rho_e \exp[-\Sigma^A_{eff}(E_2)(x_A + y_A) - \Sigma^B_{eff}(E_2)(x_B + y_B)] \tag{7.33b}$$

Now the problem has three unknowns: ρ_e, $d_A = x_A + y_A$, and $d_b = x_B + y_B$. The sum of the last two unknowns, $d = d_A + d_B$ is a geometry-dependent but is a measurable constant, d, of Eq. (7.32c). Now, one has three equations that can be solved iteratively for the three problem unknowns: ρ_e, d_1 and d_2.

Although the dual-energy method discussed in this section depends only on two measurements, to solve what is essentially a three-unknown problem, it relies on geometric parameters that need to be known and measured in advance. Its use is also limited to the case of one material enclosed within another.

7.3.9. Dual-Energy: General Case

The restrictions of the dual-energy method of section 7.3.8 can be relaxed, by recording scattering for each source energy at two different locations, as schematically shown in Figure 7.9 [200]. For such an arrangement, one would obtain four measurements, which can be modeled using Eq.(7.12) as follows:

$$S_{1,A} = K_{1,A} f(a \to P, E_1)\rho_e g(P \to b, E'_1) \tag{7.34a}$$

$$S_{2,A} = K_{2,A} f(a \to P, E_2)\rho_e g(P \to b, E'_2) \tag{7.34b}$$

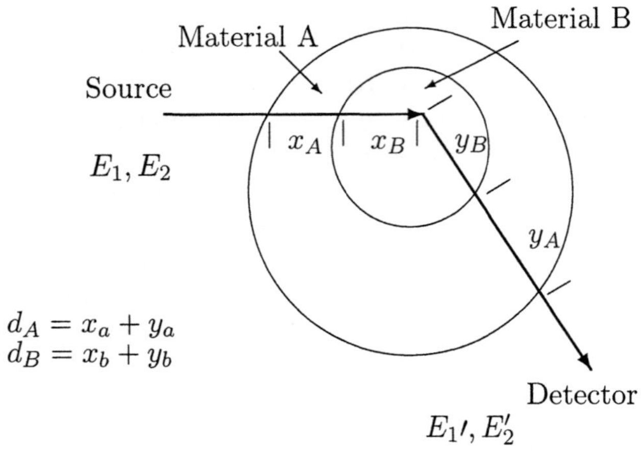

Figure 7.8. Schematic of the dual-energy probing method in an object consisting of two segregated materials.

$$S_{1,B} = K_{1,B} f(a \to P, E_1) \rho_e g(P \to c, E_1'') \qquad (7.34c)$$

$$S_{2,A} = K_{2,B} f(a \to P, E_2) \rho_e g(P \to c, E_2'') \qquad (7.34d)$$

where the usual convention is used, with A and B designating the two detectors, and E' and E'' referring to the scattering energies of the photons reaching these detectors, respectively, in accordance with Eq. (3.37).

Assuming that Compton scattering is dominant in the attenuation process, which is a valid assumption, unless the photon energy is too low and/or the material is very dense, see section 3.4, then $\Sigma(E)$ becomes proportional to the electron-density. The attenuation factors, f and g defined by Eqs. (7.2) and (7.7), respectively, can now be expressed, using Eq. (3.49), as:

$$\begin{aligned} f(a \to P, E_1) &= \exp\left(-\sigma_e(E_1) \int_a^P \rho_e(x) dx\right) \\ &= \exp[-\sigma_e(E_1) \Lambda_{a \to P}] \qquad (7.35a) \end{aligned}$$

$$\begin{aligned} f(a \to P, E_2) &= \exp\left(-\sigma_e(E_2) \int_a^P \rho_e(x) dx\right) \\ &= \exp[(-\sigma_e(E_2) \Lambda_{a \to P}] \qquad (7.35b) \end{aligned}$$

$$g(P \to b, E_1') = \exp\left(-\sigma_e(E_1') \int_P^b \rho_e(y) dy\right)$$

$$\begin{aligned}
&= \exp[-\sigma_e(E_1')\Lambda_{P\to b}] && (7.35c)\\
g(P\to b, E_2') &= \exp\left(-\sigma_e(E_2')\int_P^b \rho_e(y)dy\right) &&\\
&= \exp[-\sigma_e(E_2')\Lambda_{P\to b}] && (7.35d)\\
g(P\to b, E_1'') &= \exp\left(-\sigma_e(E_1'')\int_P^c \rho_e(y)dy\right) &&\\
&= \exp[-\sigma_e(E_1'')\Lambda_{P\to c}] && (7.35e)\\
g(P\to b, E_2'') &= \exp\left(-\sigma_e(E_2'')\int_P^c \rho_e(y)dy\right) &&\\
&= \exp[-\sigma_e(E_2'')\Lambda_{P\to c}] && (7.35f)
\end{aligned}$$

where σ_e is the microscopic cross-section per electron and Λ is equal the integral of the electron-density over a chord, the *areal electron-density*. The relationships of (7.35a) to (7.35f) enable the reformulation of the model equations of (7.34a) to (7.34d) as:

$$\begin{aligned}
S_{1,A} &= K_{1,A}\exp[-\sigma_e(E_1)\Lambda_{a\to P}]\rho_e\exp[-\sigma_e(E_1')\Lambda_{P\to b}] && (7.36a)\\
S_{2,A} &= K_{2,A}\exp(-\sigma_e(E_2)\Lambda_{a\to P}]\rho_e\exp(-\sigma_e(E_2')\Lambda_{P\to b}] && (7.36b)\\
S_{1,B} &= K_{1,B}\exp[-\sigma_e(E_1)\Lambda_{a\to P}]\rho_e\exp(-\sigma_e(E_1'')\Lambda_{P\to b}] && (7.36c)\\
S_{2,A} &= K_{2,B}\exp[-\sigma_e(E_2)\Lambda_{P\to c}]\rho_e\exp(-\sigma_e(E_2'')\Lambda_{P\to c}] && (7.36d)
\end{aligned}$$

The microscopic cross-section, σ_e's, are material independent and vary only with the source energy, thus they can be determined in advance. Therefore, the problem has four unknowns: ρ_e, $\Lambda_{a\to P}$, $\Lambda_{P\to b}$ and $\Lambda_{P\to c}$. The four equations of (7.36a) to (7.36d) can now be used to fully-determine the problem and find the four unknowns. However, the two source energies and the two detection energies, have to be sufficiently different, so that the four equations are independent of each other; otherwise one would be in essence repeating the same set of measurements twice. It should also be noted that transmission measurements, at the two source energies, could have replaced the scattering measurements provided by one of the detectors. Alternatively, the transmission measurements, and any other scattering measurements, can be used to over-determine the problem to reduce the effect of measurement uncertainties. Multiple energies can also be applied, provided that they result in independent measurements. In an over-determined problem, the well-known least-squares methods can be applied; after a matrix combining all the equations is formulated in a manner similar to that of Eq. (6.31).

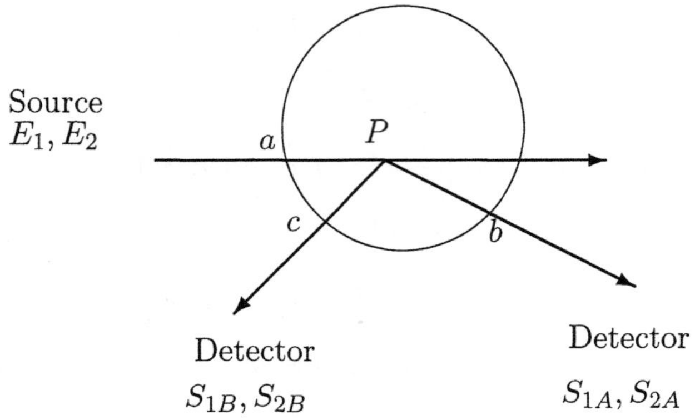

Figure 7.9. Schematic for the dual-energy general-configuration method.

7.3.10. Coherent-Scatter Probing

At small angle of scattering and at low photon-energy, photon coherent (Rayleigh) scattering can take place, as discussed in section 3.4.3.1. At a small angle of scattering, the path of a beam of photons from the source to the detector will be very close to that of the transmitted photons (which is at zero angle of scattering). While transmission will record photons which have been subjected to attenuation, most likely by Compton scattering, small-angle scattering will be caused primarily by Rayleigh scattering. Therefore, the *Rayleigh-to-transmission ratio* of photons can be indicative of the atomic-number [201], as it is a function of the Rayleigh and Compton cross-sections, given respectively by Eqs. (3.51) and (3.63). The probing volume is defined by the intersection of the field-of-view of the scattering detector with that of the incident radiation beam. This method was used for measuring the mineral content of bones [202]. A single-detector with some energy discrimination can be used to measure both coherently (Rayleigh) and incoherently (Compton) scattered photons (at small angles), since coherent scattering causes no energy loss, while Compton scattered photons have a reduced energy. Note, however, the need to use low-energy, hence low-penetration, photons to observe Rayleigh scattering, which limits this technique to small-thickness samples.

The effect of material density and radiation attenuation on the Rayleigh-to-transmission ratio can be accounted for using an approach similar to that described in section 7.3.7, manifested by Eq. (7.29d). A scattering measurement, S_1, and a corresponding transmission measurement, I_1, are obtained, then the object is rotated by 180° to record another scattering measurement, S_2, and a transmission measurement, I_2. Then the Rayleigh-to-transmission ratio can be expressed as:

$$\frac{\text{Rayleigh Scattering}}{\text{Transmission}} = \left[\frac{S_1 S_2}{I_1 I_2}\right] \tag{7.37}$$

This ratio was claimed to be more sensitive to changes in the composition than the explicit Rayleigh-to-Compton ratio, at least when the effective atomic-number was changed by changing the concentration of a potassium orthophosphate (K_2HPO_4) solution in deionized water [201]. The *Rayleigh-Compton ratio* was measured for elements of an effective atomic-number, Z_{eff}, from 10 to 25 [203], as well as for materials with $Z_{eff} < 10$ [204]. A number of industrial applications are given in section 12.2.2.6.

7.3.11. Probing with Neutrons

Probing with neutrons is more challenging than probing with photons. Fast neutrons are more difficult to collimate than photons, see section 15.1, while thermal-neutrons are not very penetrating and are cumbersome to produce as they require the slowing-down of fast-neutrons, as discussed in section 2.3. Moreover, the probability of neutron scattering, as explained in section 3.5, varies from one element to another, unlike the Compton scattering of photons which is dependent primarily on the electron-density of the material. Nevertheless, when low mass-number materials are encountered, the probability of scattering of photons becomes quite low, while the scattering probability and energy change of neutrons is quite prominent in such materials. Therefore, neutrons are useful in probing hydrogen-rich materials, and was considered for measuring the local density, or void fraction, in the two-phase flow of boiling water [205], as discussed below.

7.3.11.1 Local Void-Fraction Measurement

The density of a water-vapor mixture is determined by the amount of vapor, since the density of vapor is considerably less than that of liquid and can be neglected. Therefore, for a void-fraction, α, (void volume-fraction), the density of the mixture would be equal to $(1-\alpha)\rho_f$, where ρ_f is the density of the liquid phase. In addition, the density of the liquid-phase depends on the density of its hydrogen, since in water every two hydrogen atoms must be accompanied by an oxygen atom to form an H_2O molecule. Therefore, neutron scattering, which is very susceptible to collisions with hydrogen, as indicated by Eq. (3.80), can be used to measure the hydrogen density, hence the liquid density. The amount of slowing-down (thermalization) of fast-neutrons should then depend on the hydrogen density, consequently the liquid density, $(1-\alpha)\rho_f$. In a boiling water system, ρ_f is the density of saturated liquid and is dependent only on pressure. Therefore, within a small section of an enclosure containing boiling water, the pressure can be considered constant, and the amount of neutron scattering will be proportional directly to $(1-\alpha)$.

A method for measuring the local void-fraction at a point in a vapor-water two-phase flow was proposed in reference [205]. This method is described here with the aid of Figure 7.10. By directing a collimated beam of fast-neutrons and measuring the neutrons side (90°) scattering with a collimated detector, the location of the point of interest is determined. This side-scatter measurement, S, can be simulated using a scattering model analogous to that of Eq. (7.12) as:

$$S = K_s(1-\alpha)\exp[-X_i\Lambda_i - Y_s\Lambda_s] \qquad (7.38)$$

where K_s is a system constant, α is the void-fraction at the point of interest, X and Y are the total cross-section per unit density for the incident (fast) and scattered (slow) neutrons, respectively, and Λ_i and Λ_s are integrated flow-material densities ($\int \rho dx$, with ρ being density and x is distance) along the path of the incident and side-scattered neutrons, respectively.

The system of Figure 7.10 has the usual three unknowns associated with radiation scatter probing: $(1-\alpha)$, Λ_i and Λ_s, since Σ and Σ' can be determined in advance for a full-of-liquid pipe. An additional scattering, or transmission measurement, will add with it an additional chordal density. Solving this dilemma with any of the methods described earlier in this section for photon scattering is not very practical with neutrons, particularly in a phenomenon that evolves with time, such as flow boiling. However, with neutrons, one can make use of backscattering (also referred to as *albedo*, as well as of forward scattering [206]). The scattered neutrons can be detected with a slow-neutron detector, since they lose energy after scattering with the hydrogen of water. Therefore, energy discrimination can be applied to discard source neutrons reaching the backscattering detector. In addition, neutron transmission can be distinguished from that of forward scattering by energy since the neutron detection efficiency is quite high at low energy (at which scattering is measured), and is low at the higher energy of the source neutrons (see section 4.4).

For a backscattering (180°) measurement, B, and a forward scattering (at $\approx 0°$ scattering) measurements, F, expressions similar to Eq. (7.38)

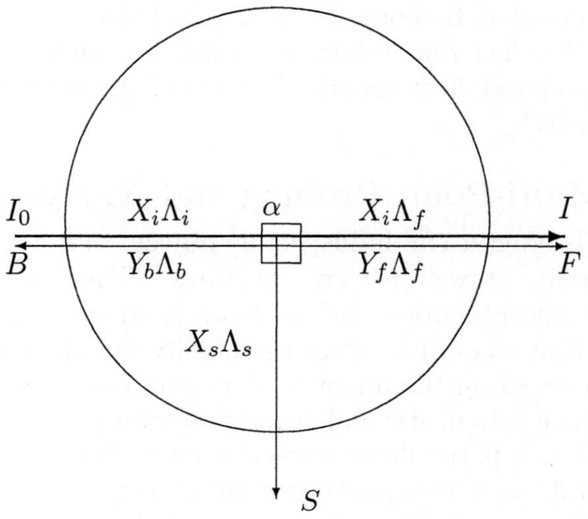

Figure 7.10. Schematic for determination of local void-fraction with neutron scattering.

can be written as:

$$B = K_b(1-\alpha)\exp[-X_i\Lambda_i - Y_i\Lambda_i] \quad (7.39)$$
$$F = K_f(1-\alpha)\exp[-X_i\Lambda_i - Y_f\Lambda_f] \quad (7.40)$$

with the notations analogous to those used in Eq. (7.38). Note that although a detector and a source cannot coexist at the same location, they can be placed in the proximity of each other, allowing the path of backscattered neutrons to be approximately equal to that of the incident neutrons. The fast-neutron transmission measurement, I, can be modeled with the aid of Eq. (6.1) as:

$$I = I_0 \exp[-X_i\Lambda_i - X_f\Lambda_f] \quad (7.41)$$

where I_0 is the intensity of the scattered neutrons for an empty test-section. Note the use of X_f instead of Y_f in Eq. (7.41), since the former is for the incident fast-neutrons, while the latter is for the scattered slow-neutrons. Nevertheless, the X and Y values are those for liquid water and can be determined, along with the system constants, K_s, K_b, K_f and I_0, in advance. Now the four equations, (7.38) to (7.41), have four unknowns, α, Λ_i, Λ_b, and Λ_f, and thus can be solved to obtain these unknowns, and in particular the local void-fraction, α. Note that any additional scattering measurement will add only one unknown, the integrated density along the path of the scattered neutrons, and thus can be used to overdetermine the problem, and reduce the effect of measurement error. It should be emphasized that this approach is made possible by the fact that values of X and Y change only with density, since the two-phase flow material is made of the same matter (water at varying density).

7.4. Multi-Point Probing and Analysis

Probing is often useful in inspecting objects that are made from uniform materials, or well-defined structures. Then one can detect the presence of discontinuities, such as flaws (voided or corroded space) or inclusions, that can either affect the quality or safety of a product. In addition to detecting the presence of an anomaly, it is also desirable to determine its location, size and density. Scatter probing, namely Compton scattering, is particularly attractive for such applications, as shown in section 7.3, as it is capable of determining the material density at isolated points in a structure. Location and size information can also be deduced from Compton scattering, as discussed in this section.

Scatter probing can determine the presence and location of discontinuity by scanning the object, and observing abrupt changes in the profile of the scattering signals. A single radiation beam can be used to simultaneously probe many points within a structure, if the energy of the scattered beam is measured [207, 208, 209]. To demonstrate this concept, let us consider a collimated beam of monoenergetic photons, of energy, E, and a detector capable of measuring the energy of the scattered photons, E', as schematically shown in Figure 7.11. Then for every measured energy, E', one can calculate a scattering angle, θ, using relationship (3.37), and hence determine the direction from which the photons are scattered. The intensity of the scattered radiation at that energy can be related to the density at the point on the beam defined by the angle of scattering corresponding to that energy.

With the line scanning process schematically shown in Figure 7.11, one can determine the presence of any material discontinuity along the incident beam by observing any unusual abrupt changes in the energy spectrum of the scattered radiation. The energy at which such distorsion is observed will correspond to a particular direction of scattering that determines the location of the discontinuity along the incident beam. One of course should keep in mind that the edge of an object, or holes or inclusions, present in the object by design, will also disrupt the detected spectrum. Moreover, a discontinuity present along the path of the scattered radiation will also disturb the profile of the energy spectrum. It is, therefore, advisable to scan the object starting with the

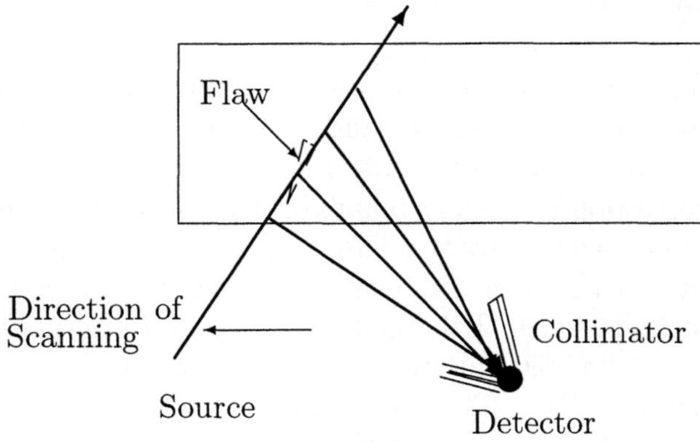

Figure 7.11. Schematic for multi-point probing with a collimated beam and an energy-sensitive detector.

beam at a location nearest to the detector, then move away from the detector. Then discontinuities in a scan will cause predictable disturbances in subsequent scans.

Soft Collimation. A detector equipped with a narrow collimator that focuses on a single direction of scattering can be used. However, such a collimator would define one angle of scattering, enabling the inspection of one-point at a time (point probing), but it eliminates the need for energy-spectrum measurement. On the other hand, by measuring the energy of the scattered photons and employing relationship (3.37), many scattering points can be simultaneously monitored. The latter process is called here "*virtual or soft collimation*", as it can be accomplished without equipping the detector with a physical "hard" collimator. Successful application of this soft-collimation process requires two conditions to be satisfied:

1 Single-scattered photons are recorded.

2 The detector is close to being a point detector.

The first condition is necessary to enable the use of the energy-angle relationship of Eq. (3.37), and can be satisfied by using a source of a sufficiently high energy, so that the length of the monitored line-segment (along the incident beam) is equivalent to, or less than, one mean-free-path of the incident radiation in the inspected material. This condition is also helped by the fact that photons that encounter more than one collision lose much more energy than once-scattered photons. Thus discarding low-energy photons in the detected signal can alleviate this problem.

The second condition is needed in order to uniquely define an angle of scattering, by the intersection of an incident and scattered beams of radiation. In reality, a detector has a volume, and a single-detected energy in the detector would correspond to many points on the incident beam. In other words, photons can scatter by the same angle, hence same energy, from various points on the incident beam but reach the detector at different locations on its surface. This makes it difficult to decipher the direction of scattering; though some workers have attempted to do so, see for example reference [210]. The more practical approach is to equip the detector with a collar made of a heavy metal, that converges into a small opening, resembling a point, at the detector's surface, as schematically shown in Figure 7.11. The opening of this collar should be sufficiently wide so that the field-of-view of the detector covers the desired inspection length along the incident beam. The collar also acts as a shield against multiscattered photons that tends to hit either the inside walls of the collar's opening or its outside surface. Such photons are either absorbed within the collimator's material, or are further reduced in energy and can be discarded by low-energy discrimination. The collar can also shield the detector from the direct exposure to photons streaming from the source shielding.

Size of Discontinuity. When inspecting for discontinuities with collinear scanning, one objective of the inspection process is to determine the size of the discontinuity. The width of a discontinuity along the path of the incident beam, ℓ, can be deduced from the width of the energy window, ΔE, of the observed disruption in the measured energy-spectrum of the scattered photon. If a disruption in the spectrum is observed between the energies, E_1 and E_2, then $\Delta E = E_1 - E_2$. The energies E_1 and E_2 each define an angle of scattering, in accordance to Eq. (3.37), and subsequently two bounding directions of scattering. The two points of the intersection of these two directions with the incident beam define the limits of ℓ. The other dimensions of the discontinuity, perpendicular to ℓ, cannot be determined from a single-beam measurements. However, the boundaries of discontinuities can be demarcated as the beam scans the object along its width and depth.

Density. If the discontinuity is a voided flaw, one would be able to observe a discontinuity in the measured spectrum distinguished by a decrease in intensity to a level corresponding to the radiation background. However, if the discontinuity happens to correspond to a corroded (lowered density) region in the object or to an inclusion (of lower or higher density than that of the object), then it may be required to determine the density of the material within the discontinuity. For a discontinuity of length ℓ along the incident beam, the measured detector response, S, can be expressed by a model of photon scattering analogous to that of Eq. (7.10) as:

$$S = K\ f\ \exp\left[-\Sigma\rho_r\frac{\ell}{2}\right]\Sigma_s\ell\rho_r \exp\left[-\Sigma'\rho_r\frac{\ell}{2}\right] g \qquad (7.42)$$

where K is a system constant, f and g are the attenuation factors of, respectively, the incident and scattered radiation within the interrogated object but outside the volume of the discontinuity (assumed to be known), Σ and Σ' are the total cross-sections of the scattered radiation before and after scattering, respectively, for some reference material (can be taken as the object base material), Σ_s is the scattering cross-section of the incident beam for the reference material at the energy at which the discontinuity is detected, ρ_r is the density of the discontinuity relative to that of the reference material. Note that in arriving at Eq. (7.42), Compton scattering is assumed to be dominant, thus enabling weighting the cross-sections of the material by the density of the discontinuity to arrive at its cross-section. Unlike the models used in section 7.3, attenuation within the inspection-volume (the volume of the discontinuity) is not ignored, but the scattering events are all lumped into a point at the middle of the discontinuity, hence the use of the distance $\frac{\ell}{2}$ in the attenuation factors. The determination of ρ_r from Eq. (7.42) requires the solution of a non-linear equation, which can be done iteratively. However, the solution for a given value of S can result into two different solutions, because of the nonlinearity of the equation. In most practical cases, however, the size of a discontinuity is usually small so that the exponential terms are reasonably close to unity, and one solution of Eq. (7.42) becomes possible. In addition, the second solution, if obtained, is usually too large to be practically acceptable.

Difference Indications. The change in the spectrum of the scattered radiation caused by the presence of a discontinuity is often too small in comparison to the underlying measurement produced by radiation scattering from the rest of the object. However, such a small change can be elucidated by subtracting the measured spectrum from that corresponding to a flawless reference object. The shape of the obtained indication in this *difference method* can also reveal information on the nature of the anomaly, e.g. a sharp crack or a smooth flaw [209, 211]. Then, in solving for density, ρ_r, one should also consider the difference indication,

$S - S_0$, where S_0 is the indication for a flawless object. The reference indication can be expressed as:

$$S_0 = Kf \exp\left[-\Sigma\frac{\ell}{2}\right] \Sigma_s \ell \exp\left[-\Sigma'\frac{\ell}{2}\right] g \tag{7.43}$$

which is similar to Eq. (7.42), except that the relative density ρ_r is set equal to unity, since no discontinuity is present in the inspected region. The difference indication, using Eqs. (7.42) and (7.43) is:

$$S - S_0 = K \; f \; \Sigma_s \; g \; \left\{ \exp[-(\Sigma + \Sigma')(\rho_r - 1)\frac{\ell}{2}] - 1 \right\} \tag{7.44}$$

This equation can be solved iteratively to obtain the value of ρ_r.

7.5. Scatterometry

Bulk scattering, that is measuring the overall scattering from an object, can provide a useful means of measuring or "metering" the overall nature of the inspected medium. This process is termed here "scatterometry", to distinguish it from the qualitative methods of scatterography, discussed in section 7.6 which provides an image. A bulk indication requires exposing as much volume of the object as possible to radiation so that the recorded signal is indicative of the overall nature of its material, while suppressing the contribution of local, but not important, effects. Scattering is particularly suited for this purpose, since scattering from many points in the object, including those that are not directly exposed to the radiation source, can be simultaneously recorded, while transmission indication arises only from the portion of the object exposed to a radiation source.

Scatterometry can be seen, as schematically shown in Figure 7.12, as the process of generating a "virtual source" of radiation within the object by the scattering of the primary source (incident radiation). The intensity of the virtual source, measured by the radiation it emanates, i.e., the intensity of the scattered radiation, is indicative of the macroscopic cross-section for radiation in the encountered material. If the source radiation scatters many times in the object, it will cover a large volume of the object. These many scatters tend to "average out" local effects, thus providing an overall bulk indication of a large portion of the object. Scatterometry is also attractive in that it can be performed from one-side of the object, by monitoring backscattered radiation, or for that matter any side of the object; unlike transmission which requires access to two opposite sides of the object.

The strength of the virtual source depends on the scattering cross-section, Σ_s, of the material for the employed radiation type. However,

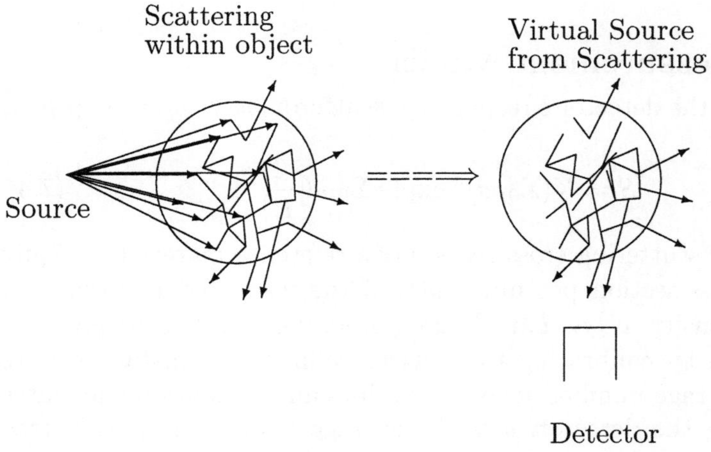

Figure 7.12. Schematic showing the generation of a virtual source in scatterometry.

the intensity of the signal reaching the detector, as schematically shown in Figure 7.12, depends not only on the strength of the virtual source, but also on the ability of the scattered radiation to penetrate the object and reach the detector. Therefore, the scattering signal has features similar to those of the single-scattering model of Eq. (7.10), at least to the first order. However, such a simple model is usually not adequate. Many collisions can occur within the medium before radiation escapes the medium and reaches the detector, and the distribution of the virtual source can be altered by radiation leakage from the boundaries of the medium. Therefore, measurements or more detailed simulation, using the Monte Carlo method (see section 16.2), are needed to more accurately model the scatterometry problem. However, the model of Eq. (7.10) can still be used to predict the overall behavior of the problem. There are two competing parameters in this model, a scattering contributor, the scattering cross-section, Σ_s, and signal dampening parameters, the attenuation factors f and g. In scatterometry, one may aim at enhancing the effect of Σ_s over that of f and g, so that the recorded indicator is dominated by the scattering effect. This can be done by proper choice of the radiation source and type for the material inspected. In some other applications, one may use scattering for the purpose of bringing the radiation energy down so that it can be absorbed by the material of the object, then one would want to enhance the contribution of the removal terms f and g over that of the scattering term. Both types if scatterometry applications are utilized, as discussed in chapter 11.

7.5.1. Measurement Model

Empirically, the detector's response in scatterometry can be expressed as:

$$S = K(\Sigma_0 \rho d)^n \exp[-\Sigma_0 \rho d] \frac{1}{R^2} \tag{7.45}$$

where Σ_0 is the scattering cross-section of a reference material, or equivalently the cross-section per unit mass of the inspected material, ρ is the material density, K, n, d are R are system-dependent constants that are determined by calibrating the device. While the constant n is related to the average number of collisions the source radiation encounters before reaching the detector, d is the average distance radiation travels within the inspected volume before radiation reaches the detector, R is a distance reflecting the reduction in radiation intensity by radiation divergence, and K is an overall-system parameter that incorporates the source strength, detector efficiency, etc. The distances d and R are related, since in essence they describe the size of the virtual source created by radiation scattering. Therefore, d can be used instead of R in Eq. (7.45), with the relative difference in their value incorporated into the system constant, K.

Further Approximations. For an extended medium with backscattering, that is when the source and detector are on the same side, d in Eq. (7.45) can be equated to the source-to-detector distance, since this distance is approximately equal to the travel distance of radiation within the inspected object in its way to the detector. If the inspected medium does not have a high removal (total) cross-section for the radiation used, then the exponential term in Eq. (7.45) can be approximated by $[1 - \Sigma_0 \rho d]$. Since the exponential term in the model of Eq. (7.45) incorporates in effect the attenuation factors f and g of Eqs. (7.2) and (7.7), the cross-section used in the exponential term should be a combination of that of the incident radiation (f term) and the scattered radiation (g term). Therefore, the model reported in reference [212], see section 11.5.11, uses the approximation $[1 - b\Sigma_0 d]$, where b is a constant so that $b\Sigma_0 \rho d = \Sigma_0 \rho x + \Sigma'_0 y$, where Σ_0 is macroscopic cross-section of the incident radiation (that travels an average distance x before scattering), Σ' is that for the scattered radiation (which travels an average distance y in its way to the detector). The value of b can also be designated as a measurable system constant that can be evaluated along with other system-dependent constants by calibration.

7.5.2. Linear Response

If many collisions occur within the medium so that the scattered radiation "forgets" its point of origin, the response of the detector can be assumed to be linear. Lowering of the radiation energy by the many collisions has the effect of reducing the argument of the exponential term in Eq. (7.45), so that it reaches its plateau value (becomes nearly constant). Also these multiple collisions will tend to spread the virtual source over a large volume, making the divergence term $\frac{1}{R^2}$ nearly independent of location of collisions, i.e. becoming nearly constant for a given source-detector arrangement. An added advantage of these many collisions is that they smear out any spatial variations within the inspected object, thus providing an overall averaging of the scattering cross-section of the encountered material. Therefore, by careful design, the response of a scatterometer can be made linear, at least within a certain range. Reference [213] shows an example of such linear scatterometer for measuring the void fraction of boiling water in a pipe, see also section 11.5.11.

7.5.3. Variable Source-to-Detector Distance Method

In soil density measurement with photon multiple scattering, a *variable source-to-detector technique* is used to avoid the use of calibration curves. In the photon energy above 300 keV or so, where Compton scattering dominates, the scattering model of Eq. (7.45) can used. By bringing both the source and detector close to the soil so that the distance within which most photon interactions take place, d, and the distance of divergence, R, are equal, then the measurement model of Eq. (7.45) becomes:

$$S = K(\Sigma_0\rho)^n \exp[-\Sigma_0\rho d]d^{n-2} \tag{7.46}$$

The function $S\ d$ reaches a maximum value when

$$\frac{\partial(S\ d)}{\partial d} = 0 \tag{7.47}$$

$$\frac{\partial}{\partial d}\left\{K(\Sigma_0\rho)^n \exp[-\Sigma_0\rho d]d^{n-1}\right\} = 0 \tag{7.48}$$

which leads to:

$$\rho d_m = \frac{n-1}{\Sigma_0} \tag{7.49}$$

where d_m is the distance at which Sd reaches a maximum value. By considering the virtual source to have a width, d, equal to the source-to-detector distance, and finding the distance at which the detector's response multiplied by d, i.e. Sd, reaches a maximum value, then one can find the material density using Eq. (7.49). The constants Σ_0 and n can be evaluated experimentally, and subsequently used in further evaluation of the soil density [214]. This approach can also be used for measurements performed on any other extended structures, such as a concrete wall.

7.5.4. Ratio Method

Another approach utilized in soil density measurements, suggested in reference [215], is to use the ratio of the counts, $\frac{S_1}{S_2}$, at two distances, d_1, d_2, which according to Eq. (7.46) is given by:

$$\frac{S_1}{S_2} = \left(\frac{d_1}{d_2}\right)^{n-2} \exp[-\Sigma_0 \rho (d_1 - d_2)] \tag{7.50}$$

This ratio approach provides a unique value for ρ, unlike the direct solution of Eq.(7.46), and does not require many measurements to observe the maximum trend of Sd as required by the procedure of Eq. (7.49). It also has the advantage of eliminating the dependence on water content in soil, as it remove the $\frac{Z}{A}$ dependence of the scattering response, where Z is the atomic-number and A is the mass-number of elements in the soil. The assumption, see section 7.2.1, that the Compton scattering cross-section is proportional to the material density, ρ, is valid for most materials and elements, where $\frac{Z}{A} \approx 0.5$. For hydrogen, $\frac{Z}{A} = 1$, and thus in the presence of hydrogen, the scattering response, S, will be higher than that for a dry material. However, the use of the ratio of Eq. (7.50) tends to reduce this effect, by removing the scattering cross-section in the term $(\Sigma_0 \rho)^n$ that appears in the model of Eq. (7.46), though leaving some weak dependence via the exponential term in Eq. (7.50). On the other hand, the enhancement in the scattering cross-section introduced by hydrogen can be used to measure the water content of soil by comparing the count rate recorded at certain source-to-detector distance, d, to that of dry soil [216].

7.5.5. Saturated Scattering

If the source energy is low, or the object is large in size, or the scattering cross-section of material is sufficiently large, the scattered radiation would eventually begin to disperse in a manner that is almost independent of the original source. Radiation "leaking" from the object, as a result of this dispersion process, can provide information on the nature of the object's material. Since this dispersion process requires low-energy radiation, the leaked radiation will also tend to be low in energy and would be quite sensitive to variations in the medium properties, since for most materials and radiation types the material cross-section is highest at low energy. Therefore, *dispersive techniques* are useful in applications where detecting small variation in the medium's *overall* properties is required. The use of the word "overall" is necessary, since a large volume is required for the leaked radiation to become independent of the source energy and direction of emission. The intensity of the scattered radiation in this dispersive process corresponds to the so-called "saturated scattering" as it arises from a virtually infinite medium, i.e. any further increase in the size of medium will not result in an increase in the response of the scattering detector. This saturated response, S_∞, can be evaluated from the integral:

$$S_\infty = K \int_0^\infty \Sigma_s \exp[-\Sigma_t x] dx = K \frac{\Sigma_s}{\Sigma_t} \tag{7.51}$$

where K is a system constant and Σ_s and Σ_t are the scattering and total cross-sections of the materials in the medium, evaluated at some "effective" energy that reflects the change in radiation energy from that of the source to the detection energy. In arriving at the measurement model of Eq. (7.51), say from the multiple scattering model of Eq. (7.45), radiation divergence was lumped into the system constant K, to simplify the integration process. However, in the model of Eq. (7.51), unlike that of Eq. (7.45), distinction was made between the scattered cross-section, Σ_s and the total cross-section, Σ_t, since at low energy, the two cross-sections can differ significantly. It is evident, however, from Eq. (7.51) that any change in the overall material properties of the medium will affect in turn the value of S_∞.

7.5.6. Energy Spectrum

The energy spectrum of scattered photons can also provide chemical composition indications. The kinematics of Compton scattering, Eq. (3.37), dictates that a monoenergetic source of photons must produce a unique energy when scattered by a certain angle. However, as discussed in section 3.4.2, bound electrons in the atom tend to affect the energy spectrum of the scattered photons by broadening it, particularly at low energy. The degree of broadening depends on how the electrons are bound to the atom, and hence is affected by the chemical bonding, and in turn composition, of the scattering medium. This approach was explored for measuring the fat/muscle ratio in tissue and bone mineral content, using an ^{241}Am source [217]. The energy spectrum of neutrons scattered from a medium also offers features that can be utilized for the identification of its material. This approach was suggested for landmine detection using the effect of the resonances in the cross-sections of various elements in the explosive material of a landmine [218].

7.5.6.1 Multivariate Analysis

Reference [219] presented a general *spectral perturbation analysis* method for extracting information from the energy spectrum of scattered radiation, by studying the pattern of perturbation in the spectrum. If the scattered spectrum is recorded in N channels (energy bins), then the total count rate, S_t, is given by:

$$S_t = \sum_{i=1}^{N} S_i \tag{7.52}$$

where S_i is the intensity of the scattered radiation in channel i. Now if some physical parameter (such as medium density), P_k changes to P'_k, then the recorded spectrum will change to:

$$S'_t = A_k S = \sum_{i=1}^{N} S'_i = \sum_{i=1}^{N} A_{k,i} S_i \tag{7.53}$$

where A_k and $A_{k,i}$ are perturbation factors that account for the change in the scattered intensities from S to S' and S_i to S'_i in channel i, respectively. The change in the shape of the spectrum due to the perturbation can be quantified using an autocorrelation function that relates the intensity in one channel to that of the others as follows:

$$Auto_r = \frac{1}{(N-r)\bar{S}^2} \sum_{i=1}^{N-r} (S_i - S_{i+r})^2 \tag{7.54}$$

where r is an integer defining the autocorrelation step, and $\bar{S}$ is the average density of the scattered spectrum. However, a more sensitive parameter for measuring changes in the spectral shape relates one segment of the spectrum to another segment. This spectral parameter can be defined in terms of energy by spectral parameters, X_i, as:

$$X_j = \frac{\int_{E_i}^{E_j} S(E)dE}{\int_{E_k}^{E_m} S(E)dE} \tag{7.55}$$

where the limits of the integration (energies) are chosen to provide a certain number of spectral parameters, say J. If there are K physical parameters, P_k, $k = 1$ to K, then the dependence of a spectral parameter, X_j, on the kth physical parameter is quantified by the partial derivative $S_{j,k} = \frac{\partial X_j}{\partial P_k}$, which represents the sensitivity of X_j to the change in P_k. There are $K \times J$ of such partial derivatives, that can be expressed by the matrix:

$$[S_{J,K}] = \begin{bmatrix} \frac{\partial X_1}{\partial P_1} & \frac{\partial X_1}{\partial P_2} & \cdots & \frac{\partial X_1}{\partial P_K} \\ \cdots & \cdots & \cdots & \cdots \\ \cdots & \cdots & \cdots & \cdots \\ \frac{\partial X_J}{\partial P_1} & \frac{\partial X_J}{\partial P_2} & \cdots & \frac{\partial X_J}{\partial P_K} \end{bmatrix} \tag{7.56}$$

One can correspondingly write the following set of equations:

$$\begin{bmatrix} X_1 \\ \cdots \\ \cdots \\ X_J \end{bmatrix} = \begin{bmatrix} \frac{\partial X_1}{\partial P_1} & \frac{\partial X_1}{\partial P_2} & \cdots & \frac{\partial X_1}{\partial P_K} \\ \cdots & \cdots & \cdots & \cdots \\ \cdots & \cdots & \cdots & \cdots \\ \frac{\partial X_J}{\partial P_1} & \frac{\partial X_J}{\partial P_2} & \cdots & \frac{\partial X_J}{\partial P_K} \end{bmatrix} \begin{bmatrix} P_1 \\ \cdots \\ \cdots \\ P_K \end{bmatrix} \tag{7.57}$$

The elements of matrix $[S_{J,K}]$ can be determined from calibration measurements, by varying one physical parameters while keeping all other constant, and evaluating the corresponding change in the spectral parameters. With the matrix $[S_{J,K}]$ known, the change in the physical parameters can be determined by inverting the matrix equation, (7.57), using a least-squares procedure since the matrix is not a square matrix.

Now, assume one is interested in determining the change in one particular physical parameter, P_k, then for that parameter the influence of the other $K-1$ physical parameters can be considered as undesired noise. The total influence of these $K-1$ parameters can then be lumped into a single hypothetical parameter, P_h. Then the system of Eq. (7.56) can be simplified to any set of 2×2 matrices such as:

$$[S_{m,j}] = \begin{bmatrix} \frac{\partial X_j}{\partial P_K} & \frac{\partial X_j}{\partial P_h} \\ \frac{\partial X_m}{\partial P_K} & \frac{\partial X_m}{\partial P_h} \end{bmatrix} \tag{7.58}$$

Note that:

$$\frac{\partial X_j}{\partial P_h} dP_h = \sum_{i=1, i \neq j}^{K} \frac{\partial X_j}{\partial P_i} dP_i$$

Naturally, $|S_{m,j}| \neq 0$ is a necessary condition for extracting changes from the special parameters, X_j and X_m, otherwise there is no indicated change in the perturbed spectrum, or that there is an interdependence between the system parameters and P_k and P_h. Reference [219] suggested the use of the following criteria for choosing the spectral parameters Z_j and Z_m for determining the change in the physical parameter P_k:

$$J(X_m, X_j) = \left| \frac{S_{m,j}}{\max\left(\frac{\partial X_{j,j}}{\partial P_j}, \frac{\partial X_{m,m}}{\partial P_m}\right)} \right| \tag{7.59}$$

That is, one should select the pair of spectral parameters, X_j and X_m, that provide the largest value for the expression of Eq. (7.59). Reference [219] applied this method, using the scattering of the 662 keV photons of ^{137}Cs in rock, for the determination of the following rock properties: heavy element content (divided into two variables), density, distance from source/detector to investigated medium, and grain size of heavy elements. The use of a multivariate approach to analyze the energy spectrum of scattered photons was also proposed as a method for determining the electron-density of a large object [220].

7.5.7. Combined Bulk and Probing Measurements

In the presence of erratic distribution of stones or boulders, large voids or pore spaces, or unevenly distributed and stratified materials, gamma-backscattering can lead to anomalous results. Reference [221] suggested the use of single-scatter probing to scan the soil for non-uniformity, since multiple scattering (scatterometry) measurements smear out such non-homogeneity. Single-scatter measurements can be registered with the same setup as that of the scatterometer by narrowing the window of the detector energy to correspond to the energy of single-scattering according to Eq. (3.37).

7.6. Scatterography

Similar to transmission radiography, discussed in section 6.3, scattering radiography can be performed by recording the scattered radiation on an extended radiation-sensitive screen, such as a film or an array of detectors, as schematically shown in Figure 7.13. The raw image obtained on the screen, without any collimation of the source or the detector, will be too convoluted to be useful for direct deduction of any useful indications, except for prominent features. However, with the aid of a small hole in a shielding material in front of the beam, a pinhole camera will be created, enabling the recording of a two-dimensional image on the film; as schematically shown at the top of Figure 7.13. Slit collimators, see section 14.6.3 can be used to further confine the image to within the solid-angles defined by the slits. Instead of collimating the detector, the source can be confined into a narrow beam or a slit, enabling the imaging of one segment of the object at a time. A broad-area uncollimated detector or a film can then be used to record the scattering imaging. A flying-spot beam can be generated either electronically, by triggering successive targets in an x-ray machine, or mechanically with the aid of a chopper wheel [222].

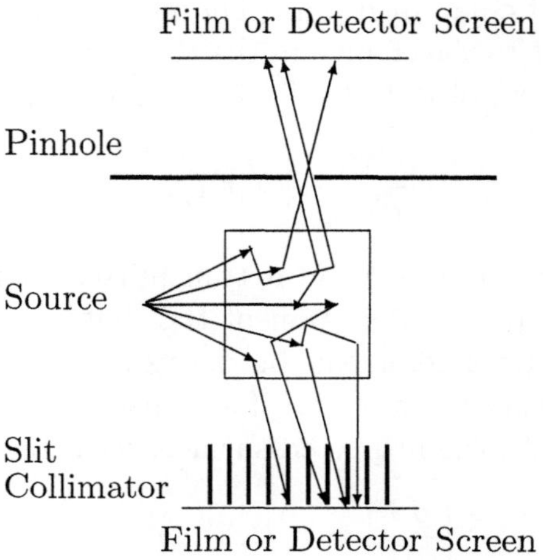

Figure 7.13. A schematic representation of a scatterography system, with pinhole or slit collimators .

Images produced by scattering are less sensitive to material deeper in the object, due to the attenuation and divergence of radiation, as the measurement model of Eq. (7.10) indicates. In scatterography, unlike in conventional radiography, the receiving screen is not shielded from direct exposure to the source by the interrogated object. Therefore, the detection screen is vulnerable to direct exposure to source radiation. The detector needs then to be shielded from such direct exposure. The presence of the shielding can, however, deflect some of the stray radiation towards the screen, thus contaminating the image with a background signal. These factors and the fact that the radiation signal is more difficult to interpret than the transmission signal are the reasons behind the relatively limited use of scatterography.

Scatterography is useful for detecting near-surface discontinuities, and in applications where conventional radiography is impractical, such as when access to two opposing sides of the object is not possible. Moreover, scatterography is also advantageous for objects that are too thin to cause any significant change in the transmission signal, or too thick that they fully absorb the incident radiation before it succeeds in traversing the medium.

Sensitivity. In order to demonstrate the difference in sensitivity and contrast between transmission radiography and scatterography, let us consider the change in the response of transmission and scattering to a small change in the cross-section of a material, caused by a discontinuity within a particular medium. The change in the transmission response with respect to the change in the total cross-section is given by Eq. (6.14), which for an object of a fixed thickness, t, is:

$$\left|\frac{\partial I}{\partial \Sigma_{total}}\right| = t\, I \tag{7.60}$$

where I is the measured intensity of the transmitted radiation, and Σ_{total} is the total macroscopic cross-section of the material. For scattering, a similar sensitivity parameter can be derived using Eq. (7.10), for a change in the scattering cross-section for a discontinuity at some fixed location, d, underneath the inspected surface, resulting in:

$$\left|\frac{\partial S}{\partial \Sigma_s}\right| = K f(\Delta) V g \tag{7.61}$$

where S replaces ψ_d in Eq. (7.10), Σ_s is the material's scattering macroscopic cross-section, K is a constant that incorporates other non-varying parameters and the arguments of the other parameters are dropped for simplification. Note that in differentiating Eq. (7.10), the attenuation factors f and g were considered constant, as they are determined by the position of the discontinuity, and thus are not affected by its scattering cross-section. However, the values of f and g determine the sensitivity of the scattering technique as defined by the left-hand-side of Eq. (7.61).

A good radiography system should result in a large change in the measured indication for a small change in the material property, i.e. a large value of the derivatives of Eqs. (7.60) and (7.61). Therefore, as Eq. (7.60) indicates, transmission is not sensitive to discontinuities in materials of small thickness, since t becomes very small. On the other hand, for thick objects, I becomes small, as dictated by Eq. (6.1), and the right-hand-side of Eq. (7.60) in turn becomes small, leading again to poor sensitivity to changes in Σ_{total}. For scattering, Eq. (7.61) indicates that the sensitivity parameter becomes small when the attenuation factors, f and g, defined by Eqs. (7.2) and (7.7), are small in value, which occurs when the discontinuity is buried deep within the object. On the other hand, for shallow (near-surface) discontinuities, the sensitivity of scattering is highest, and is only dependent on the volume of the discontinuity (ΔV). This gives scattering a clear advantage in imaging thin objects, such as plates, and near-surface flaws in thick objects.

Contrast. The contrast of transmission radiography and scatterography can be compared by considering the relative change in the indication measurement caused by a change in material property. Consider two slabs, each of thickness, t, made of the same material (material A) except that slab b has a discontinuity of small volume, ΔV, made of material B and of a thickness Δt along the direction of an incident radiation beam. A transmission radiograph along t will produce, according to the transmission model of Eq.(6.2), a measurement of $I_a = I_0 \exp[-\Sigma_A t]$ for the flawless (reference) slab and a measurement of $I_b = I_0 \exp[-\Sigma_A(t - \Delta t) - \Sigma_B \Delta t]$ for the slab with a discontinuity of thickness Δt, where I_0 is the radiation intensity in air at the receiving screen, and Σ is the total cross-section. Then the material contrast for transmission, according to Eq. (6.16), is:

$$C_{\text{transmission}} = \frac{I_a - I_b}{I_a} = |1 - \exp[-(\Sigma_B - \Sigma_A)\Delta t]| \approx |(\Sigma_B - \Sigma_A)|\Delta t \tag{7.62}$$

For scattering, the contrast can be similarly defined using the measurement model of Eq. (7.10), for scattering measurements, S_a for the flawless slab and S_b for the slab with the flaw, so that:

$$C_{\text{scattering}} = \frac{S_a - S_b}{S_a} = \left|1 - \frac{\Sigma_B}{\Sigma_A}\right| \tag{7.63}$$

In arriving at Eq. (7.63), it is assumed that the attenuation factors, f and g, are the same for both the reference slab and the one containing the discontinuity, which implies that the discontinuity is not large enough to significantly alter the values of f and g. The latter assumption also indicates that the transmission contrast, which is attenuation-based, is quite small. This is obvious from Eq. (7.62), which shows that, unless Σ_B differs considerably from that of the discontinuity, Σ_A, the transmission contrast will be very close to zero. On the other hand, the contrast for scattering, Eq. (7.63), is not dependent on the size of the discontinuity, provided of course that the discontinuity is sufficiently significant to produce a measurable difference between S_a and S_b.

Contrast for a Void. To further illustrate the above point, let us consider the case of a voided discontinuity, i.e. $\Sigma_B = 0$, then using Eqs. (7.62) and (7.63), the contrasts for transmission radiography and scatterography will be, respectively, $\Sigma_A \Delta t$ and 1. That is scatterography can provide a 100% contrast for a voided discontinuity, while transmission radiography would tend to provide a much lower contrast. Note, however, the approximation of Eq. (7.62) is only valid if $\Sigma_A \Delta t << 1$, i.e., with the discontinuity thickness $\Delta t << \frac{1}{\Sigma}$. Therefore, by lowering the energy of the incident radiation one can increase the value of Σ_A and hence decrease the size of the detectable discontinuity, provided of course that transmitted radiation is still detectable. However, in general, for small size discontinuities, scatterography can provide a much higher contrast.

Intensity. The challenge with scatterography is to be able to record a detectable signal with a moderate source strength and within a reasonable exposure time. Transmission measures mainly the uncollided radiation, the non-interacting radiation, while in scatterography the measured signal is the result of radiation collisions. In other words, transmission measures radiation that has not been removed from the beam, while scattering measures radiation that has been removed from the beam and succeeds in reaching the detector. This leads to a weak but sensitive-to-changes signal. If the fraction of radiation removed in transmission is Δ, then a relative indication of $1 - \Delta$ is measured, with unity referring to the normalized intensity of the radiation source obtained in the absence of the object. Of the amount Δ removed from the beam, a fraction is absorbed and the rest is dispersed by scattering to various directions. Of the latter fraction, a portion would be recorded in scatterography, say a fraction δ of the incident beam, with $\delta < \Delta$. Therefore, while the recorded intensity in transmission would be $1 - \Delta$, that for scattering would be only equal to δ. Thus, the recorded intensity in transmission would always be greater than that of scattering. However, in transmission one contrasts the recorded signal to the signal obtained in the absence of the object, i.e. that of the incident beam. Therefore, in transmission, unless Δ is a reasonably large fraction, the presence of the radiation source can overwhelm the indication signal. In scatterography, the absence of the object will lead to a zero indication (or more accurately a background-level indication), then the contribution of the object δ, though small, is contrasted with zero (or the background signal). This is what gives scatterography a good contrast, in spite of its low intensity signal. A word of caution: unless the background signal is reduced to a low level, scatterography can lose this contrast advantage. Methods to reduce the radiation background level are discussed in chapter 17. In general, however, one can state that although the strength of the measured signal in transmission is generally high, the contribution of the interrogated object is blurred by the direct contribution of the radiation source. On the other hand, while the small contribution of the object in scatterography competes with that of the background radiation, a carefully recorded scatterograph should be more sensitive to changes in the interrogated object.

Multiplicity of Measurement. One other advantage of scatterography is that, for the same source, one can measure scattering at many directions, while in transmission only one measurement is possible, the one downstream of the incident radiation. This advantage compensates to some extent for the weakness of the scattering signal, by providing multiple scatterographs for the same amount of radiation.

Summary. In general, scatter imaging has an inherently lower resolution than conventional transmission imaging. Scatter imaging is also subject to superposition artifacts caused by attenuation and multiple scattering [160], particularly when the depth of the imaging plane (or volume) within the object is large. The weakness of the scattering signal has also limited the use of scatter imaging. But, because of its higher contrast, scatterography is useful for imaging thin objects and near-surface discontinuities. Of course, when access to two opposite sides of an object in not possible, scatterography may be the only viable means of inspection. These attractive features have led to a number of applications of scatterography; see section 13.1.9.

7.7. Reconstructed Scatter-Imaging

Unlike scatterography (section 7.6) which produces an analog qualitative image, reconstructed imaging produces, with the aid of a mathematical algorithm, an image in the form of numerical values at various points (voxels) in the object. Scatter imaging, like photography and unlike transmission tomography (see section 6.4), lends itself to point-by-point imaging. Like digital photography, the quality of an image depends on the number of points sampled for the image. Point-by-point imaging, being differential in nature, is sensitive to small changes in the material property, unlike the line-integrated process of transmission which tends to smear out such changes (see chapter 6). Therefore, such scatter imaging is more suited for handling abrupt discontinuities, such as sharp boundaries. Another advantage of scatter imaging is that it is not restricted to two-dimensional imaging, which is inherent in the geometric projection of transmission measurements, and thus can in principle be used in three-dimensional images. Scatter imaging also eliminates the need for the elaborate rotational scanning process required in transmission tomography (see section 6.4), since the intersection of many imaging projections is not needed. Indeed, what is required in scatter imaging is that each point (or voxel) in the image contributes to at least one scattering detector, while in transmission tomography each point needs to affect a number of transmission measurements.

While point-by-point imaging is easily achieved in photography for exterior surfaces, scatter imaging by penetrating radiation is complicated by scattering from the interior of the interrogated structure. As the measurement model of Eqs. (7.10) indicates, the scattering signal originating from a point is attenuated by the surrounding media before reaching the detector. Accounting for this attenuation effect is complicated by the fact that the magnitude of attenuation depends on the characteristics of the surrounding medium; which consists of points that are part of the image one is attempting to reconstruct. In other words, reconstruction of interior images is essentially a nonlinear problem. This perhaps explains the relatively slow progress of point imaging systems, in comparison to CT transmission imaging systems. Methods developed for dealing with the attenuation problem in scatter imaging are discussed in the following subsections.

7.7.1. Point-by-Point Scanning

Since scatter imaging is a point-by-point imaging process, it is in effect equivalent to performing point probing, see section 7.3, on each point (or voxel) in the object. However, by sequential probing of successive points in the object, the imaging process can be greatly simplified by utilizing the information obtained from preceding voxels to determine the attenuation factors, f and g for the examined voxel. This imaging process is facilitated by the use of a narrow beam to scan one row of voxels at a time [223], as schematically shown in Figure 7.14. Using a set of collimated detectors, voxels along the path of the incident beam are defined by the intersection of the field-of-view of the sources and the detectors. The beam is first positioned closest to the detector array and the response of the detectors is used to reconstruct image information in each of the voxels along the beam, one-by-one. If the voxel is sufficiently small, radiation attenuation suffered by the incident and scatter radiation within the voxel (self-attenuation) can be ignored. Then the response of the scattering detector corresponding to this voxel can be used to define the scattering cross-section, Σ_s in Eq. (7.10), or in the case of Compton scattering, the electron density, ρ_e in Eq. (7.12). Therefore, using the model of Eq. (7.12), for Compton scattering, the detector responses for the first beam, beam 1, can be modeled as:

$$\begin{array}{lcl} S_{1,1} & = & K\rho_{1,1} \\ S_{1,2} & = & K\rho_{1,2}\exp[-\Sigma\rho_{1,1}\Delta x] \\ S_{1,3} & = & K\rho_{1,3}\exp[-\Sigma(\rho_{1,1}+\rho_{1,2})\Delta x] \\ \cdots\cdots\cdots & & \cdots\cdots\cdots\cdots\cdots\cdots\cdots\cdots \\ \cdots\cdots\cdots & & \cdots\cdots\cdots\cdots\cdots\cdots\cdots\cdots \\ S_{1,N} & = & K\rho_{1,N}\exp\left[-\Sigma(\sum_i^{N-1}\rho_{1,i}\Delta x)\right] \end{array} \tag{7.64}$$

where the first subscripts designates the position of the beam and the second subscript refers to the detector's position, so that $S_{i,j}$ is the detector response when the incident beam covers row i, at detector position j, and $\rho_{i,j}$ is the material density in the voxel defined by beam i and detector j, with voxel 1,1 being closest to the source and voxel N being farthest from the source, Σ is the scattering cross-section at the incident photon energy for some reference material, ρ is the voxel density relative to that of the reference material, and Δx is the width of the voxel along the incident beam. Note that the attenuation coefficient, f, of Eq. (7.2), is expressed in the relationships of Eq. (7.64) in terms of the scattering cross-section, assuming that scattering is the dominant radiation removal process, which is usually the case with photons for most materials in the energy range of gamma-ray source. With this formulation, $S_{1,1}$ gives directly the value of $\rho_{1,1}$. Having determined $\rho_{1,1}$, $S_{1,2}$ can be used to calculate the value of $\rho_{1,2}$, and so on, until the densities of all the N voxels in the beam are determined.

Moving the beam to position 2, the scattering measurements can then be modeled as:

$$\begin{array}{lcl} S_{2,1} & = & K\rho_{1,2}\exp[-\Sigma'\rho_{1,1}\Delta y] \\ S_{2,2} & = & K\rho_{2,2}\exp[-\Sigma\rho_{2,1}\Delta x]\exp[-\Sigma'\rho_{1,2}\Delta y] \\ S_{2,3} & = & K\rho_{2,3}\exp[-\Sigma(\rho_{2,1}+\rho_{2,2})\Delta x]\exp[-\Sigma'\rho_{1,3}\Delta y] \\ \cdots\cdots\cdots & & \cdots\cdots\cdots\cdots\cdots\cdots\cdots\cdots \\ S_{1,N} & = & K\rho_{2,N}\exp\left[-\Sigma(\sum_i^{N-1}\rho_{2,i}\Delta x\right]\exp[-\Sigma'\rho_{1,N}\Delta y] \end{array}$$

where Σ' is the cross-section of the scattered radiation for the reference material. With the densities of the voxels in the first row now known, the set of relationships of Eq. (7.64) can be used to determine the density of the voxels along the second beam. The beam is moved to position 3, and so on, until the process is repeated until the density of all voxels in the same section are determined. The beam can then be moved to another section, and the process is repeated, if multi-section imaging is required.

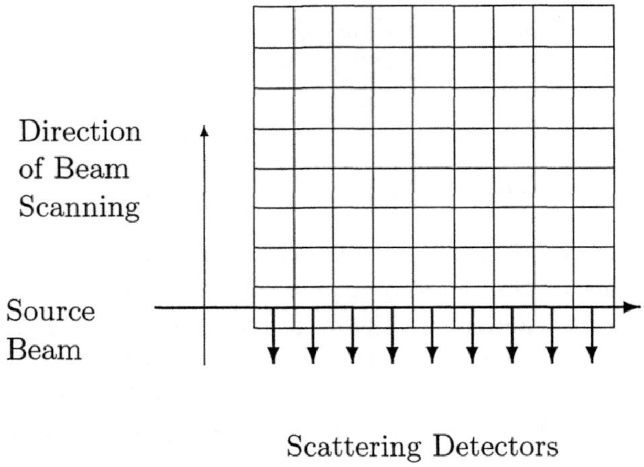

Figure 7.14. A schematic for point-by-point scatter imaging.

This point-by-point scanning process is vulnerable to error propagation, since an error caused in evaluating the density at one voxel affects the densities of all subsequent voxels. However, this problem can be alleviated by using another set of detectors, for example at the other side of the beam for the arrangement of Figure 7.14. The average of the density values obtained from each set of detectors should provide a more accurate estimate of the image density. Moreover, transmission measurement can also be used to spread the error evenly. For example, for beam 1 in Figure 7.14, a transmission measurement can be expressed, using Eq. (6.2) as:

$$I_1 = I_0 \exp[-\sum_i^N \Sigma \rho_{1,i} \Delta x] \tag{7.65}$$

Therefore, if the summation in the right-hand-side of Eq. (7.65) is evaluated using the values obtained by solving the set of equations (7.64), it should match the transmittance value, $-\ln \frac{I_1}{I_0}$, determined from the transmission measurement, within a reasonable range of accuracy. If this matching is not achieved, the densities in the voxels within the beam can be corrected, before proceeding to solving the second set of equations (7.64), say in proportion to their estimated density.

7.7.2. Integration Method

With a point isotropic source and a point detector, the locus of photons scattered by the same angle θ, on a certain plane, is a circle, with center and radius determined by the angle of scattering and the distance between the source and the detector [224], see Figure 7.15. The portion of the circle intersecting the interrogated object is called an isogonic (same-angle) line [225]. If the source is monoenergetic, and assuming single scattering, the energy of scattered photons is uniquely related to the angle of scattering, θ, by the Compton scattering kinematics, Eq. (3.37), for photons or by the elastic-scattering kinematics of neutrons, Eq. (3.80). Therefore, each radiation energy measured by the detector defines an isogonic line within the object. The intensity of the scattered radiation can then be determined by integrating the scattering from all points along the corresponding isogonic line. A finite number of points can be designated along the isogonic line, at which the image is reconstructed, and the integration is approximated by a summation. If attenuation is ignored, the summation would resemble the ray-sum of transmitted radiation in computed tomography, and the same image reconstruction techniques can be utilized, see section 6.4. Attenuation can be accounted for iteratively [224]. Alternatively, the attenuation factors along the incident and scattered radiation beams can be considered themselves to be unknown, and the problem can be solved by acquiring sufficient measurements to deal with these unknowns [225]. Note, however, that in the latter method, some relationships can be established between adjacent points on the same isogonic line and between adjacent points on neighboring isogonic lines, which reduces the total number of problem unknowns [225].

7.7.3. Nonlinear Solution

When scattering is the dominant process of removing radiation, then the attenuation factors of the incident and scattered radiation, f and g of Eqs. (7.2) and (7.7), respectively, for one voxel in the image, become a function of the scattering cross-sections of other voxels. Since the objective of reconstructed imaging is to determine the scattering cross-sections at all voxels of the image, then there is no need to determine the attenuation factors independently, as they are determined once the image is reconstructed. This, however, makes the problem a nonlinear one, that may be solved iteratively [226, 227]. Using the model of Eq. (7.12), a scattering measurement, S_i, can be expressed as:

$$S_i = \sum a_{i,j}\rho_i \tag{7.66}$$

Scattering circle:

- Center at $(\frac{d}{2}, \frac{d}{2}\frac{\cos\theta}{\sin\theta})$

Radius of circle $= \frac{d}{2\sin\theta}$

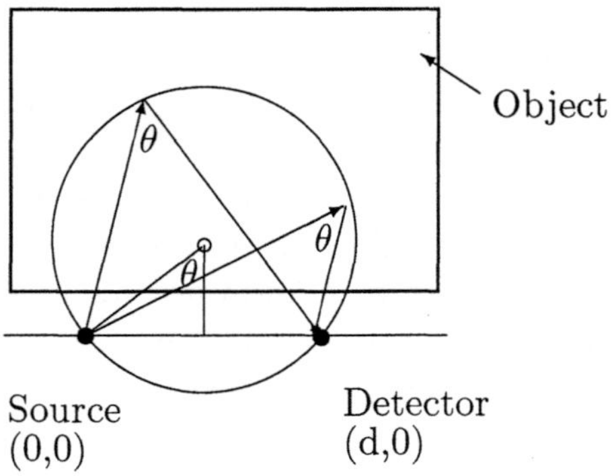

Figure 7.15. Scattering circle in single scattering.

where,

$$a_{i,j} = K f_{i,j} g_{i,j} \delta_{i,j} \tag{7.67}$$

The subscript i, j refers to the contribution of image voxel j to the detector response S_i, and $\delta_{i,j}$ is equal to unity if voxel i contributes to S_i and is zero otherwise. For an image containing N voxels, and with M scattering measurements ($M \geq N$), then following matrix equation can be written:

$$\{S\} = [A(\{\rho\})]\{\rho\} \tag{7.68}$$

where $\{S\}$ is an M-dimensional vector of the measurements, $\{\rho\}$ is an N-dimensional vector of the unknown element densities, and $[A]$ is an $M \times N$ matrix whose elements are the $a_{i,j}{'s}$ defined by Eq. (7.67). The elements of $[A]$ are function of the unknown densities, hence the expression $[A(\{\rho\})]$ in Eq. (7.68). The elements of $[A]$ can be evaluated only if $\{\rho\}$ is known. However, if one starts with an initial guess for $\{\rho\}$, the matrix $[A]$ can be constructed, albeit with approximate values, and the matrix equation (7.68) can be solved to obtain a new approximation for the vector $\{\rho\}$, which in turn can be used to reconstruct a new matrix $[A]$, and so on. This iterative process can be expressed mathematically as:

$$\{\rho\}_{k+1} = [A(\{\rho\}_k)]^{-1}\{S\} \tag{7.69}$$

where k designates the iteration number.

The conditions for convergence of this iterative process are given in reference [226], and its application to neutron scattering and gamma-ray Compton scatter imaging are described in references [228, 229, 230], respectively. Note that when the number of measurements M is greater than the number of voxels, N, the problem becomes overdetermined and the solution of matrix equation (7.69) needs to be found using least-square inversion, similar to that described in section 6.31 for transmission tomography. This solution process also allows the introduction of solution smoothing and bounding (limiting within physical range). Direct matrix-inversion need not be performed, as some of the other common non-direct matrix inversion methods can be applied, as discussed in references [229, 228]. Reference [231] also proposes the imposition of an additional criterion on the solution that provides global smoothing of the image by minimizing a negative entropy, E, defined by the function:

$$E(\rho) = \sum_i \rho_i \ln \rho_i \tag{7.70}$$

where summation is performed over density in all voxels of the image.

The formulation of Eqs. (7.67) and (7.68), unlike the algorithm of section 7.7.1, does not require the use of a narrow beam to accumulate and resolve the image point-by-point. Therefore, the matrix $[A]$ in these equations can be formulated for a wide beam, or a beam that covers the entire object at once. This enables single-exposure tomography using a wide beam, see for example reference [228]. The same formulation can also be used with partial exposure, as the $\delta_{i,j}$ term in Eq. (7.67) is simply set to zero, for voxels that are not exposed to radiation, or do not contribute to the corresponding detector. The only condition necessary for the solution of Eq. (7.68) is that each voxel in the image must contribute to at least to one measurement, otherwise no information can be extracted from the voxel.

7.7.4. Coherent-Scatter Imaging

Projections of diffraction patterns can be used to provide radiographs containing composition information. Diffraction occurs, as Eq. (3.68) indicates, when $E(\text{keV}) \ \sin\frac{\theta}{2} \leq 6.2$, where E is the energy of the scattered photons and θ is the angle of scattering. Since coherent (Rayleigh) scattering dominates at small angles, see Eq. (3.64), and at low energies, Eq. (3.63), the photons produced by common x-ray tubes, 60 to 150 kV, can be used for this purpose [232]. Also the wide spectrum of photons produced by an x-ray tube provides a large range of energies, different values of E, to allow the measurement of diffraction spectrum at a constant value of θ, i.e. at a fixed detector position, with an energy resolving detector. As the value of E changes characteristic diffraction patterns of constructive and destructive interference can be observed. A ring of detectors centered around an incident collimated photon beam can be used to record the diffraction pattern at some small angle θ, to allow averaging of variability in scattering at the same angle. Concentric rings can also be used to record scattering at different, but still small, angles. A detector, at the center of the ring, can be employed to measure the transmitted radiation. Since at small scattering angles, the path of the transmitted and scattered radiation are about equal in length, and because the energy of coherently scattered radiation and uncollided (transmitted) photons is the same, the response of the scattering detector can be used to correct for the attenuation of the scattered radiation. The *scattering-to-transmission* ratio of the detector response can, therefore, be indicative of coherently scattered radiation at the point defined by the intersection of the detector field-of-view and the scattered radiation. The scattering detectors can be collimated so that each detector, in a ring, focuses on a voxel along the incident radiation beam, so that the set of detectors covers all points in the object crossed by the incident beam [233]. By moving the object in a raster scan fashion, three-dimensional images of the object can be obtained, with each imaging point containing a diffraction profile (energy spectrum) that can be used to identify crystalline materials that exhibit characteristic profiles, such as plastic explosives.

The Rayleigh-to-Compton scatter ratio method of probing, described in section 7.3.10, was also used to produce tomographs of the atomic-number [234]. With one detector, positioned to monitor forward scattering and cover the entire section, Rayleigh scattering is distinguished from Compton scattering by energy, since the former takes place without any change in energy. Such discrimination is easily achievable when a monoenergetic source of radiation is used; a 60 keV x-ray (synchrotron source) was used in the work reported in reference [234]. By lateral movement and angular rotation of the object, radiation scattering from different pixels along the incident beam was recorded with the aid of a single detector equipped with a multi-slit collimator. Each slit in the collimator defines a field-of-view that intersects with the field-of-view of the incident beam to define a pixel. Therefore, the detector records the integrated scattering of all pixels along the incident beam. If the attenuation of radiation before and after scattering is neglected, the detector response will be indicative of the pixel summation of the Rayleigh or Compton scattering probabilities, depending on the detector's energy range. Then, with attenuation neglected, the image reconstruction problem resembles that of transmission tomography (see section 6.4). Two images, corresponding to the Compton and Rayleigh contributions, can then be reconstructed. The ratio between these two images will eliminate the pre-scattering attenuation factor, which is equal for both Rayleigh and Compton scattering, and lessens to a great extend the effect of post-scattering attenuation. The ratio image also becomes indicative of the effective atomic-number of the material in each pixel. This method was used to differentiate between polyethylene and an aqueous solution containing a low concentration of iodine [234].

7.8. X-Ray Diffraction and Refraction

7.8.1. X-Ray Diffraction

X-ray diffraction is commonly used in the macroscopic analysis of small samples, see for example references [65] and [235], using low energy x-rays to permit occurrence of the diffraction phenomenon. The Laue diffraction pattern technique, see section 3.4.3.2, is conventionally used in the examination of single crystals. This technique does not require monoenergetic photons, enabling the use of conventional (continuous energy) x-ray machines, equipped with a collimator to provide a narrow beam. These patterns can be photographed on a film either in a transmission-through arrangement with the film behind the specimen downstream of the beam, or in a back-reflection setup with the film positioned upstream of the specimen and the radiation passing through a hole in the film. Obviously the transmission method is limited to thin samples, while the back-reflection method enables only near-surface examination.

When a monoenergetic collimated beam of soft (low energy) x-rays is made incident on a single crystal at the proper angle, its photons are diffracted according to Bragg's law, Eq. (3.66). If the diffracted beam is recorded on a film, it produces an image known as a topograph: "topos" means "to describe a place". The image shape is related to the geometry of the irradiated portion of the crystal, while the intensity of the image depends on the density of the diffraction centers (atoms and their distribution). Therefore, x-ray topographic images are used to detect crystal defects, dislocations, surface damage, etc. [235].

7.8.2. Refraction

As indicated in section 3.4.3.3, the refraction of soft x-rays (less than 20 keV in energy) on a surface depends on the refractive index of material. Small particles, such as those encountered in powder, represent interfaces that function as refractive surfaces; due to the density difference between the particles and the surrounding medium (air). Pores (filled with air or gas) in a porous medium also provide interfaces for refraction. Therefore, the intensity of the refracted radiation is indicative of the refractive intensity of the interfaces presented by powder particles or pores, and consequently on the particle's, or pore's, size [67]. Since a very low photon-energy, and consequently a very short wavelength (about 0.1 nm) is required for refraction to take place, x-ray refraction enables the imaging of the microstructure of ceramics, fibers, porous materials and other low-density materials.

7.9. Neutron Diffraction

Diffraction of cold and thermal neutrons offers the ability to study objects of industrial and engineering interest, due to the higher (than photons) penetration depth of neutrons in metals. Neutron diffraction, as discussed in section 3.5.5.1, produces diffraction patterns governed by Bragg's law, Eq. (3.66). That is, if the diffraction angle (see Figure 7.16) changes, a peak in neutron diffraction occurs when the conditions for Bragg's law are satisfied. This enables the determination of crystal structure. From the diffraction pattern, one can deduce the lattice spacing (distance between atoms in a lattice), and in turn characterize the chemical or physical state of the material. A neutron beam is only diffracted from the crystal planes perpendicular to the diffraction vector (the bisector of the incident and diffracted beams, as shown in Figure 7.16). Therefore, the intensity of a neutron diffraction pattern is proportional to the number of crystal grains oriented to produce the diffraction pattern. This is helpful in studying crystallographic alignment or texture, i.e. the non-random distribution of crystallite (initiation of crystallization) orientations that usually accompany plastic deformations.

For a certain material, broadening of the lattice spacing is indicative of material strain, and consequently stress [236]. If a stress-free material results in a diffraction pattern at a lattice spacing, d_0, and a stressed sample produces a patterns at a spacing d, then the material's elastic strain is given by:

$$\epsilon = \frac{d - d_0}{d_0} \tag{7.71}$$

From the value of the strain, the elastic stress can be calculated using Hooke's law (stress-strain relationship). Usually, the variation of stress (or strain) within a material is studied by observing the diffraction pattern at different locations in the inspected object. As indicated in section 3.5.5.1, small-angle diffraction is useful in characterizing particle size, distribution and shape. Therefore, the technique is useful for

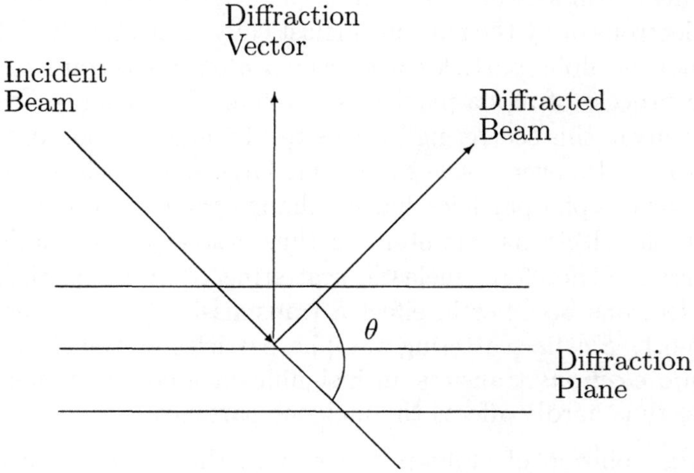

Figure 7.16. A schematic of beam diffraction.

studying small inclusions, precipitate, voids and macro-molecules [73]. Neutron reflection, see also section 3.5.5.2, is affected by changes in the surface geometry and density, and thus is used in studying thin films and surface layers, in a technique known as neutron reflectometry [73].

7.10. Scattering of Charged-Particles

The discussion in this section is based in part on reference [237]. As explained in section 3.3, charged-particles such as alpha and beta particles are deflected continuously by the electrostatic fields of the atom and the nucleus; unlike photons and neutrons which only change direction upon collision with another particle. This continuous scattering process limits, on one hand, the penetrability, of charged-particles in matter, but on the other hand gives them some unique features for use in nondestructive testing and material analysis. While it is customary to ignore the effect of air in the transport of photons and neutrons, air can have a measurable effect on the transport of charged-particles, due to their limited range and their continuous loss of energy. Therefore, the effect of the surrounding air has to measured, when the measurements are not performed in vacuum.

7.10.1. Scattering of Alpha-Particles

As indicated in section 3.3.1, alpha-particles can be scattered by the atomic electrons or by the nucleus, inelastically or elastically. The *inelastic* collision of alpha-particles with *bound atomic electrons* is the most probable process of alpha-particle scattering. The energy acquired by the electrons in this scattering process results in excitation or ionization of the atom. However, because of their large mass, relative to that of the electrons, alpha-particles hardly change their direction upon collision with the electrons, maintaining thus a straight-line path through the material. Therefore, inelastic scattering of alpha-particles by the atomic electrons becomes in effect a transmission process, discussed in section 6.6.1. *Elastic* scattering of alpha-particles by the electric field of the *atomic electrons* transfers an insignificant amount of energy to the electrons, thus hardly affects the incident particles.

Inelastic collision of alpha-particles with the *nucleus* can cause the particle to deflect from its original direction, or produce nuclear excitation. The deflection process results in the release of an electromagnetic radiation (bremsstrahlung) and a corresponding reduction in the kinetic energy of the incident alpha-particle. However, the amount of bremsstrahlung produced by an alpha-particle is very small, due to its large mass, and this effect is usually neglected. Also the likelihood of non-capture collisions resulting in nuclear excitation interaction is very low in most materials. On the other hand, *elastic* collision of alpha-particles with the *nucleus* can result in a considerable energy loss that can be used to obtain useful nucleus-specific information.

Therefore, alpha-particle scattering techniques rely mainly on nuclear scattering of alpha-particles. As shown in section 3.3.1.1, this process, known also as *Rutherford scattering*, provides an indication of the atomic-number, Z, of a material, as a result of the dependence of its microscopic cross-section on Z. Scattering information may, therefore, be utilized to identify the elemental composition of the target material. For a given material composition, the amount of scattered alpha-particles depends on the number of nuclei (or atoms) per unit volume, the thickness of the object and the distance between the detector and the object. The latter two parameters have to be such that the total distance of travel of alpha-particles before and after scattering is sufficiently small so that the scattered particles are not absorbed within the scatterer before reaching the detectors. This limitation makes the backscattering of alpha-particles useful for measuring the thickness of a coating material, on a more dense substrate. Most of the scattered alpha-particles reaching the detector would arise from the coating material, while alpha-particles reaching the substrate would most likely be absorbed.

The kinematics of nuclear elastic scattering is governed by Eq. (3.17), which depicts a unique relationship between the energy and angel of scattering, as a function of the mass of the incident charged-particle (which is known in advance) and the mass of the target nucleus. Therefore, from the energy of the scattered particle recorded at a fixed scattering angle, one can determine the mass-number of the scattering nucleus, hence identify the nucleus. The intensity of the scattered radiation at that energy determines the concentration of the scattering element. Note also that the same type of information can be obtained by measuring the energy spectrum of the recoil nucleus, since its energy is simply the difference between the incident energy and the energy of the scattered particle. For light elements, contribution from inelastic scattering becomes significant, especially as the energy of the alpha particle increases. This makes interpretation based on Eq. (3.17), which is applicable only to elastic scattering, less relevant. Techniques for chemical, or composition analysis, relying on elastic scattering need, therefore, to be designed carefully, perhaps with prior calibration for a known set of elements.

Due to their heavy weight and large charge, alpha-particles are easily stopped within a small distance in any material. The only practical scattering signal that can be obtained from alpha-particles in a thick object is backscattering. Forward scattering can be detected only if the object is very thin, and this requires the preparation of a thin special sample. Examination of liquids and gases with Rutherford scattering can be difficult since they require a container, which will absorb the charge-particles before they can reach the sample.

7.10.2. Scattering of Beta-Particles

The scattering of beta-particles is a more complex process than that of alpha-particles. Beta-particles change significantly their direction after scattering. This straggling effect makes it difficult to define a precise range for beta-particles through a straight line. Moreover, the broad energy spectrum of beta-particles complicates the relationship between energy and angle of scattering. At higher energy, the relativistic effect, discussed in appendix C, can change the particle mass. There are also more many competing scattering mechanism, when dealing with beta-particles.

Beta-particles encounter both elastic and inelastic scattering, as discussed in section 3.3.2. Due to their limited penetrability, the backscattering of beta-particles is often the most viable indication modality. This technique is most useful in measuring sheet and coating thickness, when access to only one side of the object is possible. The amount of backscattered of beta-particles can be modeled by the following empirical relationship [238]:

$$S = S_0 + S_\infty[1 - \exp(-kx)] \tag{7.72}$$

where x is the material thickness, k is an empirical attenuation-coefficient that depends on the source energy and the density and composition of the scattering materials, S_0 refers to the scattering response at a zero thickness (equal to zero for a sheet material and equal to the response of the backing material of a coating substance) and S_∞ is the response at an infinite (saturation) thickness. The value of k needs to be obtained from calibration measurements. The optimum thickness for backscattering measurements is such that $kx = 0.65$ [238].

As Eq. (3.28) indicates the probability of scattering of beta-particles is proportional to $\frac{1}{2}\sin^{-4}\phi$, where ϕ is the scattering angle. This indicates that the probability of scattering in the forward direction (small ϕ values) is much higher than that for backscattering (large values of ϕ). Therefore, when practical, it is more effective to utilize forward scattering. With a collimated beam, forward scattering can be monitored by positioning a detector downstream of the source beam, but off its direct path. With an isotropic source, the direct passage of the source particles to the detector must be blocked, using a shielding material, so that only scattered particles are measured by the detector. In either case, the material thickness must be sufficiently small to allow the scattered particles to escape from the object and reach the detector, without being removed by absorption or scattering away from the detector. This limits the use of the technique to very thin samples or to gases.

7.10.3. Scattering of Ions

Although Rutherford scattering is synonymous with alpha-particle scattering, it is also equally applicable to other accelerated charged particles (ions). The Rutherford backscattering technique is quite useful for obtaining information on the elemental composition near the surface of material [239], as discussed in section 12.1.5. When a low energy ($<$ 10 keV) beam of monoenergetic light ions bombards a surface made of much heavier nuclei, most of the backscattered ions are likely to arise after suffering a single scatter with the nuclei on the material's surface, since those scattered forward, or scattered more than once, are likely to be absorbed within the medium. For a beam of ions of energy E and mass m bombarding a surface made of nuclei of mass M, the energy of the scattered ions, E', obtained from conservation of energy and momentum (assuming $M >> m$) [240], is given by:

$$E' = E \left[\frac{m \cos\theta + (M^2 - m^2 \sin^2\theta)^{\frac{1}{2}}}{m + M} \right]^2 \tag{7.73}$$

where θ is the angle of scattering. As Eq. (7.73) indicates, at a given angle θ, the energy of the scattered ions depends on the mass, M, of the nuclei with which the ions collide. Therefore, the energy spectrum of the scattered characterizes the surface nuclei. This feature is used to obtain information on the elemental composition and structure of surfaces [240]. Note that the assumption $M >> m$ makes Eq. (7.73) applicable only to backscattering. However, the general form of Eq. (7.73) is [241]:

$$E' = E \left[\frac{m \cos\theta}{m + M} + \left\{ \left(\frac{m \cos\theta}{m + M} \right)^2 + \frac{M - m}{M + m} \right\}^{\frac{1}{2}} \right]^2 \tag{7.74}$$

Eq. (7.74) also shows that energy of the scattered ions, at a certain angle is also indicative of the mass of the target nuclei. However, the number of backscattered particles increases rapidly with decreasing ion energy [241].

One interesting application of ion scattering is for the determination of the crystalline structure of a material. If the incident beam is directed along the axis over which the atoms (nuclei) of the lattice are aligned, it would provide a smaller number of backscattering particles, than if it were randomly directed [242]. The reason for this reduction in intensity of scattering is explained by the screening effect of a row of nuclei, which tend to deflect the incident beam away from nuclei aligned along a row.

Chapter 8

EMISSION METHODS

Emission methods, as the name implies, rely on radiation originating from inside the object, for subsequent detection outside the object. In this regard, the radiation scattering methods, discussed in chapter 7, may be considered as emission methods. However, emission methods usually refer to techniques in which the detected radiation is induced by another type of radiation, or when the emitted radiation is due to an internal source introduced into the object. Tables 8.1 lists the various

Table 8.1. Summary of emission methods.

Induced Emissions		Section Number
Radiation Type		
Emitted	Incident	
Gamma-rays	Neutrons	8.1
	Charged-particles	8.2
	Photons	8.3
Charged-particles	Photons	8.5.1
	Neutrons	8.5.2
	Charged-particles	8.5.3
Neutrons	Gamma-rays	8.6.1
	Charged-particles	8.6.2
	Neutrons	8.6.3
X-rays	Photons	8.7.1, 8.7.2
	Charged-particles	8.7.3
Internal-Source Emissions		
Emitted	Source	
Photons	Photons	8.8
Photons	Positrons	8.4, 8.8.3

possible radiation emission modalities, along with the section number in this chapter in which they are discussed. The chapter begins by presenting a generalized measurement model for the radiation emission process.

8.1. Gamma-Ray Emission by Neutron Activation

The ability of a neutron to interact directly with a nucleus makes it an ideal tool for elemental analysis, particular when the neutron converts the nucleus of the target element into a radionuclide. Since each nucleus has its own characteristic radiation emission and half-life, the energy and decay-time of the resulting radionuclide can be used as a unique identifier of the parent nucleus. Gamma-ray emission is the most common mode of radioactive decay, and is quite useful in NDE applications because photons, unlike charged-particles, are subjected to less attenuation within the inspected object. The compound nucleus formed by neutron-capture is almost always in an excited state, and breaks-down almost instantaneously in a number of ways, depending on the energy of the incident neutron and the nature of the formed nucleus. In general, however, the following reactions occur, in order of increasing neutron energy, noting that some reactions are only possible if the neutron energy is above a certain threshold-energy required for the released particle to escape from the nucleus:

$$\begin{aligned}
(\mathrm{n},\gamma)&: \quad \mathrm{n} + {}^{A}\mathrm{X} \rightarrow {}^{A+1}\mathrm{X}^{*} \longrightarrow {}^{A+1}\mathrm{X} + \gamma \\
(\mathrm{n},\mathrm{n}'\gamma)&: \quad \mathrm{n} + {}^{A}\mathrm{X} \rightarrow {}^{A+1}\mathrm{X}^{*} \longrightarrow {}^{A}\mathrm{X} + \mathrm{n} + \gamma \\
(\mathrm{n},\mathrm{p})&: \quad \mathrm{n} + {}^{A}\mathrm{X} \rightarrow {}^{A+1}\mathrm{X}^{*} \longrightarrow {}^{A}(\mathrm{X}-1) + p \\
(\mathrm{n},\alpha)&: \quad \mathrm{n} + {}^{A}\mathrm{X} \rightarrow {}^{A+1}\mathrm{X}^{*} \longrightarrow {}^{A-3}(\mathrm{X}-2) + \alpha \\
(\mathrm{n},2\mathrm{n})&: \quad \mathrm{n} + {}^{A}\mathrm{X} \rightarrow {}^{A+1}\mathrm{X}^{*} \longrightarrow {}^{A-1}\mathrm{X} + 2\mathrm{n}
\end{aligned}$$

where X^* designates a compound nucleus and $\mathrm{X}-1$ indicates a decrease in the atomic number of the product nucleus by one, and so on. The compound nucleus, ${}^{A+1}\mathrm{X}$, has a high excitation state due to the binding of the neutron to the parent nucleus, ${}^{A}\mathrm{X}$. Therefore, immediately after its formation, the compound nucleus decays, usually by emitting photons, called *prompt gamma-rays.* The product nucleus formed by the decay of the compound nucleus may itself be radioactive, and subsequently decays with a certain half-life, emitting *delayed gamma-rays.*

Radiative Capture. The reaction (n,γ), known as *radiative-capture*, is in all cases an exoergic reaction, i.e. it provides more kinetic-energy than that of the incident particle. In fact, radiative-capture is most probable with thermal-neutrons, as its cross-section tends to decrease with increasing neutron energy. For high-energy neutrons, this reaction is usually ignored as the other reactions become more dominant. Therefore, radiative-capture, is always associated with thermal-neutron activation analysis, as discussed in section 8.1.2.

Inelastic Scattering. The inelastic scattering, (n,n$'\gamma$) reaction, is slightly endoergic, i.e. requires incident neutrons with some energy. This reaction is possible at neutron energies corresponding to the first excited state or higher. For elements of medium and high mass-number, a threshold neutron-energy of 0.1 to 1 MeV is required for the reaction to take place, and for light elements the threshold-energy is higher; about 6 MeV for oxygen. Inelastic scattering of neutrons with hydrogen does not occur at all. However, since the emitted radiation, a neutron and a photon, have no electric charge, there no coulomb electrostatic barrier to overcome. The (n,n$'\gamma$) reaction is, therefore, more probable than other reactions emitting charged-particles. Since the reaction does not alter the composition of the original nucleus, the (n,n$'\gamma$) reaction is only useful in prompt activation. Even in the cases where the resulting nucleus is radioactive, in a metastable state, called *isomeric state*, the half-life of such state is too short (microseconds) and considered in practice to be a prompt emission process. There are, however, some heavy elements for which the isomeric state has a sufficiently long half-life, such as ^{107m}Ag, with a half-life of 44.2 s, see reference [13].

Charged-Particle Emitting Reactions. Almost all (n,p) reactions are endoergic, with an interaction probability that depends on the height of the coulomb barrier, hence the atomic-number of the emitting nucleus. For fast-neutrons, the cross-section for the (n,p) reaction has an increasing trend for atomic-numbers up to $Z = 16$, and after reaching a maximum in the region from $Z = 16$ to 24, it decreases for heavy elements [243]. The (n,α) reaction faces a stronger coulomb barrier than the (n,p) reaction, and its probability of occurrence tends to decrease with increasing atomic-number. Although there is no coulomb barrier to overcome in the (n,2n) reaction, the release of two neutrons requires the deposition of a large amount of energy into the nucleus. This makes the (n,2n) reaction predominately endoergic, with a threshold-energy in the order of 7 to 13 MeV [243]. Therefore, the reaction is highly improbable for mass-numbers below 40, even with 14 MeV neutrons, but its cross-section is considerable for heavier elements. The (n,p), (n,α) and (n,2n) reactions are utilized in delayed fast-neutron activation, as discussed in section 8.1.4.

Sources. Activation cross-sections are generally quite low necessitating the use of intense neutron sources produced by nuclear reactors or accelerators. Nuclear reactors are suited for use in delayed-gamma neutron analysis, as the irradiated sample can be removed from the high-background environment of the reactor core to a separate low-background laboratory for counting. In prompt-gamma analysis, neutron generators are preferred over isotopic sources. Neutron generators can be turned off, thus do not produce a background component by the continuous excitation of the surroundings and shielding material as in the case of isotopic sources. Moreover, isotopic sources do not provide the high intensity required to produce a significant amount of activation within a reasonable period of time. However, isotopic sources offer some advantages. First, compared to common 14 MeV neutron generators, isotopic sources produce neutrons at a slightly lower energy, see section 2.3, making it possible to produce thermal-neutrons with a smaller amount of moderating material. Moreover, the small size of the source enables the formulation of reasonably compact moderating assemblies. Therefore, isotopic sources are often used for thermal-neutron activation. Also, the lower neutron-energy of isotopic sources makes them attractive for use in inelastic-scattering activation, where the neutron energy tends to match the reaction energies for many nuclei. Since the cross-sections of the (n,γ) reaction increases with decreasing neutron energy, the lower energy portion of the spectrum of neutron-sources makes it possible to induce this reaction with isotopic sources. Also, some nuclides exhibit resonance absorption at certain neutron energies that lie within the continuous energy range of isotopic sources. The (n,p) and the (n,α) reactions may also be induced by isotopic sources, as they require neutrons in the energy range from 1 to 10 MeV. The (n,2n) reaction, on the other hand, has a high threshold-energy, and is mainly observed with 14 MeV neutrons produced by neutron generators.

8.1.1. Measurement Model

In gamma-activation analysis, the presence of a particular nuclide is determined by the characteristic gamma-ray energy it produces. This requires the use of detectors capable of measuring the photon energy, and the performance of spectroscopic analysis, see section 4.5.4. A number of commercial package are available or this purpose [244, 245]. The concentration of the activated element is determined from the intensity of the emitted characteristic gamma-rays. However, the relationship between intensity of the radiation emitted and concentration of the element producing it is not a direct one. First, the incident neutrons have to penetrate the activated material, say by distance x from the surface of the object, where the activation reaction takes place. If the neutron flux incident on the object is ϕ_0, the neutron flux at the interaction point would then be equal to:

$$\phi = \phi_0 \exp[-\Sigma_n x] \tag{8.1}$$

where Σ_n is the total cross-section of the incident neutrons. The interaction rate, see Eq. (3.120), at x is then equal:

$$\text{Interaction rate} = \sigma_{act}(x)\phi N(x)\Delta V \tag{8.2}$$

where $\sigma_{act}(x)$ is the activation microscopic cross-section at x, N is the atomic density of material at x and ΔV is an elementary volume around x. The resulting photons have to succeed in penetrating the object to reach a detector located outside the object. If these photons travel a distance y within the object, to reach its surface, and a total distance R to reach the detector, see Figure 8.1, the flux of the emitted photons at the surface is then given by:

$$\phi_d = \sigma_{act}(x)\phi N(x)\Delta V \, \frac{\exp[-\Sigma_\gamma y]}{4\pi R^2} \tag{8.3}$$

where Σ_γ is total macroscopic cross-section (attenuation coefficient) of the emitted gamma-rays, and the $4\pi R^2$ term accounts for photon divergence, see section 3.6.2.

A number of time-dependent factors also affect the magnitude of the recorded intensity. While the sample is being irradiated, the produced gamma-emitting nuclide decays. If the irradiation time is sufficiently long, an equilibrium between the production rate by activation and the removal rate by decay would be established, and the intensity of radiation would be reduced by a factor of $[1 - \exp(-\lambda t_a)]$, where t_a is the activation (irradiation) time, and λ is the decay-constant of the radioactive isotope. If the object (or sample) is removed from the irradiation site to the laboratory to measure the activity, the gamma intensity would be further reduced by the decay factor $\exp(-\lambda t_d)$, where t_d is the time elapsed between irradiation and measurement of activity. During the counting process itself, the sample would continue to decay, reducing the activity by a factor of $\exp(-\lambda t_c)$, where t_c is the counting period. Some isotopes decay by emitting several gamma energies with various probabilities, thus the measured activity needs to be corrected by the probability of photon emission, p, at the monitored energy (also known as the branching factor, as in branching to various photon energies). The measured intensity is also reduced by the detector's efficiency, η, at the monitored energy. Since different isotopes of the same element

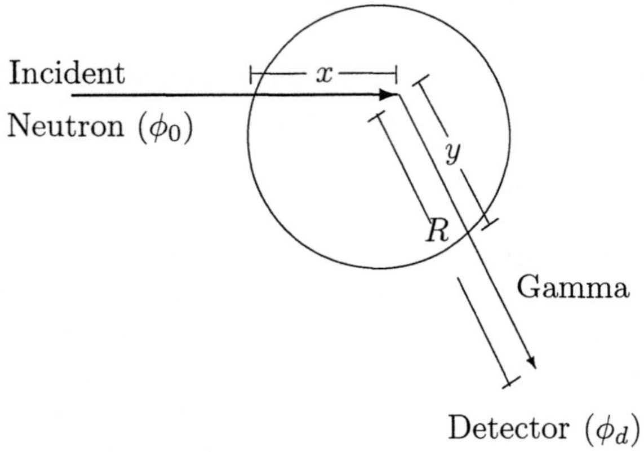

Figure 8.1. Schematic showing a neutron/gamma path in an object.

may have been present in the activated material, but only one of its activation products is measured at a certain activation energy, the relationship between concentration and intensity needs to be adjusted by the abundance, a, of the monitored isotope in the activated sample. Therefore, the intensity, (measure count rate), also known as activity, $\mathcal{A}$, of the irradiated isotope can be expressed by the following measurement model:

$$\mathcal{A} = \int_V \phi_0 \sigma_{act}(x) \; \exp[-\Sigma_n x] \; N(x) \; \exp[-\Sigma_\gamma y] \; \frac{A_d}{4\pi R^2} \; dV \\ [1 - \exp(-\lambda t_a)] \; \exp[-\lambda(t_d + t_c)] \; p \; a \; \eta \tag{8.4}$$

where A_d is the detector area, introduced to convert the flux ϕ_d of Eq. (8.3) to a count rate (or activity), and the other parameters are as defined above. The integral in Eq. (8.4) is over the volume exposed to neutrons. Note that if the sample is small in size, the attenuation (exponential) factors for the incident neutrons and emitted photons can be ignored. For prompt-gamma analysis, the exponential decay terms in Eq. (8.4) are not needed, since activation and measurements are performed on-line and simultaneously.

The measurement model is further complicated by the fact that the activation cross-section is dependent on the neutron energy, which may not be monoenergetic. Moreover, the value of the irradiation flux, ϕ_0, may not be known. In addition, the finite-resolution of a detector spreads the energy of the detected photons, and may cause two neighboring energy peaks to overlap. Interference between neighboring peaks can result in some uncertainty in determining the activity at a particular energy. Because of these factors, if possible, a standard sample of the same volume, and known concentration, of the element of interest should be irradiated at the same time as the inspected sample. The same measurement procedure should also be used simultaneously for both the standard and the test sample. This relative (or *comparator*) method of measurement eliminates the need to explicitly determine any of the factors included in Eq. (8.4), and enables direct measurement of the unknown concentration, with respect to that of the standard sample, using the ratio between the measured activities.

8.1.2. Thermal-Neutron Activation

Thermal-neutron activation (commonly known as TNA) is widely employed in elemental analysis, since the neutron absorption cross-section tends to be highest at the thermal energy. As discussed in section 8.1, low-energy (thermal) neutrons are most likely to cause activation by thermal-neutron capture. The energy of the emitted activation gamma-rays are characteristic of the parent nucleus, and hence can be used as nucleus identifiers. The *National Data Center* provides, on the *Internet*, provides a list of the thermal-neutron capture gamma-rays by target nucleus and by energy of the emitted gamma-rays [246]. Thermal-neutrons are not as readily available outside a nuclear reactor, as isotopic neutron sources and neutron generators produce fast-neutrons, as indicated in section 2.3.3. Nevertheless, a number of out-of-reactor thermal-neutron activation systems (called instrumental methods) have emerged.

Delayed versus Prompt Analysis. Delayed-gamma emission is the mode used in conventional activation analysis inside a nuclear reactor. Usually a small sample, less than a milligram, is irradiated inside the reactor for a sufficient period of time. The activated sample is then removed from the reactor to a laboratory where the released gamma-rays are measured and analyzed. Obtaining such a small sample may be considered non-destructive, in the sense that it is unlikely to affect the integrity of the object from which it is extracted. However, there are cases in which the product nuclei, ^{A+1}X, may not be active at all, may have too short a half-life that it considerably decays before reaching the laboratory for analysis, or may decay by emitting a charged-particles that are self-absorbed within the material itself. The product nucleus may also have a very long half-life, remaining radioactive long after the analysis is completed; an undesirable feature for reasons of radiological protection and disposal. In other cases, gamma-emissions from the activated nucleus may be too weak in intensity, making it difficult to detect. For the above reasons H, Li, B, C, N, F, Ne, Si, P, S and Ar are elements that cannot be detected by delayed TNA, but may be analyzed by the prompt-gamma emitted as the compound nucleus is de-excited [247]. Moreover, prompt-gamma emission is more suited for on-line and in situ analysis, or when it is not possible to remove the source, or the object, to halt the activation process and perform analysis for delayed neutrons. With the object continuously exposed to neutrons, it will also be difficult to distinguish between gamma-rays produced at different time intervals by the decay of the activated nuclei. Obviously, prompt activation has also the advantage of providing immediate indications. However, neutron activation performed outside the reactor is not as efficient as that performed within the reactor, since a limited neutron flux can be produced using neutron sources or generators. Also, in prompt analysis, neutrons are absorbed in the surrounding shielding and construction materials, and within the detector itself. This generates a background component that interferes with the monitored signal. When it is possible to monitor both prompt and delayed photon emissions, small concentrations of almost all elements can be detected, except for oxygen [247]. Oxygen is difficult to monitor with TNA due to its very low thermal-neutron activation cross-section ($\sigma_{th} = 28\ \mu$b).

8.1.2.1 Delayed Thermal-Neutron Activation

Delayed neutron-activation analysis requires removing the tested object (sample) from neutron exposure, so that the emitted delayed gamma-rays can be analyzed independently. The advantage of this approach is that the sample can be analyzed in a low-background area, and counting can be performed for as long as needed to provide an accurate measurement. This enables accurate analysis of the interrogated sample. However, it is neither desirable nor often practical to irradiate a large volume of material, since thermal-neutrons tend to be absorbed near the surface, due to their high cross-section. Moreover, the object can remain radioactive for some time due to the delayed emission of gamma-rays. Therefore, a small sample is commonly used in such analysis, a milligram or less. Such a sample is inserted inside a reactor, or a moderated-neutron assembly (see section 15.3). The efficiency of the activation process can be enhanced by *cyclic activation*, i.e. repeating the cycle of irradiation, delaying and counting, a few times to accumulate an overall increase in the total measured activity [248]. This process is also useful for measuring activation products with short half-lives that decay while being irradiated.

Short-lived isotopes produced by TNA typically require an irradiation times of five to ten minutes in a reactor, with a subsequent 5 to 15 minutes counting period [249]. Long- and medium-lived isotopes, on the other hand, can take several hours to several weeks of irradiation in a reactor, and from one-half to several hours of counting. This obviously takes delayed-activation out of the domain of rapid on-line analysis. Delayed-activation analysis offers, however, accurate information. Table 8.2 provides a list of some of the common elements that can be subjected to delayed TNA. Reference [246] gives a list of the energy of gamma-rays produced by thermal-neutron capture, ordered by energy or element. Note that in almost all cases gamma emission is the result of neutron capture, (n,γ). However, in some cases, particularly for heaver elements, the compound nucleus decays by beta-particle emission, with the resulting nuclide subsequently decaying by gamma-ray emission. In some other cases, the (n,p) reaction takes place, and the product nucleus provides the decay gamma-rays. However, both the (n,β) and the (n,p) reactions can generate some misleading indications, since the resulting isotope may be producible directly by neutron capture or by some other reaction. For example. ^{28}Al can be produced by one of the following three reactions: ^{27}Al(n,γ), ^{28}Si(n,p), or P^{31}(n,α). Therefore, when monitoring the 1.78 MeV emitted from ^{28}Al, one may not be able to definitely identify the parent isotope.

8.1.2.2 Prompt Thermal-Neutron Activation

Prompt-gamma activation analysis with thermal-neutrons eliminates the time-dependent factors, in Eq. (8.4), that influence the measured activity of delayed neutrons. The prompt emission of gamma-rays is the result of the immediate (within a picosecond or so) decay of the compound nucleus formed by neutron absorption. The compound nucleus decays to release the excitation energy that it receives during its formation. This energy is the result of the difference in binding energy between the target nucleus and the formed compound nucleus. The difference in mass, Δm (called mass defect) is converted to energy, ΔE, according to the well-known relationship $E = \Delta mc^2$, where c is the speed of light. With this relationship, each atomic mass unit (u) is equivalent to approximately 931.5 MeV in energy. For example, the reaction ^{14}N(n,γ)^{15}N, results in a $\Delta m = 15.000109 - 14.003074 - 1.008665 = 0.01163$ amu, which is equivalent to about $0.01163 \times 931.5 = 10.83$ MeV (a value very close to the measured value). That is, the small difference in mass resulting from the binding energy of different nuclei is sufficient to produce

Table 8.2. Some elements and their thermal-neutron activated isotopes* [13, 249].

Element	Isotope's Mass Number	γ Energy (keV)	Half-life
	Elements producing short-lived isotopes		
Na	24	1368.6, 2754.0	15.95 h
Mg	27	843.8, 1014.4	9.45 min
Al	28	1779.0	2.25 min
Si	(n,p) ^{29}Al	1273.0	6.6 min
Cl	38	1642.4, 2167.5	37.2 min
K	42	1524.6	12.36 h
Ca	49	3084.4	8.72 min
V	52	1434.1	3.76 min
Cu	66	1039.2	5.10 min
Ag	110	657.8	24.6 s
In	116m	1293.6, 1097.3	54.2 min
Sb	122m	61.4	4.19 min
I	128	442.9	25.00 min
Ba	139	165.9	1.396 h
U	239	74.7	23.47 min
	Elements producing medium-lived isotopes		
Zn	69	438.6	13.76 h
Ga	72	834.1, 2201.7	14.10 h
As	76	559.1	26.3 h
Mo	99	140.5	2.7476 d
Cd	(n,$\gamma\beta$)^{115}In	336.3	2.228 d
Sb	122	564.1	2.72 d
Sm	153	103.2	1.928 d
W	187	685.7	23.9 h
Au	198	411.8	2.6952 d
Hg	197	77.3	2.672 d
U	(n,$\gamma\beta$)^{239}Np	277.6	2.355 d
	Elements producing long-lived isotopes		
Fe	59	1099.2, 1291.6	44.51 d
Co	60	1173.2, 1332.4	5.271 y
Ni	(n,p)^{58}Co	810.8	70.88 d
Zn	65	1115.5	243.8 d
Zr	95	756.7	64.02 d
Ag	110	657.8	249.8 d
Cs	134	604.7, 795.8	2.065 y
Hf	181	482.1	42.4 d
Hg	203	279.2	46.61 d

*Produced via neutron-capture, unless otherwise mentioned.

an energy in the MeV range. This energy is usually released in the form of photons (gamma-rays). Prompt gamma-rays are in general higher in energy than delayed gamma-rays (resulting from the long-term decay of the product nucleus), since for most nuclei the binding energy released by capturing a neutron is about 8 MeV or so. Not all the binding energy is liberated as prompt gamma-rays, as part of the energy usually remains within the formed compound nucleus, bringing it to an excited state. The compound nucleus can promptly emit photons at various energies, as it is "de-excited". Therefore, a complex spectrum of prompt gamma-rays can be emitted, with a maximum photon energy equal to that equivalent to the mass defect of the reaction. In the above discussed example of the ^{14}N(n,γ)^{15}N reaction, more than 50 prompt gamma-rays have been reported [247]. However, the primary (maximum) prompt gamma-ray is usually monitored.

Since the mass defect in the formation of a compound nucleus is always positive, prompt gamma-rays can be produced from almost any element. However, the probability of formation depends on the thermal capture cross-section. Table 8.3 provides a list of the prompt gamma-rays produced from some common elements and the neutron cross-section needed for their formation. Reference [250] lists 1915 prompt gamma-rays emitted in the thermal-neutron activation of 84 elements. Obviously, elements with very low cross-sections are difficult to activate, as they will require a strong neutron source to produce measurable prompt activation photons.

Measurement Model. Dropping the time-dependent terms in Eq. (8.4), the measurement model for the intensity of prompt gamma-radiation, $\mathcal{A}_{pr}$, emitted at energy E_γ, is expressed as:

$$\mathcal{A}(E_\gamma)_{pr} = \int_V \phi_0 \sigma_{act}(x)\, Y\; \exp[-\Sigma_n x]\; N(x)\; \exp[-\Sigma_\gamma y]\; \frac{A_d}{4\pi R^2}\, dV\; p\; a\; \eta \tag{8.5}$$

where the terms are as defined in Eq. (8.4) and the yield factor, Y, defined in Table 8.5, is introduced, since as mentioned earlier, not all photons are necessarily emitted at energy E_γ. As Table 8.3 shows, the photon energy is in the MeV range, thus, the attenuation-coefficient of photons, Σ_γ, is likely to be low for most materials. Moreover, since most activation will tend to occur near the surface of the object, due to the low value of Σ_γ, the value of y tends also to be small. These two factors lead to a value of $\exp[-\Sigma_\gamma y]$ close to unity. Nevertheless, the attenuation factor of neutrons, $\exp[-\Sigma_n x]$, can still affect the measured activity. Thus, the measured gamma activity is not a direct indication of the atomic-density, N, of the activated nuclei. Therefore, some *a priori* knowledge,

Table 8.3. Some elements producing prompt gamma-rays by thermal-neutron capture [247].

Element	σ_{th} (barn)	E_γ (MeV)	% Yield (Y)
H	0.332	2.2232	100
Li	0.036	2.0325	89.33
B(n,α)^{7}Li	755	0.4776	93
N	0.0747	10.8293	14.12
S	0.520	5.4205	59.08
Cl	33.2	6.1109	20
Ar	0.678	4.7450	55
K	2.10	0.7703	51.48
Ca	0.43	1.9420	72.55
Mn	13.3	7.2438	12.31
Fe	2.55	7.631	28.51
Ni	4.43	8.5334	16.98
Cu	3.6366	7.6366	15.71
Zn	1.10	1.0774	18.93
As	4.30	1.5343	7.18
Ag	63.6	0.1984	22.88
Cd	2450	0.5586	72.73
Gd	49000	1.1865	10.83
Hf	102	1.2064	4.60
Au	98.8	1.2018	12.17
Hg	376	0.3681	81.35

σ_{th} = thermal neutron-capture cross section.
Y= fraction of photons of energy E_γ produced per neutron absorbed.

or estimate, of the other factors in Eq. (8.5) must be made. This can be a difficult task if the object is not homogeneous, and if its material distribution is not known in advance. Moreover, highly neutron-absorbing elements can mask the presence of other less absorbing elements, due to the attenuation effect (self-shielding) of the former elements. Of course, for small objects, radiation attenuation can be neglected altogether, and the divergence factor ($\frac{1}{4\pi R^2}$) can be easily determined. In addition, the incident neutrons usually have an energy distribution that affects the value of σ_n. In the presence of some hydrogen, the neutron energy may also significantly change, by scattering, as they travel through the object, which further complicates the measurement model. In addition, the incident flux may not be known or may vary with time. The measurement process itself requires a detector capable of resolving the photon energy, see sections 4.3 and 4.5.4. The detector's efficiency and energy resolution must also be known. Because of all the above factors, some reference measurements using a "comparator" object, of geometry and material content and distribution similar to the objects to be inspected, is used.

Background Radiation. Background radiation is another complicating factor in quantifying prompt-gamma emissions and in obtaining measurements at a high level of statistical confidence (see Appendix G). Thermal neutrons are absorbed by almost all elements, releasing either prompt or delayed photons in many cases. Therefore, absorption of neutrons in the surrounding shielding materials, source and detector collimators, or even structure materials and floors or the detector's material itself, can result in the emission of gamma-rays at different levels. Some of the emitted background photons would be low-energy photons that are easy to discard by energy discrimination, but some may arise at energies close to that emitted by the element(s) of interest. Neutron interactions within a photon detector's material can also damage it, if the exposure is prolonged or is intense, and deteriorate the detector's performance. Higher photon energies may also interfere with the energy of interest by the Compton continuum (see section 4.3) that they produce within the detector, as they lose energy by Compton scattering within the inspected object. The 0.511 MeV photons of pair-production (see section 3.4), arising from the subsequent interaction of the emitted photons and their scattering, can also contribute to the background signal for measurements below 0.511 MeV. Background radiation can also be emitted at high energy from naturally radioactive elements, such as ^{40}K (which emits 1.46 MeV photons), but the activity of such elements is usually low and can be ignored. In general though, the gamma-ray background tends to be relatively high below 2 MeV.

Residual Activity. Prompt-activation can also result in long-lived residual activities that can be of radiological concern in some cases. For example, ^{23}Na is activated by thermal-neutron absorption ($\sigma_{th} = 0.53$ b) to ^{24}Na; the latter emits 2.7541 keV photons, with a half-life of 15 hours. Sodium is also present in NaI(Tl) gamma detectors, and in foodstuff, and can result in some residual activity that can last for a few days following activation. The natural iodine (^{127}I) in a NaI(Tl) detector can also be activated ($\sigma_{th} = 6.2$ b) to ^{128}I, which emits 442.9 keV photons with 25 min half-life. The presence of possible residual activities in the inspected object, and its radiological impact, must be assessed prior to activating a material.

8.1.3. Epithermal-Neutron Activation

Neutrons slowed-down to just above the thermal energy are subjected to strong absorption by elements that have resonances in their cross-sections, as explained in section 3.5.6.2. The resonance integrals, defined by Eq. (3.103) and widely reported (see for example references [84] and [85]), indicate how amenable some elements are to this type of activation. Epithermal-activation is particularly attractive when elements with large thermal cross-sections (such as Na, Al, Mn, Fe, and Sc) are present, and it is desirable to detect other elements. Table 8.4 lists some of the elements that can be activated with epithermal-neutrons. These reactions are usually employed in delayed-neutron activation. Thus, for this technique to be nondestructive, small samples of the inspected material are irradiated and monitored in a manner similar to that described in section 8.1.2.1. Therefore, the measurement model of the activity of epithermal-neutron activation, for a small sample, can be expressed, similar to Eq. (8.4) as:

$$\mathcal{A}_{epi} = I_r\ \phi_{epi}\ N\ V\ [1 - \exp(-\lambda t_a)]\ \exp[-\lambda(t_d + t_c)]\ p\ a\ \eta \tag{8.6}$$

with the resonance integral, I_r replacing the absorption cross-section and the neutron flux designated as the epithermal flux, while other parameters are as in Eq. (8.4). The comparator method described in section 8.1.2.1, can be also used to eliminate the unknown parameters in the measurement model. Since thermal-neutrons are also likely to be present along with epithermal-neutrons, the sample needs to be wrapped in a good thermal-neutron absorbing material, such as cadmium or boron (the former in the from of a metal and the latter usually in powder form).

Although, some of the elements listed in Table 8.4 do not have large values of I_r, their thermal cross-sections are even lower. For example, while the thermal-energy cross-section, σ_{th}, for ^{100}Mo is only 0.195 barns, its resonance integral, I_r, is equal to 3.8 barns. All other elements in Table 8.4 have a high value of $\frac{I_r}{\sigma_{th}}$, and are thus easier to detect with epithermal-neutrons, than with thermal-neutrons. In addition, activation at the epithermal-energy can suppress the effect of other materials that are present at high concentration but are not of interest, such as Na (0.13, 0.32), Al (0.233, 0.17), Mn (13.3, 14.0), Fe (2.8,1.4), Sc (17.0,7.0), where the values in brackets are, respectively, σ_{th} and I_r for each of these elements in barns. These elements have low $\frac{I_r}{\sigma_{th}}$ values, thus their presence is not amplified by epithermal activation.

Table 8.4. Some elements activated by epithermal activation*

Element	I_r (barn)	Isotope Produced	E_γ (keV)	Half-life
Mn	14	^{56}Mn	846.8, 1810.8, 2113.1	2.578 h
Ga	21	^{72}Ga	834.1, 2201.7, 630.0	14.10 h
As	61	^{76}As	559.1	26.3 h
Br	89	^{80}Br	665.9	17.66 min
Rb	6	^{86}Rb	1076.7	18.65 d
Mo	25	^{101}Mo	191.9, 590.9, 1012.5	14.61 min
Ag	750	^{110}Ag	657.8, 884.7	248.8 d
Cd	70	^{115}Cd	336.3, 527.9	2.228 d
In	3200	^{116}In	1293.6, 1079.3	54.2 min
Sb	170	^{122}Sb	564.1	2.72 d
Cs	420	^{134}Cs	604.7, 795.9	2.065 y
La	12	^{140}La	1596.5, 487.0, 815.8	1.678 d
Sm	2,140	^{153}Sm	103.2, 69.7	1.928 d
Tb	430	^{160}Tb	879.4, 298.6, 966.2	72.3 d
Ta	660	^{182}Ta	67.8, 1121.3,	114.43 d
W	360	^{187}W	685.7, 479.6	23.9 h
Au	1550	^{198}Au	411.8	2.694 d
Th	85	^{233}Th	86.5, 29.4, 459.3	22.3 min d
U	278	^{239}U	74.7, 43.5	23.5 min

* Compiled from data given in reference [13].

8.1.4. Fast-Neutron Activation

A nucleus can also be activated by fast-neutrons. Although thermal-neutron activation using nuclear reactors is the most effective tool for elemental analysis, due to the high activation cross-section of thermal-neutrons and the high flux provided by reactors, fast-neutron activation is useful as it enables activation outside the reactor. Fast-neutron sources are readily produced by isotopic sources and neutron generators, see section 2.3. Moreover, the ability to turn-off and pulse a neutron generator facilitates in-situ performance of delayed-neutron activation without having to remove the object to a separate laboratory for analysis. One other advantage of using fast-neutrons directly in activation is the elimination of the need to slow-down neutrons to the thermal or epithermal energy. The materials used in slowing-down usually contain hydrogen, which when activated by slow-neutrons promptly produces 2.2232 MeV photons, see Table 8.3. These photons produce, directly or after scattering, a gamma-ray background component. This background component is reduced to a great extent when the slowing-down process is eliminated. However, materials used for neutron shielding tend also to contain hydrogen, and generate their own gamma-background as the neutrons are slowed-down and absorbed within the shielding material. Nevertheless, proper arrangement of the system, by keeping the shielding away from the gamma detector, can reduce this background component in fast-neutron activation.

The low neutron-yield of radioisotopes limits their use in activation analysis. Therefore, neutron generators or accelerators are often used for this purpose. The neutron generators employed in such applications are most often those that generate 14-MeV neutrons via the d-T reaction (see section 2.3.1.3), due to their high-neutron yield. Irradiation by generators tend also to produce a lower gamma-radiation background, since the surrounding materials are not continuously exposed to neutrons, and subjected to activation, at all time, but only when the generator is turned on. Therefore, the discussion in this section is focused on activation by 14-MeV neutrons. Delayed-activation is typically the result of transmutation of the irradiated isotope to another isotope of the same element, via the (n,2n) reaction , or due to the production of an isotope of another element via the (n,p) or (n,α) reactions. On the other hand, prompt-activation of fast-neutrons results from the inelastic scattering, (n,n$'$), of fast-neutrons, in which the nucleus absorbs the incident neutron, becomes excited and immediately releases a neutron and a photon carrying an energy characteristic of the parent nucleus. Both modes of activation are discussed in the following subsections.

8.1.4.1 Delayed Fast-Neutron Activation

Oxygen. Although fast-neutron activation, in almost all cases, has a lower activation cross-section than that of thermal-neutrons, there is one notable exemption in the activation of oxygen. While the activation cross-section of ^{16}O by thermal-neutrons is only 28 μb, the corresponding value for 14-MeV neutrons is much larger (42 mb), see Table 8.5. The activation reaction ^{16}O(n,p)^{16}N by 14-MeV neutrons produces also high-energy photons, 6.129 and 7.115 MeV, well above the background gamma-ray energy resulting from the activation of other elements. The high-energy photons are also less subject to attenuation by the material, making it possible to examine large objects. Oxygen content can even be measured at trace amounts, due to the lack of a competing background component. It should be noted, however, that fluorine also produces ^{16}N by the ^{19}F(n,α)^{16}N reaction, thus can interfere with the accurate determination of oxygen content [251]. However, the concentration of fluorine, if present, can be determined independently using one of its other reactions, listed in Table 8.5, enabling correction for its interference with the oxygen-produced signal. Oxygen is in general a difficult element to measure even with classical chemical techniques, which makes activation by fast-neutrons one of the uniquely qualified methods for measuring the elemental content of oxygen. Oxygen is, however, difficult to activate with neutrons of lower energy, as the threshold-energy for the $^{16}O(n,p)^{16}N$ reaction is 9.63 MeV [243] (while that for prompt activation is above 6 MeV, see section 8.1.4.2), thus in practice, it is only activated by 14-MeV neutrons.

Nitrogen. Nitrogen is another element that is difficult to analyze with delayed thermal-neutron activation analysis. The element produced by the (n,γ) reaction of the most abundant isotope (^{14}N) is ^{15}N, which is a stable isotope. Although the cross-section for the $^{14}N(n,p)^{14}C$ reaction with thermal-neutrons is reasonably high (1.83 b), the resulting isotope, ^{14}C, produces no gamma emission. The only isotope of nitrogen that can be activated with thermal-neutrons is ^{15}N, leading to the production of ^{16}N. However, the abundance of ^{15}N in natural nitrogen is very low, only 0.37 atom percent. With a cross-section of 0.02 mb, and with the low abundance of ^{15}N, it is very difficult to activate nitrogen with thermal-neutrons for use in delayed-activation analysis; although prompt activation of nitrogen with thermal-neutron is possible (see section 8.1.3). However, fast-neutron activation of ^{14}N enables delayed analysis of nitrogen, via the $^{14}N(n,2n)^{13}N$ reaction. The resulting isotope decays by positron emission. Each of the produced positrons is annihilated by one of the many atomic electrons available in the medium, producing two 511 keV photons that are emitted in two opposite directions to conserve the zero momentum of the interacting particles, see section 3.4. Counting of these 511 keV photons can then be used to detect the presence and concentration of ^{14}N. Interference with these annihilation photons may result from incomplete energy deposition (Compton continuum, see section 4.3) of higher-energy photons, produced by the activation of other elements. However, this interference can be eliminated by coincidence counting (described in section 4.5). But, there are some other elements that produce positron emitting isotopes, with similar half-lives to ^{13}N; namely ^{38}K, ^{53}Fe, ^{62}Cu, ^{78}Br, and ^{143}Sm. Some of the common positron emitting isotopes produced by 14-MeV neutron activation are listed in Table 8.5. Counting the activity of the 511 keV photons at different time intervals and monitoring the change in intensity may be used to quantitatively distinguish between these isotopes, but this can be a complicated process, if many positron emitters happen to coexist. If nitrogen is present in a hydrocarbon material, the recoil protons produced by the elastic-scattering of fast-neutrons with hydrogen can interact with the carbon present in the material via the reactions $^{12}C(p,\gamma)^{13}N$

Table 8.5. Some elements activated by 14-MeV neutrons [243, 251, 252, 253].

Reaction	σ (mb)	E_γ (Yield) (MeV)	Half-life
^{14}N(n,2n)^{13}N	19	0.511*(200%)	9.97 min
^{16}O(n,p)^{16}N	42	6.129 (69%), 7.115 (5%)	7.13 s
^{19}F(n,2n)^{18}F	61	0.511*(97%)	1.830 h
^{19}F(n,α)^{16}N	23	6.129 (69%), 7.115 (5%)	7.13 s
^{19}F(n,p)^{19}O	14	0.197 (64%), 1.357 (39%)	29.9 s
^{23}Na(n,p)^{23}Ne	37	0.440 (33%)	37.2 s
^{23}Na(n,α)^{20}F	170	1.634 (100%)	11 s
^{23}Na(n,2n)^{22}Na	13.8	0.511*(181%)	2.605 y
^{24}Mg(n,p)^{24}Na	180	1.369 (100%), 2.754 (100%)	14.96 h
^{25}Mg(n,p)^{25}Na	60	0.975 (15%), 0.585 (15%), 0.390 (15%)	60 s
^{26}Mg(n,α)^{23}Ne	3	0.440 (33%)	37.2 s
^{27}Al(n,p)^{27}Mg	70	0.842 (68%), 1.014 (29%)	9.45 min
^{27}Al(n,α)^{24}Na	110	1.369 (100%), 2.754 (100%)	14.96 h
^{28}Si(n,p)^{28}Al	235	1.779 (100%)	2.25 min
^{29}Si(n,p)^{29}Al	100	1.273 (85%), 2.43 (15%)	6.5 min
^{30}Si(n,α)^{27}Mg	46	1.014 (29%)	9.45 min
^{30}Si(n,p)^{30}Al	60	2.235 (61%)	3.68 s
^{31}P(n,α)^{28}Al	150	1.779 (100%)	2.25 min
^{31}P(n,p)^{31}Si	83	1.266 (1%)	2.62 h
^{31}P(n,2n)^{30}P	11	0.551*(100%)	2.50 min
^{35}Cl(n,2n)^{34}Cl	8	511*(100%), 2.128 (43%)	32.2 min
^{37}Cl(n,p)^{37}S	25	3.104 (100%)	5.05 min
^{37}Cl(n,α)^{34}P	50	2.128 (43%)	12.5 s
^{39}K(n,2n)^{38}K	27	0.511*(200%)	7.63 min
^{41}K(n,p)^{41}Ar	70	1.293 (99%)	1.82 h
^{41}K(n,α)^{38}Cl	30	2.168 (47%), 1.642 (31%)	37.2 min
^{56}Fe(n,p)^{56}Mn	103	0.847 (99%)	2.578 h
^{56}Fe(n,2n)^{53}Fe	10	511*(196%), 0.378 (30%)	8.51 min
^{63}Cu(n,2n)^{62}Cu	550	0.511*(195%)	9.74 min
^{65}Cu(n,2n)^{64}Cu	1000	0.511*(38%)	12.7 h
^{65}Cu(n,p)^{65}Ni	25	1.482 (24%)	2.52 h
^{65}Cu(n,γ)^{66}Cu	6	1.039 (9%)	5.1 min
^{90}Zr(n,2n)^{89m}Zr	770	0.511*(44%), 0.909 (99%)	4.18 min
^{90}Zr(n,α)^{87m}Sr	194	0.388 (80%)	2.80 h
^{94}Zr(n,p)^{94}Y	7	0.919 (43%)	18.7 min
^{92}Mo(n,2n)^{91}Mo	119	511*(200%)	15.5 min
^{97}Mo(n,p)^{97}Nb	110	0.658	1.23 h
^{134}Ba(n,2n)^{133}Ba	940	0.276 (17%)	1.621 d
^{136}Ba(n,2n)^{135}Ba	700	0.268 (16%)	1.2 d
^{138}Ba(n,2n)^{137}Ba	1250	0.662 (89%)	2.552 min

* Annihilation photons (positron emitters).

and ^{13}C(p,n)^{13}N, producing the same isotope as that generated by the ^{14}N(n,2n)^{13}N reaction. A similar interfering reaction can occur if oxygen is also present in the medium via the reaction ^{16}O(n,α)^{13}N. Interference from these (n,p) reactions with the signal produced by activating ^{14}N must be accounted for, say by prior calibration against a similar, but nitrogen- and oxygen-free (high purity), hydrocarbon target.

Aluminum. Another element that is detectable by 14-MeV neutrons is aluminum. For 14-MeV neutrons, as Table 8.5 shows, the activation cross-section of Al is not as high as that for thermal-neutrons[1], and the half-lives of the resulting isotopes is also longer. However, the 14-MeV activation cross-section for Al is still quite high. The half-life of ^{27}Mg, produced by the (n,p) reaction, is reasonably short and the gamma energy is sufficiently high, to make it possible to detect Al with 14-MeV neutrons. One should pay attention, however, to the interfering reactions resulting from the ^{30}Si(n,α)^{27}Mg and ^{56}Fe(n,p)^{56}Mn reactions. While, ^{56}Mn emits 0.846 MeV gamma-rays that interfere with that emitted by ^{27}Mg, it also emits 1.81 and 2.113 MeV gammas, enabling the implementation of some correction procedure to compensate for the iron content. To correct for interference from Si, one can rely on the higher gamma energies produced by the ^{27}Al(n,α)^{22}Na reaction.

Fluorine. This element is also favored for analysis by 14-MeV neutrons. While its capture cross-section for thermal-neutron is only 9.6 mb, it has considerably higher activation cross-sections for 14-MeV neutrons, particularly for the ^{19}F(n,2n)^{18}F reaction, as shown in Table 8.5. However, the latter reaction produces ^{18}F, a positron emitter that gives rise to 511 keV annihilation photons. Other likely positron emitters that may interfere with ^{18}F are ^{45}Ti, ^{86}Ge, ^{77}Cr, ^{95}Ru, ^{105}Cd, ^{123}Xe, ^{129}Ba, ^{161}Er and ^{190}Ir [243]. The ^{16}N produced in the (n,α) reaction with ^{19}F faces competition from the ^{16}O(n,p)^{16}N reaction, but there are no annihilation gamma-rays produced with ^{16}N. Activation of ^{19}F with the (n,p) reaction is also possible, though with a lower cross-section, but it produces ^{19}O that may be used in spite of its short-have life as a confirmation of the presence of fluorine. The only competing reaction that produces ^{19}O is ^{22}Ne(n,p)^{19}O, but since the presence of neon in nature is quite rare, its effect may be neglected.

[1]The cross-section is 233 mb for the reaction ^{27}Al(n,γ)^{28}Al; the product decays with a 2.25-minute half-life, emitting 1.779 MeV gammas.

Magnesium. Another element that is not too sensitive to activation by thermal-neutrons, but is detectable by 14-MeV activation, is magnesium. Two of its natural isotopes have reasonably high cross-sections for 14-MeV activation by the (n,p) reaction. However, one of the reactions, ^{24}Mg(n,p)^{24}Na, has a product that is also formed by the reaction ^{27}Al(n,α)^{24}Na. Interference with Al needs, therefore, to be accounted for and that can be done by monitoring both activation reactions (the other being ^{25}Mg(n,p)^{25}Na) simultaneously. Note also that the third natural isotope of magnesium, ^{26}Mg, though has a lower cross-section, produces ^{23}Ne via the (n,α) reaction, and that can be further used to correct for the above interferences.

Zirconium. Hafnium accompanies zirconium in nature. While Zr is almost transparent to thermal-neutrons, Hf is a good neutron absorber for such neutrons. This makes it difficult to analyze Zr with thermal-neutrons in the presence of Hf. For 14-MeV neutrons, the activation cross-section for Zr is significant enough, see Table 8.5, to make it possible to detect Zr in the presence of Hf, though the latter has still high activation cross-sections. The (n,2n) reaction with ^{90}Zr is often utilized. It produces ^{89m}Zr, which emits both 511 and 909 keV gammas. The 511 MeV photons are those of positron annihilation, which can be interfered with by those being produced from other positron emitters, in particular ^{38}K, ^{78}Br, ^{108}Ag and ^{140}Pr [243]. However, yttrium, via the reaction ^{89}Y(n,n$'$)^{89m}Y, with a cross-section of 400 mb, produces the most significant interference with Zr by generating 909 keV photons. However, the half-life of ^{89m}Y is only 15.7 s, making it possible to discriminate against it by waiting for some time for ^{89m}Y to decay. Note also that ^{89m}Y can be produced by the reaction ^{93}Nb(n,n$'\alpha$)^{89m}Y, but with a cross-section of only 2.5 mb, thus can cause interference only if present at high concentrations.

Other Elements. Activation with 14-MeV neutrons is also advantageous for Si, P, Cu and Ba [252]. While there is no gamma emission from thermal-neutron absorption by P, there is little gamma emission when Si absorbs a thermal-neutron. Copper and barium result in isotopes with shorter half-lives via the (n,2n) reaction . Fast-neutron activation of ^{63}Cu produces ^{62}Cu via the (n,2n) reaction , which has a half-life of about 9.8 min, in comparison to the 12.7 h half-life of the ^{64}Cu (4.5 b) produced by the (n,γ) reaction induced by thermal-neutrons. Also, the fast-neutron activation of ^{138}Ba by the (n,2n) reaction produces ^{137}Ba, which has a half-life of 2.55 minute; shorter than the 1.4 h of ^{139}Ba, that results from thermal-neutron activation. The shorter half-lives are advantageous as they enable replicate analysis of the same material within a short period of time, or the analysis of many samples within a short time. As Table 8.5 shows, 14-MeV neutron activation of ^{27}Al, ^{31}P, and ^{28}Si can yield the same isotope, ^{28}Al, thus the analysis of Al, P and Si in the presence of each other will cause interferences that must be corrected for.

8.1.4.2 Prompt Fast-Neutron Activation

Prompt release of gamma-rays by fast-neutrons is the result of the excitation of a nucleus and its immediate (within microseconds) return to the ground state. The de-excitation of the nucleus can take place from different nuclear levels, leading to several gamma-energies, the strongest of which can be used to identify the emitting nucleus. It is important, therefore, for the user to determine ahead the energies of the inelastic gamma-rays for the elements of interest. Elements with high inelastic cross-sections include C, Mg, Al, Si, and Fe. The excitation process can be produced as a direct result of the neutron interaction with the nucleus, or indirectly if the reaction product decays rapidly to a nucleus in an excited state. An example of the latter is the reaction ^{16}O(n,p)^{16}N, which occurs at neutron energy of 9.63 MeV. The resulting ^{16}N decays (with a 68% probability) by beta-emission with a half-life of 7.2 s producing ^{16}O in an excited state that emits 6.13 MeV photons upon its de-excitation. The transmutation of the nucleus by neutron absorption can also result in the production of an excited nucleus; as in the case of the reaction ^{16}O(n,$\alpha\gamma$)^{13}C, with ^{13}C releasing 3.09, 3.68 and 3.86 MeV photons as the excited nucleus returns to its ground state. However, neutron excitation of a nucleus is more often than not the result of inelastic scattering, (n,n$'\gamma$).

Inelastic scattering requires a minimum energy at just about the excitation energy of the target nucleus, typically over 5 MeV. Above 10 MeV neutron-energy, the other non-elastic scattering reactions become competitive. This makes it possible to use a variety of neutron sources in this technique, including neutron generators and isotopic sources. Reference [254] suggested that ^{13}C(α,n)^{16}O is an effective neutron-producing reaction for a source suitable for bulk analysis, to a low level of detection, of a wide variety of elements (Na, Mg, Si, P, Cl, K, Ca, Fe, Cu, An, Sn, Ce, Ta, and Au). With an alpha-energy of 3.5 MeV, 5.6 MeV neutrons are generated at zero angle with respect to the incident particle.

Table 8.6 lists the energies of the gamma-rays promptly emitted from the first excited levels of some nuclei that can be activated by inelastic scattering. The microscopic cross-sections for the reaction are also given in Table 8.6 at an incident neutron energy of 14 MeV. It should be kept in mind, however, that the cross-section of inelastic scattering for some elements can be higher at lower neutron energies, provided that the energy is above the first excitation energy of the considered nucleus. However, given the high neutron yield of 14-MeV neutron generators, attention is paid here to neutrons at this energy. Table 8.6 also lists the number of possible excitations levels up to an energy 14 MeV. The user should examine various energy levels to determine which ones have the highest cross-section, and which other elements present in the inspected object can generate photons of similar energies that interfere with the energies of interest. The evaluated nuclear data files (ENDF) cross-section library, available on the *Internet* via reference [78], can be helpful in this regard. The large number of energy levels shown in Table 8.6 for each nuclide can make element identification with this reaction often quite difficult. These many excitation levels also lead to background interference problems, as other elements in the surroundings or the examined object itself are excited by inelastic scattering. Prompt inelastic-activation is, therefore, used for detecting the presence of a relatively large amount of material, or high concentration of the element(s) of interest. Although the cross-section for inelastic activation is generally lower than that of the capture cross-section of thermal-neutrons, the ability to use unmoderated fast-neutrons, emitted directly from the source, compensates to some extent for the lower cross-section. Moreover, the lower degree of attenuation of fast-neutrons enables the examination of larger objects. Notice that the prompt emission of gamma-rays makes it unnecessary to include the exponential decay terms in applying the measurement model of Eq. (8.4) to prompt-activation analysis.

8.2. Gamma-Ray Emission by Charged-Particle Activation

Photon emission by activating the nucleus is also possible by charged-particles, such as protons, deuterons, tritons, or ions of heavier atoms. Obviously, for such positively charged-particles to penetrate the nucleus, they have to have sufficient energy to overcome the coulomb electrostatic barrier of the nucleus and to bring the nucleus to an excited state. This is achieved by accelerating charged-particles to an energy of few MeV for low atomic-number materials. Since charged-particles travel a fixed range, that depends on their incident energy and the material type, see section 3.3, the activation rate depends on the particle's range, R, in the material. Therefore, a measurement model similar to that used for neutron activation analysis, Eq. (8.4), can be used, with the infinitesimal activation volume, dV, replaced with $R\ dA$, where dA is the cross-sectional area of the incident beam, since the activation process takes

Table 8.6. Some elements activated by inelastic-scattering emitting prompt gamma-rays [78].

Element	σ at 14 MeV (mb)	E_γ First Excitation Level (MeV)	Number of Excitation Levels (up to 14 MeV)
^{12}C	179	4.4391	14
^{14}N	8	2.3128	21
^{16}O	101	6.1299	22
^{23}Na	81	0.4400	27
^{24}Mg	117	1.3686	11
^{27}Al	8	0.8438	39
^{28}Si	117	1.7790	13
^{29}Si	19	1.2730	15
^{35}Cl	6	1.2193	20
^{37}Cl	4	1.7266	27
^{56}Fe	74	0.8470	25

place only within the range of incident particles. If the thickness of the inspected sample is so thin that it is smaller than the range of the particle, in modeling the emission, the range R should be replaced by the thickness of the object. More accurate modeling should include the effect of the reduced particle energy on the activation cross-section, as the incident particles propagate through the object. The comparator method, discussed in section 8.1.2.1, can be used to avoid the need for determining all other parameters in the measurement model.

Gamma-rays emitted from the activation process have the advantage of being penetrating. The emitted photons could be delayed or prompt. Table 8.7 lists some of the common activation reactions used in charged-particle delayed activation, see reference [255] for more details. Due to the weaker coulomb electrostatic field of lighter elements, these elements are relatively easier to activate thus charged-particle activation is particularly useful for detecting lighter elements within metals or semiconductors. However, distinguishing such light elements from each other, if they simultaneously exist, may be difficult, since as Table 8.7 shows, with the same incident particle, the same photon energy (mainly 511 keV) can result from different elements. The 511 keV gamma-rays are emitted by the subsequent annihilation of positrons produced by the decay of the product nuclei. Although the difference in the half-life of the various nuclei emitting 511 keV gammas can be used to distinguish between them, two different elements can produce the same product nuclei even when bombarded by the same incident particle, as in the case of the

Table 8.7. Some charged-particle activation reactions [255].

Incident Particle	Target Element	Activation Product			
		Particle	Element	Gamma Energy (keV)	Half-Life
Proton	^{10}B	α	^{7}Be	477.6	52.28 d
	^{11}B	n	^{11}C	511	20.2 min
	^{14}N	α	^{11}C	511	20.2 min
	^{14}N	n	^{14}O	511, 2312.7	70.60 s
	^{16}O	α	^{13}N	511	9.97 min
	^{19}F	n	^{19}Ne	511	17.22 s
Deuteron	^{10}B	n	^{11}C	511	20.2 min
	^{12}C	n	^{13}N	511	9.97 min
	^{14}N	n	^{15}O	511	122.2 s
Alpha	^{31}P	n	^{34m}Cl	2127.8	32.2 s
	^{40}Ca	p	^{43}Sc	511, 372.8	3.89 h

production of ^{11}C by the proton activation of ^{11}B and ^{14}N. Prompt charged-particle activation is also possible. Examples of such reactions include the prompt activation of Li, Be, F, Na, Mg, Al and Si with protons [256].

Resonance Activation. If the incident charge-particle is chosen to have an energy that corresponds to one of the excitation energies of the target nucleus, a strong emission of gamma-rays takes place due to the *resonance absorption* of the incident radiation. For elements of high Z values, the energy levels in the nucleus become adjacent to each other, making it difficult to identify a unique resonance energy. However, some elements, such as ^{48}Ti, ^{52}Cr and ^{58}Ni, have some dominant states of resonances that can be used for resonance activation [257]. The correspondence between the incident energy and the resonance energy cannot be maintained over a large penetration distance within the target material, since charged-particles continuously lose energy. However, this hindrance can be turned into an advantage by selecting the incident charged-particle energy so that it has an energy higher than the resonance energy of the nuclide of interest. Then, the charged-particle reaches the resonance energy after penetrating a certain depth within the object. Thus by gradually increasing the energy of the incident beam from a value equal to that of the resonance energy to higher values, one will be gradually obtaining resonance emissions from the surface of the object and progressively from deeper regions within. This enables depth profiling of the element of interest, provided that the source energy is not sufficiently increased to reach the next excitation resonance

Table 8.8. Some charged-particle resonance reactions [257].

Element	Reaction	Resonance Energy (MeV)	Reaction Product	Photon Energy MeV
^{1}H	(^{7}Li, $\alpha\gamma$)	3.073	^{4}He	17.64, 14.70
	(^{15}N,$\alpha\gamma$)	6.385, 13.350	^{12}C	4.439
	(^{19}F,$\alpha\gamma$)	6.420, 16.440	^{16}O	6.130, 6.912, 7.117
^{4}He	(^{7}Li,γ)	1.680, 1.430	^{11}B	2,313
^{12}C	(p,γ)	0.456	^{13}N	2.365
		1.699		3.511 (95%)
^{14}N	(p,γ)	1.058, 1.550	^{15}O	0.511
^{15}N	(p,$\alpha\gamma$)	0.429, 0.897, 1.210	^{12}C	4.439
^{18}O	(p,$\alpha\gamma$)	0.633	^{15}N	0.429, 0.710
	(p,γ)	1.169	^{19}F	1.261
^{19}F	(p,$\alpha\gamma$)	0.872	^{16}O	6.130, 6.921, 7.117
^{23}Na	(p,γ)	0.309	^{24}Mg	1.369
	(p,$\alpha\gamma$)	1.011	^{20}Ne	1.634
^{24}Mg	(p,γ)	0.223	^{25}Al	1.612
^{26}Mg	(p,γ)	1.548	^{27}Al	0.720
^{27}Al	(p,γ)	0.992	^{28}Si	1.778

energy, unless emissions from more than one resonance are accounted for. Some of the employed nuclear-resonance activation reactions are listed in Table 8.8. These reactions are usually employed for studying the distribution of light elements implanted in metals. The use of heavy-ions to investigate the presence of light nuclei, such as ^{1}H and ^{4}He, can be viewed as an inverse reaction of protons and alpha-particles, respectively, with the nuclei of the incident heavy ions, i.e. the observed ions, being those of the nuclei of the heavy ions.

Prompt Activation. Photon-producing prompt activation analysis with charged-particles is also used in material analysis. Reference [257] catalogues the photon energy and the yield of photons induced by 4.5 MeV proton bombardment and alpha bombardment at 5, 11, and 16 MeV energy. The measurement model for this prompt activation process is similar to that discussed at the beginning of this section, but with the exclusion of the terms accounting for radiation decay. The method is capable of analyzing minor components in the range of mg/g of many elements.

8.3. Gamma-Ray Emission by Photon Activation

For photons to penetrate a nucleus and activate it, they must possess a high energy. The photonuclear cross-section is also much lower than that of neutrons. Unlike charged-particles, which be can accelerated to high energy, photon energy cannot be increased. Nevertheless, when a photon source is readily available, or when other activation techniques do not provide the required sensitivity or are hindered by interferes between various elements in the examined object, the use of photon activation can be advantageous. Photon activation requires, therefore, a high-energy and a high-intensity source, which can only be produced by high-energy electron accelerators. When highly energetic electrons bombard a high atomic-number material, they produce high-energy photons (x-rays) by the bremsstrahlung process (section 3.3).

A number of photonuclear activation reactions can take place, the most common of which is the (γ,n) reaction. This (γ,n) reaction is equivalent, in terms of the reaction products to the (n,2n) reaction, see section 8.6.3. However, the ability to produce intense beams of photons with an accelerator makes it more likely to induce activation with the (γ,n) reaction. Neutrons emitted from such reaction can themselves be used to identify a few elements, see section 8.6.1. The other main attractions of the (γ,n) reaction is its ability to produce radioactive nuclides in elements where neutron activation, by the (n,γ) reaction, produces a stable isotope, a very low activation rate (long half-life), or a too-short half-life for delayed activation analysis. Examples of the (n,γ) reactions that produces stable isotopes are those with ^{12}C, ^{14}N, and ^{16}O, producing, respectively, ^{13}C, ^{15}N, and ^{17}O. However, the photoneutron activation of these elements produces ^{11}C, ^{13}N, and ^{15}O; all are positron emitters, that subsequently emit 511 keV photons by positron annihilation. Note that emissions from these elements can interfere with each other. Another advantage of photoneutron activation is the ability to activate lead (^{204}Pb), which with neutrons produces a stable isotope (^{205}Pb).

Table 8.9 lists some of the elements which are candidates for photoneutron activation. A comprehensive list is given in reference [258]. Note the high threshold photon-energy required for the reaction, which necessitates the use of powerful accelerators. The measurement model of Eq. (8.4), with the photon cross-sections replacing the neutron cross-sections for the incident radiation, can be used to model this activation process. However, the need to use highly energetic photons makes it possible in some cases to ignore altogether the attenuation of the incident radiation. In addition to the (γ,n) reaction, Table 8.9 also shows that the (γ,p) reaction can be employed, as well as well as other multiplicity reactions: (γ,2n), (γ,3n), and (γ,np). These reactions produce nuclides that cannot be formed by other activation methods, and in turn offer unique indications that further help identify some elements.

Table 8.9. Some candidate nuclides for photonuclear activation [12, 259].

Nuclide	Threshold Incident Energy (MeV)	Reaction Product Isotope	Half-Life	Photon Energy (keV)	Yield (%)
		Via the (γ,n) Reaction			
^{12}C	18.7	^{11}C	20.39 min	511	$\leq$ 199.52
^{14}N	10.6	^{13}N	9.965 min	511	$\leq$ 199.61
^{16}O	15.7	^{15}O	122.24 s	511	$\leq$ 199.80
^{19}F	10.4	^{20}F	109.77 min	511	$\leq$ 193.46
^{58}Ni	12.2	^{57}Ni	35.60 hour	511	$\leq$ 87.17
				1377.63	81.70
^{64}Zn	11.9	^{63}Zn	38.47 min	511	$\leq$ 185.49
^{90}Zr	11.9	^{89}Zr	4.161 min	587.80	89.64
^{123}Sb	8.9	^{122}Sb	2.724 day	564.24	70.67
^{133}Cs	9.1	^{123}Cs	6.479 day	667.718	97.52
^{204}Pb	8.4	^{203m}Pb	6.3s	825.2	71.40
		^{203}Pb	51.873 hour	279.20	80.80
		Via the (γ,p) Reaction			
^{43}Ca	10.7	^{42}K	12.360 hour	1525	18.08
^{44}Ca	12.2	^{43}K	22.3 hour	617.5	79.20
^{48}Ti	11.5	^{47}Sc	3.3492 day	159.38	68.30
^{49}Ti	11.4	^{48}Sc	43.67 hour	983.53	100.10
				1037.52	97.60
				1312.12	100.10
^{68}Zn	10.0	^{67}Cu	61.83 hour	93.311	16.10
				184.58	48.7
^{74}Ge	10.9	^{73}Ga	4.86 hour	297.32	79.8
				325.70	11.2
		Via (γ,multiplicity) Reactions			
^{54}Fe	20.9	^{52}Mn	5.591 day	1434.09	100.00
				935.54	94.50
				744.23	90.00
				511	$\leq$ 59.20
^{91}Zr	19.2	^{89}Zr	78.41 hour	909.15	99.04
				511	$\leq$ 45.48
^{204}Pb	15.2	^{202}Pb	3.53 hour	960.70	92.00
				422.12	86.00
				786.99	49.80
				657.49	32.4
^{204}Pb	14.4	^{202}Tl	12.23 day	439.56	91.4
^{206}Pb	25.1	^{204m}Pb	67.2 min	899.15	99.17
				911.74	90.69
				374.74	89.00
^{209}Bi	22.3	^{206}Bi	6.243 day	803.10	98.9
				881.01	66.2
				516.18	40.70
				1718.70	31.80
				537.45	30.50

Since the incident flux and the activation cross-sections may not be completely known, it is necessary to calibrate a tested sample against a reference sample of known composition, using the comparator method described in section 8.1.2.1. Alternatively, the internal standard method can be used. This method, as explained in reference [260], requires knowing the content of one element in both the reference material and the examined sample. This element can be then used as an internal standard.

8.4. Gamma-Ray Emission by Positronium Decay

When a positron passes through matter it loses energy, and is subsequently captured by an electron, forming a short-lived hydrogen-like atom, called a positronium, in which the electron and the positron revolve around each other with an equal speed. As the two particles rotate around each other, each one spins around its own axis. If the positron and the electron of this atom spin in opposite directions, the "atom". is called a parapositronium, and is unstable. It decays within 0.1252 ns (mean life) to two photons of equal energy (511 keV each), emitted in opposite directions to conserve momentum. On the other hand, if the positron and electron spin in the same direction, the positronium (called then an orthopositronium) is more stable and survives for a longer period (0.1421 ms mean life) before it decays into three photons (each carrying an energy of $\frac{1022}{3} = 0.341$ keV), emitted in a plane at equally spaced directions (120°) to conserve momentum. The two-photon annihilation process is about 370 times more probable than the three-photon annihilation process [49], thus is the process utilized in positron emission imaging, discussed in section 8.8.3. However, the longer life-time of the orthopositroniums is indicative of their lower rate of decay, compared to that of parapositronium, and is what makes the latter useful for obtaining some material characteristic information. The life-time of orthopositronium, given above, are attained in dilute gases and highly porous materials, but is only a few nanoseconds in condensed matter, due the interaction of the positronium with the electrons. In metals and semiconductors, where the electron density is high, the available electrons make it difficult for positroniums to form, unless some void space (defects or dislocations) is present to isolate the positroniums from the surrounding electrons. This is the basis of using parapositron annihilation decay-time for material characterization.

The lifetime distribution of positrons and positroniums is monitored by measuring gamma-decay emission with fast-fast coincidence timing electronics, see reference [261], in which the start-time is marked by the prompt gamma-rays normally associated with a positron source, e.g. the

1.28 MeV associated with the decay of ^{22}Na. The stop-time of this decay process is marked by the annihilation of slowed-down (thermalized) positrons as they are being captured by the electrons of the medium producing two coincident 511 keV photons. Between the start-time and the decay-time, a decay curve is observed within 15 ns or so. The fastest component of this decay curve corresponds to the fast decay of the para-positronium, while the slowest component relates to the decay of the orthopositroniums. The intermediate component corresponds to the direct annihilation of species formed by the reaction of the positroniums with certain types of molecules within the medium [261]. Therefore, the decay curve of positronium's can be represented by the following measurement model:

$$N(t) = \sum_{i=1}^{3} I_i \exp(-\lambda_i)t \tag{8.7}$$

where $N(t)$ is the positron intensity at time t, I_i is the intensity arising from component i, and λ_i is the decay constant of that component. Typical values of I_i (normalized to total counts) and λ_i are: $I_1 = 3$ to 20%, $I_2 = 10$ to 60%, $I_3 = 10$ to 70%, $\frac{1}{\lambda_1} = 0.125$ ns, $\frac{1}{\lambda_2} = 0.3$ to 0.5 ns, $\frac{1}{\lambda_3} = 0.5$ to 10 ns [261].

8.5. Charged-Particles Emission

The emission of charged-particles is seldom utilized in multi-elemental analysis, but rather for detecting a single element. This particular element is designated by the nature of the monitored reaction, which is often caused by a single element in the examined material.

A unique feature of charged-particles is that they continuously lose energy, as discussed in section 3.3. Therefore, the emission of charged-particles can be used to determine the distribution with depth (depth profile) of an element from which the charged-particles are emitted. The concept relies on the fact that the energy of the emitted charged-particle depends on its depth of burial within the object. The deeper the location of emission, the lower is the particle's energy, in accordance to the stopping power, $\frac{-dE}{dx}$, of the charged-particle of interest in the inspected material. The energy, E, of a charged-particle emitted after traveling a distance x can be expressed as:

$$E = E_0 - \frac{x}{\cos\theta}\frac{dE}{dx} \tag{8.8}$$

where E_0 is the initial energy of the emitted particle and θ is the angle the exiting particle makes with the normal to the surface of the sample.

The penetration depth of charged-particles is limited to typically less than 100 μm below the surface of a material.

While the number of emitted charged-particles relates to the concentration of the affected material in the object, the energy distribution of emitted particles is indicative of the depth of emission. Owing to the short range of charged-particles, this depth profiling technique is limited to subsurface or thin-layer analysis. Moreover, since the range of a charge-particle is material dependent, the technique is used to determine the amount of impurity in an otherwise uniform material, so that a constant value of the range is maintained. To detect charged-particles, it is necessary to position the inspected object in a vacuum chamber, to avoid energy losses in air.

8.5.1. Charged-Particle Emission by Photon Activation

High-energy photons, as discussed in section 8.3, can penetrate a nucleus and release some of its protons. Some of the possible reactions are: the (γ,p) reaction, as in those listed in Table 8.9, the (γ,np) reaction, as in ^{56}Fe(γ,np)^{54}Mn, and the (γ,2p) reaction, as in ^{207}Pb(γ,2p)^{205}Hg. The emitted protons can in principle be used for depth profiling, employing the principle exemplified by Eq. (8.8). The emitted protons can exit a sample for subsequent detection, but only if the sample is quite thin. For such a thin sample, the incident high-energy photons will hardly suffer any interactions, for the activation to take place. Nevertheless, photonuclear activation can produce reaction products that decay by emitting high-energy beta-particles, either in the form of electrons or positrons. For example, ^{7}Li(γ,p) produces ^{6}He which decays by emitting beta-particles with a maximum energy of 3.51 MeV and 0.81 s half-life. Similarly, the reaction ^{9}Be(γ,p)^{8}Li produces a beta-emitter that decays with half lives of 0.84 s with maximum beta-particle energy of 13 MeV, while the reaction ^{32}S(γ,n)^{31}S gives rise to a positron emitter that decays with a 2.56 s half-life and a maxim positron energy of 4.39 MeV [13]. These reactions have threshold-energies of 12 MeV for ^{7}Li, 19 MeV for ^{9}Be and 15.24 MeV for ^{32}S [262]. The high-energy of emitted beta-particles and their low mass give them higher penetrability than heavier charged-particles, thus can be used for the analysis of lithium, beryllium and sulfur [259]. The short-half life of the reaction products either enables on-line (immediate) detection, or requires a rapid transfer system for delayed analysis away from the photon source.

Electrons can also be emitted from the atom by Compton scattering and the photoelectric effect. Therefore, electrons emitted by a material exposed to x-rays can be used to radiograph its surface. Most of the emitted electrons would arise from near the surface exposed to radiation, due to the poor penetrability of electrons in matter. A film placed on the surface, on the same side as the x-ray source, can then be used to radiograph the surface of the object exposed to x-rays. This method is known as electron-emission radiography [134]. The amount of electrons emitted depends on the electron-density of the material, thus the method can be used to image variations in material and their concentration near the surface of an object. This method is similar in principle to the electron-transmission method discussed in section 6.6.3. Therefore, the same requirements of high-energy x-rays, close proximity of the film to the object and the use of slow (fine grain) films, are also applicable to electron-emission radiography, for the same reasons discussed in section 6.6.3. However, electron-emission can be applied to image the surface of a thick object, and to detect the presence of metallic content in a non-metallic objects.

8.5.2. Charged-Particles Emission by Neutron Activation

Neutron-induced charged-particle emission relies on the (n,α) or (n,p) reactions introduced by thermal-neutrons in some elements [263]. These are the same reactions used in neutron detection, see chapter 4; that is, the ^{3}He(n,p)^{3}H, ^{6}Li(n,α)^{3}H and ^{10}B(n,α)^{6}Li reactions. Table 8.10 lists the main properties of these reactions, along with other reactions that have a reasonably high cross-section. Theses reactions are exothermic, releasing energy (Q-value) that is distributed among the emitted charged-particle and the heavier product nucleus, in reverse proportion to their mass. For example, the Q-value of the ^{10}B(n,α)^{6}Li is 2.310 MeV (96% of the time), $\frac{7}{11}$ of which is carried by the alpha-particle (1.470 MeV) and the rest (0.840 keV) by ^{6}Li.

The technique is primarily used for analyzing boron content, which is widely employed in the semiconductor industry. Boron is used to dope silicon wafers to create p-type semiconductors, and is also used in manufacturing borophosphislicate glass, employed as insulating layers on integrated circuits. The high abundance of the highly neutron-absorbing ^{10}B (19.9 % in natural boron, the rest being ^{11}B), makes boron a natural candidate for direct use in neutron depth-profiling. However, the technique requires a neutrons flux in the order of 10^{8} neutrons/ cm^{2} s to produce high sensitivity to changes in concentration. Therefore, this technique is usually employed in conjunction with research reactor facilities [264].

Table 8.10. Reactions useful for neutron depth profiling [264].

Target Nuclide	Product Nuclide	Q-Value (MeV)	Cross Section (barn)
	Via the (n,p) Reaction		
^{3}He	^{3}H	0.763	5,333
^{7}Be	^{7}Li	1.645	48,000
^{14}N	^{14}C	0.626	1.8
^{22}Na	^{22}Ne	2.350	31,000
	Via the (n,α) Reaction		
^{6}Li	^{3}H	4782	940
^{10}B	^{7}Li	2.310	3,837
^{40}K	^{40}Ar	2.287	4.4
^{59}Ni	^{56}Fe	5.097	12.3

8.5.3. Charged-Particle Emission by Charged-Particle Activation

The short range of the incident charged-particles limits their use to the analysis of thin targets. However, the dependence of the energy of charged-particles exiting a medium on the distance traveled by the emitted particles makes the technique useful for *depth profiling* of elements, particularly light elements. Table 8.11 lists some of the useful reactions, along with their Q-value. The Q-value affects the energy of the emitted particle, which also depends on the angle of emission. In the interaction between an incident charged-particle of incident energy E_{in} and a nucleus, one can consider the nucleus to be at rest. Conservation of energy dictates that the kinetic energy of the incident particle, E_{in}, plus the Q value of the reaction, be equal to the kinetic energy of the emitted charged-particle, E_{em}, plus the kinetic energy given to the produced nucleus, E_{pr}. That is:

$$E_{in} + Q = E_{em} + E_{pr}$$

If a charged-particle is emitted at an angle θ, with respect to the direction of the incident particle, and the product nucleus emerges at an angle ϕ, conservation of momentum in the direction of the incident beam and normal to it results in:

$$\begin{aligned} \sqrt{m_{in}E_{in}} &= \sqrt{m_{em}E_{em}}\cos\theta + \sqrt{m_{pr}E_{pr}}\cos\phi \\ 0 &= \sqrt{m_{em}E_{em}}\sin\theta - \sqrt{m_{pr}E_{pr}}\sin\phi \end{aligned}$$

where m refers to the mass of the interacting species. Eliminating E_{pr} and ϕ from the above equation yields:

$$E_{em} = [b \pm \sqrt{b^2 + c}]^2 \tag{8.9}$$

where,

$$b = \frac{\sqrt{m_{in} m_{em} E_{in}}}{m_{em} + m_{pr}} \cos\theta$$

$$c = \frac{m_{pr}(Q + E_{in}) - m_{in} E_{in}}{m_{em} + m_{pr}}$$

Therefore, for the energy of the emitted particle to be uniquely known, one must fix the direction of emission. Similarly, the energy of the recoil product can be shown to be equal to:

$$E_{pr} = \left[B \pm \sqrt{B^2 + C}\right]^2 \tag{8.10}$$

where,

$$B = \frac{\sqrt{m_{in} m_{pr} E_{in}}}{m_{em} + m_{pr}} \cos\phi$$

$$C = \frac{m_{em}(Q + E_{in}) - m_{in} E_{in}}{m_{em} + m_{pr}}$$

Therefore, if both E_{em} and E_{pr} are measured in coincidence at the proper θ and ϕ angles, a unique identification of the reaction can be obtained. This coincidence-measurement technique is not quite selective and specific for the reaction products of interest, but is an efficient counting process since each activation event is counted twice (once for each emitted particle). However, since the recoil-particle is usually a heavy charged-particle, its penetration depth is quite low, which restricts the use of coincidence-measurement to very thin objects of light elements [257]. As Eq. (8.10) indicates, the largest value for E_{pr} is at $\psi = 0$, i.e. in the same direction of the incident beam. Therefore, it is customary to measure the recoil particles in the forward direction where it has the largest penetration depth. At $\phi = 0$, $\sin\phi = 0$, and conservation of momentum dictates that $\sin\theta = 0$, with the two emerging particles moving in opposing directions, that is, θ has to be equal to 180°. Therefore, heavy-particles can be monitored in the forward direction, while light-particles have to be detected in the backward direction. The reactions listed in Table 8.11 for light nuclei can be used in this coincidence measurement technique; reference [257] lists other suitable reactions.

Resonances. Use can also be made of resonances in the activation cross-sections of charged-particles, at energies corresponding to the ex-

Table 8.11. Some reactions of charged-particle emission by other charged-particles [257]

Reaction	Q-Value (MeV)
^{1}H(^{11}B, α)^{8}Be $\longrightarrow \alpha\alpha$	8.59
^{2}H(^{3}He, p)^{4}He	18.354
^{2}H(d, p)^{3}H	4.033
^{3}He(d, p)^{4}He	18.534
^{6}Li(d, α)^{4}He	22.374
^{9}Be(p, d)^{8}Be	0.560
^{9}Be(p, α)^{6}Li	2.125
^{9}Be(d, p)^{10}Be	4.587
^{11}B(p, α)^{8}Be	8.591
^{10}B(d, p)^{11}B	9.231
^{10}B(d, α)^{8}Be	17.822
^{12}C(d, p)^{13}C	2.719
^{14}N(d, p)^{15}N	8.609
^{14}N(d, α)^{12}C	13.575
^{16}O(d, p)^{17}O	1.918
^{16}O(d, α)^{14}N	3.111
^{18}O(p, α)^{15}N	3.970
^{28}Si(d, p)^{29}Si	6.249
^{52}Cr(d, p)^{53}Cr	5.716
^{58}Ni(d, p)^{59}Ni	6.775
^{60}Ni(d, p)^{61}Ni	5.950

citation levels of the nucleus. Examples of these resonance activations reactions are [257]: ^{10}B(α,p)^{13}C at 1.507 MeV, ^{16}O(α,α')^{16}O at 3.048 MeV, ^{18}O(p,α)^{15}N at 1.165 MeV, and ^{27}Al(p,γ)^{24}Mg at 1.183 MeV, where the reported energy is the resonance energy for the reaction. Activation at these energies leads to an enhanced emission of charged-particles.

8.6. Neutron Emission

8.6.1. Neutron Emission by Gamma-Ray Activation

With a sufficiently high incident photon-energy (threshold on the order of 10 MeV, see Table 8.9), all neutron-containing nuclei will emit neutrons. It is difficult, therefore, to perform multi-elemental analysis by monitoring neutron emissions. However, a couple of nuclides have a relatively low threshold-energy for the photoneutron reaction; namely the ^{2}H(γ,n)^{1}H reaction with a threshold-energy of 1.67 MeV and the ^{9}Be(γ,n)^{4}He^{4}He with a 2.23 MeV threshold energy. This makes it possible to detect deuterium and beryllium using an isotopic gamma-source, such as ^{24}Na (2.754 MeV, 99.94%), for both elements and ^{124}Sb (2.091 MeV, 5.51%) for ^{9}Be only, where the values in brackets are the photon-energy emitted by the source and its relative yield. Table 2.17 also shows how neutron sources are formed by combining deuterium or beryllium with a gamma source; the same gamma sources can be used separately to detect the presence of these two nuclides within an object.

The fissionable elements (^{238}U and ^{239}Pu) are another set of nuclides that have a threshold-energy to photons that is lower (less than 10 MeV) than that of all other elements (with the exception of ^{2}H and ^{9}Be). Neutrons are emitted as a result of the usual photon emission process, (γ,n), or by photofission (γ,fission). The result is a significant neutron yield that makes it possible to detect the presence of these elements, even if they are embedded in, or surrounded by, other materials. Delayed-neutrons are also produced, by the decay of fission products, thus providing another unique indication of the presence of fissionable materials. It is also possible to determine the relative abundance of these fissionable elements (with respect to each other), by varying the energy of the incident photons to take advantage of the fact that the threshold-energy for photofission is element-dependent (5.26 MeV for ^{235}U, 6.0 MeV for ^{239}Pu, 7.8 MeV for ^{238}U and 9.28 MeV for ^{232}Th [262]).

The following (γ,n) reactions also produce delayed-neutron emitters [259]: ^{11}B(γ,2p)^{9}Li, ^{18}O(γ,2p)^{16}C, ^{18}O(γ,p)^{17} and ^{19}F(γ,2p)^{17}N. The reaction products have half-lives of 177 ms (^{9}Li), 0.75 s (^{16}C) and 4.17 s (^{17}N). These short half-lives make it possible to measure in real-time (almost promptly) the boron, oxygen or fluorine content.

8.6.2. Neutron Emission by Charged-Particle Activation

Neutron emission due to charged-particle bombardment is useful in concentration profiling of some elements in thin samples. Although neutrons are hardly attenuated in such a thin sample, the energy of emitted neutrons depends on the energy of the incident charged-particle. Since the latter changes with depth as the beam of charged-particles moves across the sample, the energy spectrum of emitted neutrons will depend on the concentration profile of the activated element. The measurement model of Eq. (8.8), applied this time to the incident particle, can be employed to determine the energy of the incident charged-particle at some depth x within the beam. Table 8.12 listed some of the useful reactions. Note that the Table contains mainly light elements, since it

Table 8.12. Some useful reactions for neutron emission by charged-particles [257].

Reaction	Q-Value (MeV)
^{2}H(d, n)^{3}He	3.269
^{7}Li(p, n)^{7}Be	−1.644
^{6}Li(d, n)^{7}Be	3.383
^{7}Li(d, n)^{9}Be	15.031
^{9}Be(α, n)^{12}C	5.702
^{11}B(d, n)^{12}C	13.733
^{11}B(^{3}He, n)^{13}N	10.182
^{12}C(d, n)^{13}N	−0.281
^{13}C(p, n)^{13}N	−3.003
^{14}N(d, n)^{15}O	5.067
^{15}N(p, n)^{15}O	3.542
^{16}O(d, n)^{17}F	−1.624
^{18}O(α, n)^{21}Ne	−0.699
^{43}Ca(p, n)^{43}Sc	−3.003
^{48}Ca(p, n)^{48}Sc	−0.510

is easier for charged-particles to penetrate lighter nuclides, because of their low coulomb barrier. However, some heavier nuclides, such as ^{48}Ca, ^{51}V, ^{35}Cr, ^{55}Mn, and ^{59}Co can be penetrated by beams of energy lower than their coulomb barrier [257]. The energy spectrum of the emitted neutrons can be measured either with a proton-recoil scintillator (such as BC-501, NE213) see section 4.4, or with the time-of-flight method discussed in section 4.5.5.

8.6.3. Neutron Emission by Neutron Activation

Neutron emission induced by incident neutrons can result from the (n,n′), (n,2n) and (n, fission) reactions, as they all produce emissions of neutrons that are different from the incident neutrons, due to interactions with the nucleus. However, in order to be able to utilize such neutron emissions, there must be a way to distinguish between incident and emitted neutrons. In the case of the (n,fission) reaction, the fission process can be induced in fissile isotopes, such as ^{235}U by thermal-neutrons, and since the fission process results in the emission of fast-neutrons, energy can be used to distinguish between the incident and emitted neutrons. Instead of energy discrimination, one can rely on timing information by monitoring delayed-neutron emissions. Very few reactions are suited for this purpose [265]; namely ^{9}Be(n,p)^{9}Li (178.3 ms) and ^{17}O(n,p)^{17}N (4.174 s), with the half-lives given in brackets. The neutron fission of ^{235}U, with thermal-neutrons, and ^{232}Th, ^{238}U and ^{239}Pu, with fast-neutrons, also produce delayed-neutrons. Measuring these delayed neutrons requires that the detection process take place after the object ceases to be exposed to source neutrons. This can be achieved by either removal of the exposed object away from the neutron source, or using a pulsed source. Since delayed-neutron emission occurs only for a few elements, it offers a high level of specificity for detecting the affected elements, while not creating interfering emissions by activating background material. The activation model of Eq. (8.4) is also applicable to delayed-neutron activation, with the gamma-ray parameters replaced by the corresponding appropriate neutron parameters.

8.7. X-Ray Emission

Photons (produced from x-ray tubes or isotopic sources), or charged-particles, can be used to excite an atom, by moving its electrons out of their orbits. If a bound atomic-electron, say in the shell closest to the nucleus, the K shell, receives an energy greater than its binding energy, it will be liberated from the atom, carrying a kinetic energy equal to the difference between the energy it receives and its binding energy. The liberated electron will leave a vacancy in the atomic shell that will be filled by an electron moving from an outer shell. This movement is accompanied by the immediate (in less than a picosecond) emission of a photon (x-ray) with an energy equal to the difference between the binding energies of the two levels. This process is known as x-ray fluorescence (XRF), and results when an atom is excited by an external source of an energy slightly greater than the electron's binding energy. Electrons moving to the K shell produce the so-called K x-rays, those moving to the L shell are called L x-rays, and so on. The energy needed to remove K-shell electrons varies from 13.4 eV for hydrogen to 133.6 keV for californium [238]. Each transition can arise from any of the angular momentum and spin states of higher energy levels. Such transitions are differentiated according to the notation of Siegbahn by adding a Greek letter and a subscript number after the transition line, with α designating transition from the closest electron shell, and α_1 referring to the strongest line, see for example reference [266]. Therefore, the line K_{α_1} corresponds to transition of an electron from the L_{III} energy level to the K shell, where the subscript III refers to an electron with an orbital angular quantum number equal 2 and a total (orbital plus spin) angular quantum number of $\frac{3}{2}$. Transition-level notation is also used, as it identifies directly the two levels between which electron transition takes place, and as such is a more straightforward and easier notation to use. The energies, E, in keV of the emitted x-ray corresponding to the K, L and M main levels can be approximately expressed as [267]:

$$E(\mathrm{K}) = \frac{Z^2}{100} \quad \mathrm{keV} \tag{8.11a}$$

$$E(\mathrm{L}) = \frac{Z^2}{750} \quad \mathrm{keV} \tag{8.11b}$$

$$E(\mathrm{M}) = \frac{Z^2}{3000} \quad \mathrm{keV} \tag{8.11c}$$

The K and L x-rays are the most distinguishable x-rays, due to the large energy difference between these two levels and the energy levels of outer electron orbits. Table 8.13 lists the energy of the K_α emissions from some atoms. Since the energy levels of an atom depends on its atomic number, the emitted x-rays are characteristic of the atom, and can be detected and used for elemental identification. Si(Li) detectors, see section 4.3.3, are usually used for detecting these low-energy photons. However, Si(Li) detectors are sensitive to photons of energy below 30 keV, or so. Therefore, they can detect K x-ray emissions from elements of atomic-number, Z, from 11 (Na) to 55 (Cs), but are insensitive to K x-rays generated from elements of Z from 70 (Yb) to 82 (Pb). However, L x-rays of these heavier elements are detectable by a Si(Li) detector. Elements of Z between 56 (Ba) and 69 (Tm) are difficult to analyze with a Si(Li) detector, since their K x-ray emissions are above the upper energy limit of the detector, while they produce L x-rays with energies that overlap with the K x-rays of lighter elements. However, planar high purity germanium (HPGe) detectors, see section 4.3.3, are sensitive to photons in the 30 to 125 keV range. This allows the detection of the K x-rays of many elements. Since chemical bonds occur mainly between electrons in the outer shell of the atom, emissions from the inner K and L shells are not affected greatly by chemical bonds, except for light elements. The following sections discuss the three different methods used for generating XRF by atomic excitation.

8.7.1. Excitation by Isotopic Sources

Isotopic sources can be used for atomic excitation, if they produce photons with energies that are as close as possible (but higher than) the binding energies of the K or L shells of the atoms of interest. Table 8.14 lists some radioisotopic sources and the elements they can excite.

As Table 8.14 indicates, the source energy used in XRF is quite low for most elements, since the electron binding energy is generally low, as indicated by the K_α energies of the elements listed Table 8.13. These low excitation energies introduce a number of factors that should be taken into account when arriving at a measurement model for XRF. First, as

Table 8.13. Energy of K_α emissions for some elements.

Element	Ti	V	Cr	Mn	Fe	Co	Ni
K_α energy (keV)	4.511	4.952	5.415	5.899	6.404	6.930	7.478
Element	Cu	Zn	Ga	Ge	As	Se	Br
K_α energy (keV)	8.045	8.639	9.258	9.886	10.544	11.222	11.924
Element	Kr	Rb	Sr	Zr	Nb	Mo	Cd
K_α energy (keV)	12.649	13.395	14.165	15.775	16.615	17.479	23.174
Element	Sn	I	Hf	W	Hg	Pb	
K_α energy (keV)	25.271	28.612	55.790	59.318	70.819	74.969	

* For a complete list for K and L emissions see references [22], [268] and [269].

Table 8.14. Some elements excited by x-ray fluorescence using radioisotopes [270].

Source	Emission Energy, keV		Some XRF Excited
	X-rays (yield)	γ-rays (yield)	Elements
^{241}Am	11.9 , 21.3 (37%)	59.5 (36%)	Cu, Zn, Cd, Sn, Sb, Au, U, Th, Pu
^{109}Cd	22.1, 24.9 (73%)	392 (64%)	Fe, Cu, Zn, Mo, Pb, U, Th
^{55}Fe	5.9, 6.5 (25%)		Al, P, S, Cl, K, Ca, Cr
^{125}I	27.5, 31.0 (141%)	35.4 (7%)	Cr, Mn, Fe, Co, Ni, Cu, Zn, Br, Rb,Sr, Zr, Mo, Cd
^{192}Ir/^{133}Ba	^{192}Ir:	296 (30%), 308 (32%), 316 (87%), 468 (52%)	Pb, U, Th, Pu
	^{133}Ba:	356 (62%), 303 (19%)	
^{57}Co	6.4, 7.0 (48%)	121.9 (85%)	Pb, U, Th, Pu
^{170}Tm	52.4, 57.5 (5%)	84 (3%)	Zr
^{3}H	β emitter, max. energy = 18.6 keV		Mg, Cl, K, Sr

explained in section 3.4, the *Auger effect* competes with fluorescence emission, as the emitted photons may be absorbed in the outer shells of the atom. Therefore, the fluorescence yield, ω, which is the ratio of the number of emitted x-ray photons to the number of vacancies created by the excitation source, should be taken into account. The value of ω for K emissions is much higher than that of L fluorescence. Also, the fluorescence yield increases as the atomic number (Z) increases, which explains why no low-Z elements are listed in Table 8.14. In general, elements of atomic-number less than 12 are very difficult to detect with XRF, since the energy of the emitted x-rays is so low that they are easily self-absorbed within the object [238].

The low particle-energy required for XRF makes it also necessary to use sources with thin windows, generally made of Be, to minimize absorption of the emitted radiation within the source itself. The low energy of the emitted x-rays also causes a significant amount of attenuation within the inspected object. If the object is quite thick, x-rays at the characteristic XRF energy may not be detectable, and the technique can become useless. Therefore, XRF is limited to the analysis of small samples, or low-density gases or liquid solvents. Even if the sample is sufficiently thin, the intensity of emitted x-rays will depend on the sample thickness, due to the attenuation of both the incident and emitted radiation. In order to demonstrate this effect let us construct a measurement model for XRF.

8.7.1.1 Measurement Model

Consider the case of excitation by photons. Assuming that the detector is shielded from direct exposure to the source photons, the detector count rate, $C(E_K)$, at a certain energy corresponding to an excitation level, say E_K, can be expressed as [238]:

$$C(E_K) = G \int \exp(-\Sigma_i x) \; N_i \; \sigma(E_K) \; \omega_K \; \exp(-\Sigma_o x) \; dx \qquad (8.12)$$

where G is a system constant that takes into account the source strength, detection efficiency and system geometry, Σ_i and Σ_o are the total cross-section of the incident and emitted photons, respectively, N_i is the atomic density of the element i that produces XRF photons at the E_K energy, with a fluorescence yield of ω_K, and σ_{E_K} is the microscopic cross-section for producing these XRF photons. The integration in Eq. (8.12) is carried out over the path of radiation within the sample. Because of the small size of the sample, the distance traveled by the source photons within the sample is assumed to be equal to the distance XRF x-rays travel in their way out of the sample. The exponential terms in Eq. (8.12) account for the attenuation of the incident and emitted photons, while the other terms represent the probability of interaction once the photons reach a distance x within the object. The microscopic cross-section, σ_{E_K}, for a photon source depends on the photoelectric cross-section, $\sigma_{pe}(E_K)$. The latter cross-section, as shown in section 3.4, can be expressed by the relationship of Eq. (3.30), which shows strong, but continuous dependence, on photon energy, and the atomic-number. However, as schematically shown in Figure 8.2, σ_{pe} exhibits strong jumps at the energies corresponding to the energies of XRF emission, as conditions become most favorable for the XRF process. Owing to these jumps

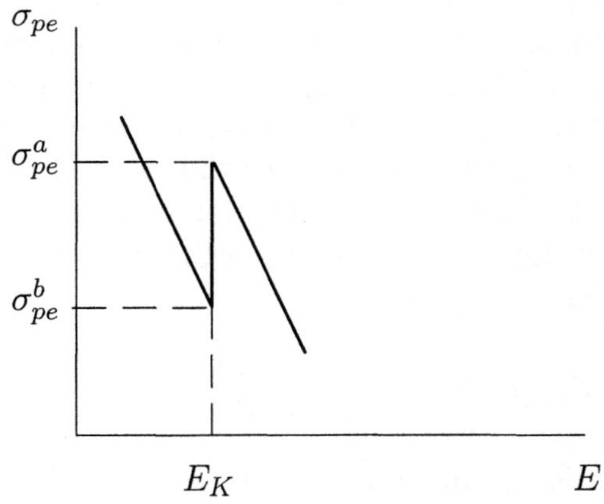

Figure 8.2. Schematic showing an absorption K-edge in the photoelectric cross-section.

in σ_{pe}, $\sigma(E_K)$, can be expressed as:

$$\sigma_{pe}(E_K) = \sigma_{pe} \frac{\sigma_{pe}^a - \sigma_{pe}^b}{\sigma_{pe}^a} \tag{8.13}$$

where σ_{pe} is the value of the photoelectric cross-section at the source energy (an energy that has to be slightly higher than E_K), σ_{pe}^a is the cross-section value just above the absorption edge at E_K and σ_{pe}^b is the value just before the edge.

Integrating Eq. (8.12) over the sample thickness, t, results in the relationship:

$$C(E_K) = G\ \{N_i\ \sigma(E_K)\ \omega_K\ \}\ \left\{\frac{1 - \exp[-(\Sigma_i + \Sigma_o)t]}{(\Sigma_i + \Sigma_o)}\right\} \tag{8.14}$$

The model of Eq. (8.14) can be generalized to other elements in the object by measuring the spectrum of various XRF photons expected in the sample, using one of the spectroscopy detectors described in section 4.5.4.

For small values of $(\Sigma_i + \Sigma_o)t$ ($<< 1$), i.e. for a thin sample or a low density target, the count rate, C, becomes:

$$C(E_K) = G\ N_i\ \sigma(E_K)\ \omega_K\ t \quad \text{for} \quad (\Sigma_i + \Sigma_o)t << 1 \tag{8.15}$$

The count rate becomes linearly proportional to the atomic-density of the element of interest, N_i, for a given sample thickness. Also, since $\sigma(E_K)$ is evaluated for the element of interest, then as Eq. (8.15) indicates, the count rate is also independent of the composition of the sample, that is independent of the presence of other elements, since the total cross-sections Σ_i and Σ_o do not appear in the measurement model.

For larger values of $(\Sigma_i + \Sigma_o)t$, as Eq. (8.14) indicates, the count rate becomes nonlinearly dependent on N_i, since the value of N_i also affects the value of Σ_i and Σ_o. However, for very large values of $(\Sigma_i + \Sigma_o)t$, the exponential term in Eq. (8.14) becomes negligible and the count rate reaches a *saturation value*, C_∞, of:

$$C_\infty(E_K) = \frac{G\, N_i\, \sigma(E_K)\, \omega_K}{(\Sigma_i + \Sigma_o)} \tag{8.16}$$

This saturation value is the maximum count rate obtainable, and the thickness at which this value is reached is known as the *critical thickness* or the "infinite thickness". Then, the count rate, though independent of the thickness, is influenced by the composition and density of the object, due to the presence of Σ_i and Σ_o in Eq. (8.16). If one considers the critical thickness to be attained at a value of $(\Sigma_i + \Sigma_o)t$ equal to, say, six, with $e^{-5} = 6.7 \times 10^{-3} << 1$, then the critical thickness can taken as equal to $\frac{5}{\Sigma_i + \Sigma_o}$. This is still a small value for most materials, given the low energy of the photon source needed to cause excitation. Therefore, even for a large sample, only a skin-deep layer, about $\frac{5}{\Sigma_i + \Sigma_o}$ in thickness, on the surface of the source is analyzed, by monitoring emissions from near the surface.

The model of Eq. (8.12) does not include the effect of Compton scattering, which competes with the photoelectric-effect as explained in section 3.4. Since the photon energy of the excitation source has to be slightly higher than the binding energy of the electrons in the K shells of the atom, Compton scattering can reduce the energy of the incident photons, according to Eq. (3.37), to values equal or below the required excitation-energy. The scattered photons can also reach the detector measuring the XRF emissions. The detected photons at the energy of interest will then contain an XRF component, and a background component from Compton scattering. However, by proper design of the setup, Compton-scattered photons can be made to scatter at a large angle, so that their energy is well below the monitored XRF photon energy.

Self-excitation is another complicating factor that is not considered in the model of Eq. (8.12). The emitted XRF photons resulting from one element can themselves be absorbed by other elements within the object, causing excitation of one or more of these elements in the sample. For this excitation process to occur, the energy of the original XRF photons has to match the binding energies of electrons of these elements. This in turn implies that the excited elements have binding energies close to that of the element of interest. The result would be an observed enhancement of the XRF emission, that is difficult to account for. Fortunately, however, the radiation intensities used with radioisotopic sources is too low (typically in the range of 370 MBq) to produce sufficient secondary emission to interfere with the measurements [270].

Matrix Effect. The attenuation coefficients in the measurement model of Eq. (8.14) must be correctly accounted for, if accurate quantitative measurements are to be obtained. This attenuation effect is refereed to as the *matrix effect*, with the matrix being the material within which the emissions arise. Reference [20] provides a number of approaches for dealing with this effect, which can be summarized as follows:

Iteration. Make an initial estimate of the material composition, calculate the relative emission intensities for that composition, compare that with the measured values, correct the initial estimate and repeat the process until calculation-convergence is attained.

Influence Coefficients. These coefficients define in advance the matrix-effect on emissions of present elements for a given excitation source. The coefficients can be calculated or measured for a typical reference sample.

Measured Attenuation. The attenuation coefficient of the specimen is measured separately for a thin sample, by measuring the transmission of the source radiation within the sample, since the incident radiation energy is close to the fluorescence energy.

Internal Standard. Adding a known proportion of a compound containing an element with known fluorescence characteristics and using its measured intensity to correct for the matrix effect.

Dilution. Measure the intensity at two different dilution levels and use the corresponding measurements to correct for the attenuation effect, or dilute the sample to a high level to minimize the matrix effect.

8.7.2. X-Ray Excitation

X-rays can be used to excite atoms and produce x-ray fluorescence. X-ray tubes, as mentioned in section 2.2.1, have two main features; (1) they can produce intense radiation fields, and (2) they generate a wide energy-spectrum of photons. The first feature is clearly attractive in XRF, as it increases the incident flux, and hence results in the production of more x-ray fluorescence. The increased x-ray emissions provide strong indications, by increasing the signal-to-background ratio and improving statistical confidence in the indication. In addition, the wide spectrum enables monitoring more than one element at a time, as it can cover a wide range of K and L emissions in many atoms. However, these advantages are countered by the complexities introduced by the simultaneous analysis of the resulting many emissions. Also, a wide spectrum of photons incident on an object will produce Compton scattered photons at many energies, overlapping the XRF energy peaks. This increases the background signal, and acts against the gain attained by using an intense source. The scattering background signal can be reduced by placing a filter in front of the x-ray machine that has a strong absorption edge just above the energy edge for the elements of interest, to remove the higher-energy photons that will scatter to lower energies. The placing of such filter comes, however, at the expense of reduced intensity of the excitation radiation. Alternatively, a special target, aside from the common high atomic-number targets (tungsten and gold), can be used to produce characteristic x-rays at the desired energy (close but slightly higher in energy than the monitored fluorescence energy). This comes, however, at the expense of a reduced overall intensity, due to the lower atomic-number of the target material. These so-called characteristic x-ray tubes employ elements such as silver, rhodium, molybdenum, copper and chromium [269]. In addition, a secondary target can also be used [269]. That is, the photons of a primary x-ray (with a high atomic-number target) can be utilized to excite an external secondary target. The fluoroscopic emission from the secondary target, after filtering to cut-off low-energy radiation, becomes essentially monoenergetic with an energy equal to the excitation energy of the material of the secondary target. The material of the secondary target can, therefore, be chosen to match the element to be analyzed. While the use of such changeable external target offers unlimited flexibility in tailoring the x-ray energy to the desired energy, it reduces the intensity of the incident flux, because of the lower relative yield of the secondary target.

X-rays, as an excitation source, can also be produced indirectly using charged-particles. When charged-particles are mixed with the material of the object, they produce their own photons with the bremsstrahlung process, described in section 3.4. The bremsstrahlung photons, in turn, can cause the excitation of atoms, and subsequent XRF emissions. Beta-emitting sources are used for this purpose, since the amount of bremsstrahlung radiation produced by heavier charged-particles is very small, due to their large mass.

8.7.3. Charged-Particle Excitation

High-energy charged-particles are used as external sources for atomic excitation, in a process refereed to as PIXE (particle-induced x-ray emission) [271, 267]. Electron beams are not preferred for such applications, since their bremsstrahlung effect produces a strong continuous background signal that makes it difficult to segregate XRF signals. Protons or alpha-particles (0.1 to 10 MeV in energy) are usually used in PIXE, as they can deposit significant amount of energy within a short distance. While a positively-charged particle with a few MeV in energy can penetrate the coulomb electrostatic field of a light nuclei (low atomic-number) causing activation, see section 8.2, it is unlikely for such particles to overcome the field associated with a heavier nuclei. Instead, charged-particles as they are slowed down, will impart their energy to the electron of the atom creating vacancies in the inner electronic shells. This excitation of the atom leads to the release of x-rays, as electrons from higher orbits lose energy by moving into vacated lower shells. This is the same x-ray fluorescence (XRF) process triggered by photon sources, as explained in sections 8.7.1 and 8.7.2. However, charged-particles are particularly advantageous when dealing with small samples (on the order of a few millimeters), or with aerosols or gases, where the cross-section for charged-particle interaction is much higher than that of photons.

Measurement Model. The x-ray count rate at energy E_k from a particular element i by PIXE can be represented by a measurement model similar to that of Eq. (8.12):

$$C(E_K) = G \int_0^x N_i \, \sigma(E_K) \, \omega_K \, \exp\left(-\int_0^y \Sigma_\gamma \, dy\right) dx \qquad (8.17)$$

where G is a system constant that takes into account the source strength, detection efficiency and system geometry, N_i is the atomic-density of the element i that produces the XRF photons at an energy E_K, with a fluorescence yield of ω_K, $\sigma(E_K)$ is the microscopic cross-section for producing these XRF photons, Σ_γ is the total cross-section of the emitted photons, x is the distance traveled by the charged-particle in the inspected material, and dx and dy are incremental distances in the direction of the incident beam and the emitted photons, respectively. The exponential term in

Eq. (8.17) accounts for the attenuation of the emitted photons. Since the number of incident-charge particles remains constant until they are absorbed at the end of their range, no attenuation term is included in the model of Eq. (8.17). For charged-particles, the stopping power $\frac{-dE}{dx}$ can be determined in advance, see section 3.3. Therefore, the integral along the path of the incident charged-particle along x in Eq. (8.17), can be performed over energy, leading to the model:

$$C(E_K) = G \int_{E_i}^{E_x} N_i \; \sigma(E_K) \; \omega_K \; \exp\left(-\int_0^y \Sigma_\gamma \; dy\right) \; \frac{1}{-\frac{dE}{dx}} dE \quad (8.18)$$

where E_i is the energy of the incident beam of charged-particles, and E_x is the beam's energy after traveling a distance x in the inspected material. The production cross-section, $\sigma(E_K)$ in the model of Eq. (8.18), can be evaluated, see reference [272]. Other parameters in Eq. (8.18) can be determined in advance from physical properties or calibration, enabling the determination of the atomic-density of the element of interest from a count rate measured at a photon energy, E_K, characteristic of the element of interest (see Table 8.13 for the K_α of some elements). Two limits of the measurements model of Eq. (8.18) have some useful detection implications.

If the object to be inspected is small, less than 1 mg in mass, then one can neglect the energy-loss charged-particles encounter in the object. Then, all energy-dependent terms in Eq. (8.18) can be assigned constant values, equal to their values at the incident energy. Also photon attenuation can be neglected in such a small target. Eq. (8.18) is then be simplified for a small thickness, Δx, to the model:

$$C(E_K) = GN_i\sigma(E_K)\omega_K\Delta x \qquad \text{for a thin target} \qquad (8.19)$$

Then, the x-ray count rate is directly a linear function of the concentration of the element, N_i, producing x-rays at energy E_K. This is obviously a simple but crude model. The other extreme is to consider a thick target in which the incident beam loses all its energy. Then the upper limit of the first integral in Eq. (8.17) becomes equal to its range, R, and the upper energy limit of integration in Eq. (8.18) becomes equal to zero. This makes it possible to calculate the integrals in Eq. (8.18) without determining the target thickness, since beam's interactions will all occur within a distance less than its range. Therefore, while the advantage of a thin target is the ability to linearly relate count rate to concentration by Eq. (8.19), the thick target eliminates the need to measure the target thickness. However, one should keep in mind that a thin target is merely a target of a thickness less than the range of the employed charged-particle. A material of thickness of about 10 μg/mm^3 (multiply by density to obtain thickness in μm) can be considered to be a thin target for commonly used beam energies [267]. Therefore, the PIXE method is limited to examining thin samples (1 mg or less) and aerosol droplets.

Background. As charged-particles lose energy they also emit photons by the bremsstrahlung process discussed in section 3.3. This creates a continuous background photon spectrum underlying the fluorescence radiation of interest. The background is higher for lighter charged-particles, since heavy ones do not produce much bremsstrahlung. In addition, electrons ejected and subsequently accelerated by the incident charged-particles, also produce secondary bremsstrahlung radiation that further adds to the background component. The recoil-electrons that are produced by the Compton scattering of the emitted x-rays can also contribute to the background continuum. In light elements, nucleus activation can occur, producing its own photon emission, as for example in the 511 keV photons produced by fluorine in teflon via the ^{19}F(p,n)^{19}Ne reaction, see Table 8.7. The gamma-rays emitted in such a reaction produce a continuous background component by scattering within the target and the surroundings. Every effort should made to reduce these, and other possible sources of background effects, to increase the signal-to-background ratio and obtain good indications for fluorescence emission.

Ionoluminescence. X-rays emitted by PIXE are simultaneously accompanied by luminescent light, which has a frequency extending from that of infrared to that of ultraviolet [273]. This light, called *ionoluminescence*, can be used for the detection of rare-earth elements, as the luminescence emission lines from such elements are sharp, stable and have a high yield in a variety of elements [274].

8.8. Emission from Internal Sources

In industrial applications involving liquids, gases, or particulates, an isotopic source can be directly inserted into a medium. If a radioactive source is inserted, it has to be chosen so that it has chemical and physical properties similar to those of the medium in which it is introduced. If the medium or the internal source are in motion, the internal sources can be used as radiotracers. If the radioactive material is incorporated into a specific phase (gas, liquid, or solid) or into a part of the system, it can then be used as a "marker" or "label" of that phase, or portion. Radiation emitted from the tracers can then be used to track the motion of the medium. Internal sources can be those of a radioactive material, and the radiation emitted can be used to detect the presence of radioactivity and identify the nature of the emitting isotopes. Internal sources can also be used for imaging purposes. Obviously in fluids, radioactive sources that can be injected into a fluid have to be open sources that dissolve easily into the liquid. The sources are either beta, positron or gamma emitters. Although, beta-particles have a short penetration depth, compared to photons, they are safer to handle as they can be easily shielded against, but as shown in section 4.2 their detection requires special attention. Due to their limited penetrability, the detection of beta-particles within a system will require either a thin window in the walls of the system, or the taking of a sample for off-line detection. Gamma-rays are attractive as internal sources, not only because of their penetration power, but also due to their distinct energies that allow the introduction of more than one internal source in the system for simultaneous tracking (by energy), say of various species in a mixture of material. Positron-emitters are used to produce annihilation (511 keV) photons, as the positron combine with the electrons of the medium. Since these annihilation photons are produced in opposite directions, they are often used for imaging purposes, rather than in radiotracer applications. In solids, a radiotracers can be introduced by thermal-neutron activation. For the produced radioactive source to act as a tracer, it has to be subjected to some movement, e.g. as in the particulate movement produced by wear or severe erosion. Both gauging (radiotracing and measurement of radioactivity) and imaging methods with internals sources are discussed below.

8.8.1. Radiotracing

A radioactive tracer can be externally monitored by direct measurement of the count rate of emitted radiation, or by timing its arrival [2, 238, 275]. The following are useful radiotracer isotopes [276]: ^{3}H, ^{14}C, ^{32}P, ^{35}S,^{41}Ar, ^{45}Ca, ^{68}Ga, ^{76}As, ^{82}Br, ^{85}Kr, ^{90}Y, ^{133}Xe, ^{140}La, ^{198}Au, ^{203}Hg. The measurement model presented below demonstrates the type of information obtained from radiotracer indications.

Measurement Model. An internal source of a specific activity $\mathcal{A}$ at some point within an object, will produce a count rate C at some external position x, see Figure 8.3, that can be expressed as:

$$C = K \int_V \mathcal{A} \, \exp\left[-\int_0^R \Sigma(x)\, dx\right] \frac{1}{R^2}\, dV \tag{8.20}$$

where K is a system constant that takes into account geometry and detection effects, $\Sigma(x)$ is the total cross-section at some distance x along the radiation path towards the detector, R is the distance between the

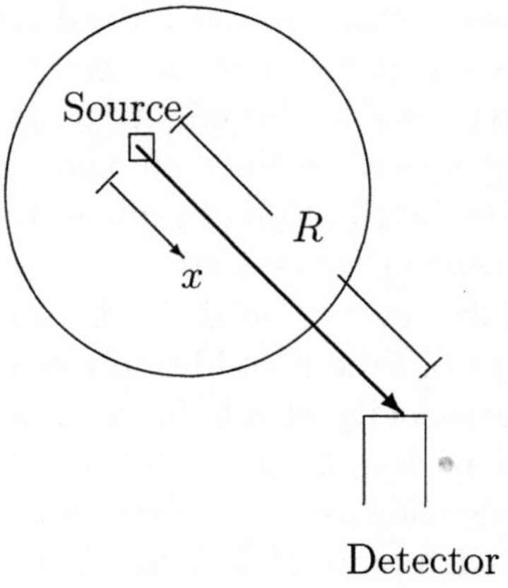

Figure 8.3. A schematic showing detected radiation emission from a point in an object.

emission point and the detector, and the integration is over the volume, V, of the object containing the radiotracer.

With the concentration of the radiotracer material being small so that it does not affect the overall system properties (a necessary condition for a nonintrusive measurement), the model of Eq. (8.20) indicates that the count rate C will change mainly with the amount of radioactivity in the system, $\int \mathcal{A}\, dV$. If the specific activity, $\mathcal{A}$, is constant, say determined by the solubility of the radiotracer in a liquid, then C becomes proportional to the volume, V, and as such can be used to determine the volume within which the radiotracer material is embedded. This can be useful, for example, for the determination of the amount of flow leakage from an otherwise closed system. If a tracer is flowing in a duct at a volumetric flow rate of Q, then $dV = Qdt$, where dt is equal to some time which can be taken as the measurement time. Then, at a constant $\mathcal{A}$, the count rate, as Eq. (8.20) shows, becomes proportional to Q, and can be used to measure the volumetric flow rate.

When the volume, V, of an internal source is fixed, the value of C becomes an indication of the concentration of the radiotracer in the material. Then, if a number of measurements are taken at different positions, a profile of the count rate will be obtained. This spatial profile of the count rate can be used to monitor the distribution of the tracer within the inspected volume. Mixing, blending, deposition, or loss (leakage) information can be obtained. If the volume of the internal source and its specific activity are both fixed, then C will be mainly dependent on the position of the radiotracer, which can be used, for example, to locate a pipe buried underground. For such fixed geometry and activity, the spatial profile of the count rate can be used to determine the area over which the radioactivity is distributed. A reduction in the count rate and/or a widening of the spatial profile can also be taken as an indication of a disruption in the distribution of radioactivity due to leakage or obstruction of flow, or some other reasons.

The motion of a radiotracer can be timed to provide information on its velocity or the amount of time it resides in a particular location. For example, to measure the velocity of a fluid in a pipe, a radiotracer can be injected over a short period, to represent a pulse, and its arrival is monitored at some distance downstream of the flow. The time, t, it takes for the pulse signal to reach a point at distance L from the injection point is used to determine the flow velocity, v as $v = \frac{L}{t}$. This value of v is a good estimate of the flow velocity, if the radiotracer is well-mixed in the flow, i.e. if the flow is turbulent and the distance L is sufficiently long to ensure complete mixing by the flow turbulence. A time profile, change of count rate with time at some distance from the point of injecting a pulse of a radiotracer, is also a measure of the time a chunk of fluid takes to travel between the injection point and the measurement point, called the residence-time. The residence-time and its distribution can be used, for example, to examine the effectiveness of stirring in mixing materials in a chemical reactor [277].

The interpretation of information obtained from radiotracer measurements depends greatly on the nature of the system examined and on the nature of the radiotracer, i.e. whether the radiotracer material behave identically as the material of the medium. Many studies are devoted to understanding the behavior of radiotracers in chemical, geological, and agricultural systems, and in biology and medicine.

8.8.2. Radioactive Materials

Radiation emission can be used to detect and identify materials that are naturally radioactive or known to be radioactive. Due to their high penetrability, gamma-ray photons are used to detect the presence and concentration of elements emitting radioactivity. The decay of ^{40}K results in the emission of single-energy 1.461 MeV photons, enabling detection of potassium, and potassium-rich, natural materials. Uranium and thorium can be detected in nature or in radioactive materials, by the many gamma-ray energies resulting from the complex decay series of ^{232}Th, ^{235}U and ^{238}U, see section 2.4. Some prominent gamma-rays are used to identify these decay chains. The ^{232}Th decay-series has a distinct photon peak at 2.615 MeV from the decay of ^{208}Tl [107]. Other energy peaks characteristic of gamma-emissions from the ^{232}Th series (in equilibrium with all its daughters) include 1.630 MeV (^{228}Ac), 1.621 MeV(^{212}Bi) and 1.588 MeV (^{228}Ac), 0.969, 0.965, and 0.911 MeV (^{228}Ac), and other lower energy peaks. The main gamma-emissions from ^{235}U series are all below 1 MeV, with ^{211}Pb giving the highest energy of emission at 0.832 MeV. Since the abundance of ^{235}U in natural uranium is quite low, as indicated in Table 2.19, its contribution to gamma-ray emission is negligible in comparison to that of the ^{238}U series. The most distinguishable gamma peaks for the ^{238}U series in the MeV range are those emitted by ^{214}Bi at 1.765, 1.120 and 0.609 MeV. There are also lower energy emissions. At higher energies ^{214}Bi produces weaker gamma-emissions at 2.448 and 2.204 MeV. The spontaneous-fission of ^{235}U and ^{238}U produce gamma-rays over a wide range of energy, as the fission products decay. Neutrons emitted from the spontaneous fission process can also activate the surroundings, producing background gamma-rays. However, given that the probability of spontaneous fission is quite low, their effect on gamma-ray emission can be ignored. Note, however, that higher energy peaks (above 1.5 MeV) can produce double-escape and single-escape peaks, see section 4.5.4, at energies of $E - 2m_0c^2$ and $E - m_0c^2$, where m_0c^2 is the electron rest-mass energy (0.511 MeV). These escape peaks should not be interpreted as identifiers of some other elements. Reference [40] shows also a 2.20 MeV gamma peak in the decay of the ^{232}Th series. In general, the gamma-peak at 2.615 MeV is a good indicator of the thorium decay products, while the 1.765 MeV energy characterizes the uranium decay series, and the unique gamma-energy of ^{40}K (1.461 MeV) directly identifies its presence [40]. The lower-energy emissions of gamma-rays from the daughters of ^{232}Th and ^{238}U are more difficult to analyze, due to interference by higher energy photons, as they lose energy by Compton scattering.

A number of radionuclides are also present in nature and can be used for example in environmental studies, such as groundwater studies. Secondary cosmic emissions produce the following isotopes in nature (half-lives indicates in brackets): ^{3}H (12.4 y), ^{14}C (5730 y), ^{32}Si (105 y), ^{36}Cl (301 000 y), ^{39}Ar (269 y) and ^{81}Kr (210 000 y). Atmospheric tests of thermonuclear devices produce the following radionuclides: ^{3}H, ^{14}C and ^{36}Cl, ^{90}Sr (29 y) and ^{137}Cs (30 y). Tritium, ^{14}C and ^{85}Kr(10.7 y) may also be released from nuclear plants and some nuclear reprocessing facilities. Natural nuclear reactions (subterranean) produce ^{36}Cl and ^{39}Ar. An indication of the amount of rainfall can also be determined from the concentration of ^{7}Be.

When a material is suspected to be radioactive, when searching for a missing radioactive material, or when the nature of a radioactive material is to be determined, radiation emission can be monitored and analyzed. Although alpha and beta emissions can be used for this purpose, they are often limited to the detection of radioactive contamination of surfaces, due to their limited penetrability. Gamma-radiation is the preferred emitter, not only because of its high penetrability, but also due to the ability to identify elements with their characteristic emission energy. Indeed, gamma-ray spectroscopy is widely used for identifying radioactive materials, to the extent that computer commercial packages are available for analyzing measured gamma-ray spectra for routine analysis of radioactive substances, see reference [244] for comparison between these packages. References [278] and [279] provide a comprehensive list of pulse-height spectra experimentally measured with scintillation and solid-state detectors, respectively.

Measurement Model. The amount of gamma-ray emissions from a radioactive material can be modeled in a manner similar to that of Eq. (8.20), with the concentration of the radioactive isotope replacing activity. Therefore, for a radioactive source emitting photons at a certain characteristic energy, E, the count rate can be expressed as:

$$\mathcal{A} = KC(E) = K \int_V N \,\exp[-\Sigma_{total}(E)x] \frac{1}{R^2}\, dx \qquad (8.21)$$

where $C(E)$ is the count rate at energy E, N is the atomic-density of the isotope producing the gamma-rays with energy E, $\Sigma_{total}(E)$ is the photon total cross-section at E of the material containing the radioactive material, x is the distance the radiation travels in its way towards the detector, R is the distance from the emission point to the detector site, and K is a system constant that depends mainly on the detection efficiency. The exponential term in Eq. (8.21) accounts for the attenuation of photons within the object (self-attenuation) and the $\frac{1}{R^2}$ factor incorporates the divergence of radiation with distance. These two factors detract from the direct linearity of the count rate, $C(E)$, with the concentration, N. The attenuation-factor makes it possible for a dense object to produce less counts than a lighter object containing the same concentration of the radioactive isotope, since the total cross-selection of photons increases with increasing energy, see section 3.4. The radiation divergence term, on the other hand, reduces the contribution of radiation present in the object at locations far away from the source. It may be necessary, therefore, to monitor emitted radiation from an object at different orientations and take the average value of these measurements to obtain a more representative indication of the object.

For dense materials, and low photon-energy emission, the attenuation term in Eq. (8.21) can be so dominant that only a small amount of radiation, confined to the surface of the material, can be detected. This problem is particularly encountered in nuclear fuel assay. For instance, for the detection of ^{235}U in an enriched nuclear fuel, where equilibrium conditions between the decay daughters are not attained, the only viable emission is that of the 185.7 keV gamma emitted from the decay of ^{235}U itself. The attenuation cross-section for photons in metallic uranium at this photon energy is about 25 cm^{-1}, leading to a mean-free-path of about 0.4 mm. With this high cross-section, radiation emitted from beyond about a few millimeters would be completely attenuated. This also makes it possible to deal with the attenuation problem by comparing the measurements to a standard sample, of known enrichment, and with a thickness that would result in almost complete attenuation of the emitted radiation, a few millimeter thick [107]. However, the low emission-rate, due to the long half-life of ^{235}U, see Table 2.19, makes it necessary to minimize the background signal from uranium in the environment, and to watch and correct for the interference of the 186.1 MeV gamma-rays emitted by ^{226}Ra, a daughter of ^{238}U. The closeness in energy between the gamma-rays of ^{235}U and ^{226}Ra necessitates the use a detector with a good energy resolution.

Gamma-emission from plutonium is also used to assay nuclear fuel, an important aspect of nuclear safeguards. Plutonium can be present in irradiated nuclear fuel at various concentration of its isotopes: ^{238}Pu (87.7 y), ^{239}Pu (2.41×10^4 y), ^{240}Pu ($6. \times 10^3$ y), ^{241}Pu (14.4 y) and ^{242}Pu (3.75×10^5 y), with the half-life given in brackets. The relative abundance of these isotopes depends on the neutron flux and duration of exposure of the uranium fuel (i.e. the neutron fluence). However, ^{239}Pu is usually the most dominant isotope, as it is directly formed by the neutron absorption of ^{238}U, usually the most abundant isotope in nuclear fuel, transforming it to ^{239}U, followed by two consecutive, beta decays, to ^{239}Np then to ^{239}Pu. The other heavier isotopes of plutonium are produced by sequential steps of neutron absorption of ^{239}Pu to ^{240}Pu, then to ^{241}Pu, and the latter to ^{241}Pu, while ^{238}Pu results from ^{235}U, by multiple neutron capture and beta decay (of ^{238}Np).

Many photon-energy emissions accompany the decay of plutonium isotopes, see for example [268]. In addition, plutonium is usually present in conjunction with other fission products. A plutonium assay is, therefore, usually conducted after several years of its production to allow for the decay of fission products. Also, the relative, not absolute, abundance of the plutonium isotopes are determined to avoid the complicated process of calculating the detector efficiency for each monitored energy. The relative content of plutonium isotopes are indicative of the total plutonium content of a nuclear material, since it is a function of the neutron

Table 8.15. Energy pairs for determination of relative concentration of plutonium isotopes [107].

Isotope Pair	Energy Pair (keV)
^{238}Pu & ^{239}Pu	743 & 646
^{238}Pu & ^{240}Pu	153 & 160
^{238}Pu & ^{241}Pu	153 & 149
^{239}Pu & ^{240}Pu	161 & 160
	646 & 642
^{239}Pu & ^{241}Pu	144 & 149
	146 & 149
	161 & 165 (^{237}U)
	204 & 208 (^{237}U)
	345 & 332 (^{237}U)
	375 & 371 (^{237}U)
^{240}Pu & ^{241}Pu	160 & 149
	160 & 165
^{241}Am & ^{239}Pu	125 & 129
	335 & 345
	337 & 375
	653 & 646
	662 & 646

fluence to which the fuel was exposed, as indicated above. By performing relative measurements at adjacent energies, one also avoids the need to correct for photon attenuation factors that appear in the measurement model of Eq. (8.21), by assuming that the monitored radiation is emitted from the same location within the object. Therefore, the relative concentration of plutonium isotopes is determined by comparing the intensity of pairs of adjacent energy gamma-rays, see Table 8.15. In this process, gamma-rays emitted from ^{237}U, a short-lived isotope (6.75 days) produced by the alpha decay of ^{241}Pu, is used as an indicator of the presence of the latter. Notably absent from Table 8.15 is ^{242}Pu, as it is the longest-lived (3.7×10^5y) isotopes produced by uranium irradiation. This along with its low abundance, as it requires absorption of multiple neutrons, makes it difficult to detect ^{242}Pu. Being a source of spontaneous fission neutrons, along with ^{240}Pu, makes the determination of the concentration of ^{240}Pu important for criticality considerations. The concentration of ^{242}Pu is, therefore, estimated from the concentration of other plutonium isotopes.

8.8.3. Emission Imaging

Industrial applications of emission imaging using radiation emitted from internal sources is limited to liquids or gases, where a radioactive source can be injected. Gamma-ray emission is usually employed due to the higher penetrability of photons, in comparison to charged-particles. Emission imaging can also be combined with activation analysis (see section 8.1), with the activated elements acting as the internal source. High-energy photon interactions can also be utilized to produce positrons, that subsequently annihilate to a pair of 511 keV photons, as discussed in section 8.4. These monoenergetic photons can be used for composition imaging, as discussed in section 12.2.2.7. Note that in imaging with external induced emissions, corrections should be made for the attenuation of the external radiation before it induces emission, as the reduced intensity of the incident radiation directly affects the intensity of emission; the discussion in this section does not include correction for this attenuation effect.

Measurement Model. Emission imaging is best performed when emission is monitored at two opposite directions, for reasons that will become apparent after the measurement model for this imaging process is discussed. Let us start first by assuming that the activity inserted within an object is distributed within a volume V inside the object, and is detected by two collinear detectors, 1 and 2, located at two opposite directions of the object, as schematically shown in Figure 8.4. The two detectors are assumed to be well-collimated so that they only detect photons emitted along the line connecting the location of the two detectors. Also, it is assumed that the measurement time is much smaller than the half-life of the internal source, so that the source intensity does not change significantly during the measurement time. Now using the idealized narrow-beam model of radiation transmission (the limitations of which are discussed in section 6.1), the count rate at detector 1, ignoring attenuation in air, can be modeled as:

$$C_1 = K \int_{x_a}^{x_b} \int \mathcal{A} \, \exp[-\Sigma(x - x_1)] \, dA \, dx \tag{8.22}$$

where K is a system constant that takes into account the detector's efficiency, its collimator width, length, etc., $\mathcal{A}$ is the source activity per unit volume within V, x is the direction along which the radiation is detected, dA is an infinitesimal area of V normal to x, Σ is the total cross-section of the source photons within the region where the internal source is distributed (assumed to be constant), and x_a and x_b designate the location of the edge points of the object along the field-of-view of

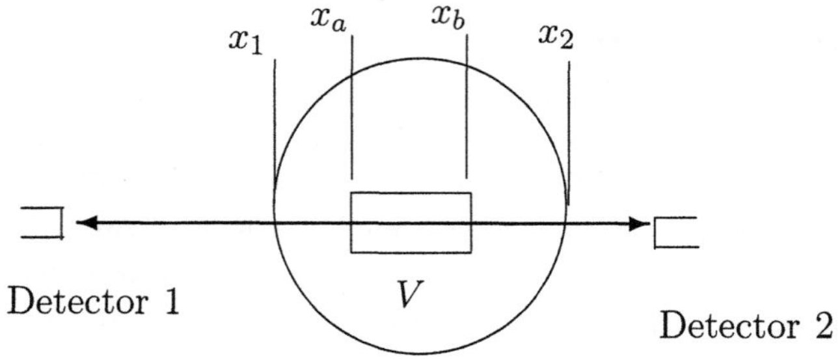

Figure 8.4. Schematic for simultaneous detection of emitted photons from a source with a volume V inside an object.

the detector, and x_1 corresponds to the point of the object boundary closest to the detector along its detection line, see Figure 8.4.

If the activity is distributed evenly within V, then the integration in Eq. (8.22) can be easily performed, leading to:

$$\begin{aligned} C_1 &= K \int \frac{\mathcal{A}}{\Sigma} \exp[-\Sigma(x_a - x_1)] \; \{1 - \exp[-\Sigma(x_b - x_a)]\} \; dA \\ &= K \left[\int \mathcal{A} \; (x_b - x_a) \; dA\right] \; \exp[-\Sigma(x_a - x_1)] \; f(\Sigma, x_b - x_a) \\ &= K \left[\int \mathcal{A} \; dV\right] \; \exp[-\Sigma(x_a - x_1)] \; f(\Sigma, x_b - x_a) \end{aligned} \quad (8.23)$$

with $V = A(x_b - x_a)$, and

$$f(\Sigma, \Delta x) = \frac{1 - \exp[-\Sigma \; \Delta x]}{\Sigma \; \Delta x} = \frac{\sinh(\Sigma \frac{\Delta x}{2})}{\exp[\Sigma \frac{\Delta x}{2}](\Sigma \frac{\Delta x}{2})} \quad (8.24)$$

where sinh is the hyperbolic sine function. Similarly, the count rate at detector 2, C_2, can be expressed as:

$$\begin{aligned} C_2 &= K \int_{x_a}^{x_b} \int \mathcal{A} \; \exp[-\Sigma(x_2 - x)] \; dA \; dx \\ &= K \int \frac{\mathcal{A}}{\Sigma} \exp[-\Sigma(x_2 - x_b)] \; \{1 - \exp[-\Sigma(x_b - x_a)]\} \; dA \\ &= K \left[\int \mathcal{A} \; (x_b - x_a) \; dA\right] \exp[-\Sigma(x_2 - x_b)] \; F(\Sigma, x_b - x_a) \\ &= K \left[\int \mathcal{A} \; dV\right] \exp[-\Sigma(x_2 - x_b)] \; F(\Sigma, x_b - x_a) \end{aligned} \quad (8.25)$$

where x_b and x_b are as shown in Figure 8.4. The relationships (8.23) and (8.25) can be combined to obtain the geometric average, or the conjugate, count rate, of the two counts, $\sqrt{C_1C_2}$, given by:

$$\sqrt{C_1C_2} = K\left[\int A\, dV\right] \exp\left[-\frac{1}{2}\Sigma[(x_2-x_1)-(x_b-x_a)]\right] \times f(\Sigma, x_2-x_1) \quad (8.26)$$

The conjugate count rate, as Eq. (8.26) shows, is independent of the source depth, i.e. independent of x_a and x_b, but is dependent on the thickness of the object along the line of detection, $(x_2 - x_1)$. This count rate, $\sqrt{C_1C_2}$, can be made independent of the source thickness, if multiplied by a correction factor, F_c, defined such that:

$$F_c = \frac{\exp\left[\Sigma(x_2-x_1)-(x_b-x_a)\right]]}{f(\Sigma, \Delta x)} \quad (8.27)$$

then,

$$F_c\sqrt{C_1C_2} = K\left[\int A\, dV\right] \quad (8.28)$$

becomes proportional to the total activity within the activated volume, $\int AdV$, provided the system constant K is known in advance (by calibration). The corrected measurement provides then a direct indication of the source activity, $\int AdV$, at the point.

One limit of the model of Eq. (8.26) is for emission from a single point, i.e., when the width dV is sufficiently small so that $\Sigma(x_b - x_a) << 1$. Then, $x_a = x_b = x$ and $\Delta x \rightarrow 0$. Then, from Eq. (8.24), f becomes equal to unity, on the limit of $\Delta x \rightarrow 0$, and Eq. (8.26) becomes:

$$\sqrt{C_1C_2} = K\left[\int AdV\right] \exp\left[-\frac{\Sigma(x_2-x_1)}{2}\right] \quad \text{for point emission} \quad (8.29)$$

Then, the correction factor of

$$F_c = \exp\left[\frac{\Sigma(x_2-x_1)}{2}\right] \quad \text{for point emission} \quad (8.30)$$

can be applied to the conjugate measurement, $\sqrt{C_1C_2}$. The correction factor, F_c of Eq (8.30), can be obtained from assumed values of Σ along with a thickness measurement for $(x_2 - x_1)$, or by a transmission measurement with an external source of the same energy as that of the internal source [280].

The other limit for the model of Eq. (8.26) is when emission occurs from all points of the object. Then, $x_a = x_1$ and $x_b = x_2$. The exponential factor in Eq. (8.26) becomes equal to unity, and the correction factor, F_c, required for the conjugate measurement, $\sqrt{C_1 C_2}$, becomes equal to $\frac{1}{f}$. Using Eq. (8.24), F_c can be expressed as:

$$\begin{aligned} F_c &= \frac{1}{f(\Sigma, x_2 - x_1)} \\ &= \frac{[\frac{1}{2}\Sigma(x_2 - x_1)] \exp[\frac{1}{2}\Sigma(x_2 - x_1)]}{\sinh(\frac{1}{2}\Sigma(x_2 - x_1))} \\ &\quad \text{for emission from all points} \end{aligned} \tag{8.31}$$

The right-hand-side of Eq. (8.31) is the expression often found in literature on medical imaging, see for example references [281] and [282]. The correction term in Eq. (8.26) requires knowing the value of $\Sigma(x_2 - x_1)$. This value can be determined from either assumed values and measurement of the thickness, $(x_a - x_2)$, or deduced from independent transmission measurements along the line defined by the two detectors using a source emitting photons with the same energy as that of the internal source.

Indications. The corrected value of the conjugate count, as the right-hand-side of Eq. (8.28) indicates, provides a quantity that is linearly proportional to $\int \mathcal{A}\, dx$, since $dV = \mathcal{A}dx$, if the area of the detected beam is assumed to be constant (i.e. if the effect of divergence of the field-of-view of a collimated detector is ignored). The integrated activity is analogous to the integrated attenuation-coefficient obtained from the logarithm of a transmission measurement, see discussion following Eq. (6.3) in section 6.2. Therefore, each conjugate measurement, after being corrected by the factor, F_c, represents a *projection* of the integrated activity. Consequently, a corrected (pre-filtered) single conjugate measurement can be used for line probing, along the direction defined by the two detectors. If the count rate of an uncollimated detector, viewing the entire object, is utilized, it will provide an indication of the total activity of the internal source. This approach can be viewed as *bulk gauging* that determines the location and overall intensity of the inserted radioactivity within the object. Alternatively, if a number of intersecting measurements are taken around the object, with two collimated but opposite detectors, then the values of $F_c\sqrt{C_1 C_2}$ provide a set of activity projections that can be used to reconstruct an image of the distribution of the activity, $\mathcal{A}$, within V. Any of the image reconstruction methods used in transmission imaging, discussed in section 6.4, can be employed for reconstructing these emission-activity images.

Positrons. Positron-emitting internal sources, such as those listed in Table 2.4, present a special and unique category of gamma-ray emitters. If positrons are introduced into an object, they would be annihilated by the electrons of the medium, typically within 1 to 3 mm from their emission site, producing two 511 keV photons that are emitted into two opposite directions (to conserve the zero momentum of the combined positron and electron). These two photons can be detected in coincidence (see section 4.5.5.5), as they travel at the speed of light, and as such can be detected instantaneously by two detectors located at two opposite sides of the direction of emission of the pair of photons. Coincidence measurements provide two main attractive features to emission imaging, in the process known as PET (positron emission tomography). The first attractive feature is that it acts as a virtual collimator, that is, a pair of photons detected in coincidence at two detectors would have been emitted somewhere along a line (or a cylinder) connecting the two detectors. The second feature of coincidence measurements in PET is that in essence each annihilation event is detected independently. These unique features offered by positrons led the medical imaging community to distinguish PET from photon-emission imaging employing direct photon sources. Imaging by the latter method is refereed to as *single-photon emission computed tomography* (SPECT), or *scintigraphy*, to signify that only singly-emitted photons (not a pair of photons) are counted.

Coincidence measurements allow the object to be surrounded by a number of detectors. Then, coincidence counting, by any pair of detectors, represents collimated counting by the two detectors. This makes it possible to provide flexible virtual collimation that changes according to the occurrence of coincidence. This electronic virtual ("soft") collimation process eliminates the need to use photon elimination with physical ("hard") collimators made of a dense material such as lead. The soft collimation process is also more accurate in defining the combined fields-of-view of the two detectors. It does not suffer as much from the divergence problem of the field-of-view of hard collimators, since the field-of-view of the electronic collimators is defined by two opposite detectors, rather than the shape and dimensions of a hard collimation (see section 14.6.3). Note, however, that some detector-collimation is usually applied to define a tomographic plate (section) by shielding the detector from radiation emitted or scattered from the section of interest.

Within the line connecting any two detectors in coincidence, the conjugate measurement of the two detectors during a certain coinciding interval (from about 4 to 20 ns) will correspond to a single annihilation event at a point within the line. With a counting interval consisting of many coincidence intervals, the total count will be the time integral of individual annihilations at all points on the line connecting the sites of the two detectors. Note that photons travel at a speed of about 0.3 m per ns, so that coincidence intervals of a few nanoseconds, will allow the coincidence detection of events occurring over a distance in the order of a meter. Since every pair of annihilation radiation detected in coincidence must travel the total path-length between the detectors, the attenuation factor of conjugate measurement has to be equal to that obtained using Eq. (8.29) for emission from a single point, i.e. $\exp[-\Sigma(x_2 - x_1)]$. Since, the total count for any conjugate measurement is now a time integral, the same attenuation factor is maintained for all detected pairs of photons. Therefore, the correction factor for PET is simply that given by F_c, as defined by Eq. (8.30). That is, the value of F_c depends only on the object thickness traversed by the coincidence photons, which can be measured, and the total cross-section of the 511 keV photons which can be estimated in advance. Alternatively, a 511 keV gamma-source, such as ^{22}Na, can be used to perform a transmission measurement along the direction connecting the coincidence detectors, and the value of F_c can be deduced from such measurement.

The use of a conjugate measurement corrected for attenuation, Eq. (8.28), has the advantage of lending the image-reconstruction problem into a linear form analogous to that of transmission imaging. However, the introduced correction is susceptible to error, if the total cross-section is assumed to be known in advance, and requires tedious thickness, or radiation transmission measurements. Therefore, methods have been developed to solve the problem of reconstructing images from emission imaging iteratively, without supplementary radiation transmission or thickness measurements [281, 283]. This process relies on providing an initial guess of the distribution of activity within the imaged volume. With this initial guess, the attenuation factors of radiation are estimated, a corresponding set of detector count rates are calculated, using the model of Eq. (8.22), applied to the entire section of the object. The calculated count rates, for the assumed distribution, and the measured ones are then compared, and the difference is used to correct the assumed image. The process is repeated until the problem converges to a solution. The iterative techniques discussed in section 6.4 can be used for this purpose.

8.8.4. Gamma Cameras

The intensity of gamma-ray emission can be observed using a gamma camera, discussed in section 4.3.2, or a Compton-scatter camera (also known as the Compton gamma-camera [284]). While a conventional gamma-camera employs an array of collimated scintillation detectors, a

Compton camera eliminates the collimation process, which reduces the number of detected photons. A Compton camera employs instead two detector arrays in parallel, as schematically shown in Figure 8.5. The first array is usually a thin array of a semiconductor detection material that allows the scattering of photons, and the subsequent loss of some energy. This array is, therefore, called the scatter array. The second array, called the absorption array, is designed to measure the photons incident on it, and can be a usual gamma camera without the associated collimator.

Let us consider a monoenergetic photon of energy E_0 that loses, hence deposits, an energy ΔE on the scatter detector at location A in the scatter array. The position of incidence of that photon, B, on the absorption array will then be determined by the energy $E_0 - \Delta E$ that it must deposit in the absorption array. The direction AB then determines the direction the source photon has undertaken after being scattered in the first array. The energy of that scattered photon is equal to $E_0 - \Delta E$, and if E_0 is known, then according to the kinematics of Compton scattering, Eq. (3.37), its polar angle of scattering, θ, is uniquely determined. However, the kinematics of scattering are also satisfied by an azimuthal angle of scattering anywhere on the surface of a cone whose axis is along AB, its tip at A and its semi-angle (opening angle) is θ, as schematically shown in Figure 8.5. The intersection of these scattering points of an arbitrary image plane defines an ellipse; along its circumference a source point could have existed. A point source will produce another scattering angle θ', another cone along another axis, $A'B'$, and another ellipse on the image plane, and so on. All these ellipses should intersect at one point, the source point, within a range determined by the energy resolution of the detectors. For an extended source, each point on an ellipse is a possible source location, thus the intensity of the image is defined by the direct summation of all ellipses in the image plane. Obviously, this imaging process involves a numerical image reconstruction algorithm, but is a powerful imaging tool, since every source particle reaching the two detector arrays will be detected.

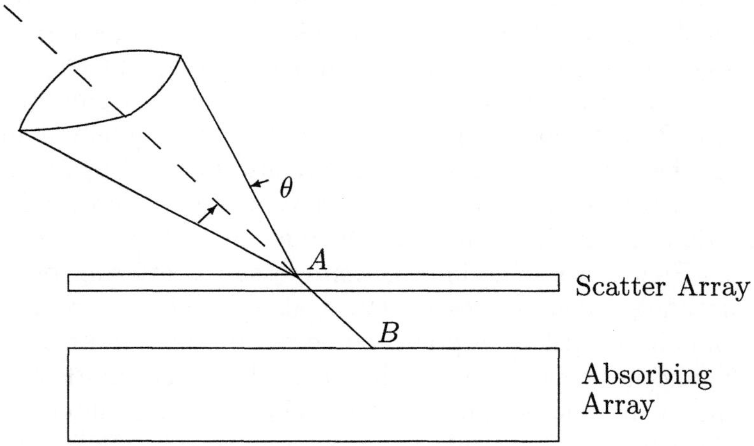

Figure 8.5. Schematic of a Compton scatter camera.

Chapter 9

ABSORPTION METHODS

9.1. Absorption of Charged Particles

As explained in section 3.3, charged-particles can cause ionization in matter, as the atomic electrons are liberated from the atom. In gases, because of their low density, it is possible to gather the liberated ions by applying an external electric potential before the ions recombine with available electrons. This is the essence of gas-filled radiation detectors, discussed in section 4.2.2. The same concept can also be utilized for obtaining useful gas properties, such as pressure, density, and material content, as demonstrated by the measurement model presented below.

Ionization Cross-Section. The probability of ionization can be defined in terms of an interaction cross-section, Σ_{ion}. An effective ionization cross-section can be defined using the stopping power, $\frac{dE}{ds}$ defined in section 3.3, and the mean ionization-energy, ϵ, (energy required to produce an ion pair) as [24] :

$$\Sigma_{ion} = \frac{1}{\epsilon}\left|\frac{dE}{ds}\right| \tag{9.1}$$

where ϵ is the energy required to ionize an atom in a particular gas. By incorporating both ϵ and $\frac{dE}{ds}$ (the stopping power), the cross-section of Eq. (9.1) represents the susceptibility of a certain gas to ionization by a particular type of charged-particles. The stopping power, as shown in section 3.3, depends on the atomic-density of the gas, as well as on its atomic-number, see for Eqs. (3.20), (3.27) and (3.29). This dependence on atomic-density enables the determination of the gas density (or pressure, as the two are related at a constant temperature), while the dependence on the atomic-number is useful for distinguishing between two types of gases, if present together. The value of ϵ also depends on the nature of the gas, but is typically about 30 eV, as indicated in section 4.2.2, being as high as 42.3 eV for helium and as low as 22.0 keV for xenon [24]. For a mixture, or a compound, the ionization cross-section can be determined in a manner similar to that used for calculating the macroscopic cross-section, see appendix E.

The ionization current produced by applying an external electric-field generally increases with increasing Σ_{ion}. However, the value of the ionization current depends on the ability to collect ionized atoms before they recombine. The current increases with increasing voltage until all ionized atoms are collected where the current stays constant over a voltage range (saturation voltage, or plateau), as in the case with gas-detectors (Figure 4.2). When the voltage is sufficiently high, electrons liberated by radiation ionization acquire an energy greater than the ionization energy of the gas, and begin to introduce their own ionization. In this multiplication zone, the magnitude of ionization is elevated, increasing the amount of collected ionization current. Measurements are performed in either the plateau region where instabilities in the applied voltage do not affect the current, or in the multiplication zone where the magnitude of the current is increased for improved detection.

Measurement Model. Collecting the ionization current requires two electrodes through which the applied voltage is applied. The electrodes can be housed within an ionization chamber, which also incorporates a charged-particle source, alpha or beta particles. When the gas to be analyzed passes through the chamber, it generates an ionization current, the magnitude of which, i, can be expressed by the following Measurement model:

$$i = K\Sigma_{ion}r \quad \text{for} \quad r < R \tag{9.2}$$

where K is a system constant that depends on the source's activity and applied voltage, r is the radius of the ionization zone around the radiation source as determined by the physical size of the chamber, and R is the range of the charged-particle in the material considered. The applicability of Eq. (9.2) is limited to a value of $r < R$, since no direct ionization takes place beyond the range of the particle. Note that the model of Eq. (9.2) does not take into account ion recombination, which occurs at low voltage, nor gas multiplication which happens at high voltage, and thus is best suited for application within the zone of plateau voltage, shown in Figure 4.2. Even then calibration curves are required, due to the inherent complexity of the model of Eq. (9.2), as it does not for instance include the effect of the chamber geometry on the distribution of the electric field, or secondary ionization.

9.2. Photon Absorption Methods

Absorption of photons is usually measured indirectly via the transmission technique, discussed in chapter 6. When photon absorption results in the release of secondary radiation, e.g. fluoroscopic emission at low source energy, or 511 keV photons, by positron annihilation at high energy, emission techniques are applied, see chapter 8. However, there are occasions where photon absorption is measured within the bulk of the interrogated medium itself, to characterize its composition. This is commonly done in subsurface analysis of the Earth's crust [285]. The basics of this approach are discussed below.

Using the approximation for the photoelectric cross-section, given by Eq. (3.31), the "average" cross-section per electron, τ_e of Eq. (3.33), can be written in terms of two independent functions of the photon energy, E, and the atomic-number, Z, as follows:

$$\tau_e = K(E)P_e(Z) \tag{9.3}$$

where,

$$K(E) = \frac{4.8 \times 10^4}{E^{3.15}} \tag{9.4}$$

and

$$P_e(Z) = \left(\frac{Z}{10}\right)^{3.6} \tag{9.5}$$

with E in keV and τ_e in barns. While K is an energy-dependent coefficient, P_e is an energy-dependent parameter that reflects the ability of an "average" electron in atom with an atomic number, Z, to absorb a photon. The latter parameter is often referred to as the *photoelectric absorption index* [40], or simply the photoelectric factor [285]. This index indicates clearly that the ability of an atomic electron to absorb photons depends on the atom in which it resides, and increases rapidly with Z. Therefore, P_e is characteristic of the atom causing photoelectric absorption and can be used as an element identifier. While P_e is a microscopic quantity (per electron), an analogous macroscopic value (per unit volume) can now defined as:

$$U = P_e N_e \tag{9.6}$$

where N_e is the electron-density. The parameter, U of Eq. (9.6), is the volumetric photoelectric index. For a mixture of materials,

$$U = \sum_i \alpha_i U_i \tag{9.7}$$

where U_i is the volumetric photoelectric index for component i and α_i is its volume fraction. For such a mixture, an effective P_e value can be estimated as $\frac{U}{N_e}$, with N_e being the electron-density of the medium.

By placing a photon (x- or gamma-ray) source within the medium, and measuring the number of low-energy photons reaching a detector far away from the source, one can obtain an indication of U. The lower the value of U, the less absorption, the higher the detectors count rate at the detector. Therefore, if the medium's detector density is known, or measured by some other means, one can obtain an estimate of P_e of the medium, hence determine its general composition indication character. The value of P_e varies from a fraction for oil and water, to a value between 1.8 to 5 for sedimentary minerals [285]: sandstone (SiO_3), limestone ($CaCO_3$) and dolomite ($CaMgC_2O_6$), and for high-Z elements the value of P_e is in the hundreds [40].

9.3. Neutron Flux Depression Method

In some applications, its is possible to insert an object within a larger medium in which a neutron source is enclosed. Such an arrangement can be useful, for example, in establishing an inspection station to which small objects can be brought in for examination, or for constructing a facility within which many objects can be inspected simultaneously. Within the inspection facility, one can create a neutron "cloud", inside which the inspected object(s) can be placed. This will require the placement of a source at the center of the assembly, or at one of its ends, while performing the inspection at some distance from the source. The source neutrons will then spread within the inspection assembly, creating the desired neutron cloud. The inspection can be then done with minimum interference by neutrons directly emitted from the source. Minimizing the direct source contribution enables modeling the neutron cloud, created away from the source, by the diffusion theory, discussed in section 3.6.4. If the inserted material has an absorption cross-section that is significantly different from that of the material within which the neutron cloud is created, the inserted material will disrupt the distribution on the radiation flux in its surroundings. The disruption in flux can be sensed by a nearby detector, as schematically shown in Figure 9.1. The amount of flux disruption will be indicative of the nature of the inspected object.

Measurement Model. The change in flux distribution caused by the introduction of a small object within a larger medium is similar to the flux change caused by a fuel rod inside a nuclear reactor. The same concept is used here to introduce the basic aspects of this gauging method.

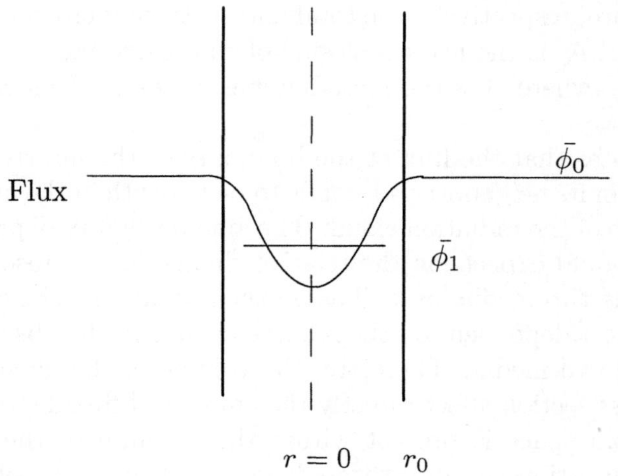

Figure 9.1. Schematic for neutron-flux depression due to the introduction of an absorbing cylindrical object of radius r_0.

Let us consider a cylindrical rod of radius r_0, with an absorption cross-section $\Sigma_{a,0}$ inserted inside the medium of the neutron cloud, which has an absorption cross section, $\Sigma_{a,1}$. With a coordinate system with one axis coinciding with the rod's axis, as schematically shown in Figure 9.1, two diffusion equations can be written for the rod's medium and for the medium of the cloud. The solution of these equations, with the proper boundary conditions, produces the following relationship [83]:

$$\phi(r_0) = \bar{\phi}_1 \left[1 - \frac{\Sigma_{a,0} V_0 \bar{\phi}_0}{\Sigma_{a,1} V_1 \bar{\phi}_1} X \right] \tag{9.8}$$

where $\bar{\phi}$ designates the average flux in the medium identified by the subscript, V refers to the volume of the corresponding medium and X is a geometry-dependent parameter given by:

$$X \approx \frac{R^2 \Sigma_{a,1}}{2D_1} \left[\frac{R^2}{r_0^2 - R^2} \ln\left(\frac{r_0}{R}\right) - \frac{3}{4} + \frac{r_0^2}{4R^2} \right] \tag{9.9}$$

where D_1 is the diffusion coefficient in the medium of the cloud and R is a radial distance from the source at which there is no net flow of neutrons into or out of the medium. The diffusion coefficient, D, as given by the transport theory, is such that:

$$D_1 = \frac{1}{3(\Sigma_t - \Sigma_s \bar{\mu}_0)} \tag{9.10}$$

where Σ_t and Σ_s are, respectively, the total and scattering cross-sections of the medium and $\bar{\mu_0}$ is the average cosine of the scattering angle (for neutrons: $\bar{\mu_0} = \frac{2}{3A}$ where A is the mass-number of the nucleus causing scattering) [28].

Eq. (9.8) indicates that the flux at the boundary of the inserted rod, and consequently in its neighborhood, tends to be lower than the average flux in the medium of the radiation cloud. The amount of flux-depression depends, as one would expect, on the relative change in the absorption cross-section of the two media, as well as on their volumes. The change in the flux is in effect dependent on the relative change in the absorption rate, $\Sigma\phi V$, in the two media. Therefore, the volume of the object and its absorption cross-section affect directly the amount of flux depression. Note that if a void space is present within the medium of the cloud to facilitate the insertion the interrogated object, then in the absence of the object $\Sigma_{a,0} = 0$, and as Eq. (9.8) indicates, the flux at the boundary of void gap stays equal to the average flux of the cloud medium. The effect of a flux perturbation extends to about $3L_1$, where L_1 is the diffusion-length in the medium of the flux cloud ($L_1 = \sqrt{\frac{D}{\Sigma_a}}$). Therefore, the detector should be placed within a distance less than $3L_1$ from the introduced object.

The geometric coefficient X of Eq. (9.9) indicates that the relative size of the inserted object and the inspection facility affect also the amount of flux depression. This parameter is dependent on the geometry of the object and the inspection facility, and thus it can be optimized for maximum change in flux distribution.

It should, however, be kept in mind that the formulation of Eqs. (9.8) and (9.9) is based on the diffusion theory, which has limitations as indicated in section 3.6.4, and is derived for monoenergetic radiation. Note also that the detector itself will introduce its own flux depression which will overlap with that of the inspection object, and was not taken into account in the above formulation. However, an empirical measurement model that takes into account the effect of the detector, estimates the flux measured by the detector, ϕ_d, as [24]:

$$\phi_d = \frac{\bar{\phi_1}}{\Sigma_{a,d} + \Sigma_{a,1} + \Sigma a, 0} \tag{9.11}$$

where $\Sigma_{a,d}$ is the macroscopic absorption cross-section of the detector. Therefore, the above analysis should be seen only as a means of understanding the physical process associated with flux depression. Monte Carlo simulations, described in section 16.2, can be helpful in designing the inspection assembly, while experimental calibration should be used to produce absorption curves relating changes in the properties of the inspected objects to the measured radiation flux.

The flux-depression method discussed below usually applies to the detection of thermal-neutron absorbers, such as B, Li, Ag, Cd, Gd, In and Hg. Elements with strong resonances in their absorption cross-section can also be examined by this method. The following are some of these elements [24]: Mn, Co, RH, Ag, Cd, In, Sm, Eu, Gd, Dy, H, Ta, Au and U. For theses elements, the neutrons in the flux cloud need not be well-thermalized, since resonances occur at energies well above the thermal-neutron energy.

9.4. Decay-Time of Neutrons

The half-life of a free-neutron in vacuum is about 13 minutes, as it decays into a proton, an electron, and an antineutrino. This time is, however, quite large that neutrons usually interact with matter before they disintegrate; a thermal-neutron would survive only 0.1 ms in the free state before encountering an interaction. The lifetime of a neutron in matter depends on the ability of the material to absorb neutrons. Timing the survival time of a neutron requires the use of a pulsed source. Pulsing can be achieved, if neutrons are emitted by a generator, see section 2.3, which emits fast-neutrons, or with the aid of a mechanical chopper made of a strong neutron-absorbing material. In the presence of hydrogenous materials (water or oil), these fast-neutrons will be readily slowed-down to the thermal energy. The resulting thermal-neutrons will eventually dissipate ("die-away") at a rate that is directly proportional to the macroscopic absorption cross-section of the thermal-neutron in the medium, Σ_a, as shown below. Therefore, by monitoring the population of the generated thermal-neutrons and their "decay-time", an estimate of Σ_a can be determined. The thermal-neutron population can be measured, at some location within the medium, directly by their count rate or indirectly by the gamma-rays associated with their capture. Since gamma-rays are generally more penetrating than thermal-neutrons, a gamma-ray detector covers photon emission from a wider surrounding area, than a thermal-neutron detector. This in turn increases the size of the domain of analysis, which can be advantages if one desires to survey a wide area, since the inspection domain is the volume encompassed between the source and the detector.

Measurement Model. After the short time taken for the buildup of the thermal-neutron population due to the introduction of a fast-neutron pulse, the thermal-neutron population will start to collapse by neutron absorption, and the lack of supply of new neutrons by the cessation of the fast-neutron pulse. This exponential decay is due to the fact that, between pulses, the thermal-neutron population will decrease at a rate, $\frac{dN}{dt}$, proportional to both the available thermal-neutron population, $N(t)$ at the time, t, the macroscopic thermal-neutron cross-section of the medium, Σ_a in which the thermal-neutrons are created, and the speed of thermal-neutrons, v. That is:

$$-\frac{dN(t)}{dt} = \Sigma_a N(t) v \tag{9.12}$$

which leads to

$$N(t) = N_0 \exp(-\Sigma_a v t) \tag{9.13}$$

where N_0 is the number of neutrons at an arbitrary time, $t = 0$. The velocity of thermal-neutrons, v (= 2200 m/s at room temperature) appears in Eq. (9.13), not only to balance the units, but also to indicate that neutrons are removed from the measurement volume by simply moving out of it. Now, if the neutron population is measured at two different times, t_1 and t_2, the measurement model can be applied at both times and the value of Σ_a can then be derived from these measurements, N_1 and N_2 as:

$$\Sigma_a = \frac{1}{v(t_1 - t_2)} \ln \frac{N_1}{N_2} \tag{9.14}$$

Eq. (9.13) also indicates that the time needed for the thermal-neutron population to decrease to $\frac{1}{e}$ (= 0.3678) of its value at time zero is $\tau = \frac{1}{\Sigma_a v}$. The τ value is known as the *die-away time*. The importance of this die-away time is that it is independent of the temperature of the medium, though still dependent on its neutron-absorption ability. This is due to the fact that thermal-neutron cross-section has a $\frac{1}{v}$ dependence (see section 3.5 in chapter 3), making the multiplication $\Sigma_a v$ independent of v, and consequently independent of temperature (since v for thermal-neutrons is temperature-dependent). From Eq. (9.12), one can show that τ can be determined from the decay rate as:

$$\tau = -\frac{1}{N}\frac{dN}{dt} \tag{9.15}$$

Therefore, τ is equal to the normalized decay rate, which can be determined from two measurements at adjacent time intervals.

The model of Eq. (9.12) and its derivatives, Eqs. (9.13) to (9.15), neglect the transport (or diffusion) of neutrons into and away from the measurement volume around the detector, by neutron scattering, see section 3.6. However, the obtained indications, either of Eq. (9.14) or Eq. (9.15), are still good indicators of the neutron-absorption ability of the medium.

Appendix A
Basic Units and Constants

Units

The *Systéme International* (SI)[1] units are widely used, legislated by most jurisdictions, and required by almost all scientific journals. In radiation units, the use of "Curie" for measuring activity is replaced with the SI unit, "Bq". The "rem" and "rad" units for absorbed radiation dose and dose equivalent, respectively, are also replaced by "Sv" and "Gy", respectively; some workers use centi-gray (cGy) and centi-sievert (cSv) in place of the Gy and Sv, respectively. Although, the units of "eV" for radiation energy and "barn" for the microscopic cross-section are not strictly SI units, their use is permitted [1361].

Commonly used radiation units are defined below[2]:

Mass: 1 u (atomic mass unit or amu) $= 1.6605 \times 10^{-27}$ kg

$$= 931.493 \text{ MeV} = \frac{\text{mass of } ^{12}\text{C}}{12} = \frac{\text{g/mol}}{\text{Avogadro's number}}.$$

[1]The SI system is based on the units of meter (m) for distance, kilogram (kg) for mass, second (s) for time, ampere (A) for electric current, Kelvin (K) for temperature and mole (mol) for amount of substance; with radian (rad) for plane angle and steradian (sr) for solid angle. Derived units include: newton (N = kg m/s^2) for force, joule (J = N m) for energy, watt (W = J/s) for power, pascal (Pa = N/m^2) for pressure, volt (V = W/A) for electric potential, coulomb (C = A s) for electrical charge, weber (Wb = V s) for magnetic flux and tesla (T = Wb/m^2) for magnetic flux density. The SI allowed prefixes are: yotta (Y = 10^{24}), zetta (Z = 10^{21}), exa (E = 10^{18}), peta (P = 10^{15}), terra (T = 10^{12}), giga (G = 10^9), mega (M = 10^6), kilo (k = 10^3), hecto (h = 10^2), deka (da = 10), deci (d = 10^{-1}), centi (c = 10^{-2}), mili (m = 10^{-3}), micro ($\mu = 10^{-6}$), nano (n = 10^{-9}), pico (p = 10^{-12}), femto (f = 10^{-15}), atto (a = 10^{-18}), zepto (z = 10^{-21}), and yocto (y = 10^{-24}); see references [1360] and [1361] for more details.

[2]The on-line reference [1362] (http://physics.nist.gov/cuu/Reference/unitconversions.html) provides access to a number of on-line links for conversion between units.

Microscopic cross-section: 1 b (barn) $= 10^{-24}$ cm^2 $= 10^{-28}$ m^2.

Energy: 1 eV (electron volt) $= 1.602 \times 10^{-19}$ J.

Activity: 1 Bq (becquerel) = 1 disintegration/s $= \frac{1}{3.7\times10^{10}}$ Ci.

Exposure: 1 C/kg of produced charge in air $= \frac{1}{2.58\times10^{-4}}$ R (roentgen).

Absorbed dose: 1 Gy (gray) = 1 J/kg $= 6.2415 \times 10^{12}$ MeV/kg = 100 rad.

Dose equivalent: 1 Sv (sievert) = absorbed dose in Gy × radiation weighing factors = 100 rem; see chapter 5 for weighting factor.

Constants

The following are useful constants in radiation physics [13][3]:

Symbol	Name	Value
N_A	Avogadro's number	6.022×10^{23}/mol
c	speed of light	2.9979×10^{8} m/s
h	Planck's constant	4.136×10^{-15} eV s
k	Boltzmann's constant	8.617×10^{-5} eV/K
e	elementary charge	1.602×10^{-19} C
		1.5189×10^{-14} m$^{\frac{3}{2}}$kg$^{1/2}$s^{-1}
m_0 or m_e	electron rest-mass	0.511 MeV $\equiv 5.4858 \times 10^{-4}$ u
m_n	proton rest-mass	1.007 u
m_p	neutron rest-mass	1.0087 u
m_α	alpha particle rest-mass	4.001 u
1 y	one year	3.155760×10^{7} s

[3]See reference [1363] (http://physics.nist.gov/cuu/Constants/index.html) for up-to-date, on-line, exact values.

Appendix B
List of Elements and Natural Isotopes

For information on nuclear and atomic properties, visit the web site of the National Nuclear Data Center of the Brookhaven National Laboratory: "http://www.nndc.bnl.gov/".

Element	Symbol	Atomic Number (Z)	Natural Isotopes A (%)[a]
Actinium	Ac	89	...
Aluminum	Al	13	27 (100)
Americium	Am	95	...
Antimony	Sb	51	121 (57.21), 123 (42.79)
Argon	Ar	18	36 (0.3365), 38 (0.0632), 40 (99.6003)
Arsenic	As	33	75 (100)
Astatine	At	85	...
Barium	Ba	56	130 (0.106), 132 (0.101), 134 (2.417) 135 (6.592), 136 (7.854), 137 (11.232), 138 (71.698)
Berkelium	Bk	97	...
Beryllium	Be	4	9 (100)
Bismuth	Bi	83	209 (100)
Boron	B	5	10 (19.8), 11 (80.2)
Bohrium	Bh	107	...
Bromine	Br	35	79 (50.69), 81 (49.31)
Cadmium	Cd	48	106 (1.25), 108 (0.89), 110 (12.49) 111 (12.80), 112 (24.13) 113 (12.22), 114 (28.73), 116 (7.49)

[a] A: mass number, % natural abundance, atom percent, compiled from reference [1364].

Element	Symbol	Atomic Number (Z)	Natural Isotopes A (%)
Calcium	Ca	20	40 (96.94), 42 (0.647), 43 (0.135), 44 (2.09), 46 (0.004), 48 (0.187)
Californium	Cf	98	...
Carbon	C	6	12 (98.89), 13 (1.11)
Cerium	Ce	58	136 (0.185), 138 (0.251), 140 (88.450), 142 (11.114)
Cesium	Cs	55	133 (100)
Chlorine	Cl	17	35 (75.77), 37 (24.23)
Chromium	Cr	24	50 (4.345), 52 (83.789), 53 (9.501), 54 (2.365)
Cobalt	Co	27	59 (100)
Copper	Cu	29	63 (69.17), 65 (30.83)
Curium	Cm	96	...
Dubnium	Db	105	...
Dysporsium	Dy	66	156 (0.06), 158 (0.10), 160 (2.34), 161 (18.91), 162 (25.51), 163 (24.90), 164 (28.18)
Einsteinium	Es	99	...
Erbium	Er	68	162 (0.139), 164 (1.601), 166 (33.503), 167 (22.869), 168 (26.978), 170 (14.910)
Europium	Eu	63	151 (47.81), 153 (52.19)
Fermium	Fm	100	...
Flouroine	F	9	19 (100)
Francium	Fr	87	...
Gadolinium	Gd	64	152 (0.20), 154 (2.18), 155 (14.80), 156 (20.47), 157 (15.65), 158 (24.84), 160 (21.86)
Gallium	Ga	31	69 (60.108), 71 (39.892)
Germanium	Ge	32	70 (20.37), 72 (27.31), 73 (7.76), 74 (36.73)
Gold	Au	79	197 (100)

Element	Symbol	Atomic Number (Z)	Natural Isotopes A (%)
Hafnium	Hf	72	174 (0.16), 176 (5.26), 177 (18.60), 178 (27.28), 179 (13.62), 180 (35.08)
Hassium	Hs	108	...
Helium	He	2	3 (0.000137), 4 (99.999863)
Holmium	Ho	67	165 (100)
Hydrogen	H	1	1 (99.985%), 2 (0.015)
Indium	In	49	113 (4.29), 115 (95.71)
Iodine	I	53	127 (100)
Iridium	Ir	77	191 (37.3), 193 (62.7)
Iron	Fe	26	54 (5.845), 56 (91.754), 57 (2.119), 58 (0.282)
Krypton	Kr	36	78 (0.35), 80 (2.28), 82 (11.58), 83 (11.49), 84 (57.00), 86 (17.30)
Lanthanum	La	57	138 (0.090), 139 (99.910)
Lawrencium	Lr	103	...
Lead	Pb	82	204 (1.4), 206 (24.1), 207 (22.1) 208 (52.4)
Lithium	Li	3	6 (7.59), 7 (92.41)
Lutetium	Lu	71	175 (97.41), 176 (2.59)
Magnesium	Mg	12	24 (78.99), 25 (10.00), 26 (11.01)
Manganese	Mn	25	55 (100)
Meithnerium	Mt	109	...
Mendelevium	Md	101	...
Mercury	Hg	80	196 (0.15), 198 (9.97), 199 (16.87), 200 (23.10), 201 (13.18), 202 (29.86), 204 (6.87)
Molybdenum	Mo	42	92 (14.84), 94 (9.25), 95 (15.92), 96 (16.68), 97 (9.55), 98 (24.13) 100 (9.63)
Neodymium	Nd	60	142 (27.2), 143 (12.2), 144 (23.8), 145 (8.3), 146 (17.2), 148 (5.7), 150 (5.6)
Neon	Ne	10	20 (90.48), 21 (0.27), 22 (9.25)
Neptunium	Np	93	...
Nickel	Ni	28	58 (68.077), 60 (26.223), 61 (1.140), 62 (3.634), 64 (0.926)
Niobium	Nb	41	93 (100)
Nitrogen	N	7	14 (99.634), 15 (0.366)
Nobelium	No	102	...
Osmium	Os	76	184 (0.02), 186 (1.59), 187 (1.6), 188 (13.29), 189 (16.21), 190 (26.36) 192 (40.93)

Element	Symbol	Atomic Number (Z)	Natural Isotopes A (%)
Oxygen	O	8	16 (99.762), 17 (0.038), 18 (0.200)
Palladium	Pd	46	102 (1.02), 104 (11.14), 105 (22.33), 106 (27.33), 108 (26.46), 110 (11.72)
Phosphorus	P	15	31 (100)
Platinum	Pt	78	190 (0.014), 192 (0.782), 194 (32.967), 195 (33.832), 196 (25.242), 198 (7.163)
Plutonium	Pu	94	...
Polonium	Po	84	
Potassium	K	19	39 (93.2581), 40 (0.0117), 41 (6.7302)
Praseodymium	Pr	59	141 (100)
Promenthium	Pm	61	...
Protactinium	Pa	91	...
Radium	Ra	88	...
Radon	Rn	86	...
Rhenium	Re	75	185 (37.40), 187 (62.60)
Rhodium	Rh	45	103 (100)
Rubidium	Rb	37	85 (72.17), 87 (27.83)
Ruthenium	Ru	44	96 (5.54), 98 (1.87), 99 (12.76), 100 (12.60), 101 (17.06), 102 (31.55) 104 (18.62)
Rutherfordium	Rf	104	...
Smarium	Sm	62	144 (3.07), 147 (14.99), 148 (11.24), 149 (13.82), 150 (7.38), 152 (26.75), 154 (22.75)
Scandium	Sc	21	45 (100)
Seaborgium	Sg	106	...
Selenium	Se	34	74 (0.89), 76 (9.37), 77 (7.63), 78 (23.77), 80 (49.61), 82 (8.73)
Silicon	Si	14	28 (92.230), 29 (4.683), 30 (3.087)
Silver	Ag	47	107 (51.839), 109 (48.161)
Sodium	Na	11	23 (100)
Strontium	Sr	38	84 (0.56), 86 (9.86), 87 (7.00), 88 (82.58)
Sulfur	S	16	32 (95.02), 33 (0.75), 34 (4.21), 36 (0.02)
Tantalum	Ta	73	180^m (0.012), 181 (99.988)
Technetium	Tc	43	...
Tellurium	Te	52	120 (0.09), 122 (2.55), 123 (0.89), 124 (4.74), 125 (7.07), 126 (18.84), 128 (31.74), 130 (34.08)
Terbium	Tb	65	159 (100)
Thallium	Tl	81	203 (29.524), 205 (70.476)
Thorium	Th	90	232 (100)
Thulium	Tm	69	169 (100)

Element	Symbol	Atomic Number (Z)	Natural Isotopes A (%)
Tin	Sn	50	112 (0.97), 114 (0.66), 115 (0.34), 116 (14.54), 117 (7.68), 118 (24.22), 119 (8.59), 120 (32.58), 122 (4.63) 124 (5.79)
Titanium	Ti	22	46 (8.25), 47 (7.44), 48 (73.72), 49 (5.41), 50 (5.18)
Tungsten	W	74	180 (0.12), 182 (26.50), 183 (14.31), 184 (30.64), 186 (28.43)
Uranium	U	92	234 (0.0054), 235 (0.7204), 238 (99.2742)
Vanadium	V	23	50 (0.250), 51 (99.750)
Xenon	Xe	54	124 (0.095), 126 (0.089), 128 (1.910), 129 (26.40), 130 (4.071), 131 (21.232), 132 (26.909), 134 (10.436), 136 (8.857)
Ytterbium	Yb	70	168 (0.13), 170 (3.04), 171 (14.28), 172 (21.83), 173 (16.13), 174 (31.83), 176 (12.76)
Yttrium	Y	39	89 (100)
Zinc	Zn	30	64 (48.63), 66 (27.90), 67 (4.10), 68 (18.75), 70 (0.62)
Zirconium	Zr	40	90 (51.45), 91 (11.22), 92 (17.15), 94 (17.38), 96 (2.80)

Appendix C
Relativistic Mechanics

Small nuclear and atomic particles can reach a speed close to the speed of light. At such high-speed their corpuscular behavior begins to blur with that of wave properties. This in turn affects particle kinematics (mass, momentum and energy). Relativistic mechanics is applied to describe a particle's behavior when its speed approaches the speed of light. Since radiation particles usually move in a straight line, until they encounter an interaction, the special theory of relativity is applied. This appendix provides a short summary of the basics of relativistic analysis; for more details consult a basic physics textbook, such as references [48] and [1365].

The special theory of relativity is based on the premises that:

1 A physical law maintains the same mathematical form in coordinate systems that are in uniform relative translational motion.

2 The speed of light, c, is the same for any two observers who are in a uniform rectilinear relative motion, and is independent of the motion of the source.

In order to accommodate these two postulates, space and time are viewed together as forming a four-dimensional Euclidean space whose basic axes are, for any given observer, orthogonal and independent.

In this four-dimensional space, the time t is an independent variable, much like the position variables x, y and z. This necessitates the introduction of a fourth component of force and momentum, and some other time-like variable called the "proper time", τ. The proper-time interval between two events whose space coordinates are a distance Δr apart from each other is defined by [1365]:

$$(\Delta\tau)^2 = (\Delta t)^2 - \frac{(\Delta r)^2}{c^2} \qquad \text{(C.1)}$$

The time interval, $(\Delta\tau)^2$ can be either positive, negative or zero, so that $\Delta\tau$ can be real, imaginary, or singular. When $\Delta\tau$ is real, it resembles a "time-like" interval, while when it is imaginary it is a "space-like" interval.

In the four-dimensional space, a four-vector velocity, and a corresponding four-momentum components, can be defined as:

$$U_1 = \frac{dx}{d\tau} = \frac{dx}{dt}\frac{dt}{d\tau} = \frac{\dot{x}}{\sqrt{1-\frac{v^2}{c^2}}}; \quad P_1 = P_x = \frac{m_0\dot{x}}{\sqrt{1-\frac{v^2}{c^2}}} \qquad \text{(C.2)}$$

$$U_2 = \frac{dy}{d\tau} = \frac{dy}{dt}\frac{dt}{d\tau} = \frac{\dot{y}}{\sqrt{1-\frac{v^2}{c^2}}}; \quad P_2 = P_y = \frac{m_0\dot{y}}{\sqrt{1-\frac{v^2}{c^2}}} \qquad \text{(C.3)}$$

$$U_3 = \frac{dz}{d\tau} = \frac{dz}{dt}\frac{dt}{d\tau} = \frac{\dot{z}}{\sqrt{1-\frac{v^2}{c^2}}}; \quad P_3 = P_z = \frac{m_0\dot{z}}{\sqrt{1-\frac{v^2}{c^2}}} \qquad \text{(C.4)}$$

$$U_4 = i\frac{cdt}{d\tau} = \frac{ic}{\sqrt{1-\frac{v^2}{c^2}}}; \quad P_4 = i\frac{m_0c}{\sqrt{1-\frac{v^2}{c^2}}} \qquad \text{(C.5)}$$

where v^2 is the square of the ordinary 3-vector velocity, and $i = \sqrt{-1}$, and m_0 is the "rest" (at zero speed) mass of the object.

The physical meaning of P_4 becomes apparent if one considers the low-speed case, $v << c$, where relativistic effects are not important. Then the last equation of (C.5) can be expanded[1] as:

$$P_4 = im_0c\left[1 + \frac{1}{2}\left(\frac{v}{c}\right)^2 + \frac{3}{8}\left(\frac{v}{c}\right)^4 + \frac{5}{16}\left(\frac{v}{c}\right)^6 + \cdots\right] \qquad \text{(C.6)}$$

The dominant term in this expression is precisely $\frac{i}{c}$ times the well-known non-relativistic kinetic energy, $\frac{1}{2}m_0v^2$. Analogously, $P_4\frac{c}{i}$ is an energy term, which is called the total energy of the particle. This term defines the "total" energy, E:

$$E = \frac{cP_4}{i} = \frac{m_0c^2}{\sqrt{1-\frac{v^2}{c^2}}} \qquad \text{(C.7)}$$

[1] Taylor series expansion: $f(x) = f(0) + \sum_{n=1}^{\infty} \frac{x^n}{n!}\frac{\partial^n}{\partial x^n} f(x)\,|_{x=0}$

This total-energy consists of an intrinsic constant component, m_0c^2, called "self-energy" or "rest-energy", and a variable component acquired by motion, the kinetic energy, T. That is, $E = m_0c^2 + T$. The kinetic energy is equal to $\frac{1}{2}m_0v^2$ only for speeds that are small compared to c. The kinetic energy, $T = E - m_0c^2$, can be approximated using Eqs. (C.7) and (C.6) as:

$$T = m_0c^2\left[\frac{1}{2}\left(\frac{v}{c}\right)^2 + \frac{3}{8}\left(\frac{v}{c}\right)^4 + \frac{5}{16}\left(\frac{v}{c}\right)^6 + \cdots\right] \quad \text{(C.8)}$$

The total energy, E, is also expressed in terms of a "relativistic mass", m, so that $E = mc^2$. Then from Eq. (C.7), one obtains:

$$m = \frac{m_0}{\sqrt{1 - \frac{v^2}{c^2}}} \quad \text{(C.9)}$$

The introduction of the relativistic mass simplifies the expressions for the four-vector momentum of Eqs. (C.5) to $P_1 = m\dot{x}$, $P_2 = m\dot{y}$, $P_3 = m\dot{z}$, $P_4 = imc$.

The ordinary momentum of a particle along its direction of motion is equal to $p = mv$, then by squaring Eq. (C.9) and multiplying it by c^2 one obtains:

$$(pc)^2 = (mc^2)^2 - (m_0c^2)^2 \quad \text{(C.10)}$$

Therefore, in a potential-free medium, the difference between the total-energy of a particle, mc^2, and its rest-mass energy, m_0c^2, must be equal to its kinetic energy, T. That is, $T = mc^2 - m_0c^2$. Then one can obtain the relationship:

$$p^2c^2 = T^2 + 2m_0c^2T \quad \text{(C.11)}$$

Therefore, for a "particle" that has a zero rest-mass (e.g. a photon or a neutrino), the momentum is simply $p = \frac{T}{c}$. The "equivalent" mass[2], m, is then equal to $\frac{T}{c^2}$, and the velocity must be equal to c. Note that solving Eq. (C.11) for T produces two states:

$$T + m_0c^2 = \pm\left[(p^2c^2 + (m_0c^2)^2\right]^{\frac{1}{2}} \quad \text{(C.12)}$$

The concept of negative energy is utilized to explain some physical phenomena, see section D.4.

Using $E = m_0c^2 + T$, Eq. (C.11) can be rewritten as:

$$E^2 = p^2c^2 + (m_0c^2)^2 \quad \text{(C.13)}$$

[2]Recall that mass is resistance to motion.

The rest-mass of the electron is so small that the relativistic increase in the mass of an electron is 1% of the rest-mass for each 5 keV of kinetic energy [48]. It is a must, therefore, to treat the kinematics of electron motion in a medium relativistically [48].

Appendix D
Quantum Mechanics

D.1. Preliminaries

Conventional mechanics is concerned with the conservation of mass, energy and momentum. In radiation physics, these conservation laws, along with relativistic mechanics (appendix C), are used to describe the kinematics of an interaction, i.e. the change in energy and direction. The probability of occurrence of a particular interaction is not addressed by conservation principles. Quantum mechanics deals with this problem by providing a framework for determining the cross-section of a particular reaction. This is done by examining the effect of the potential of a target nucleus on incoming radiation, with the latter viewed as an incoming wave even for radiation with corpuscular nature. Atomic and nuclear systems can exist only in discrete quantized states of internal motion. The uniqueness of these states is what makes such systems inherently stable and reproducible. This quantum (discrete) nature becomes apparent once the wave mechanics of the interaction process are formulated. Although exact calculations with wave mechanics are not possible, they provide some insight into the nature of the cross-section and their behavior with energy. This appendix provides a short introduction to the concepts of wave mechanics. For more details, the reader should consult one of the many available textbooks on quantum mechanics.

Duality of Particle and Wave. A particle moving with a momentum p can be assigned an equivalent wavelength λ, called the *de Broglie wavelength*, determined by the relationship:

$$\lambda = \frac{h}{p} \tag{D.1}$$

where h is Planck's constant. For elementary particles (electrons, protons, neutrons, etc.), the wavelength is usually of the order of nuclear dimensions, due to the small value of h ($= 6.626 \times 10^{-34}$ J s). Therefore, this duality principle is applicable only to interactions between particles with dimensions comparable to the de Broglie wavelength.

Duality also means that waves can assume corpuscular properties. Thus, a photon of energy $E = h\nu$ can be given a momentum $p = \frac{E}{c} = \frac{h\nu}{c} = \frac{h}{\lambda}$, and a relativistic mass $m = \frac{E}{c^2} = \frac{h\nu}{c^2}$, where ν is the wave frequency, λ is the wavelength and $c = \nu\lambda$ is the speed of electromagnetic waves (speed of light).

Classical Wave Equation. The one-dimensional form of the wave equation, in the z-direction, is given by:

$$\frac{\partial^2 \Psi}{\partial t^2} = c^2 \frac{\partial^2 \Psi}{\partial z^2} \tag{D.2}$$

where Ψ is the amplitude of the wave motion, and t refers to time. A general solution of Eq. (D.2) can be obtained by assuming that the spatial behavior of Ψ is completely independent of its temporal behavior, so that Ψ can be expressed as:

$$\Psi = Z(z)\Gamma(t) \tag{D.3}$$

This enables the separation of Eq. (D.2) into to two equations, one in z and the other in t:

$$\frac{1}{Z(z)} \frac{\partial^2 Z}{\partial z^2} = \frac{1}{c^2 \Gamma(t)} \frac{\partial^2 Z}{\partial t^2} = -k^2 \tag{D.4}$$

where k^2 is a constant, since each function is dependent on a separate variable. The function $Z(z)$ has one of the following solutions:

$$Z(z) = \sin(kz), \qquad Z(z) = \cos(kz), \qquad \text{or} \quad Z(z) = \exp(\pm ikz) \tag{D.5}$$

This is the solution for a simple harmonic motion. The function $\Gamma(t)$ has the solutions:

$$\Gamma(t) = \sin(ktc), \qquad \Gamma(t) = \cos(ktc) \qquad \text{or} \quad \Gamma(t) = \exp(\pm iktc) \tag{D.6}$$

All the these solutions have a periodical behavior as their arguments change by 2π. Therefore, from Eqs. (D.5) and (D.6), one can conclude that k is the wave-number of a wave of wavelength λ: $k = \frac{2\pi}{\lambda}$.

One can choose a particular solution of the wave equation, Eq. (D.2), as:

$$\Psi(z,t) = Z(z) \exp[-i(2\pi\nu t)] \tag{D.7}$$

where use is made of the fact that $c = \lambda\nu$, with ν being the wave frequency. Substituting the above solution into the general wave equation (D.2) leads t:

$$\frac{\partial^2\psi}{\partial z^2} + \left[\frac{2\pi}{\lambda}\right]^2 \psi = 0 \tag{D.8}$$

This is the classical wave equation when expressed in space only, assuming periodic behavior in time.

D.2. Schrödinger Equation

Consider a free particle, with a kinetic energy $\frac{p^2}{2m}$, entering a system of a potential energy U. The free particles are elementary particles, such as electrons, neutron, protons, while the potential is that of an atom or a nucleus. Conservation of energy requires that total energy, $W = \frac{p^2}{2m} + U$, remains constant. Therefore, the momentum of the particle at any state must be equal to:

$$p^2 = 2m(W - U) \tag{D.9}$$

The wave associated with this particle according to the duality principle, Eq. (D.1), results in a wave number, k, such that:

$$k^2 = \left(\frac{2\pi}{\lambda}\right)^2 = \left(\frac{2\pi p}{h}\right)^2 = \frac{p^2}{\hbar^2} = \frac{2m(W-U)}{\hbar^2} \tag{D.10}$$

where $\hbar = \frac{h}{2\pi}$. Eq. (D.8) can now be expressed as:

$$\frac{\partial^2\psi}{\partial z^2} + \frac{2m(W-U)}{\hbar^2}\psi = 0 \tag{D.11}$$

This is the Schrödinger's equation.

When two particles, say a neutron and a nucleus, approach each other, a wave equation can be written taking into account their relative motion and center-of-mass. The three-dimensional wave equation for two particles of mass, m and M, is given by [48]:

$$\nabla^2\psi + \frac{\frac{2mM}{m+M}}{\hbar^2}[W - U(r)]\psi = 0 \tag{D.12}$$

where r is the relative separation distance between the two particles. The value $\frac{mM}{m+M}$ is known as the reduced mass, and is equal to m if $M >> m$. In the case of a neutron and a heavy nucleus, the reduced mass would be equal to the mass of the neutron indicating that the center-of-mass of the two particles almost coincides with that of the heavy nucleus. The wavelength of relative motion is then given by:

$$\lambda = \frac{h}{\sqrt{2\frac{mM}{m+M}[W - U(r)]}} \tag{D.13}$$

The solution of Eq. (D.12) determines the states this two-particle system can presume.

The simplest form of a potential field is that of a square-well model, where the potential energy of the nucleus is constant, ($= -U_o$), over a distance extending from the center of the nucleus to its radius. Outside the nucleus, the potential energy follows the coulomb value $\frac{(Ze)(ze)}{r}$, where Ze is the charge of the nucleus, ze is the charge of the incoming particle and r is the distance to the center of the nucleus.

Solution. Assuming that the solutions of the wave Ψ in the radial, polar and azimuthal directions, see Figure 3.1, are not interdependent, then a solution for Eq. (D.12) can be found using the separation of variables[1]:

$$\psi(r,\theta,\phi) = R(r)Y(\theta,\phi) \qquad \text{and} \quad Y(\theta,\phi) = \Theta(\theta)\Phi(\phi) \tag{D.14}$$

Analogous to Eq. (D.4), one can find separation constants for the differential equations for the functions R and Y, and for the differential equations for the functions Θ and Φ. The separation constant for R and Y is designated as $\ell(\ell+1)$, for reasons that will become apparent shortly, while m_ℓ^2 is used as the separation constant for the functions Θ and Φ for the solution of Y at a given value of ℓ. Then, applying Eq. (D.14) to Schrödinger's equation, Eq. (D.12), leads to:

Radial:

$$\frac{1}{r^2R}\frac{d}{dr}(r^2\frac{dR}{dr}) + \frac{2\frac{mM}{m+M}}{\hbar^2}\left[W - U(r) - \frac{\ell(\ell+1)}{2\frac{mM}{m+M}r^2}\hbar^2\right] = 0 \tag{D.15}$$

Polar:

$$\frac{1}{\Theta\sin\theta}\frac{d}{d\theta}\left(\sin\theta\frac{d\Theta}{d\theta} + \ell(\ell+1\right) - \frac{m_\ell^2}{\sin^2\theta} = 0 \tag{D.16}$$

Azimuthal:

$$\frac{d^2\Phi}{d\phi^2} + m_\ell^2\Phi = 0 \tag{D.17}$$

Magnetic Quantum Number. The differential equation for Φ, Eq. (D.17), is the well-known equation for simple harmonics motion, and has possible solutions analogous to those of Eq. (D.5), as $\sin(m\phi)$

[1]The process of separation of variables, is only valid if the potential, $U(r)$ is spherically symmetric and if the separation constants are suitably chosen.

or $\cos(m\phi)$, or a combination of. However, only integer values of m_ℓ (positive or negative) are acceptable solutions, since the wave function Φ, and consequently Ψ, can assume only single-values for each value of ϕ. The integer m_ℓ is known as the magnetic quantum number. The value of m defines the angular momentum of a particle along an external (uniaxial) magnetic field. When applied to an atom under the effect of an external field, the energy levels of the atom split into new levels, each with a different value of m_ℓ, although the original energy of the electron's orbit does not depend on the value of m_ℓ. This effect is known as the Zeeman effect.

Angular-Momentum Quantum Number. Eq. (D.16) is also a mathematically familiar second-order differential equation, known as Legendre's equation. It has two general independent solutions, each of which can be written as power series in $\cos\theta$. However, when $\cos\theta = \pm 1$, i.e. along the polar axis, both of these solutions for Θ become infinite, except for particular values of ℓ. When ℓ is a positive integer or zero, and when $|m| \leq \ell$, the resulting solutions are finite, single-valued, and continuous for all values of θ, making them physically acceptable solutions for Θ. In the special case of $m_\ell = 0$, Φ becomes constant, and the solutions of Legendre's equations are called the Legendre Polynomials The Legendre Polynomials $P_\ell(\cos\theta)$ of degree ℓ, where ℓ is an integer, are defined such that:

$$P_\ell(\cos\theta) = \frac{1}{2^\ell \ell!} \frac{d^{|\ell|}}{dx^{|\ell|}} (\cos^2\theta - 1)^n \tag{D.18}$$

The values of the first few Legendre polynomials are: $P_0(\cos\theta) = 1$; $P_1(\cos\theta) = \cos\theta$; $P_2(\cos\theta) = \frac{1}{2}(3\cos^2\theta - 1)$, $P_3(\cos\theta) = \frac{1}{2}(5\cos^3\theta - 3\cos\theta)$. When $m_\ell \neq 0$ and $\ell \geq |m_\ell|$, the solutions of Legendre's equation are called the associated Legendre functions, $P_n^{m_\ell}(x)$. These can be expressed in terms of Legendre polynomials as:

$$P_\ell^{m_\ell}(\cos\theta) = (1 - \cos^2\theta)^{|\frac{m_\ell|}{2}} \frac{d^{|m|}}{d(\cos\theta)^{|m_\ell|}} P_\ell(\cos\theta) \tag{D.19}$$

The radial part of the solution of the wave equation, Eq. (D.15), can be best examined by considering an incident particle, of mass m, at its closest distance of approach, r_0, measured from the center-of-mass of the potential field of a target nucleus of mass M. Then, non-relativistic conservation of energy leads to:

$$W = \frac{1}{2} \frac{mM}{m + M} v_0^2 + U(r_0) \tag{D.20}$$

where v_0 is the particle's speed at r_0 and the total energy, W, is the initial kinetic energy of the particle (when the rest-mass energy is not included), at some distance far way from the field. At r_0, the particle is expected to be moving normal to r_0, i.e. tangential to the radius of the sphere around the potential field. The angular momentum, J, of the particle, around the center-of-mass is then:

$$J = \frac{mM}{m+M} v_0 r_0 \tag{D.21}$$

The combination of Eqs. (D.20) and (D.21), leads to:

$$W = U(r_0) + \frac{J^2}{2\frac{mM}{m+M} r_0^2} \tag{D.22}$$

Therefore, at the distance of closest approach, the total energy is the sum of the potential energy and the rotational kinetic energy. Note that $\frac{mM}{m+M} r_0^2$ is equal to the moment of inertia of the system at r_0. The angular momentum, J, is a constant of motion, and thus must have the same value away from the potential field and when it approaches it. Therefore, from Eqs. (D.22), and by comparison with the last term in Eq. (D.15), one can conclude that:

$$J = \sqrt{\ell(\ell+1)}\hbar \tag{D.23}$$

which is the angular momentum of the system about an axis through its center-of-mass and normal to the polar axis. The integer ℓ is called the angular-momentum quantum number; or more accurately the orbital-angular-momentum quantum number, to accommodate bound states, such as those of the atomic electrons around a nucleus or the nucleons within a nucleus.

Nuclear collisions are classified using the ℓ values, with a terminology borrowed from Rydberg's notation in atomic spectroscopy [48], see Table D.1. The magnetic quantum number m_ℓ can be viewed as corresponding to the numerical value of the projection of the angular momentum, associated with ℓ, on an external magnetic field. Hence m_ℓ, according to Eq. (D.23), is an integer, and may have only any of the $(2\ell+1)$ values from $-\ell$ to $+\ell$, including zero, as shown in Table D.1. Therefore, at $\ell = 0$ the collisions between a free particle and a potential field of a target are called the s-wave collisions. The wave represented by any value of ℓ is called a partial wave. The summation of all these partial wave represents the wave "packet" associated with the particle. The microscopic cross-section is proportional to the intensity of waves, i.e. to the square of the amplitude of the wave obtained by the solution

Table D.1. Quantum states at angular momentum, ℓ and magnetic moment m_ℓ [48].

ℓ	Letter	Rydberg notation	m
0	s	sharp	0
1	p	principal	-1, 0, +1
2	d	diffuse	-2, -1, 0, +1, +2
3	f	fundamental	-3, -2, -1, 0, +1, +2, +3
4	g	g	-4, -3, -2, -1, 0, +1, +2, +3, +4

of Eq. (D.12), using the particular potential of the nucleus. Note that in coherent scattering, amplitudes from different partial waves are added while preserving their relative phases, then squared to obtain the scattering cross-section. On the other hand, in incoherent scattering, partial waves are first squared, then combined to define the cross-section.

With neutral particles, such as neutrons, the potential of the nucleus is the only potential that needs to be considered. With charged particles, the coulomb electrostatic field, $\frac{(Ze)(ze)}{r}$, must also be considered. Note that away from a potential field, the wave packet becomes a plane wave. S-wave collisions tend to dominate, in most collisions.

Spin. Free particles and nuclei have also a spin angular momentum. That is, a particle spins around an axis passing through itself. This spin accounts in part for the splitting up of the angular momentum of a particle along an external axis. The spin states are assigned a quantum number, s. The spin of a free particle can be in one of two opposite directions (spin-up or spin-down), and hence assumes the values of $\pm\frac{1}{2}$. The angular momentum corresponding to s, in analogy to Eq. (D.23), is equal to $\sqrt{sl(s+1)}\hbar$. The nucleons of a nucleus tend to form pairs so that their individual spins and magnetic moments cancel out. Thus the spin of a nucleus with even-even nucleons (even number of protons and neutrons) is always zero, while the spin of a nucleus with an odd mass-number is in many cases entirely due to the motion and spin of the single unpaired nucleon.

The spin of individual electrons forming an atom combine to produce an overall angular momentum vector, with a corresponding spin quantum number $S = s_1 + s_2 + \cdots$, where s_i is the spin of the ith electron. The number of permitted orientations of S with respect to an external field direction, in analogy to m_ℓ in Table D.1, is $(2S + 1)$.

The nucleons of a nucleus also combine to form an overall spin I, which is similar in nature to that of the atomic electrons. When atoms combine to from a molecule their angular moments also combine. For example, in the case of a hydrogen gas, H_2, the spin of each of the hydrogen nuclei (protons) is $\frac{1}{2}$. It is then possible that some molecules may have two protons spinning in the same direction, giving a total spin $S = 1$, while others spinning against each other, resulting in a total spin of zero. Since there are $(2S+1)$ orientations with respect to an external field direction, $S = 1$ can have three different orientations (states), $+1, 0, -1$, while $S = 0$, has only one state. Therefore, the proton combination of $S = 1$ is three times more probable than that of $S = 0$. Molecules in the state of $S = 0$ are known as the otrho-hydrogen molecules, while molecules with $S = 0$ are known as the para-hydrogen molecules. A similar situation arises when a neutron interacts with a proton, where either triplet, $S = 1$, or singlet, $S = 0$, spin states can exist for the combined neutron-proton system, with a 3:1 probability of the former to the latter. In general for a nucleus of a spin I interacting with a neutron, two spins states can exists: $I + \frac{1}{2}$ and $I - \frac{1}{2}$, with a probability $\frac{I+1}{2I+1}$ and $\frac{I}{2I+1}$, respectively.

D.3. Concept of Cross-Section

Free Particles. A beam of free particles of wavelength λ can be visualized as a set of co-cylinders whose radii are $\frac{\ell}{k}$, where k is the wave-number, $\frac{2\pi}{\lambda}$. Particles in the beam with an angular-momentum quantum number ℓ can be found somewhere in the annulus with an inner radius of $\frac{\ell}{k}$ and an outer radius of $\frac{\ell+1}{k}$. The area of this annulus is:

$$\frac{\pi(\ell+1)^2}{k^2} - \frac{\pi\ell^2}{k^2} = \frac{\pi(2\ell+1)}{k^2} \tag{D.24}$$

A free beam can be occupied by particles of continuous (not discrete) values of angular moment. The number of free particles that can be found between angular moments represented by ℓ, $\ell + 1$ is given by equation (D.24). When the beam is subjected to the field of a nucleus, these particles will occupy orbits defined by ℓ, $\ell+1$, etc. An upper limit of the *absorption cross-section* can be represented by the area of the zones, i.e. by the maximum number of particles that can be absorbed by the potential field. For the ℓth partial wave, this upper limit is:

$$\sigma^{\ell}_{a,max} = \frac{\pi(2\ell+1)}{k^2} \tag{D.25}$$

However, as shown later, Eq. (D.40), the maximum cross-section for the *potential scattering cross section* is $4\pi\frac{(2\ell+1)}{k^2}$.

The incident wave satisfies the wave equation, Eq. (D.12), when $r \rightarrow \infty$, i.e. when the incident particles are far away from the target nucleus, so that the potential of the nucleus has no effect on the wave and can be set to zero, leading to:

$$\nabla^2\psi + k^2\psi = 0 \tag{D.26}$$

with $k = \frac{2\pi\sqrt{\mu T}}{h}$, where E is the kinetic energy of an incident particle, which is then equal to its total energy, W, since the particle is away from the potential field. One solution for Eq. (D.26) is $\exp(ikz)$. Another solution can be obtained in terms of the partial waves, by the separation of variables, analogous to Eqs. (D.15) to (D.17). The radial part of the solution, similar to that of Eqs. (D.15), is the *spherical Bessel function*:

$$j_\ell(kr) = \sqrt{\frac{\pi}{2kr}} J_{\ell+\frac{1}{2}}(kr) \tag{D.27}$$

The plane wave for free particles has a zero magnetic quantum number m, since it is not subjected to the potential field of the nucleus. For $m = 0$, the Legendre equation, Eq. (D.16), gives the Legendre polynomials $P_\ell(\cos\theta)$. The wave equation (D.26) has, therefore, the solution

$$\psi_{incident} = \sum_{\ell=0}^{\ell=\infty} A_\ell j_\ell(kr) P_\ell(\cos\theta) \tag{D.28}$$

where A_ℓ's are constants and can be shown to be given by [48]:

$$A_\ell = (2\ell + 1) i^\ell \tag{D.29}$$

in order to satisfy the fact that the probability of finding a particle everywhere in the spatial space is equal to unity.

Given the fact that $\exp(ikz)$ is also a solution of the wave equation (D.26), then

$$\exp(ikz) = \sum_{\ell=0}^{\ell=\infty} A_\ell j_\ell(kr) P_\ell(\cos\theta) \tag{D.30}$$

which shows how the plane wave can be expressed in terms of elementary spherical partial waves.

Potential Scattering. A beam of free particles can also be represented by a plane wave $\exp(ikz)$, where $k = \frac{2\pi\mu v}{h} = \frac{2\pi}{\lambda}$, with v being the velocity of the particles, and z is their direction of propagation. When this beam approaches a nucleus, it can be scattered by its potential well, $U(r)$. This form of scattering is, therefore, called potential scattering. Note that v is the relative velocity of approach of the incident particles in the center-of-mass system and its value is not altered by collision, see chapter 3.

The probability of scattering by an angle θ can be represented in terms of the square of the amplitude of the scattered spherical wave emanating from the center of the nucleus. The scattered wave can be given the form $f(\theta)\frac{\exp(ikr)}{r}$, where $f(\theta)$ is a function of the wave number of the incoming wave, k, and the potential of the target nucleus, $U(r)$. The term $\exp(ikr)$ is used, instead of $\exp(-ikr)$, since the incoming wave is moving in a direction opposite to $\vec{r}$, which originates from the center of the nucleus. The scattering wave is a function of $\frac{1}{r}$, not $\frac{1}{r^2}$, since by squaring the magnitude wave one obtains the intensity of the scattered particles, which then follows the law of divergence, section 3.6.2.

The scattered wave combined with the incoming wave at a large distance, r, from the target nucleus, i.e. for $kr >> \ell$, can be expressed as:

$$\psi_{total} = \exp(ikz) + f(\theta)\frac{\exp(ikr)}{r} \tag{D.31}$$

Subtracting the incident wave, $\exp(ikz)$, from the total wave gives the scattered wave at $kr >> \ell$.

Total Disturbance Wave. A solution for the total wave function, Ψ_{total}, must include the effect of $U(r)$. By analogy with Eq. (D.28), for $m = 0$, a solution for Ψ_{total} can be expressed as:

$$\psi_{total} = \sum_{\ell=0}^{\ell=\infty} B_\ell R_\ell(kr) P_\ell(\cos\theta) \tag{D.32}$$

where B_ℓ are constants that satisfy Eq. (D.31) and R_ℓ is a solution of the radial wave equation, Eq. (D.15), with a potential $U(r)$. The constants B_ℓ are found to have the values [48]:

$$B_\ell = (2\ell + 1)i^\ell \exp(i\delta_\ell) \tag{D.33}$$

where δ_ℓ is *phase shift* of the ℓth partial wave caused by the scattering potential $U(r)$. This phase shift is due to the presence of the potential $U(r)$, and therefore is independent of the angle θ.

Scattered Wave. Now, subtracting Eq. (D.28) from equation (D.32), a solution for the scattered wave, $\frac{f(\theta)\exp(ikr)}{r}$, can be expressed as:

$$f(\theta) = \frac{1}{2ik}\sum_{\ell=0}^{\ell=\infty}(2\ell + 1)[\exp(2i\delta_\ell) - 1]P_\ell(\cos\theta) \tag{D.34}$$

which is a valid solution when $U(r)$ decreases faster than $\frac{1}{r}$ for large r [48]. Note that $f(\theta)$ is a complex function, with real and imaginary components.

Differential scattering cross-section. Normalizing the intensity of incident particles per unit volume to unity, that is $|\exp(ikz)|^2 = 1$, the corresponding fluence becomes equal to $|\exp(ikz)|^2 v$ particles per unit area. The number of scattered particles crossing an element of area dS per unit time, at a location designated by the spherical coordinate (r, θ, ϕ), is the product of the probability of scattering, the area dS, and the velocity v of the elastically scattered particles, and thus is equal to $v|f(\theta)\frac{\exp(ikr)}{r}|^2 dS$. Dividing this number by the incident flux, $|\exp(ikz)|^2 v$, one obtains the *differential cross-section* for elastic scattering (into the solid angle $d\Omega = \frac{dS}{r^2}$). That is:

$$\frac{d\sigma}{d\Omega} = |f(\theta)|^2 \tag{D.35}$$

The involves only the square of the absolute value, which can be written as

$$\frac{d\sigma}{d\Omega} = |f(\theta)|^2 = C^2 + D^2 \tag{D.36}$$

where

$$C = \frac{1}{2k} \sum_{\ell=0}^{\ell=\infty} (2\ell + 1)(\cos 2\delta_\ell - 1) P_\ell(\cos\theta) \tag{D.37}$$

and

$$D = \frac{1}{2k} \sum_{\ell=0}^{\ell=\infty} (2\ell + 1)(\sin 2\delta_\ell) P_\ell(\cos\theta) \tag{D.38}$$

The cross-section for elastic scattering is the integral of the differential cross-section at all possible angles:

$$\sigma = \int \frac{d\sigma}{d\Omega} d\Omega = 2\pi \int_0^\pi |f(\theta)|^2 \sin\theta \, d\theta \tag{D.39}$$

which gives:

$$\sigma = \frac{4\pi}{k^2} \sum_{\ell=0}^{\ell=\infty} (2\ell + 1) \sin^2 \delta_\ell \tag{D.40}$$

where the phase-shift δ_ℓ is a functions of k and $U(r)$, and $U(r)$ must decrease faster than $\frac{1}{r}$ for very large r. The actual value of the phase-shifts can be obtained, at a given incident particle energy, only if the potential of the target nucleus is known and is well-defined. This is where experiment and theory meet in the process of determining the cross-sections.

Using Eq. (D.34), one can show that the differential cross-section for elastic s-wave scattering is given by:

$$d\sigma = \frac{\sin^2 \delta_0}{k^2} d\Omega \tag{D.41}$$

Since this differential cross-section is not dependent on the angle of scattering, the scattering of the s-wave is isotropic in the center-of-mass system (at which the above analysis is performed). The integral of Eq. (D.41) can also be shown to be in agreement with Eq. (D.40).

D.4. Quantum Electrodynamics

The probability of interaction of charged-particles with electromagnetic fields, including that of photons, can be evaluated by representing the electromagnetic forces by "virtual" photons that are emitted and absorbed [48]. In other words, the energy of an electromagnetic field can be translated into a number of "equivalent" photons, or quanta, the number of which is small, a few times a, where:

$$a = \frac{2\pi e^2}{hc} = \frac{1}{137} \tag{D.42}$$

The constant, a is known as the fine-structure constant. The strength of coupling between a charged-particle and an electromagnetic field is, therefore, represented by a. Virtual photons can be seen as representing transitions between negative (Dirac holes) and positive energy states; see Eq. (C.12). The negative states are considered to be normally filled and are only observed when a vacancy is created in them.

In bremsstrahlung, a charged-particle loses energy by undergoing transitions between two positive states, emitting real photons (photons in a free state). This process is accomplished by the scattering of a charged-particle first by a virtual photon from the electromagnetic field of the nucleus, then by the free photon created in the process. In pair-production, the energy of a photon is used to lift an electron from a hole to a positive energy, by crossing an energy barrier equal to $2m_0c^2$, where m_0 is the electron's rest-mass [48]. The "hole" becomes then an observable positron. Pair-production can also be seen as the result of a positron scattered by the virtual photon, then by the real photon creating an electron. Therefore, both the bremsstrahlung of electrons and the pair-production process of photons have the same nuclear cross-section, both of the order of $Z^2ar_0^2$ [48], where $r_0 = \frac{e^2}{m_0c^2}$ is a constant known as the classical electron radius ($= 2.818 \times 10^{-15}$m). Exchange of more than one virtual photon introduces another a term in the cross-section, requiring the so-called radiative corrections to the cross-section. This correction is associated with the emission and reabsorption of virtual photons, and with the emission of low and high energy real photons [68].

Compton scattering can also be described by quantum electrodynamics as a combination of two processes [48]:

1 The incident photon of energy E_0 is absorbed virtually by the electron, then a real photon of energy E' is emitted to ensure energy and momentum conservation.

2 The electron virtually emits a photon of energy E', then a real photon of energy E_0 is absorbed to conserve energy and momentum.

This representation is the essence of the Klein-Nishina differential cross-section of Eq.(3.42) [48]. The involvement of more than one virtual photons introduces radiative and double Compton corrections, in the order of a, to the cross-section [51]. The emission and reabsorption of virtual photons leads to the radiative correction, while the double Compton effect is associated with the emission of a real photon, usually very low in energy.

Appendix E
Nuclear/Atomic Parameters for Compounds and Mixtures

This appendix presents a number of approaches for calculating the atomic and nuclear parameters of a chemical compound or a mixture of elements. The application of these methods is demonstrated for water (H_2O), as an example of chemical compound, which can also be viewed as a mixture of hydrogen and oxygen atoms.

E.1. Atomic Density

Consider a molecule of a molecular weight M consisting of various species (types) of atoms, each has mass-number of A_i. The number of molecules per unit mass, N_m, is:

$$N_m = \frac{1}{M\text{u}} \tag{E.1}$$

where u is the atomic mass unit ($1.6605402 \times 10^{-27}$ kg). The number of molecules per unit volume, N, is

$$N = \frac{\rho}{M\text{u}} \tag{E.2}$$

where ρ is the mass-density of the material. For a single element, M in Eqs. (E.1) and (E.2) is replaced by the mass number, A.

For a mixture of known density, ρ, and weight fractions, w_i, of components, the atomic-density N_i, of component i, can be expressed as:

$$N_i = \frac{\rho}{\text{u}} \frac{w_i}{A_i} \tag{E.3}$$

Note that density of the mixture can be calculated from the density of individual components, ρ_i's, as:

$$\frac{1}{\rho} = \sum \frac{w_i}{\rho_i} \tag{E.4}$$

When atomic fractions, f_i's are given, as often the case in chemistry handbooks, the atomic density of the mixture can be expressed as:

$$N = \frac{\rho}{\mathrm{u}} \frac{1}{\sum f_i A_i} = \frac{\rho}{\mathrm{u} A_a} \tag{E.5}$$

where A_a defines an effective mass-number based on the weight fractions as:

$$A_a = \sum f_i A_i \tag{E.6}$$

The atomic density, N_i, of component, i, is calculated as $N_i = f_i N$. Note that, the atomic fraction, f_i is related to the weight fraction, w_i, by:

$$f_i = w_i \frac{A_w}{A_i} \tag{E.7}$$

where A_w is an effective mass-number defined as:

$$A_w = \left[\sum \frac{w_i}{A_i} \right]^{-1} \tag{E.8}$$

The effective mass-number defined by Eq. (E.8) is different from that of Eq. (E.6), yet different from that defined later by Eq. (E.18), since the definition of the effective value depends on its use.

E.2. Electron Density

For a compound, the number of electrons per unit volume, or *electron density*, N_e, making use of Eq. (E.2), is equal to:

$$N_e = ZN = \frac{Z\rho}{M\mathrm{u}} \tag{E.9}$$

where M is the molecular-number of the compound (number of electrons per molecule), and Z is the number of electron per molecule.

The electron-density for a mixture is defined as:

$$N_e = \sum_i \frac{Z_i \, \rho \, w_i}{A_i \, \mathrm{u}} \tag{E.10}$$

where w_i, Z_i and A_i are, respectively, the weight (or mass) fraction of species i, its atomic-number and its mass-number. The electron-density in Eq. (E.10) is derived by summing the number of electrons per unit volume for each species.

For example, H_2O has two species (H and O), with $w_H = \frac{2}{18}$, $M_H = 1$, $A_H = 1$ (for hydrogen), $w_O = \frac{16}{18}$, $M_O = 8$, $A_O = 16$ (for oxygen), $M = 2 \times 1 + 16 = 18$, $Z = 2 \times 1 + 8 = 10$, for the molecule. It is straightforward to show that for H_2O, $\frac{N_e u}{\rho} = \frac{10}{18}$ from Eq. (E.9) and equal to $1 \times (\frac{2}{18}) + 8 \times \frac{16}{18}\frac{1}{16} = \frac{2+8}{18} = \frac{10}{18}$ from Eq. (E.10), that is, as one would expect, Eqs. (E.9) and (E.10) are identical.

E.3. Macroscopic Cross-Section

The macroscopic cross-section of a compound, Σ, can be calculated from the microscopic cross-sections, σ_i's, of its constituents as:

$$\Sigma = \sum_i N_i \sigma_i \tag{E.11}$$

where N_i is the atomic-density of atoms of type i, which can be calculated as:

$$N_i = \frac{\rho}{M\mathrm{u}} n_i \tag{E.12}$$

here n_i is the number of atoms of type i in the molecule; for instance for H_2O, $n_H = 2$, and $n_O = 1$. The expression of Eq. (E.12) can be manipulated as follows:

$$N_i = \frac{n_i A_i}{M} \frac{\rho}{A_i \mathrm{u}} = w_i \frac{\rho}{A_i u} \tag{E.13}$$

where A_i is the mass-number of species i, hence the weight-fraction of that species is equal to $w_i = \frac{n_i A_i}{M}$. Therefore, the macroscopic cross-section of the compound can also be calculated, using Eqs. (E.11) and (E.13), as:

$$\Sigma = \sum_i w_i \frac{\rho}{A_i \mathrm{u}} \sigma_i = \sum_i w_i \Sigma_i \tag{E.14}$$

where Σ_i is the macroscopic cross-section of atomic species i, *if it had the density of the compound.*

If one forms a mixture of materials each has its own density, ρ_i, and after mixing had a weight fraction, w_i, then the $w_i = \frac{\rho_i V_i}{\rho V}$, where V_i is the volume occupied by species i in the mixture and V is the total volume of the mixture. Therefore, $w_i = \frac{\rho_i}{\rho}\alpha_i$, where $\alpha_i = \frac{V_i}{V}$ is the volume-fraction occupied by species i. Then, by substituting this expression of w_i in the middle side of Eq. (E.14), the macroscopic cross-section of the mixture becomes equal to:

$$\Sigma = \sum_i \frac{\rho_i \alpha_i \sigma_i}{A_i \mathrm{u}} \tag{E.15}$$

For H_2O, expressions (E.11) and (E.12) give:

$$\Sigma(H_2O) = \frac{\rho}{18u}(2\sigma_H + \sigma_O)$$

Using Eq. (E.14), one gets:

$$\begin{aligned}\Sigma(H_2O) &= \left(\frac{2}{18}\Sigma_H + \frac{16}{18}\Sigma_O\right) \\ &= \left(\frac{2}{18}\,\frac{\rho}{1u}\sigma_H + \frac{16}{18}\,\frac{\rho}{16u}\Sigma_O\right) \\ &= \frac{\rho}{18u}(2\sigma_H + \sigma_O)\end{aligned}$$

which shows that relationships (E.11) and (E.14) are equivalent, though the latter is directly applicable to mixtures. Note, however, that these two expressions are not useful when molecular vibration can affect the radiation interaction, as in the case of thermal-neutron interactions, where measured values for the molecule of interest should be used (see section 3.5).

If one considers boiling water, say under saturation conditions, with a vapor phase occupying a volume-fraction, α, then the liquid phase will have a volume fraction of $1 - \alpha$. The specific volume (inverse density) can be obtained for each phase from steam tables, under a given pressure or temperature, as v_g for the vapor phase and v_f for the liquid phase. For this two-phase flow mixture, one can use Eq. (E.15) to calculate its macroscopic cross-section, Σ, as:

$$\begin{aligned}\Sigma &= \frac{\sigma}{18u}\left[\frac{\alpha}{v_g} + \frac{(1-\alpha)}{v_f}\right] \\ &= (2\sigma_H + \sigma_O)\left[\frac{\alpha}{v_g} + \frac{(1-\alpha)}{v_f}\right] \\ &\approx \frac{(1-\alpha)}{v_f}(2\sigma_H + \sigma_O) \qquad \text{(E.16)}\end{aligned}$$

Note that since both phases are made of the same material, H_2O, they both have the same microscopic cross-section and molecular weight $M =$ 18. Therefore, σ in the above expression is equal to $\frac{\Sigma(H_2O)}{N} = (2\sigma_H + \sigma_O)$, where $\Sigma(H_2O)$ is as calculated in the above paragraph and N is the atomic-density of H_2O . Note also that the approximation in the above expression is due to the fact that usually $v_g >> v_f$ in most cases (for pressures less than about 10 MPa). The same expression of Eq. (E.16) could have been arrived at by viewing the macroscopic cross-sections as the number of barns per unit volume, and weighting its value for each phase by its volume fraction.

E.4. Effective Mass and Atomic Numbers

Since photons mainly interact with the atomic electrons, see section 3.4, it is possible to conceptualize a chemical compound as consisting of an equivalent monatomic substance of an *effective atomic number* Z_{eff} and an *effective mass number*, A_{eff}. With the equivalent single atom/nucleus entity, modeling of radiation interactions can be simplified (though approximately). A number of approaches for calculating theses effective values are described below.

E.4.1. Electron-Density Based

One can define a fictitious "effective-atom" of mass-number A_{eff} and atomic-number Z_{eff}, while maintaining the same electron density, so that, using Eqs. (E.9) and (E.10), one obtains:

$$N_e = \frac{Z_{eff}\rho}{A_{eff}u} = \sum_i \frac{Z_i \; \rho \; w_i}{A_i \mathrm{u}} \tag{E.17}$$

Based on Eq.(E.17), one can define the following effective mass-number:

$$A_{eff} = \sum_i w_i \; A_i \tag{E.18}$$

and effective atomic-number:

$$Z_{eff} = A_{eff} \sum_i \frac{Z_i w_i}{A_i} = A_{eff} \; \frac{Z_m}{M} \tag{E.19}$$

For H_2O, Eqs. (E.18) and (E.19) give, respectively, $A_{eff} = \frac{258}{18} = 14.33$, and $Z_{eff} = 7.96$. Then $\frac{Z_{eff}}{A_{eff}} = 0.555$. While $\frac{Z}{A}$ for most stable elements is approximately equal to 0.5, but $\frac{Z}{A}$ for hydrogen is equal to 1.

E.4.2. Reaction Cross-Section Based

Since the interaction probability (cross-section) of photons is linearly proportional to density only in the case of Compton scattering (see section 3.4), then the approach discussed in section E.4.1 is valid when Compton scattering is the dominant mode of interaction. In general, one should modify the expression of Eq. (E.19) to account for the non-linear dependence of the photon cross-section with atomic number. As shown in section 3.4, the atomic macroscopic cross-section for photons can be expressed as:

$$\sigma = k_r(E)Z^m \tag{E.20}$$

where $k_r(E)$ is a function that depends on the type of interaction and the photon energy, E, and m is interaction-dependent index ($m = 1$ for Compton scattering, $m = 3$ to 5 for the photoelectric effect, $m =$ 1 to 2 for pair-production and $m = 2$ to 3 for Rayleigh scattering). For a mixture, or a compound, using expression (E.14) and substituting Eq. (E.20) for the macroscopic cross-section, one obtains:

$$\Sigma = \sum_i w_i \frac{\rho}{A_i \mathrm{u}} k_r(E)Z^m = \frac{\rho}{\mathrm{u}} \sum_i \frac{w_i}{A_i} Z_i^m \tag{E.21}$$

An equivalent macroscopic cross-section can be written, assuming an effective mass and atomic numbers, Z_{eff} and Z_{eff}, as:

$$\Sigma = N\sigma = \frac{\rho}{A_{eff}\mathrm{u}} k_r(E) Z_{eff}^m \tag{E.22}$$

Equating, the right-hand-sides of Eqs. (E.21) and (E.22), one can define Z_{eff} as:

$$Z_{eff}^m = A_{eff} \sum_i \frac{w_i}{A_i} Z_i^m \tag{E.23}$$

where A_{eff} can be defined using Eq. (E.18). Obviously, for Compton scattering, ($m = 1$), Eqs. (E.23) and (E.19) become identical. However, for other types of photon interactions, one will obtain different values.

For H_2O, Eq. (E.23) gives the following values for Z_{eff}: 7.56 for the photoelectric effect ($m = 4$), 7.96 for Compton scattering ($m = 1$), 7.25 for pair production ($m = 2$), and 7.42 for Rayleigh scattering (for $m = 3$).

Other methods for calculating the effective atomic-number rely on matching the value of $\frac{\Sigma}{N_e}$ (i.e. the cross-section per electron for the compound or the nucleus) with the microscopic cross-section per electron for an individual element [1366]. For example, reference [1367] measured the effective atomic-number of an W/Cu alloy by looking for the value of Z that matches the total photon cross-section of the alloy measured by narrow-beam gamma transmission.

E.4.3. Reaction-Ratio Based

The effective atomic-number may also be deduced from the ratio of two photon cross-sections, e.g. the photoelectric-to-Compton ratio or the Rayleigh-to-Compton ratio, using the definition of those cross-sections for a compound. Using Eqs. (E.2), (3.30), (3.51) and (3.49) and (3.63), the cross-sections for the photoelectric effect, Σ_{pe}, Compton scattering, Σ_{cs} and Rayleigh scattering, Σ_{rs}, can be expressed as functions of the atomic-number of individual atoms in the mixture as follows:

$$\Sigma_{pe} = k_{pe} \sum_i N_i Z_i^n \tag{E.24}$$

$$\Sigma_{cs} = k_{cs} \sum_i N_i Z_i \tag{E.25}$$

$$\Sigma_{pr} = k_{rs} \sum_i N_i Z_i^3 \tag{E.26}$$

where k designates proportionality factors that are not explicitly expressed in the above relationships and n takes a value form 3 to 5, as indicated following Eq. (3.30). Then one can define the followings ratios:

$$\frac{\Sigma_{pe}}{\Sigma_{cs}} = \frac{k_{pe} \sum_i N_i Z_i^n}{k_{cs} \sum_i N_i Z_i} \tag{E.27}$$

$$\frac{\Sigma_{rs}}{\Sigma_{cs}} = \frac{k_{rs} \sum_i N_i Z_i^3}{k_{cs} \sum_i N_i Z_i} \tag{E.28}$$

The same ratios for a single element are expressed as follows:

$$\frac{\Sigma_{pe}}{\Sigma_{cs}} = \frac{k_{pe}}{k_{cs}} Z^{n-1} \tag{E.29}$$

$$\frac{\Sigma_{rs}}{\Sigma_{cs}} = \frac{k_{rs}}{k_{cs}} Z^2 \tag{E.30}$$

Therefore, by comparing Eq. (E.27) with Eq. (E.29) and Eq. (E.28) with Eq. (E.30), the effective atomic-number of a compound can be expressed as [203]:

$$Z_{eff}|_{pe/cs} = \left(\frac{\sum_i N_i Z_i^n}{\sum_i N_i Z_i} \right)^{\frac{1}{n-1}} \tag{E.31}$$

$$Z_{eff}|_{rs/cs} = \left(\frac{\sum_i N_i Z_i^3}{\sum_i N_i Z_i} \right)^{\frac{1}{2}} \tag{E.32}$$

where pe/cs and rs/cs refer to the two reaction cross-sections that are used in determining the value of the effective atomic-number. The rs/cs ratio was measured experimentally and the results obtained compared favorably with those calculated using Eq. (E.31) [203]. Note that these ratios can be used for elemental identification, see section 7.3.10.

Appendix F
Effective Energy

When analyzing indications obtained from a system that employs a multienergetic radiation source, such as an x-ray source, or an isotopic neutron source, it is often convenient to consider an equivalent monoenergetic source that gives the same indication. This concept is also useful in describing the overall distribution of radiation after interacting with matter. This appendix presents some of the methods that can be used to obtain this equivalent energy.

In the analysis below the source energy distribution is designated by the function, $\chi(E)$, where E is the radiation energy, so that:

$$\int_{E_{min}}^{E_{max}} \xi(E)\, dE = 1 \tag{F.1}$$

where E_{min} is the minimum energy of the radiation emitted from the source, and E_{max} is the maximum radiation energy.

F.1. Mean Energy

The most straightforward approach is to use the average (or mean) energy of the source, $\bar{E}$, as the equivalent source energy, so that:

$$\bar{E} = \frac{\int_{E_{min}}^{E_{max}} E\, \xi(E)\, dE}{\int_{E_{min}}^{E_{max}} \xi(E)\, dE} \tag{F.2}$$

The obvious disadvantage of this approach is that if the source energy range spans a few decades of magnitude, the mean value will be strongly biased toward the higher-energy radiation. For example, for radiation emissions at two energies, $E_1 = 1$ eV and $E_2 = 1$ MeV $= 10^6$ eV, with equal probability of emissions, i.e. $\chi(E_1) = \chi(E_2) = p$, then $\bar{E} = pE_1 +$

$pE_1 \times 10^6 \approx p \times 10^6$. That is, the energy E_2 will dominate the value of $\bar{E}$, although both E_1 and E_2 have the same emission probability. Therefore, the mean-energy approach to the effective-energy is only useful when the source emits energies that are not too different in magnitude from each other.

F.2. Most Probable Energy

The most-probable energy (E_p) is the energy corresponding to the maximum value of $\chi(E)dE$. If used as the equivalent source energy, E_p gives a natural bias to the radiation that has the highest probability of emission. For neutrons, the kinetic energy corresponding to the most probable velocity interval based on the Maxwell-Boltzmann distribution of thermal neutrons is used to describe the overall behavior of thermal neutrons at the thermal-energy (see section 3.5). This energy is equal to kT, where k is the Boltzmann constant and T is the absolute medium temperature. This most-probable energy approach is obviously not very adequate, if the radiation distribution does not have a prominent peak.

F.3. Cross-Section Dependent

The overall effect of radiation depends not only on the distribution of the source radiation but also on the manner it interacts with matter. It is, therefore, appropriate to define an equivalent energy, or an *effective-energy*, E_{eff}, as the energy corresponding to the average cross-section of the medium, so that:

$$\sigma(E_{eff}) = \frac{\int_{E_{min}}^{E_{max}} \sigma(E)\xi(E)\, dE}{\int_{E_{min}}^{E_{max}} \xi(E)\, dE} \tag{F.3}$$

Since material density does not change with energy, either macroscopic or microscopic cross-sections can be used in Eq. (F.3). The material for which σ is chosen obviously affects the value of E_{eff}. Therefore, the obtained value of E_{eff} should be used with materials that do not differ significantly in their radiation cross-section from that of the reference material used for calculating E_{eff}. This approach can be further expanded by multiplying $\chi(E)$ by the modifying model of the considered phenomenon, e.g. the exponential attenuation factor in transmission, see section 6.1, or the scattering probability in scattering techniques, etc.

F.4. Best Match

The measurement model of a particular NDE device can be used to define an effective-energy, by minimizing the difference between the response of the device produced by the model and that resulting from a reference evaluated experimentally or by using detailed simulations. That is,

$$E_{eff} \Longrightarrow \min_{E} |\text{Model Value at } E - \text{Reference Value}| \tag{F.4}$$

This requires one to find the energy that produces model results (obtained assuming a monoenergetic source) which are closest in value to the reference results (which incorporates all source energy and interactions). Though this is perhaps the most accurate approach, it works best when the problem configuration for which the reference results are obtained is closest to the situation in which the value of E_{eff} is to be applied.

Appendix G
Radiation Counting Statistics

The statistics of radiation counting is addressed in a number of references [15, 92, 1368, 1369, 1370, 1371]. This appendix provides a summary of methods used to deal with the statistical variability associated with radiation-counting measurements. Question addressed here include:

1 How to evaluate statistical error?

2 How good are the data?

3 How to reduce the error associated with measurements?.

4 How many times should a measurements be repeated?,

5 What is the optimum counting period?

To answer these question, it is necessary to understand the nature of the Poisson statistics of radiation counting, see section 4.5.6.1.

G.1. Poisson Statistics

In radiation counting, one is usually interested in obtaining an estimate of a count rate, (the number of radiating counts per unit time). This is obtained by recording a number of counts, C, within a finite time interval, t, with the count rate, $\dot{C}$, being equal to $\frac{C}{t}$. If the time interval, t, is measured electronically, one can assume that the time interval is accurate and precise[1]. Is is required, however, to determine the level of

[1] An accurate measurement is one with low bias or small systematic error, while a precise measurement has a low variance, i.e. small random variation; see section 14.2.

confidence[2] one has in the recorded counts, C, i.e. the number of photons, neutrons, etc. registered by a detector. If these measurements are not corrupted by spurious signals, some sources of which are discussed in section G.3, one can expect the measured count to follow a Poisson distribution, as discussed in section 4.5.6.1.

If the mean number of counts is m, then according to the Poisson distribution, the probability that a measurement will give C counts is:

$$P(C; m) = \frac{m^C}{C!} \exp(-m) \tag{G.1}$$

The variance of this distribution is equal to its mean, i.e. $m = \sigma^2$, where σ^2 is the distribution variance. The Poisson distribution approaches a normal (Gaussian) distribution, with variance equal to m, only when m is large.

The statistical problem of radiation counting can be stated as follows:

- Obtain an estimate of the distribution mean of the measurement and define a confidence interval around this estimated value.

It is necessary to determine the confidence interval in order to determine the precision of a measurement, that is to say its variability (which arises from the random nature of the problem).

G.1.1. Mean and Variance

Let us consider n measurements, $C_1, \cdots, C_n$. The distribution mean can be estimated as:

$$\bar{C} = m \approx \frac{1}{n} \sum_{i=1}^{n} C_i \tag{G.2}$$

The distribution variance is subsequently estimated as:

$$\text{Var}(C_i) = C_i \tag{G.3}$$

where $\text{Var}(C_i)$ refers to the variance of any measurement C_i sampled from a Poisson distribution. The variance of the estimated mean, $\text{Var}(\bar{C})$ is subsequently obtained by combining all variances involved in evaluat-

[2]The value of a measurement, m, should be reported along with its associated uncertainty, $\mathcal{E}$, and confidence level; e.g. $m \pm \mathcal{E}$, with a 68% confidence. Stating the value of a measurement without indicating the degree of its uncertainty is not worth reporting at all. The ANSI/ASME Measurement Uncertainty, Part I, ANSI/ASME PTC 19.1-1985, recommended using a confidence level of 95% in uncertainty analysis.

ing $\bar{C}$, according to the rules of combining errors, as follows:

$$\mathrm{Var}(\bar{C}) = \sum_{i=1}^{n} \left[\frac{\partial \bar{C}}{\partial C_i}\right]^2 \mathrm{Var}(C_i) = \sum_{i=1}^{n} \frac{1}{n^2} \mathrm{Var}(C_i) \approx \frac{1}{n^2} \sum_{i=1}^{n} C_i = \frac{n}{n^2}\bar{C} = \frac{\bar{C}}{n} \tag{G.4}$$

The confidence interval associated with the estimated mean is defined by $[\bar{C} - \mathcal{E}, \bar{C} + \mathcal{E}]$, where $\mathcal{E}$ can be taken to be equal to $\sqrt{\mathrm{Var}(\bar{C})}$, for a 68% confidence level, see section G.1.2. That is,

$$\mathcal{E} = \sqrt{\frac{\bar{C}}{n}} \tag{G.5}$$

The estimated mean can, therefore, be expressed as: $\bar{C} \pm \mathcal{E}$.

The precision of a measurement is usually defined using the percentage error[3]:

$$\text{Percentage Error} = \frac{\mathcal{E}}{\bar{C}} \times 100 = \frac{100}{(n\bar{C})^{\frac{1}{2}}} \tag{G.6}$$

This relationship confirms what one would intuitively expect: increasing the number of measurements, n, or the number of registered counts, C, diminishes uncertainty, i.e. decreases statistical error.

G.1.2. Population Statistics

One can also evaluate the variance using conventional statistics, from a population of counts, or observation. For n observations, $C_1, \cdots, C_n$, an estimate of the distribution variance can be calculated using the well-known relationship:

$$\mathrm{Var}(C_i) \approx s^2 = \frac{1}{n-1} \sum_{i=1}^{n} (C_i - \bar{C})^2 \tag{G.7}$$

The value $\mathrm{Var}(C_i) \approx \bar{C}$, obtained using Poisson statistics, Eq. (G.3), leads however to a better, more precise, estimate of the distribution variance, than that evaluated using Eq. (G.7). As the number of measurements, n, increases, both estimates will approach each other. Typically for 20 measurements or more, the two estimates of the variance are equally good. For a small number of measurements, however, it is advisable to use the variance estimates provided by the Poisson statistics.

According to the *Central Limit Theorem*, see chapter 16, any set of independent measurements will tend to resemble a normal (Gaussian)

[3]By error, it is not meant that the measurement is in error. The statistical error is a measure of the variability of the distribution from which measurements are sampled.

distribution, as the number of observations increases. For twenty observations or more, one can claim that the statistics of the normal distribution is applicable. The standard error or the variance of the mean, $\mathcal{E}$, can then be used to define a confidence level. One can then define the following confidence intervals:

1–σ : 68% of the measurements lie within $[\bar{C} + \mathcal{E}, \bar{C} - \mathcal{E}]$.

2–σ: 95% of the measurements fall within $[\bar{C} + 2\mathcal{E}, \bar{C} - 2\mathcal{E}]$.

3–σ: 99.7% is within $[\bar{C} + 3\mathcal{E}, \bar{C} - 3\mathcal{E}]$

For a small number of measurements, one cannot apply the same confidence levels to the corresponding confidence intervals. Due to the lack of a simple process for determining the confidence levels in Poisson statistics, the normal distribution confidence levels are usually adopted. One should, however, be aware that the resulting confidence levels are approximate.

G.2. Gross/Background Count Rates

G.2.1. Net Count Rate

In radiation counting, one is usually interested in the count rate, i.e. the number of counts per unit time. The background count rate is also recorded, and the net count rate is obtained as the difference between the gross (foreground) and background count rates. The background is usually taken as the reference level, usually recorded in NDE in the absence of the inspected object. This section deals with error calculations for the net count rate estimates. Since digital clocks are usually used for measuring the time interval with a great degree of precision, no error is assumed to be associated with measuring the counting period.

Let us assume that a total of k measurements, $B_1, \cdots, B_k$, are recorded for the background, each within a counting period of t_B. The ith count rate would then be equal to $b_i = \frac{B_i}{t_B}$. Let the gross count rate be measured n times; resulting in the counts $G_1, \cdots, G_n$, with a counting period t_G for each count. The ith count rate, g_i is, therefore, equal to $\frac{G_i}{t_G}$. One can then calculate the average background and gross count rates, $\bar{b}$ and $\bar{g}$, respectively, and the associated errors $\mathcal{E}_b$ and $\mathcal{E}_g$, respectively, as shown below.

For the background count rate:

$$\bar{b} = \frac{1}{k}\sum_{i=1}^{k} b_i = \frac{1}{kt_B}\sum_{i=1}^{k} B_i \qquad \text{(G.8)}$$

$$\mathrm{Var}(\bar{b}) = \sum_{i=1}^{k} \left[\frac{\partial \bar{b}}{\partial B_i}\right]^2 \mathrm{Var}(B_i) = \sum_{i=1}^{k} \frac{1}{k^2 t_B^2} \mathrm{Var}(B_i) \approx \frac{1}{k^2 t_B^2} \sum_{i=1}^{k} B_i = \frac{\bar{b}}{k t_B} \tag{G.9}$$

$$\mathcal{E}_b = \sqrt{\frac{\bar{b}}{k t_B}} \tag{G.10}$$

Similarly, for the gross count rate:

$$\bar{g} = \frac{1}{n} \sum_{i=1}^{n} g_i \tag{G.11}$$

$$\mathrm{Var}(\bar{g}) = \frac{\bar{g}}{n t_G} \tag{G.12}$$

$$\mathcal{E}_g = \sqrt{\frac{\bar{g}}{n t_G}} \tag{G.13}$$

In order to evaluate the net count rate, $\bar{r}$, the following relationships apply:

$$\bar{r} = \bar{g} - \bar{b} \tag{G.14}$$

$$\mathrm{Var}(\bar{r}) = \mathrm{Var}(\bar{g}) + \mathrm{Var}(\bar{b}) = \frac{\bar{g}}{n t_G} + \frac{\bar{b}}{k t_B} \tag{G.15}$$

$$\mathcal{E}_r = \sqrt{\frac{\bar{g}}{n t_G} + \frac{\bar{b}}{k t_B}} \tag{G.16}$$

One should avoid subtracting the background from individual gross measurements. This may lead to a distorted distribution of the net count rate, particularly if an individual gross count rate happens to be smaller than the background, while most of the others are larger than the background (or vice versa).

It is also advisable to have two estimates of the background, one at the beginning and the other at the end of an experiment, in order to ensure stability of the system. If the average of the two sets of background measurements differ significantly, this may be an indication of electronic instability. If the system is reasonably stable, one should use the average of the two sets of backgrounds and the associated combined error in the calculation of the net count rate.

G.2.2. Number of Measurements and Counting Period

The number of measurements and the counting period depend on the magnitude of error one is willing to accept, as well as the time available. It is recommended that, if possible, an error of no more than 2% be obtained. The counting period should be chosen such that at least 1000 counts are observed. This leads to about 3% error in each measurement. If then three measurements are taken, the combined error would be less than 2%.

A question which often arises is, if one has a fixed amount of time, t, available for performing measurements: should one take one single measurement over the period t, or record n measurements each over a counting period of $\frac{t}{n}$? If the data truly follow a Poisson distribution, both approaches should lead to the same mean and variance; provided of course that the n measurements are recorded sequentially with no gap in between. It is a good practice, however, to record at least three measurements for each data point, to ensure the goodness of data. Unusually large variance can provide an indication of some drifting in the electronics or some interference from undesirable effects, see section G.3.

In order for the background to not significantly affect the net count rate, according to equation (G.16), $\frac{\bar{g}}{nt_G}$ must be much greater than $\frac{\bar{b}}{kt_B}$. One should have $\frac{\bar{g}}{nt_G}$ equal at least 10 times the value of $\frac{\bar{b}}{kt_B}$. If, for example, $\bar{g}$ is twice as large as b, then kt_B must be at least 5 times larger than nt_G. Since the background measurement is in essence a reference point, it should be recorded at least twice as many times as that for foreground (gross) measurements. That is, k should be equal at least to $2n$. In this case, t_B should be at least 2.5 times longer than t_G.

G.3. Goodness of Data

Are all the obtained data points representative of the expected physical situation? This question can be answered by performing one of the following tests:

Reproducibility. Repeating an experiment as many times as possible, and observing whether the results are reproducible or not.

Deviation from Mean. Examining how the results deviate from their average value; too large or too small deviations, say outside $2\mathcal{E}$, are suspicious and may unduly influence the results.

χ^2 Test. Goodness of data can be checked using the χ^2 criterion:

$$\chi^2 = \frac{1}{\bar{C}} \sum_{i=1}^{n} (C_i - \bar{C})^2 \tag{G.17}$$

This test is actually applicable only to normal distributions. For a sufficiently large number of measurements, 20 or more, one can apply this test. To perform this test, find from the χ^2 table[4], the value of the probability, p, corresponding to the number of degrees of freedom, $n-1$, and the value of χ^2. It is desirable to have p close to 0.5, or equivalently $\chi^2 \approx n-1$, otherwise the test is considered to have been failed. If the data fail the χ^2 test, one or two of the outliers should be discarded, using the procedure described below, and the test should be repeated. If the data still fail the χ^2 test, then the collected data may be "bad". Some possible reasons are given below.

Outliers. Data points that are much smaller or larger than the mean value are called outliers. To determine whether an outlier should be rejected, use should be made of the Chauvenet's criterion, which states that a reading may be rejected if its deviation from the mean is greater than a specific number of standard deviations given in Table G.1. For example, in a series of 10 measurements, the Table shows that the number of standard deviation away from the average, beyond which a measurement may be rejected, is 1.96σ. Therefore, if $d = |C_i - \bar{C}|$ exceed 1.96σ for 10 measurements, then the reading should be rejected. Note that σ can be estimated by the value of $\sqrt{\text{Var}}$ calculated using Eq. (G.7). A new mean, and standard deviation, should be then be calculated without this measurement, since the original values were unduly influenced by the extreme observation.

A similar test for rejecting data is proposed in the document: Measurement Uncertainty, Part I, ANSI/ASME PTC 19.1-1985, using a technique called the *modified Thompson τ technique*, see for example reference [1372]. This technique is slightly more liberal in rejecting data than the Chauvenet's criterion.

Inconsistent Data. Possible reasons for statistically inconsistent "bad" include:

- Unstable equipment, e.g. spurious counts generated by a faulty component or an instrument.

- External signals that may be picked up by the apparatus and be recorded. Sparks, radio signals, welding machines, etc. produce signals that may be measurable by a pulse-type counting system.

---

[4]The χ^2 table is available on the web, see for example: "http://fonsg3.let.uva.nl/Service/Statistics/ChiSquare_distribution.html".

Table G.1. Chauvenet's criterion for rejecting a measurement [1373].

N	2	3	4	5	6	7	8
d/σ	1.15	1.38	1.54	1.65	1.73	1.80	1.87
N	9	10	12	14	15	16	18
d/σ	1.91	1.96	2.04	2.10	2.13	2.15	2.20
N	20	25	30	40	50	60	80
d/σ	2.24	2.33	2.39	2.49	2.57	2.64	2.74
N	100	150	200	300	400	500	1000
d/σ	2.81	2.93	3.02	3.14	3.23	3.29	3.48

N = Number of measurements.
d/σ = Maximum acceptable number of standard deviations from mean.

- Sufficient time for equipment warm-up may have not been allowed, or equipment overheating may have occurred.

G.4. Current-Mode Statistics

Current-mode counting, see section 4.5, provides a time-integrated indication. The integration process is accomplished electrically with the aid of an *RC* circuit, which provides an indication of the count rate, r. In order that a true response be obtained, the time-constant, τ, of a rate meter should be shorter than the observation time of the actual change in measured parameters. If pulses are incoming at a rate, r, the number of pulses, $r\Delta t$ counted within a time interval Δt would be randomly distributed around the true mean value. According to Poisson statistics, Eq. (G.3), the variance in this recorded count is $r\Delta t$. However, within Δt pulses from earlier times are also recorded and will affect the statistical variance of the recorded counts. Therefore, the variance of a signal recorded over some observation time t_0 must be integrated over the variance of the collected charge up to t_0.

The charge rate collection in an *RC* integrating circuit of a time-constant τ (= *RC*) is given by:

$$\frac{dQ}{dt} = rq - \frac{Q}{\tau} \tag{G.18}$$

where Q is the charge in the capacitor, q is the charge per pulse received by the circuit, r is the incoming pulsing rate, and t refers to time. With a initial zero charge and a constant value of r, the solution of Eq. (G.18) is:

$$Q(t) = qr\tau \left[1 - \exp\left(-\frac{t}{\tau}\right)\right] \tag{G.19}$$

The change in the collected charge, ΔQ, after some time, Δt, due to the introduction of $r\Delta t$ counts, can be expressed as:

$$\begin{aligned}\Delta Q &= \frac{dQ}{dt}\Delta t \\ &= q(r\Delta t)\exp\left(-\frac{\Delta t}{\tau}\right) \end{aligned} \tag{G.20}$$

where use is made of Eqs. (G.18) and (G.19). Consequently, the variance in ΔQ is given by:

$$\text{Var}(\Delta Q) = \left(\frac{\partial \Delta Q}{\partial (r\Delta t)}\right)^2 \sigma^2(r\Delta t) = \left[q\exp\left(-\frac{t}{\tau}\right)\right]^2 \sigma^2(r\Delta t) \tag{G.21}$$

According to Poisson's counting statistics, $\sigma^2(r\Delta t) = r\Delta t$. Therefore,

$$\text{Var}(\Delta Q) = q^2\left[\exp\left(-\frac{2t}{\tau}\right)\right](r\Delta t) \tag{G.22}$$

For a measurement at time, t_0, all pulses collected from time $t = 0$ to t_0 contribute to the variance, so that

$$\begin{aligned}\text{Var}[Q(t_0)] &= \int_0^{t_0} q^2 r\left[\exp\left(-\frac{-2t}{\tau}\right)\right] dt \\ &= q^2\frac{r\tau}{2}\left[1-\exp\left(\frac{-2t_0}{\tau}\right)\right] \approx q^2\frac{r\tau}{2}\end{aligned} \tag{G.23}$$

with the approximation obtained after performing the integration and assuming that $t_0 >> \tau$. With the same assumption, Eq. (G.19) gives $r = \frac{Q}{q\tau}$. Therefore, the variance in the reading of the rate-meter, r, observed at t_0 is given by:

$$\text{Var}(r) = \frac{\text{Var}[Q(t_0)]}{q^2\tau^2} = \frac{r}{2\tau}\left[1-\exp\left(\frac{-2t_0}{\tau}\right)\right] \tag{G.24}$$

Therefore, the relative error in r is given by:

$$\frac{\mathcal{E}(r)}{r} = \frac{1}{\sqrt{2r\tau}}\left[1-\exp\left(\frac{-2t_0}{\tau}\right)\right]^{\frac{1}{2}} \tag{G.25}$$

An actual change in the count rate from r_i to r_f is fully detectable when the signal r reaches a value of $r_f - \mathcal{E}(r_f)$, see Figure G.1. Below this value, the signal would still be evolving with a time-constant, say τ'. Therefore, to detect the correct value of r_f with the range of its statistical variability, a time t_0 must or elapse so that:

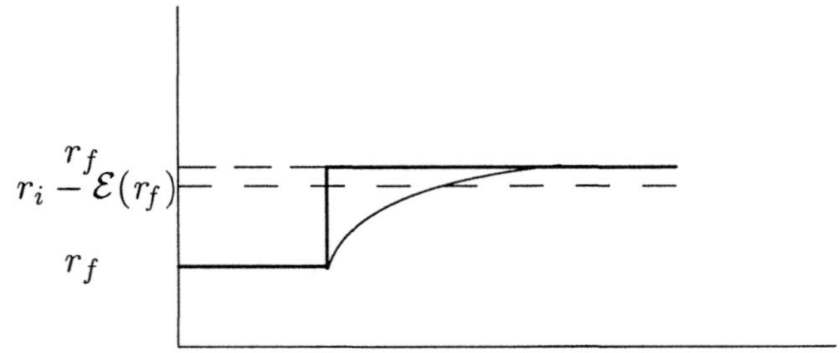

Figure G.1. A step change in count rate in a current-mode measurement.

$$r_i + (r_f - r_i)\left[1 - \exp\left(-\frac{t_0}{\tau'}\right)\right] \geq r_f - \mathcal{E}(r_f) \tag{G.26}$$

or equivalently,

$$\mathcal{E}(r_f) \geq (r_f - r_i)\exp\left(\frac{-t_0}{\tau'}\right) \tag{G.27}$$

As Eq. (G.24) indicates, the variability in r changes with a time-constant equal to $\frac{\tau}{2}$. Therefore, by squaring Eq. (G.27) to obtain the variance, one can also use the time constant $\frac{\tau}{2}$ for the variance, leading to:

$$\mathrm{Var}(r_f) \geq (r_f - r_i)^2 \exp\left(\frac{-2t}{\tau}\right) \tag{G.28}$$

Making use of Eq. (G.24) and assuming $t_0 >> \tau$, then:

$$t_0 \geq \frac{\tau}{2}\ln\left[\frac{2\tau(r_f - r_i)^2}{r_f}\right] \tag{G.29}$$

As Eq. (G.29) indicates, the time required to observe a change depends on the magnitude of the expected change. Shortening the value of τ will also reduce the value of t_0. However, with a too small time-constant, there is little smoothing of the statistical fluctuations, so that assessment of the meter's response with time becomes highly subjective [410]. By optimizing the time-constant of the meter, genuine changes in the measurements can be captured, while smoothing out statistical fluctuations. In general, however, as relationships (G.25) to (G.29) indicate, a rapid response (small value of t_0), requires a high count rate, for changes to be observable with a small uncertainty.

G.5. Elemental Error

Aside from the statistical variability, other sources of error, can occur. These are called elemental errors, and their source are identified according to the ANSI/ASME Measurement Uncertainty, Part I, ANSI/ASME PTC 19.1-1985, as:

Calibration Errors: these are errors associated with system calibration. Although the purpose of calibration is to minimize systematic (bias) error, some uncertainties may still exist. For example, uncertainty in knowing source strength, nature of calibration object, randomness in source emission and detector response, can introduce some variability.

Data Acquisition Errors: such errors can be introduced by the electronic processing units, e.g. a multichannel analyzer, a scalar, timer, etc.

Data Reduction Errors: interpolation, differentiation of data curves, peak-locating and peak-width determination schemes, etc., can introduce their own errors.

Regardless of the source of errors, their contribution should be assessed and included in the uncertainty analysis. These errors can be combined using the process of Eq. (G.4).

References

[1] J. S. Charlton and E. F. Wellman, "Quality measurement in industrial process plants - the role of radioisotopes," *Applied Radiation and Isotopes*, vol. 41, no. 10/11, pp. 1067–1077, 1990.

[2] J. S. Charlton, ed., *Radioisotope Techniques for Problem-Solving in Industrial Process Plants.* Glasgow: Leonard Hill, 1986.

[3] ASTM, "E1316-01 Standard Terminology for Nondestructive Examinations." American Society for Testing and Materials, West Conshohocken, PA, 2001.

[4] D. E. Bray and R. K. Stanley, *Nondestructive Evaluation: A Tool in Design, Manufacturing and Service.* Boca Raton: CRC Press, 1997.

[5] ASNT, *Non-Destructive Testing Handbook.* Columbus, OH: American Society for Nondestructive Testing, 2nd ed., 1982-1995. nine-volume set.

[6] J. Krautkramer and H. Krautkramer, *Ultrasonic Testing of Materials.* Berlin: Springer-Verlag, 1983.

[7] R. Halmshaw, *Industrial Radiography.* London: Champan & Hall, 1995.

[8] J. R. Matthews, ed., *Acoustic Emission.* New York: Gordon and Breach, 1983.

[9] J. Bahr, *Microwave Nondestructive Testing Methods.* New York: Gordon and Breach, 1982.

[10] H. Burger, "Nondestructive testing." U. S. Atomic Energy Commission, 1967.

[11] "Catalog of radioactive and stable isotopes." Isotope Production and Distribution, U. S. Department of Energy, December 1994.

[12] R. R. Kinsey, "The NuDat program for nuclear data on the web." National Nuclear Data Center, Brookhaven National Laboratory, August 1996. (http://www.nndc.bnl.gov/nndc/nudat/, accessed February 2000).

[13] "Nuclides and isotopes: Chart of the nuclides." GE Nuclear Energy, Fifteenth Edition, San Jose, CA, 1996.

[14] Amersham, "Catalog AC3, radiation sources." Amersham Corporation, Arlington Heights, Illinois.

[15] G. F. Knoll, *Radiation Detection and Measurements.* New York: Wiley, 3rd ed., 2000.

[16] IPL, "Radiation sources for research industry environmental applications." Isotope Products Laboratories Catalog, Burbank, California. (http://www.isotopeproducts.com/, accessed February 2002).

[17] Amersham, "Catalog S912139, radiation sources for industrial gauging and analytical, instrumentation." Amersham Corporation, Arlington Heights, Illinois.

[18] M. Borghesi, H. D. Campbell, M. Galimberti, L. A. Gizzi, A. J. Mackinnon, W. Nazarov, and A. Schiavi, "Propagation issues and fast particle source characterization in laser-plasma interactions at intensities exceeding 10^{19} W/cm^2," in *Proc. ECLIM 2000: 26th European Conference on Laser Interaction with Matter, June 12-16 2000, Prague,* vol. 4424, pp. 414–417, SPIE - International Society for Optical Engineering, 2001.

[19] M. J. Berger, J. H. Hubbell, S. M. Seltzer, J. S. Coursey, and D. Zucker, *XCOM: Photon Cross Section Database (version 1. 2), [Online].* National Institute of Standards and Technology, NBSIR 87-3597, Gaithersburg, MD, 1999. (http://physics.nist.gov/xcom, accessed, February 2002).

[20] G. R. Lachance and F. Claisse, *Quantitative X-ray Fluorescence Analysis: Theory and Application.* Chichester: John Wiley & Sons, 1995.

[21] R. C. Weast, ed., *Handbook of Chemistry and Physics.* Boca Raton, FL: CRC Press, 65th ed., 1984.

[22] W. H. McMaster, N. K. D. Grande, J. H. Mallett, and J. H. Hubbell, "Compilation of x-ray cross sections," Tech. Rep. UCRL-50174, Section II Revision I, Lawrence Livermore National Laboratory, 1969. (available on: http://cars9.uchicago.edu/~newville/mcbook/ and http://www.csrri.iit.edu/periodic-table.html, accessed October 2002).

[23] D. W. L. Hukins, *X-ray Diffraction by Disordered and Ordered Systems.* Oxford Oxfordshire: Pergamon Press, 1981.

[24] J. Tölgyessy and M. Kyrš, *Radioanalytical Chemistry,* vol. II, ch. 9. New York: Ellis Horwood, 1989.

[25] M. L., ed., *An introduction to Mössbauer Spectroscopy.* New York: Plenum Press, 1971.

[26] B. J. Micklich, C. L. Fink, L. Sagalovsky, and D. L. Smith, "Contraband detection using high-energy gamma rays from $^{16}O^*$," in *Proc. Physics-Based*

Technologies for the Detection of Contraband (L. O. Malotky and J. J. Pennella, eds.), vol. 2936, pp. 201–209, SPIE - International Society for Optical Engineering, 1997.

[27] KAERI, "Table of Nuclides at the Korea Atomic Energy Research Institute." (http://www.dne.bnl.gov/CoN/index.html, accessed February 2000), 1998.

[28] S. Glasstone and A. Sesonske, *Nuclear Reactor Engineering.* New York: Van Nostrand Reinhold, 1981.

[29] H. H. Barschall *et al.*, eds., *Neutron Sources or Basic Physics and Application.* Oxford: Pergamon Press, 1983.

[30] E. A. Lorch, "Neutron spectra of ^{241}Am/Be, ^{241}Am/F, ^{242}Cm/Be,^{238}Pu/^{13}C and ^{252}Cf isotopic neutron sources," *Applied Radiation and Isotopes*, vol. 24, pp. 585–591, 1973.

[31] J. W. Boldeman, D. Culley, and R. J. Cawley, "The fission spectrum from the spontaneous fission of ^{252}Cf," *Transactions of the American Nuclear Society*, vol. 32, pp. 733–735, 1979.

[32] J. T. Mihalczo, N. W. Hill, E. D. Blakeman, and P. V. K., "Detection-system characterization using ^{252}Cf ionization chambers," *Nuclear Instruments and Methods in Physics Research*, vol. A299, pp. 34–40, 1990.

[33] E. J. Hall and H. H. Rossi, "Californium-252 in teaching and research," Tech. Rep. 159, International Atomic Energy Agency, Vienna, 1974. Technical Report Series.

[34] W. H. Makky and T. G. Miller, "Tailoring neutron spectra for fast neutron spectroscopy," in *Neutrons, X Rays, and Gamma Rays: Imaging Detectors, Material Characterization Techniques, and Applications* (J. M. Carpenter, D. B. Cline, R. Lanza, and D. F. Mildner, eds.), vol. 1737, pp. 197–208, SPIE - International Society for Optical Engineering, 1993.

[35] E. A. Burrill, "Neutron production and protection." High Voltage Engineering Corporation, Burlington, Massachusetts, 1964.

[36] K. Inoue, Y. Kiyanagi, and H. Iwasa, "An accelerator-based cold neutron source," *Nuclear Instruments and Methods*, vol. 192, pp. 129–136, 1982.

[37] A.-J. Dianoux and G. Lander, eds., *Neutron Data Booklet.* Neutrons for Science, Grenoble: Institut Laue-Langevin, April 2002.

[38] G. E. Bacon, *Neutron Diffraction.* Oxford: Clareaden Press, 1975.

[39] M. Benedict, T. H. Pigford, and H. W. Levi, *Nuclear Chemical Engineering.* New York: McGraw-Hill Company, 1981.

[40] O. Serra, *Fundamentals of Well-Log Interpretation, the Acquisition of Logging Data.* Amsterdam: Elsevier, 1984.

[41] J. A. S. Adams and W. M. Lowder, eds., *The Natural Radiation Environment.* Chicago: The University of Chicago Press, 1964.

[42] M. Eisenbud and T. F. Gesell, *Environmental Radioactivity: from Natural, Industrial, and Military Sources.* San Diego: Academic Press, 1997.

[43] A. Brodsky, *CRC Handbook of Radiation Measurement and Protection.* West Palm Beach, Fl: CRC Press, 1978.

[44] "Accelerator mass spectrometry of ^{10}Be, ^{14}C, ^{26}Al, ^{36}Cl, ^{129}I." Purdue Rare Isotope Measurement Laboratory. (http://plsv1.physics.purdue.edu/primelab/primelab_home.html, accessed June 2002).

[45] W. R. Nelson, H. Hirayama, and D. W. Rogers, "The EGS Code-System," Tech. Rep. No. SLAC-265, Stanford Linear Accelerator Center, Stanford, 1985.

[46] J. A. Halbleib and W. H. Vandevender, "TIGER, A one-dimensional multilayer electron/photon Monte Carlo transport code," *Nuclear Science and Engineering*, vol. 57, pp. 94–94, 1975.

[47] F. J. Ziegler, "The stopping and range of ions in matter (SRIM-2000)." IBM-Research, Yorktown, NY, October 1999. (http://www.srim.org/, accessed February 2000).

[48] R. D. Evans, *The Atomic Nucleus.* New York: McGraw-Hill, 1955.

[49] A. P. Ayra, *Fundamentals of Atomic Physics.* Boston: Allyn and Bacom, 1971.

[50] J. M. A. Lenihan and S. J. Thomson, eds., *Activation Analysis: Principle and Applications.* London: Academic Press, 1965. Proc. NATO Advances Study Institute held in Glasgow.

[51] J. H. Hubbell, "Photon cross section attenuation coefficients, and energy absorption coefficients from 10 keV to 100 GeV," Tech. Rep. SNRDS-NBS 29, US National Bureau of Standards, August 1969.

[52] D. C. Creagh and J. H. Hubbell, "X-ray absorption (or attenuation) coefficients," in *International Tables for Crystallography* (A. J. C. Wilson and E. Prince, eds.), vol. C, Dordrecht: Kluwer Academic Publishers, 1999.

[53] J. H. Hubbell, W. J. Veigele, E. A. Briggs, R. T. Brown, D. T. Cromer, and R. J. Howerton, "Atomic form factors, incoherent scattering functions, and photon scattering cross sections," *Journal of physical and chemical reference data*, vol. 4, no. x, pp. 471–538, 1975. (erratum in vol. 6, pp. 615-616, 1977).

[54] R. Ribberfors, "Relationship of the relativistic Compton cross section to the momentum distribution of bound electron states," *Physical Review B*, vol. 12, no. 6, pp. 2067–2074, 1975.

[55] D. Brusa, G. Stutz, J. A. Riveros, J. M. Fernández-Varea, and F. Salvat, "Fast sampling algorithm for the simulation of photon Compton scattering," *Nuclear Instruments and Methods A*, vol. 379, pp. 167–175, 1996.

[56] F. Biggs, L. B. Mendelsohn, and J. B. Mann, "Hartree-fock Compton profiles for the elements," *Atomic Data and Nuclear Data Tables*, vol. 16, pp. 201–309, 1975.

[57] M. Cooper, P. Pattison, and J. D. Schneider, "Compton profile measurement with 412 keV gamma-rays," *Philosophical Magazine*, vol. 34, no. 2, pp. 2067–257, 1975.

[58] M. Cooper, R. Holt, P. Pattison, and K. R. Lea, "Compton-profile measurements with synchrotron radiation," *Communications on Physics*, vol. 1, pp. 159–165, August 1976.

[59] J. Datta, P. K. Bera, and B. Talukdar, "On atomic Compton profile calculations," *Journal of Physics B*, vol. 25, pp. 361–3265, 1992.

[60] S. Bandyopadhyay, S. K. Saha, A. Chatterjee, and A. K. Charrerjee, "Compton profile of CuO," *Radiation Physics and Chemistry*, vol. 4, no. 5, pp. 729–732, 1995.

[61] B. Mallick, "New x-ray scattering from polymers and its quantum profile analysis applied to materials characterization," in *Proc. Electroactive Polymer, Actuators and Devices-Smart Structures and Materials, March 2001, Newport Beach, CA*, vol. 4329, pp. 311–318, SPIE - International Society for Optical Engineering, 2001.

[62] B. Mallick, "Innovation of x-ray frequency multiplier for radiotherapy," in *Proc. Nondestructive and Evaluation of Materials and Composites V, March 2001, Newport Beach, CA*, vol. 4336, pp. 224–227, SPIE - International Society for Optical Engineering, 2001.

[63] LLNL, "Elastic photon-atom scattering." Lawrence Livermore National Laboratory. (http://www-phys.llnl.gov/Research/scattering/, accessed December 2002).

[64] C. A. Ziegler, L. L. Bird, and D. J. Chlećk, "X-ray Rayleigh scattering method for analysis of heavy atoms in low Z media," *Analytical Chemistry*, vol. 31, no. 11, pp. 1794–1798, 1959.

[65] P. E. Bertin, *Principles and Practice of X-Ray Spectrometric Analysis.* New York: Plenum Press, 2nd ed., 1975.

[66] A. H. Compton and S. K. Allison, *X-rays in Theory and Experiment.* New York: Van Nostrand, 2nd ed., 1935.

[67] K.-W. A. Harbich, M. P. A. Hentschel, and J. A. Schors, "X-ray refraction characterization of non-metallic materials," *NDT & E International*, vol. 34, pp. 297–302, June 2001.

[68] J. H. Hubbell, H. A. Gimm, and I. Øverø, "Pair, triplet, and total atomic cross sections (and mass attenuation coefficients) for 1 MeV-100 GeV photons in elements Z = 1 to 100," *Journal of physical and chemical reference data*, vol. 9, pp. 1023–1147, 1980.

[69] WHO, "Food safety aspects relating to the application of x-ray surveillance equipment: Memorandum from a WHO meeting," *Bulletin of the World Health Organization*, vol. 68, no. 3, pp. 297–301, 1990.

[70] N. M. Schaeffer, *Reactor Shielding for Nuclear Engineers.* TID-25951, National Technical Information Service, Springfield, Virginia: U. S. Atomic Energy Commission, 973.

[71] R. Senesi, C. Andreani, Z. Bowden, D. Colognesi, E. Degiorgi, A. Fielding, J. Mayers, M. Nardone, J. Norris, M. Praitano, N. Rhodes, W. Stirling, J. Tomkinson, and C. Uden, "VESUVIO: A novel instrument for performing spectroscopic studies in condensed matter with eV neutrons at the ISIS facility," *Physica B: Condensed Matter*, vol. 276, pp. 200–201, 2000.

[72] A. L. Fielding and J. Mayers, "Calibration of the electron volt spectrometer, a deep inelastic neutron scattering spectrometer at the ISIS pulsed neutron spallation source," *Nuclear Instruments and Methods in Physics Research, A*, vol. 480, pp. 680–689, March 2002.

[73] M. W. Jognson, "The industrial uses of neutrons," *Applied Radiation and Isotopes*, vol. 46, no. 6/7, pp. 673–690, 1995.

[74] K. Podurets, S. S. Shilstein, and V. A. Somenkov, "Application of very small angle neutron scattering to some problems of material science and nondestructive testing," *Journal De Physique*, vol. 3, pp. 455–458, December 1993.

[75] A. K. Freund, "Cross-sections of materials used as neutron monochromators and filters," *Nuclear Instruments and Methods in Physics Research*, vol. 213, pp. 495–501, August 1983.

[76] J. Penfold, "The use of specular reflection of neutrons to study surfaces and interfaces," *Progress in Colloid & Polymer Science*, vol. 81, pp. 198–202, 1990.

[77] NIST, "Neutron scattering lengths and cross sections." NISR Center for Nuclear Research. (http://www.ncnr.nist.gov/resources/n-lengths/, accessed June 2002, and Neutron News,Vol. 3, No. 3, 1992, pp. 29-37).

[78] BNL, "Evaluated nuclear data files." National Nuclear Data Center, Brookhaven National Laboratory Upton, N.Y. . (http://www.nndc.bnl.gov/nndc/endf/, accessed February 2002).

[79] G. Bell and S. Glasstone, *Nuclear Reactor Theory.* Malabar, Fl: R. E. Krieger Pub. Co., 1985.

[80] P. A. Egelstaff and P. Schofield, "On the evaluation of the thermal neutron scattering law," *Nuclear Science and Engineering*, vol. 12, pp. 260–270, 1962.

[81] A. Radkowsky, ed., *Naval Reactors Physics Handbook.* Washington: Naval Reactors, Division of Reactor Development, United States Atomic Energy Commission, 1964.

[82] M. E. Toms, "A computer analysis to obtain neutron spectra from organic scintillators," *Nuclear Instruments and Methods*, vol. 92, pp. 61–70, 1971.

[83] A. M. Weinberg and E. P. Wigner, *Physical Theory of Neutron Chain Reactors*, ch. XVIII. Chicago: University of Chicago, 1958.

[84] M. S. F, Diveadeenam, and N. E. Holden, *Neutron Cross Section*, vol. 1. Oralndo: Academic Press, neutron resonance parameters and thermal cross section ed., 1984.

[85] IAEA, "Handbook on nuclear activation data," Tech. Rep. 273, International Atomic Energy Agency, Vienna, 1987.

[86] R. Q. Wright, N. M. Greene, J. L. Lucius, and J. C. W. Craven, *SUPERTOG III : Data generator - fine group constants and P_N scattering matrices from ENDF/B.* Oak Ridge, TN, September 1973. ORNL-TM-2679/PSR-13.

[87] B. Davison, *Neutron Transport Theory.* Oxford: Clarendon Press, 1958.

[88] M. K. Butler and J. M. Cook, "One-dimensional diffusion theory," in *Computing methods in reactor physics* (H. Greenspan, C. N. Kelber, and D. Okrent, eds.), ch. 1, New York: Gordon and Breach, 1968.

[89] E. M. A. Hussein, "On the limitations of the diffusion equation," *Developments in Chemical Engineering & Mineral Processing*, vol. 2, pp. 149–157, 1994.

[90] R. Nath, "Monte Carlo simulations in radiation therapy," in *Monte Carlo Simulation in the Radiological Sciences* (R. L. Morin, ed.), Boca Raton: CRC Press, 1988.

[91] "The whys and wherefores of gamma- and x-ray - detectors." EG&G ORTEC, Oak Ridge, TN, 1997.

[92] N. Tsoulfanidis, *Measurement and Detection of Radiation.* Washington, D. C.: Taylor & Francis, 1995.

[93] LND, "LND products catalog." LND Inc., Oceanside, NY. (http://www.lndinc.com, accessed April 2002).

[94] A. Oed, "Position-sensitive detector with microstrip anode for electron multiplication with gases," *Nuclear Instruments and Methods in Physics Research, A*, vol. 263, pp. 351–359, January 1988.

[95] G. C. Giakos, "Emerging imaging sensor technologies: from aerospace to health care," *IEEE Instrumentation & Measurement Magazine*, vol. 1, pp. 16–19, December 1998.

[96] C. Budtz-Jorgensen, A. Bahnsen, C. Olesen, M. M. Madsen, P. Jonasson, H. W. Schnopper, and A. Oed, "Microstrip proportional counters for x-ray astronomy," *Nuclear Instruments and Methods in Physics Research A*, vol. 310, pp. 82–87, December 1991.

[97] Y. Giomataris, P. Rebourgeard, J. Robert, and C. G., "MICROMEGAS: a high-granularity position-sensitive gaseous detector for high particle-flux environments," *Nuclear Instruments and Methods in Physics Research A*, vol. 376, no. 1, pp. 29–35, 1996.

[98] B. G. Harvey, *Introduction to Nuclear Physics and Chemistry.* Englewood Cliffs, N.J.: Prentice-Hall, 1962.

[99] W. B. Rolnick and J. Huston, *The Fundamental Particles and their Interactions.* Reading, Mass.: Addison-Wesley, 1994.

[100] W. Blum and L. Rolandi, *Particle Detection with Drift Chambers.* Berlin: Springer-Verlag, 1994.

[101] L. M. Lederman and D. N. Schramm, *From Quarks to the Cosmos: Tools of Discovery.* New York: Scientific American Library, Distributed by, W. H. Freeman, 1989.

[102] R. K. Bock and A. Vasilescu, "The particle detector briefbook." Springer, Berlin, 1998. (http://rd11.web.cern.ch/RD11/rkb/titleD.html, accessed November 2000).

[103] Bicron. Brochures, Newburt, Ohio. (http://www.bicron.com, accessed April 2002).

[104] Canberra, "Reference material." Canberra Industries: Meriden, CT. (http://www.canberra.com, accessed April 2002).

[105] Gresham, "Table of material characteristics." Gresham Scientific Instruments Ltd. Globeside Business Park, Marlow, Bucks, England (http://www.gsinst.com, accessed April 2002).

[106] M. Tanaka, K. Hara, S. Kim, K. Kondo, H. Takano, KobayashiM., H. Ishibashi, K. Kurashige, K. Susa, and M. Ishii, "Applications of cerium-doped gadolinium slicate Gd_2SiO_5:Ce scintillator to calorimeters in high radiation environment," *Nuclear Instruments and Methods in Physics Research A*, vol. 404, pp. 283–94, 1998.

[107] K. Debertin and R. G. Helmer, *Gamma- and X-Ray Spectrometry with Semiconductor Detectors.* New York: Elsevier Science Publ., 1988.

[108] IEEE, "NSI/IEEE 325-1996: Standard Test Procedures for Germanium Gamma-Ray Detectors." American National Standards Institute, Washington, DC, 1996.

[109] R. Polichar, R. Schirato, D. Shreve, J. Reed, and A. Policahr, "Energy selective imaging systems based on monolithic CdZnTe arrays operated in high speed counting mode," *Nuclear Instruments and Methods in Physics Research A*, vol. 380, pp. 232–330, 1996.

[110] Takahashi and S. Watanabe, "Recent progress in CdTe and CdZnTe detectors," *IEEE Transaction in Nuclear Science*, vol. 48, pp. 950–959, August 2001.

[111] Constellation Technology Corporation, Largo, FL, "Mercuric iodide detectors." (http://www.contech.com/Mercuric_Iodide_Catalog.htm, accessed March 2002).

[112] G. L. Mathis, "The effect of thermal neutron temperature upon ^{3}He detector response," *IEEE Transactions on Nuclear Science*, vol. 36, pp. 1182–1184, February 1989.

[113] S. Pszona, "A new application of ^{3}He and ^{10}BF$_3$ proportional counters in a polyethylene moderator," *Nuclear Instruments and Methods A*, vol. 402, pp. 139–142, 1998.

[114] Reuter-Stockes, "Engineering data sheets." Reuter-Stockes Inc., Cleveland, Ohio.

[115] R. K. Crawford, "Position-sensitive detection of slow neutrons: survey of fundamental principles," in *Proc. Neutrons, X Rays, and Gamma Rays: Imaging Detectors, Material Characterization Techniques, and Applications* (J. M. Carpenter, D. B. Cline, R. C. Lanza, and D. F. Mildner, eds.), vol. 1737, pp. 210–223, SPIE - International Society for Optical Engineering, 1993.

[116] C. M. M. Mori, T. Haga, and A. Uritani, "A new position-sensitive nuclear fission counter with a convenient readout for neutron measurement," *Nuclear Instruments and Methods in Physics Research A*, vol. 310, pp. 437–440, 1991.

[117] IAEA, "Neutron moisture gauges: A guide-book on theory and practice," Tech. Rep. STI/DOC/10/112, International Atomic Energy Agency, Vienna, 1970. Technical Report Series No. 112.

[118] D. W. Vehar and F. M. Clikeman, "Neutron spectrum measurement using miniature proton recoil proportional spectrometers," *Nuclear Instruments and Methods*, vol. 190, pp. 351–364, 1981.

[119] R. S. Seymour, B. Richardson, M. Morichi, M. Bliss, R. A. Craig, and D. S. Sunberg, "Scintillating-glass-fiber neutron sensors, their application and performance for plutonium detection and monitoring," *Journal of Radioanalytical and Nuclear Chemistry*, vol. 243, no. 2, pp. 387–388, 2000.

[120] J. A. Harvey and N. W. Hill, "Scintillation for neutron physics research," *Nuclear Instruments and Methods*, vol. 162, pp. 507–529, 1979.

[121] D. M. Drake, W. C. Feldman, and C. Hurlbut, "New electronically black neutron detectors," *Nuclear Instruments and Methods in Physics Research A*, vol. 247, pp. 576–582, 1986.

[122] R. C. Byrd, C. F. Auchampaugh, C. E. Moss, and W. C. Feldman, "Warhead counting using neutron scintillators: Detectors development, testing, and demonstration," *IEEE Transactions on Nuclear Science*, vol. 39, no. 4, pp. 1051–1055, 1992.

[123] E. M. A. Hussein, *Fast Neutron Scattering Method for Local Void Fraction Measurement in Two Phase Flow.* PhD, McMaster University, Hamilton, Canada, 1983.

[124] A. Aksoy, A. Coban, A. A. Naqvi, F. Z. Khiari, J. M. Hanly, C. R. Howell, W. Tornow, P. Felsher, M. A. Al-Ohali, and R. L. Walter, "Response-function

measurement of an NE213 scintillator using the^{2}H(d,n)^{3}He reaction," *Nuclear Instruments and Methods in Physics Research A*, vol. 237, pp. 486–491, 1994.

[125] RSICC, *SPEC-4: Calculated Recoil Proton Energy Distributions from Monoenergetic and Continuous Spectrum Neutrons.* Radiation Safety Information Computational Center, Oak Ridge National Laboratory, Oak Ridge, TN. Computer Code, PSR-099, (http://www-rsicc.ornl.gov, accessed April 2002).

[126] RSICC, *FLYSPEC/SHORTSPEC, FORIST, MATXUF and NE-SPEC.* Computer Codes for Unfolding the Pulse-Height Distribution of Neutrons Measured with Organic Scintillators, Radiation Safety Information Computational Center, Oak Ridge National Laboratory. (http://www-rsicc.ornl.gov, accessed April 2002).

[127] C. Petrillo, "SI(111)/Gd - A fast and high-resolution thermal-neutron detector," *Nuovo Cimento Della Societa Italiana di Fisicad-Condensed Matter, Atomic Molecular and chemical Physics Fluids Plasmas Biophysics*, vol. 20, pp. 931–939, 1998. (http://www.hmi.de/people/wilpert-t/ensa/ensa_dwg_perugia.html, accessed April 2002).

[128] FATDUD, SAND II, S. and WINDOWS, and WINDOWS II, *Computer Codes for Calculating Neutron Spectra from Foil Activation Measurements.* Radiation Safety Information Computational Center, Oak Ridge National Laboratory. (http://www-rsicc.ornl.gov, accessed April 2002).

[129] R. T. Reiman, G. P. Lasche, R. L. Coldwell, and J. A. Nobel, "Performance comparison of a new approach to nuclear spectral analysis with commercial codes," *IEEE Transactions on Nuclear Science*, vol. 47, pp. 1522–1527, August 2000.

[130] ORTEC, "Multichannel scalers and multiple-stop time spectrometer." (http://www.ortec-online.com/, accessed November 2002).

[131] "Radiation protection and the safety of radiation sources." International Atomic Energy Agency Safety Series No. 120, Vienna, 1996. Food and Agriculture Organization of the United Nations, International Atomic Energy Agency, International Labour Organisation, OECD Nuclear Energy Agency, Pan American Health Organization, World Health Organization.

[132] ICRP, "Recommendations of the International Commission on Radiological Protection, 1990," Tech. Rep. ICRP 21 (1-3), International Commission on Radiological Protection, Pergamon Press, Oxford, 1991. ICRP Publication 60.

[133] Y. Harima, "Historical review and current status of buildup factor calculations and applications," *Radiation Physics and Chemistry*, vol. 41, pp. 631–672, April/May 1993.

[134] Kodac, "Radiography in modern industry." Kodak. (http://www.kodak.com/country/US/en/government/industrial/ndt/radiography/14.shtml,accessed July 2002).

[135] E. L. Bryant and P. McIntire, eds., *Radiography and Radiation Testing*, vol. 3 of *Non-Destructive Testing Handbook*. Columbus, OH: American Society for Nondestructive Testing, 2nd ed., 1985.

[136] T. Hibiki and K. Mishima, "Approximate method for measurement of phase-distribution in multiphase materials with small neutron-attenuation using a neutron beam as a probe," *Nuclear Instruments and Methods in Physics A*, vol. 374, pp. 345–351, June 1996.

[137] Ü. Bulubay and A. B. Tuğrul, "Quantitative evaluation of intensifying screen materials in Co-60 radiography for steel parts," *NDT & E International*, vol. 31, no. 3, pp. 193–199, 1998.

[138] Polaroid Corporation, "Company history, 1950's." Cambridge, MA. (http://instantphoto.polaroid.com/polinfo/news_ room/history/50s.html, accessed June 2001).

[139] ACEPT W3 Group, "X-Radiography." Department of Physics and Astronomy, Arizona State University, Tempe, AZ. (http://acept.la.asu.edu/PiN/rdg/visnxray/xradiography.shtml, accessed June 2001).

[140] H. E. Johns and J. R. Cunningham, *The Physics of Radiology*. Springfield, Ill: Charles C. Thomas, 4th ed., 1983.

[141] J. Lang and T. Higgins, "Recent developments in radiography at the British museum," *British Journal of Non-Destructive Testing*, vol. 35, pp. 363–368, July 1993.

[142] S. W. Hughes, A. Sofat, C. Baldock, W. Wong, K. Tonge, R. Davis, and J. Spencer, "CT imaging of an Egyptian mummy," *British Journal of Non-Destructive Testing*, vol. 35, pp. 369–374, July 1993.

[143] X. Cao and H. K. Huang, "Current status and future advances of digital radiography and PACS," *IEEE Engineering in Medicine and Biology*, vol. 19, pp. 80–88, Sep 2000.

[144] Y. Suga, K. Kojima, and T. Tominaga, "Detection of weld defects by computer aided x-ray radiography image processing," in *Proc. 4th International Offshore and Polar Engineering Conference, Osaka, Japan*, vol. 4, pp. 431–436, Int. Soc. of Offshore and Polar Engineers (ISOPE), 1994.

[145] T. W. Liao and K. Tang, "Automated extraction of welds from digitized radiographic images based on MLP neural networks," *Applied Artificial Intelligence*, vol. 11, pp. 197–218, April/May 1997.

[146] T. W. Liao and J. Ni, "Automated radiographic NDT system for weld inspection: Part I - Weld extraction," *NDT & E International*, vol. 29, pp. 157–162, June 1996.

[147] T. W. Liao and Y. Li, "Automated radiographic NDT system for weld inspection: Part II - Flaw detection," *NDT & E International*, vol. 31, pp. 183–192, June 1998.

[148] T. W. Liao, D.-M. Li, and Y.-M. Li, "Detection of welding flaws from radiographic images with fuzzy clustering methods," *Fuzzy Sets and Systems*, vol. 108, no. 2, pp. 145–158, 1999.

[149] NDTech, "NDTech Newsletter #15: Digital Radiography: A technology whose time has come." Berkeley Heights, NJ. (http://www.ndtech.net/archive/newsletter15.html, accessed June 2001).

[150] M. Rossi, F. Casali, S. V. Golovkin, and V. N. Govorun, "Digital radiography using an EBCCD-based imaging device," *Applied Radiation and Isotopes*, vol. 53, pp. 699–709, July 2000.

[151] P. Xue and S. F. Schubert, "Low contrast detectability and dose savings with an amorphous silicon detector designed for x-ray radiography," in *Proc. Medical Imaging 2000: Image Perception and Performance, San Diego, CA*, vol. 3981, pp. 194–200, SPIE - International Society for Optical Engineering, 2000.

[152] E. L. Gingold, D. L. Y. Lee, L. S. Jeromin, B. Rodricks, M. G. Hoffberg, and C. L. Williams, "Development of a novel high-resolution direct conversion x-ray detector," in *Proc. Medical Imaging 2000: Physics of Medical Imaging, San Diego, CA*, vol. 3977, pp. 185–193, SPIE - International Society for Optical Engineering, 2000.

[153] A. R. Cowen, A. G. Davies, S. M. Kengyelics, E. L. Baker, G. McLeod, and G. Parkin, "First experiences with a dry digital laser imager for computed radiography," in *Medical Imaging*, vol. 2707, pp. 498–506, SPIE - International Society for Optical Engineering, 1996.

[154] J. D. Cox, D. Langford, and D. W. Williams, "Vertically integrated electronic x-ray imager," in *Proc. X-Ray Detector Physics and Applications, San Diego, CA*, vol. 2009, pp. 22–30, SPIE - International Society for Optical Engineering, 1993.

[155] R. H. Huesman, G. T. Gullberg, W. L. Greenberg, and T. F. Budinger, *RECLBL Library Users Manual: Donner Algorithms for Reconstruction Tomography.* Lawrence Berkeley Laboratory, 1977. PUB-214, (ftp://cfi.lbl.gov/pub/reclbl/, accessed May 2002).

[156] B. A. Brooks and G. Di Chiro, "Principles of computer assisted tomography (CAT) in radiographic and radioisotopic imaging," *Physics in Medicine and Biology*, vol. 21, pp. 689–732, 1976.

[157] C. L. Morgan, *Basic Principles of Computed Tomography.* Baltimore, MD: University Park Press, 1983.

[158] H. H. Barrett, *Radiological Imaging*, vol. 1 and 2. New York: Academic Press, 1981.

[159] K. Mueller, R. Yagel, and J. J. Wheller, "Anti-aliased three-dimensional cone-beam reconstruction of low-contrast objects with algebraic methods," *IEEE Transactions on Medical Imaging*, vol. 18, no. 6, pp. 519–537, 1999.

[160] E. C. Greenawald, S. Y. Ham, and C. F. Poranski, "Artifacts in x-ray backscatter tomography of non-homogeneous objects," in *Review of Progress in Quantitative Nondestructive Evaluation* (D. O. Thompson and D. E. Chimenti, eds.), vol. 13, pp. 739–446, New York: Plenum Press, 1994.

[161] O. J. Omotosho, B. Frith, A. Plaskowski, and B. M. S., "A sensing system for non-destructive imaging using externally Compton-scattered photons," *Sensors and Actuators*, vol. 18, pp. 1–15, 1989.

[162] L. Brateman, A. M. Jacobs, and L. T. Fitzgerald, "Compton scatter axial tomography with x-rays: SCAT-CAT," *Physics in Medicine and Biology*, vol. 29, pp. 1353–1370, November 1984.

[163] P. Busono and E. M. A. Hussein, "Algorithms for density and composition-discrimination imaging for fourth generation CT systems," *Physics in Medicine and Biology*, vol. 44, pp. 1455 –1477, 1999.

[164] P. Busono and E. M. A. Hussein, "Combined transmission/scatter x-ray tomography for material identification," *Journal of the Nuclear Society of Thailand*, vol. 1, pp. 9–23, 1998.

[165] D. J. Gentle and N. M. Spyrou, "Region of interest tomography in industrial applications," *Nuclear Instruments in Physics Research A*, vol. 299, pp. 534–537, 1990.

[166] P. Engler and W. D. Friedman, "Review of dual-energy computed tomography techniques," *Materials Evaluation*, vol. 48, pp. 623–629, May 1990.

[167] R. E. Alvarez and M. A., "Energy-selective reconstructions in x-ray computerized tomography," *Physics and Medicine and Biology*, vol. 21, no. 5, pp. 733–744, 1976.

[168] L. A. Lehmann, R. E. Alvareaz, A. Macovski, W. R. Brody, N. J. Pelc, S. J. Riederer, and A. L. Hall, "Generalised image combinations in dual kV digital radiography," *Medical Physics*, vol. 8, pp. 659–687, 1981.

[169] G. Christ, "Exact treatment of the dual-energy method in CT using polyenergetic x-ray spectra," *Physics in Medicine and Biology*, vol. 29, no. 12, pp. 1501–1510, 1984.

[170] H. Tominaga, N. Wada, N. Tachikawa, Y. Kuramochi, S. Horiuchi, Y. Sase, H. Amano, N. Okubo, and H. Nishikawa, "Simultaneous utilization of neutrons and gamma rays from Cf-252 for measurement of moisture and density," *Applied Radiation and Isotopes*, vol. 34, pp. 429–436, 1983.

[171] C. M. Bartle and C. R. Purcell, "Neutron-gamma transmission (NEUGAT) based composition measurements using an accelerator," *Nuclear Instruments and Methods in Physics Research A*, vol. 254, pp. 219–221, February 1987.

[172] R. A. Schrack, J. W. Behrens, R. Jonhson, and C. D. Bowman, "Resonance neutron absorption using an electron linac," *IEEE Transactions on Nuclear Science*, vol. NS-28, pp. 1640–1643, 1981.

[173] J. C. Overley, "Element-sensitive computed tomography with fast neutrons," *IEEE Transaction on Nuclear Science*, vol. NS-30, no. 2, pp. 1677–1679, 1983.

[174] C. Nicholls, J. D. King, and F. McDaniel, "Selective gamma ray absorption for explosives detection," in *Proc. 1st Int. Conf. on Explosive Detection Technology* (S. Khan, ed.), DOT/FAA/CT-92/11, pp. 333–345, Federal Aviation Administration, Department of Transportation, May 1992.

[175] J. J. Sredniawski, "Detecting concealed explosives with gamma rays," *The Industrial Physicist*, vol. 3, pp. 24–27, March/April 1997.

[176] J. Stehn, M. D. Goldberg, B. A. Magrurno, and R. Wiener-Chasman, "Neutron cross sections," Tech. Rep. MNL 325, Brookhaven National Laboratory, 1964.

[177] P. Gokhale and E. M. A. Hussein, "A Californium-252 neutron transmission technique for detection of explosives," *Applied Radiation and Isotopes*, vol. 48, pp. 973–979, 1997.

[178] E. Kuzmann, S. Nagy, and V. A., "Mössbauer spectroscopy in chemical analysis," in *Chemical Analysis by Nuclear Methods* (Z. B. Alfassi, ed.), ch. 17, Chichester: John Wiley & Sons, 1994.

[179] G. E. Erwall and K. Forsberg, H. G. Ljunggren, *Industrial Isotope Techniques.* Copenhagen: Munksgaard, 1964.

[180] E. M. A. Hussein, "Compton scatter imaging systems," in *Bioinstrumentation: Research Development and Applications* (D. L. Wise, ed.), ch. 35, pp. 1053–1086, Storeham, M. A.: Butterworths Publ., 1990.

[181] F. A. Balogun and N. M. Spyrou, "Compton scattering tomography in the study of a dense inclusion in a lighter matrix," *Nuclear Instrumentation and Methods in Physics Research*, vol. B83, pp. 533–538, 1993.

[182] P. G. Lale, "The examination of internal tissues by high-energy scattered x radiation," *Radiology*, vol. 90, pp. 510–517, 1968.

[183] J. S. Al-Bahri and N. M. Spyrou, "Electron density of normal and pathological breast tissues using a Compton scattering technique," *Applied Radiation and Isotopes*, vol. 49, pp. 1677–1684, 1998.

[184] F. T. Farmer and M. P. Collins, "A new approach to the determination of anatomical cross-section of the body by Compton-scattering of gamma-rays," *Physics in Medicine and Biology*, vol. 16, pp. 577–586, 1970.

[185] F. T. Famer and M. P. Collins, "A further appraisal of the Compton-scattering method for determining anatomical cross-sections of the body," *Physics in Medicine and Biology*, vol. 19, pp. 808–818, 1974.

[186] R. Guzzardi and M. Mey, "Further appraisal and improvements of 90° Compton scattering tomography of the lung," *Physics in Medicine and Biology*, vol. 26, pp. 155–161, 1981.

[187] B. W. Loo, S. Goulding, and D. S. Simon, "A new Compton densitometer for measuring pulmonary edema," *IEEE Transactions on Nuclear Science*, vol. NS-33, no. 1, pp. 531–536, 1986.

[188] P. Zhu, G. Piex, and J. Muller, "In-line density measurement system using x-ray Compton scattering," *NDT&E International*, vol. 28, pp. 3–7, 1995.

[189] M. Guzzardi, R. Mey and C. Giuntini, "90° Compton scattering tomography of the lung: Detection characteristics and correction of the attenuation," *Journal of Nuclear Medicine and Allied Sciences*, vol. 24, pp. 163–169, 1980.

[190] R. L. Clarke and G. Van Dyk, "Compton-scattered gamma rays in diagnostic radiography," in *Medical Radioisotope Scintigraphy*, no. 193 in STI/PUB/, ch. vol. 1, pp. 247–260, Vienna: International Atomic Energy Agency, 1969.

[191] L. A. Zielke, C. G. Howard, R. L. Currie, and C. D. Morgan, "Rod bundle subchannel void fraction by gamma scattering," *Transactions of American Nuclear Society*, vol. 21, pp. 412–413, 1975.

[192] R. L. Clarke and G. Van Dyk, "A new method for measurement of bone mineral content using both transmitted and scattered beams of gamma-rays'," *Physics in Medicine and Biology*, vol. 18, pp. 532–539, 1973.

[193] L. Kaufman, G. Gamsu, C. Savoca, S. Swann, L. Murphey, B. Hruska, D. Palmer, and J. Ullman, "Measurement of absolute lung density by Compton-Scatter densitometry," *IEEE Transactions of Nuclear Science*, vol. NS-23, pp. 599–605, 1976.

[194] A. L. Huddleston and D. Bhaduri, "Compton scatter densitometry in cancellous bone," *Physics in Medicine and Biology*, vol. 24, pp. 310–318, 1979.

[195] E. S. Garnett, T. J. Kennett, D. B. Kenyon, and C. E. Webber, "A photon scattering technique for the measurement of absolute bone density in man," *Radiology*, vol. 106, pp. 209–212, January 1973.

[196] C. E. Webber and T. J. Kennett, "Bone density measured by photon scattering. I. A system for clinical use," *Physics in Medicine and Biology*, vol. 21, pp. 760–769, 1976.

[197] C. E. Webber and G. Coates, "Clinical system for the in vivo measurement of lung density," *Medical Physics*, vol. 9, pp. 473–477, July/August 1982.

[198] A. L. Huddleston and J. P. Sackler, "The determination of electron density by the dual-energy Compton scatter method," *Medical Physics*, vol. 12, pp. 13–19, January/February 1985.

[199] A. L. Huddleston and J. B. Weaver, "Dual-energy compton-scatter densitometry," *Applied Radiation and Isotopes*, vol. 53, pp. 997–1002, 1983.

[200] B. Achmad, "X-ray Compton scatter density measurement at a point within an object," MScE, University of New Brunswick, Fredericton, Canada, 2000.

[201] J. R. Mossop, S. A. Kerr, D. A. Bradley, *et al.*, "The use of coherent gamma-ray scattering for the characterization of materials," *Nuclear Instruments and Method in Physics Research A*, vol. 255, pp. 419–422, 1987.

[202] A. M. Ndlovu, T. J. Farrell, and C. E. Webber, "Coherent scattering and bone mineral measurement: The dependence of sensitivity on angle and energy," *Medical Physics*, vol. 18, pp. 985–989, September 1991.

[203] S. Manninen, T. Pitkänen, T. Tikanen, S. Koikkalainen, and T. Paakkari, "Study of the ratio of elastic to inelastic scattering of photons," *Applied Radiation and Isotopes*, vol. 35, no. 2, pp. 93–98, 1984.

[204] H. P. Schätzler, "Basic aspects on the use of elastic and inelastic scattered gamma radiation for the determination of binary system with effective atomic numbers less than 10," *Applied Radiation and Isotopes*, vol. 30, pp. 115–121, 1979.

[205] E. M. A. Hussein, G. R. McGee, and P. T. Wan, "A neutron technique for local void fraction measurement in two-phase flow," *Transactions of American Nuclear Society*, vol. 65, no. 1, pp. 37–38, 1992.

[206] J. H. Hubbell, "Radiation gauging," in *Encyclopedia of Materials Science and Engineering* (M. B. Bever, ed.), pp. 4040–4042, Oxford: Pergamon Press, 1986.

[207] A. M. Jacobs, "Compton profile radiography," *International Advances in Nondestructive Testing*, vol. 12, pp. 17–52, 1986.

[208] S. Anghaie, L. L. Humphries, and N. J. Diaz, "Material characterization and flaw detection, sizing and location by the differential gamma scattering spectroscopy technique: Part I Development of theoretical basis," *Nuclear Technology*, vol. 91, pp. 361–375, 1990.

[209] H. A. Jama and E. M. A. Hussein, "Design aspects of a gamma-ray energy-spectral Compton-Scatter nondestructive testing method," *Applied Radiation and Isotopes*, vol. 50, pp. 331–342, 1998.

[210] S. K. Mullin and E. M. A. Hussein, "A Compton-Scatter spectrometry technique for flow detection," *Nuclear Instruments and Methods in Physics Research A*, vol. 353, pp. 663–667, 1994.

[211] A. C. Ho and E. M. A. Hussein, "Quantification of gamma-ray Compton-scatter nondestructive testing," *Applied Radiation and Isotopes*, vol. 53, pp. 541–546, 2000.

[212] T. J. Kennett, M. V. Prestwich, and A. Robertson, "Dynamic density measurement by high energy photon scattering," *Applied Radiation and Isotopes*, vol. 27, pp. 529–532, 1976.

[213] E. M. A. Hussein, "Simulation and design of a neutron scatterometer for void-fraction measurement," *Nuclear Engineering and Design*, vol. 105, pp. 33–348, 1988.

[214] K. Lin, E. Pirie, and D. Taylor, "A moving gamma-source method of measuring soil densities," *Nuclear Instruments and Methods*, vol. 72, pp. 325–328, 1969.

[215] L. Daddi, "A gamma backscatter method of measuring soil density," *Applied Radiation and Isotopes*, vol. 24, pp. 295–295, 1973.

[216] C. Ertek and N. Haselberger, "Measurement of density and water content of soil using photon multiple scattering," *Nuclear Instruments and Methods*, vol. 227, pp. 182–185, 1984.

[217] H. M. Morgan, J. T. Shakeshaft, and S. C. Lillicrap, "Gamma-ray backscatter for body composition measurement," *Applied Radiation and Isotopes*, vol. 49, no. 5-6, pp. 555–557, 1998.

[218] E. M. A. Hussein, "Multi-indicator radiation-based identifier for landmines," in *Detection and Remediation Technologies for Mine and Minelike Targets VII*, vol. 4742, pp. 502–508, SPIE - International Society for Optical Engineering, 2002.

[219] J. Charbucinski, "A universal gamma-gamma method for simultaneous determination of rock and ore properties," *Applied Radiation and Isotopes*, vol. 34, no. 1, pp. 353–361, 1983.

[220] C. Bonifazzi, G. Di Domenico, E. Lodi, G. Maino, and A. Tartari, "Principal component analysis of larger layer density Compton scattering measurements," *Applied Radiation and Isotopes*, vol. 53, no. 571-579, 2000.

[221] G. Devlin and D. Taylor, "Improvements in gamma backscatter density gauges for non-homogeneous soils," *Applied Radiation and Isotopes*, vol. 21, pp. 32–33, 1970.

[222] L. R. Lawson, "Backscatter imaging," *Materials Evaluation*, vol. 60, pp. 1295–1316, November 2002. (See also third edition, vol. 4, of reference [5]).

[223] J. J. Battista, L. M. Santon, and M. J. Bronskill, "Compton-scatter imaging of transverse sections: Correction for multiple scattering and attenuation," *Physics in Medicine and Biology*, vol. 22, no. 2, pp. 229–244, 1977.

[224] S. J. Norton, "Compton scattering tomography," *Journal of Applied Physics*, vol. 76, pp. 2007–2015, August 1994.

[225] N. Kondic, A. Jacobs, and D. Ebert, "Three-dimensional density field determination by external stationary detectors and gamma sources using selective scattering," in *Proc. 2nd Int. Topical Meeting on Nuclear Reactor Thermal-Hydraulics* (M. Merilo, ed.), vol. II, pp. 1443–1462, American Nuclear Society, January 1983.

[226] E. M. A. Hussein, D. A. Meneley, and S. Banerjee, "On the solution of the inverse problem of radiation scattering imaging," *Nuclear Science and Engineering*, vol. 92, pp. 341–349, 1986.

[227] R. Guzzardi and G. Licitra, "Critical review of Compton imaging," *CRC Critical Reviews in Biomedical Engineering*, vol. 15, no. 3, pp. 237–268, 1988.

[228] E. M. A. Hussein, D. A. Meneley, and S. Banerjee, "Single exposure neutron tomography of two-phase flow," *International Journal of Multiphase Flow*, vol. 12, pp. 1–36, 1986.

[229] N. V. Arendtsz and E. M. A. Hussein, "Energy-spectral scatter imaging. Part I: Theory and mathematics," *IEEE Transactions on Nuclear Science*, vol. 42, pp. 2155–2165, 1995.

[230] N. V. Arendtsz and E. M. A. Hussein, "Energy-spectral scatter imaging. Part II: Experiments," *IEEE Transactions on Nuclear Science*, vol. 42, pp. 2166–2172, 1995.

[231] P. Chen and Y. Wang, "Multicriterion Compton backscatter imaging," *IEE Proceedings: Science, Measurement and Technology*, vol. 143, pp. 357–361, November 1996.

[232] G. Harding and J. Kosanetzky, "Status and outlook of coherent-x-ray scatter imaging," *Journal of Optical Society of America A*, vol. 4, pp. 933–944, 1987.

[233] G. Harding and B. Schreiber, "Coherent x-ray scatter imaging and its applications in biomedical science and industry," *Radiation Physics and Chemistry*, pp. 229–245, 1999.

[234] P. Duvauchelle, G. Peix, and D. Babot, "Rayleigh to Compton ratio computed tomography using synchrotron radiation," *NDT & E International*, vol. 33, pp. 23–31, January 2000.

[235] B. D. Cullity and S. R. Stock, *Elements of X-ray Diffraction.* Upper Saddle River, NJ: Prentice Hall, 3rd ed., 2001.

[236] M. T. Hutchings and A. D. Krawitz, *Measurement of Residual and Applied Stress using Neutron Diffraction.* Dordrecht: Kluwer Academic, 1992.

[237] E. M. A. Hussein, "Scattering of alpha, beta and gamma radiation for chemical analysis," in *Chemical Analysis by Nuclear Methods* (Z. B. Alfassi, ed.), ch. 16, Chichester: John Wiley & Sons, 1994.

[238] R. P. Gardner and R. E. Ely Jr, *Radioisotope Measurement Applications in Engineering.* New York: Reinhold, 1967.

[239] A. Climent-Font, M. T. Fernández-Jiménez, U. Wätjen, and J. Perière, "RBS: an analytical technique for elemental characterization of standards; advantages and limits of application," *Nuclear Instruments and Methods in Physics Research A*, vol. 353, pp. 575–578, 1994.

[240] H. H. Brongersma, L. C. M. Beirens, and G. C. J. van der Ligt, "Applications of low-energy ion scattering," in *Material Characterization using Ion Beams* (J. Thomas and A. Cachard, eds.), pp. 65–79, New York: Plenum Press, 1978.

[241] H. Verbeek, "Backscattering of ions with intermediate energies," in *Thomas, J.P. and Cachard, A.* (J. Thomas and A. Cachard, eds.), pp. 303–332, New York: Plenum Press, 1978.

[242] E. Rauhala, "Ion backscattering and elastic recoil detection," in *Chemical Analysis by Nuclear Methods* (Z. B. Alfassi, ed.), ch. 11, Chichester: John Wiley & Sons, 1994.

[243] S. S. Nargolwalla and E. P. Przybylowicz, *Activation Analysis with Neutron Generators.* New York: John Wiley & Sons, 1973.

[244] M. Blaauw, "The 1995 IAEA intercomparison of gamma-ray spectrum analysis software," *Nuclear Instruments and Methods in Physics Research A*, vol. 387, pp. 416–432, 1997.

[245] J. E. Cline, "A comparison of detection-limit computations for four commercial gamma-ray analysis programs," *Nuclear Instruments and Methods in Physics Research A*, vol. 286, pp. 421–428, 1990.

[246] J. K. Tuli, "Thermal neutron capture gamma-rays." National Nuclear Data Center. (http://iaeand.iaea.or.at/wallet/tnc/capgam.shtml) and (http://www.nndc.bnl.gov/wallet/tnc/capgam.shtml).

[247] C. Chung, "Prompt activation analysis," in *Chemical Analysis by Nuclear Methods* (Z. B. Alfassi, ed.), ch. 8, Chichester: John Wiley & Sons, 1994.

[248] Z. B. Alfassi, "Activation analysis with nuclear reactors," in *Activation Analysis* (Z. B. Alfassi, ed.), vol. II, ch. 1, Boca Raton: CRC Press, 1990.

[249] S. Lansberger, "Delayed instrumentation neutron activation analysis," in *Chemical Analysis by Nuclear Methods* (Z. B. Alfassi, ed.), ch. 6, Chichester: John Wiley & Sons, 1994.

[250] M. A. Lone, R. A. Leavitt, and D. A. Harrison, "Prompt gamma rays from thermal-neutron capture," *Atomic and Nuclear Data Tables*, vol. 26, pp. 511–559, 1981.

[251] L. L. Chyi, "14 MeV neutron activation analysis of oxygen and nitrogen in coal," in *Proc. 4th Conf. on the Scientific & Industrial Applications of Small Accelerators* (J. L. Duggan and I. L. Morgan, eds.), no. 1175-9 NPS in 76CH, (Piscataway, NJ), pp. 281–292, IEEE, 1976.

[252] D. A. Miller and G. E. Miller, "Activation analysis course experiments with a 14 MeV neutron generator," in *Proc. 4th Conf. on the Scientific & Industrial Applications of Small Accelerators* (J. L. Duggan and I. L. Morgan, eds.), 76CH 1175-9 NPS, pp. 270–272, IEEE, Piscataway, NJ, 1976.

[253] J. W. McKlveen, *Fast Neutron Analysis: Elemental Data Base.* Ann Arbor: Ann Arbor Science, 1981.

[254] K. M. Barfoot, C. W. Cheng, J. D. MacArthur, B. C. Robertson, S. K. Saha, and K. M. Wilson, "Bulk elemental analysis by the inelastic scattering of neutrons produced with a Vande Graaff accelerator," *IEEE Transactions on Nuclear Science*, vol. NS-28, pp. 1644–1646, April 19981.

[255] C. Vandecasteele, *Activation Analysis with Charged Particles.* Chichester: Ellis Horwood, 1988. (Charlmers, R. A. , Ed.).

[256] M. Glascock, "Nuclear reaction chemical analysis: Prompt and delayed measurements," in *Chemical Analysis by Nuclear Methods* (Z. B. Alfassi, ed.), ch. 4, Chichester: John Wiley & Sons, 1994.

[257] M. Peisach, "Prompt activation analysis with charged particles," in *Activation Analysis* (Z. B. Alfassi, ed.), vol. II, ch. 3, Boca Raton: CRC Presss, 1990.

[258] C. Segebade, H.-P. Weise, and G. J. Lutz, *Photon Activation Analysis.* Berlin: Walter de Gruyter, 1988.

[259] A. P. Kushelevsky, "Photon activation analysis," in *Activation Analysis* (Z. B. Alfassi, ed.), vol. II, ch. 4, Boca Raton: CRC Presss, 1990.

[260] J. Ni, R. C. Block, and X. G. Xu, "Photon activation analysis: A proof of principle using a NIST sediment standard and an electron accelerator at Rensselaer Polytechnic Institute," *Applied Radiation and Isotopes*, vol. 53, pp. 535–540, 2000.

[261] Y. C. Jean, "Chemical analysis by positron annihilation," in *Chemical Analysis by Nuclear Methods* (Z. B. Alfassi, ed.), ch. 18, Chichester: John Wiley & Sons, 1994.

[262] L. McLane, "Nuclear data center network, experimental nuclear reaction data file (EXFOR [CSISRS])." National Nuclear Data Center, Brookhaven National Laboratory, BNL-NCS-63330, 1996. (http://www.nndc.bnl.gov/nndc/exfor/, accessed May 2000).

[263] J. F. Ziegler, G. W. Cole, and J. E. E. Baglin, "Technique for determining concentration of boron impurities in substrates," *Journal of Applied Physics*, vol. 43, pp. 3809–3815, 1972.

[264] R. G. Downing, R. I. Fleming, J. K. Langland, and D. H. Vincent, "Neutron depth profiling at the national bureau of standards," *Nuclear Instruments and Methods in Physics Research*, vol. 218, pp. 47–51, 1983.

[265] Z. B. Alfassi, "Use of delayed neutrons in activation analysis," in *Activation Analysis* (Z. B. Alfassi, ed.), ch. 6, Boca Raton: CRC Press, 1990.

[266] R. Tertian and F. Claisse, *Principles of Quantitative X-ray Fluorescence Analysis.* London: Heyden, 1982.

[267] U. A. S. Tapper, W. J. Przybyllowicz, and H. J. Annegarn, "Particle-induced x-ray emission," in *Chemical Analysis by Nuclear Methods* (Z. B. Alfassi, ed.), ch. 13, Chichester: John Wiley & Sons, 1994.

[268] C. M. Lederer and Shirley, eds., *Table of Isotopes.* New York: Wiley Interscience, 1978.

[269] R. W. Woldseth, *X-Ray Energy Spectrometry.* Burlingame, CA: Kevex Corp, 1973.

[270] T. Biran-Izak and M. Manrel, "X-ray fluorescence analysis with radioactive sources," in *Chemical Analysis by Nuclear Methods* (Z. B. Alfassi, ed.), ch. 15, Chichester: John Wiley & Sons, 1994.

[271] A. E. Johansson and J. L. Campbell, *PIXE: A Novel Technique for Elemental Analysis.* New York: John Wiley & Sons, 1988.

[272] D. D. Cohen and M. Harrigan, "K- and L-shell ionization cross sections for protons and helium ions calculation in the ECPSSR theory," *Atomic Data and Nuclear Data Tables*, vol. 33, pp. 255–343, 1985.

[273] H. W. Leverenz, *An introduction to Luminescence of Solids.* New York: Wiley, 1950.

[274] G. H. Dieke, H. M. Crosswhite, and H. Crosswhite, *Spectra and Energy Levels of Rare Earth Ions in Crystals.* New York: Interscience Publishers, 1968.

[275] E. A. Evans and M. Muramatsu, eds., *Radiotracer Techniques and Applications.* New York: Marcel Dekker, 1977.

[276] T. L. Jones, "Radioactive tracer applications," in *Radioisotope Techniques for Problem-Solving in Industrial Process Plants* (J. S. Charlton, ed.), ch. 7, Glasgow: Leonard Hill, 1986.

[277] G. Reed, "Measurement of residence times and residence-time distribution," in *Radioisotope Techniques for Problem-Solving in Industrial Process Plants* (J. S. Charlton, ed.), ch. 9, Glasgow: Leonard Hill, 1986.

[278] R. L. Heath, "Scintillation spectrometry, gamma-ray spectrum catalogue," Tech. Rep. IDO-16880, Idaho National Engineering & Environmental Center, Idaho Falls, ID, 1964. Revised Electronic Update, February 1997.

[279] R. L. Heath, "Gamma-ray spectrum catalogue, Ge(Li) and Si(Li) spectrometry," Tech. Rep. ANCR-1000-2 (TID-4500), Idaho National Engineering & Environmental Center, Idaho Falls, ID, 1974. Revised Electronic Update September 1998.

[280] D. E. Gustafson, "Single photon emission-computed tomography," in *Three Dimensional Biomedical Imaging* (R. A. Robb, ed.), vol. II, pp. 1–39, Boca Raton: CRC Press, 1985.

[281] R. J. Ott, M. A. Flower, J. W. Babich, and P. K. Marsden, "The physics of radioisotope imaging," in *The Physics of Medical Imaging* (S. Webb, ed.), ch. 6, Bristol: Adam Higler, 1988.

[282] K. K. Shung, M. B. Smith, and B. M. W. Tsui, *Principle of Medical Imaging.* San Diego: Academic Press, 1992.

[283] T. F. Budinger and G. T. Gullberg, "Transverse section reconstruction of gamma-ray emitting radionuclides in patients," in *Reconstruction Tomography in Diagnostic Radiology and Nuclear Medicine* (M. M. Ted-Pogossian *et al.*, eds.), pp. 315–342, Baltimore: University Park, 1977.

[284] D. B. Everett, J. S. Fleming, R. W. Todd, and J. Nightingale, "Gamma-radiation imaging system based on the Compton effect," *Proc. IEE*, vol. 124, pp. 995–1000, November 1977.

[285] D. V. Ellis, "Neutron and gamma ray scattering measurements for subsurface geochemistry," *Science*, vol. 250, pp. 82–87, October 1990.

[286] J. S. Charlton, "Gamma-ray absorption techniques," in *Radioisotope Techniques for Problem-Solving in Industrial Process Plants* (J. S. Charlton, ed.), ch. 13, Glasgow: Leonard Hill, 1986.

[287] E. A. Edmonds, "Radiation scattering techniques," in *Radioisotope Techniques for Problem-Solving in Industrial Process Plants* (J. S. Charlton, ed.), ch. 14, Glasgow: Leonard Hill, 1986.

[288] P. S. Ong, V. Patel, and A. Balasubramanyan, "Quantitative characterization evaluation," *Journal of Nondestructive Evaluation*, vol. 16, no. 3, pp. 135–146, 1997.

[289] J. S. Charlton, "Neutron techniques," in *Radioisotope Techniques for Problem-Solving in Industrial Process Plants* (J. S. Charlton, ed.), ch. 15, Glasgow: Leonard Hill, 1986.

[290] J. W. Mayer, M.-A. Nicolet, and W. K. Chu, "Backscattering analysis with MeV ^{4}He ions," in *Material Characterization using Ion Beams* (J. Thomas and A. Cachard, eds.), pp. 333–366, New York: Plenum Press, 1978.

[291] K. Schmid and H. Ryssel, "Backscattering measurement and surface measurement," *Nuclear Instruments and Methods*, vol. 119, pp. 287–289, 1974.

[292] J. Salgado, F. G. Carvalho, C. Oiveira, J. Manteigas, M. Minto, A. Talefe, C. Cruz, J. Neves, and Ambrósio, "Design, construction and use of nuclear instrumentation in Portugal," *Applied Radiation and Isotopes*, vol. 41, no. 10/11, pp. 1057–1066, 1990.

[293] W. C. McHarris, R. M. Ronningen, H. J. Schock, R. Schalek, and D. S. Grummon, "Radioactive ion implantation for wear studies," *Nuclear Instruments and Methods in Physics Research A*, vol. 353, pp. 583–587, 1994.

[294] G. Albertini, F. Carsughi, F. M. Cernuschi, F. Eichhorn, F. Fiori, A. Hempel, M. Magnani, and F. Rustichelli, "Tomography of a IN738 superalloy turbine blade by small-angle neutron scattering," in *Proc. Int. Conf. Neutrons in Research and Industry, Crete, June 1996* (G. Vourvopoulos, ed.), vol. 2867, pp. 127–130, SPIE - International Society for Optical Engineers, 1997.

[295] R. P. Hjelm, W. Wesley, and M. Gerspacher, "The structure of carbon black-elastomer composites by small-angle neutron scattering and the method of contrast variation," in *Proc. 1999 1st Conference on Radiation Sources and Radiation Interactions, July 1999* (G. Vourvopoulos, ed.), vol. 2867, pp. 144–147, SPIE - International Society for Optical Engineers, 1997.

[296] K. Mergia and S. Messoloras, "Studies of the precipitation process by small angle neutron scattering," in *Proc. 1999 1st Conference on Radiation Sources and Radiation Interactions, July 1999* (G. Vourvopoulos, ed.), vol. 2867, pp. 152–155, SPIE - International Society for Optical Engineers, 1997.

[297] G. Albertini, G. Caglioti, F. Fiori, and R. Pastorelli, "SANS investigation of precipitation in heat-treated AA6082 alloy," *Physica B: Condensed Matte*, vol. 276, pp. 921–922, 2000.

[298] D. Mukherji, R. Gilles, P. Strunz, S. Lieske, A. Wiedenmann, and R. P. Wahi, "Measurement of γ prime precipitate morphology by small angle neutron scattering," *Scripta Materialia*, vol. 41, pp. 31–38, June 1999.

[299] H. Hermann, A. Wiedenmann, N. Mattern, and A. Heinemann, "SANS study of nanocrystalline Fe_3Si particles in amorphous FeSiBCuNb matrix," *Physica B: Condensed Matte*, vol. 276, pp. 704–705, 2000.

[300] J. S. Pedersen, P. Vysckocil, B. Schø'nfeld, and G. Kostorz, "Small-angle neutron scattering of precipitates in Ni-rich Ni-Ti alloys. II. Methods for analyzing anisotropic scattering data," *Journal of Applied Crystallography*, vol. 30, pp. 975–984, December 1997.

[301] G. P. Felcher, K. E. Gray, R. T. Kampwirth, and M. B. Brodsky, "Magnetic depth profiles by neutron reflection.," *Physica B: Physics of Condensed Matter & C: Atomic, Molecular and Plasma Physics, Optics*, vol. 136, pp. 59–63, January/February 1985.

[302] A. Sy Ong and G. Duravaldo, "Non-destructive detection of faults in conventional explosive fuses," *Applied Radiation and Isotopes*, vol. 26, pp. 313 –314, 1975.

[303] W. L. Dunn, M. H. van Haaren, and A. M. Yacout, "A pattern-deviation photon scanning technique for precise measurement of internal component locations," *Nuclear Instruments and Methods in Physics Research A*, vol. 353, pp. 658–662, 1994.

[304] W. L. Dunn, F. O'Foghludha, and A. M. Yacout, "Precise measurement of internal sense-wire locations in high-energy physics detectors," *Transactions of the American Nuclear Society*, vol. 65, no. Supp. 1, pp. 56–56, 1992.

[305] W. L. Dunn, F. O'Foghludha, and A. M. Yacout, "Precise measurement of internal sense-wire locations in high energy physics detectors," *Applied Radiation and Isotopes*, vol. 44, pp. 1321–1326, October/November 1993.

[306] P. Zhu, G. Peix, D. Babot, and J. Muller, "In-line density measurement system using x-ray Compton scattering," *NDT & E International*, vol. 28, no. 1, pp. 3–7, 1995.

[307] D. Babot, G. Berodias, and G. Peix, "Detection and sizing by x-ray Compton scattering of near-surface cracks under weld deposited cladding," *NDT & E International*, vol. 24, pp. 247–251, October 1991.

[308] R. S. Holt, M. J. Cooper, and D. F. Jackson, "Gamma-ray scattering techniques for non-destructive testing and imaging," *Nuclear Instruments and Methods in Physics Research*, vol. 221, pp. 98–104, 1984.

[309] R. S. Holt and M. J. Cooper, "Gamma-ray scattering NDE," *NDT International*, vol. 20, no. 3, pp. 161–165, 1987.

[310] H. A. Jama, E. M. A. Hussein, and P. Lee-Sullivan, "Detection of debonding in Composite-aluminum joints using gamma-ray Compton scattering," *NDT & International*, vol. 31, pp. 99–103, 1998.

[311] N. Kim and J. . D. Achenbach, "Quantitative characterization of multiple delaminations in laminated composites using the Compton backscatter technique," *Journal of Nondestructive Evaluation*, vol. 17, June 1998.

[312] C. R. Buill, R. Z. Zwiggelaar, and R. D. Speller, "Review of inspection technique based on the elastic and inelastic scattering of x-rays and their potential in the food and agricultural industry," *Journal of Food Engineering*, vol. 33, pp. 167–179, 1997.

[313] B. Bridge, J. M. Gunnell, D. C. Imrie, and N. J. Olson, "The use of Compton backscatter imaging for the detection of corrosion pittting in offshore structures," *Nondestructive Testing Communications*, vol. 2, pp. 103–113, 1986.

[314] M. J. Anjos, R. T. Lopes, and J. C. Borges, "Compton scattering of gamma-rays as surface inspection technique," *Nuclear Instruments and Methods in Physics Research A*, vol. 280, pp. 535–538, 1989.

[315] R. A. Fookes, J. S. Watt, B. W. Seatonberry, A. Davison, R. A. Greigand, H. W. G. Lowe, and A. C. Abbott, "Gamma-ray backscatter applied to the on-line location of termite-damages railway sleepers," *International Journal of Applied Radiation and Isotopes*, vol. 29, pp. 721–728, 1978.

[316] S. Tuzi and O. Sato, "Locating the positions of reinforcing bars in reinforced concrete using backscattered gamma rays," *International Journal of Applied Radiation and Isotope*, vol. 41, pp. 1013–1018, 1990.

[317] S. Tuzi and O. Sato, "Positioning of reinforcing bars in reinforced concrete by backscattered gamma rays - II: Experimental and Monte Carlo results," *Transactions of the American Nuclear Society*, vol. 65 (Suppl. 1), pp. 69–69, 1992.

[318] S. Tuzi and O. Sato, "Determination of positions of reinforcing bars in reinforced concrete by backscattered gamma rays - II. Experimental and Monte Carlo results," *Applied radiation and isotopes*, vol. 44, pp. 1279–1284, October 1993.

[319] E. M. A. Hussein and T. M. Whynot, "A Compton scattering method for inspecting concrete structures," *Nuclear Instruments and Methods in Physics Research A*, vol. 283, pp. 100–106, 1989.

[320] R. T. Lopes, C. M. Valente, E. F. O. de Jesus, and C. S. Camerini, "Detection of paraffin deposition inside a draining tubulation by the Compton scattering technique," *Applied Radiation and Isotopes*, vol. 48, no. 10-12, pp. 1443–1450, 1997.

[321] S. Anghaie, L. L. Humphries, and N. J. Diaz, "Material characterization and flaw detection, sizing and location by the differential gamma scattering spectroscopy technique: Part II Experiment," *Nuclear Technology*, vol. 91, pp. 376–387, 1990.

[322] W. L. Dunn and A. M. Yacout, "Corrosion detection in aircraft by x-ray backscatter methods," *Applied Radiation and Isotopes*, vol. 53, no. 4, pp. 625–632, 2000.

[323] C. G. Clayton and J. S. Schweitzer, "A review of aspects of nuclear geophysics," *Nuclear Geophysics*, vol. 7, no. 2, pp. 143–172, 1993.

[324] V. Thaveeprungsriporn, P. Swatewacharkul, K. Talerngsuk, S. Punnachiya, and D. Throng-Aram, "Microstructure characterization using compensated doppler-broadened positron annihilation spectroscopy," *Nuclear Instruments and Methods in Physics Research B*, vol. 155, pp. 459–467, 1999.

[325] H. J. Prask and P. C. Brand, "Residual stress determination by means of neutron diffraction," in *Proc. Int. Conf. Neutrons in Research and Industry, Crete, June 1996* (G. Vourvopoulos, ed.), vol. 2867, pp. 106–115, SPIE - International Society for Optical Engineers, 1997.

[326] D. J. Snoha, *X-ray Diffraction Characterization of Process-Induced Residual Stress.* Aberdeen Proving Ground, MD: U.S. Army Research Laboratory, 1996.

[327] R. J. Homicz, *Fundamentals and Basic Techniques of Residual Stress Measurements with a Portable X-ray Diffraction Unit.* New York: Society of Automotive Engineers, 1967.

[328] R. A. Winholtz and A. D. Krawitz, "The relaxation of residual stresses with post weld heat treatment in a high-performance weld measured with neutron diffraction," *Metallurgical and Materials Transactions A*, vol. 26A, pp. 1287–1295, May 1995.

[329] L. Clapham, T. W. Krause, H. Olsen, B. Ma, D. L. Atherton, P. Clark, and T. M. Holden, "Characterization of texture and residual stress in a section of 610 mm pipeline steel," *NDT & E International*, vol. 28, pp. 73–82, April 1995.

[330] P. J. Webster, G. Mills, X. D. Wang, W. P. Kang, and T. M. Holden, "Residual stresses in alumino-thermic welded rails," *Journal of Strain Analysis for Engineering Design*, vol. 32, pp. 389–400, November 1997.

[331] A. Ezeilo and G. A. Webster, "Neutron diffraction analysis of the residual stress distribution in a bent bar," *Journal of Strain Analysis for Engineering Design*, vol. 35, pp. 235–246, July 2000.

[332] G. Albertini, G. Bruno, A. Carrado, F. Fiori, M. Rogante, and F. Rustichelli, "Determination of residual stresses in materials and industrial components by neutron diffraction," *Measurement Science and Technology*, vol. 10, pp. R56–R73, March 1999.

[333] A. D. Krawitz, R. A. Winholtz, E. F. Drake, and N. D. Griffin, "Residual stresses in polycrystalline diamond compacts," *International Journal of Refractory Metals and Hard Materials*, vol. 17, no. 1, 1999.

[334] A. A. Lodini, "The recent development of neutronic techniques for determination of residual stresses," *Radiation Physics and Chemistry*, vol. 61, pp. 227–233, June 2001.

[335] A. Carradò, F. Fiori, E. Girardin, T. Pirling, P. Powell, and F. Rustichelli, "Neutron diffraction measurements of residual stresses in metal matrix composite samples," *Radiation Physics and Chemistry*, vol. 61, pp. 575–577, 2001 2001.

[336] R. Vaidyanathan, M. A. M. Bourke, and D. C. Dunand, "Stress-induced martensitic transformations in NiTi and NiTi-TiC composites investigated by neutron diffraction," *Materials Science and Engineering A*, vol. 273, pp. 404–409, 2000.

[337] M. A. M. Bourke, D. C. Dunn, C. Lawson, and H. G. Priesmeyer, "Phase transformations in engineering materials," in *Proc. Int. Conf. Neutrons in Research and Industry, Crete, June 1996* (G. Vourvopoulos, ed.), vol. 2867, pp. 131–135, SPIE - International Society for Optical Engineers, 1997.

[338] D. C. Dunand, D. Mari, M. A. M. Bourke, *et al.*, "NiTi and NiTi-TiC composites: Part IV. Neutron diffraction study of twinning and shape-memory recovery," *Metallurgical and Materials Transactions. A*, vol. 27, no. 9, pp. 2820–2836, 1996.

[339] D. Dragoi, E. Ustundag, B. Clausen, and M. A. M. Bourke, "Investigation of thermal residual stresses in tungsten-fiber/bulk metallic glass matrix composites," *Scripta Materialia*, vol. 45, pp. 245–252, July 2001.

[340] D. H. Carter and M. A. M. Bourke, "Neutron diffraction study of the deformation behavior of beryllium-aluminum composites," *Acta Materialia*, vol. 48, no. 11, pp. 2885–2900, 2000.

[341] R. U. Vaidya, P. Rangaswamy, M. A. M. Bourke, and D. P. Butt, "Measurement of bulk residual stresses in molybdenum disilicide/stainless steel joints using neutron scattering," *Acta Materialia*, vol. 46, pp. 2047–206, March 1998.

[342] R. I. Todd, M. A. M. Bourke, C. E. Borsa, and R. J. Brook, "Neutron diffraction measurements of residual stresses in alumina/SiC nanocomposites," *Acta Materialia*, vol. 45, pp. 1791–1800, April 1997.

[343] C. M. Bartle, "Chemical analysis by the thermalization, scattering and absorption of neutrons," in *Chemical Analysis by Nuclear Methods* (Z. B. Alfassi, ed.), ch. 9, Chichester: John Wiley & Sons, 1994.

[344] J. S. Michalik, Z. Bazaniak, and J. Palige, "Application of radiotracers to process optimization in polish copper metallurgy," *Nuclear Geophysics*, vol. 3, pp. 425–435, 1989.

[345] A. J. Niemi, R. Zuniga, and N. Heresi, "Tracer-based modelling of grinding systems," *Nuclear Geophysics*, vol. 3, no. 4, pp. 419–423, 1989.

[346] G. G. Eichholz and W. N. Roberts, "The adsorption of tracers on mineral particles," in *Proc. Second Int. Conf. Peaceful Uses of Atomic Energy*, vol. 19, (Geneva), pp. 240–248, United Nations, 1958.

[347] N. Heresi, P. Vega, P. Henriquez, and I. Villanueva, "Tracer techniques applied to the study of leaks in a dike in Chile," *Nuclear Geophysics*, vol. 3, no. 4, pp. 437–443, 1989.

[348] C. G. Clayton, "Some comments on the development of radiation and radioisotope measurement application in industry," *Applied Radiation and Isotopes*, vol. 41, no. 10/11, pp. 917–934, 1990.

[349] UIC, "Radioisotopes in industry: Nuclear issues briefing paper 27." Uranium Information Centre, Melbourne, Australia, November 2001. (http://www.uic.com.au/nip27.htm, accessed June 2002).

[350] M. N. Iqbal, R. P. Gardner, and K. Verghese, "A radiotracer method for the dynamic measurement of the in-process inventory of dissolved materials," *Applied Radiation and Isotopes*, vol. 41, no. 10/11, pp. 1083–1093, 1990.

[351] J. Fodor and C. Varga, "Rapid determination of phosphorus in steel by isotope dilution analysis," in *Proc. Second Int. Conf. Peaceful Uses of Atomic Energy*, vol. 19, (Geneva), pp. 231–234, United Nations, 1958.

[352] L. Bozóky and D. Vödrös, "Determination of current efficiency in aluminium furnaces," in *Proc. Second Int. Conf. Peaceful Uses of Atomic Energy*, vol. 19, (Geneva), pp. 237–239, United Nations, 1958.

[353] R. W. Perkins, W. Thomas, and J. A. Young, "Application of short-lived cosmogenic radionuclides as tracers of in-cloud scavenging processes," *Journal of Geophysics Research*, vol. 75, pp. 3076–3087, May 1970.

[354] M. Brian, "How smoke detectors work." (http://www.howstuffworks.com/smoke1.htm, accessed on February 2000).

[355] A. E. Nevissi, "Detection and measurement of radon and radon decay products," in *Environmental Radon: Occurrence, Control, and Health Hazards* (S. K. Majumdar, R. F. Schmalz, and E. W. Miller, eds.), ch. 9, Easton, PA: The Pennsylvania Academy of Science, 1990.

[356] K. Q. Lao, *Controlling Indoor Radon*, ch. 3. New York: Van Nostrand Reinhold, 1990.

[357] A. C. George, "An overview of instrumentation for measuring environmental radon and radon progeny," *IEEE Transactions on Nuclear Science*, vol. 37, pp. 892–901, April 1990.

[358] H. F. Lucas, "Improved low-level alpha-scintillation counter for radon," *Review of Scientific Instruments*, vol. 28, pp. 680–683, 1957.

[359] T. M. Semkow, P. P. Parekh, C. D. Schwenker, R. Dansereau, and J. S. Webber, "Efficiency of the Lucas scintillation cell," *Nuclear Instruments and Methods in Physics Research*, vol. 253, pp. 515–518, 1994.

[360] C. Costa-Riberio, J. Thomas, R. T. Drew, M. E. Wrenn, and M. Eisenbud, "A radon detector suitable for personnel and area monitoring," *Health Physics*, vol. 17, pp. 193–198, 1969.

[361] M. E. Wrenn and H. Spitz, "Design of a continuous digital-output environmental monitor," *IEEE Transactions on Nuclear Science*, vol. NS-22, pp. 645–647, February 1975.

[362] V. C. Negro and S. Watnick, "Funghi - A radon measuring instrument with fast response," *IEEE Transactions on Nuclear Science*, vol. NS-25, pp. 757–761, February 1978.

[363] C. V. Negro, "Radometer - A portable field instrument for the rapid measurement of environmental radon and thoron," *IEEE Transactions on Nuclear Science*, vol. 37, pp. 854–858, April 1990.

[364] J. E. Sayers, F. E. Monroe, D. D. Smith, and S. A. Wallace, "A workplace air monitor for uranium particulate detection," *IEEE Transactions on Nuclear Science*, vol. 37, pp. 2165–2170, December 1990.

[365] C. Samuelsson, "Retrospective determination of radon in houses," *Nature*, pp. 338–340, July 1988.

[366] M. Urban and E. Piesch, "Low level environmental radon dosimetry with a passive track etch detector device," *Radiation and Protection Dosimetry*, vol. 1, pp. 97–109, 1981.

[367] G. Meesen, J. Uyttenhove, A. Poffijn, K. Van Laere, and Buysse, "Passive detector for measurement of the implanted ^{210}Po activity in glass," *IEEE Transactions on Nuclear Science*, vol. 41, pp. 960–963, August 1994.

[368] P. E. Mix, *Introduction to Nondestructive Testing: a Training guide.* New York: John Wiley & Sons, 1987.

[369] "MARINE CORPS ORDER 5104.3A. MARINE CORPS RADIATION SAFETY." (http://www.usmc.mil/directiv.nsf/bf7ed869c4398a1685256517005818da/455ef43a2d5bdc118525656100535bdc/$FILE/MCO%205104.3.pdf, accessed November 2002), August 1997.

[370] V. G. Baryshevsky, A. A. Khrutchinsky, V. I. Moroz, and M. D. Dezhurko, "Gamma-ray spectroscopic system for remote detection and monitoring of fissile materials," *IEEE Transaction on Nuclear Science*, vol. 41, pp. 971–975, August 1994.

[371] A. De Volpi, C. L. Fink, G. E. Marsh, E. A. Rhodes, and G. Stanford, "Fast-neutron hodoscope at TREAT: Method of quantitative determination of fuel dispersal," *Nuclear Technology*, vol. 56, pp. 141–188, 1982.

[372] A. De Volpi, R. J. Pecina, R. T. Daly, D. J. Travis, R. R. Stewart, and E. A. Rhodes, "Fast-neutron hodoscope at REAT: Development and operation," *Nuclear Technology*, vol. 27, pp. 449–487, November 1975.

[373] A. DeVolpi, "Applications of cineradiography to nuclear-reactor safety studies," *Review of Scientific Instruments*, vol. 55, August 1984.

[374] C. E. Dickerman, R. C. Doerner, J. P. Regis, E. A. Rhodes, G. S. Stanford, D. J. Travis, and A. De Volpi, "New radiation hodoscope developments for

arms control treaty verification," *IEEE Transactions on Nuclear Science*, vol. 39, pp. 1417–1421, October 1992.

[375] M. Marseguerra and E. Zio, "Monte Carlo approach to the detection of a gamma source within a scrap-iron truckload," *Nuclear Science and Engineering*, vol. 126, June 1999.

[376] A. LaMastra, "Features - advances in monitoring scrap steel for radioactivity," *Iron and Steel Engineer*, vol. 76, no. 5, pp. 48–50, 1999. Association of Iron and Steel Engineers.

[377] P. J. Campbell, "Field application experience with K9108 "buster" contraband detector," in *Proc. Cargo Inspection Technologies* (A. H. Lawrence, ed.), vol. 2276, pp. 457–462, SPIE - International Society for Optical Engineering, 1994.

[378] J. Van Wart, "A portable device for detecting visually obscured objects hidden within extended walls using Compton scattering," MScE, University of New Brunswick, 2001.

[379] J. Hubbell, "Industrial, agricultural, and medical applications of metrology: Current status and prospects for the 1990's," in *Proc. Sixth Pacific Basin Nuclear Conference* (L. Rongguang and Z. Renkai, eds.), (Beijing, China), pp. 407–412, Chinese Nuclear Society, September 1987.

[380] J. H. Hubbell, "Survey of industrial, agricultural, and medical applications of radiometric gauging and process control," *Journal of Research of the National Institute of Standards and Technology*, vol. 95, pp. 689–699, November/December 1990.

[381] B. Sellers, H. N. Ballard, and M. Izquierdo, "Direct measurement of air in the 30-60 km region by beta-ray forward scattering," *International Journal of Applied Radiation and Isotopes*, vol. 20, pp. 341–351, 1969.

[382] B. Charlton and J. M. Newton, "Application of gamma-ray attenuation to the determination of density distributions with compacted powders," *Powder Technology*, vol. 41, pp. 123–134, February 1985.

[383] S. C. Oldham, P. F. Nolan, and M. P. Mclenan, "A gamma-radiographic study of wood and polymer combustion and the effects of flame retardants," *Nuclear Instruments and Methods in Physics Research A*, vol. 299, pp. 661–665, 1990.

[384] M. Borsaru, R. J. Holmes, and P. J. Mathew, "Bulk analysis using nuclear techniques," *Applied Radiation and Isotopes*, vol. 34, pp. 397–405, 1983.

[385] N. Adil, "The measurement of concrete density by back-scattered gamma radiation," *British Journal of Non-destructive Testing*, vol. 19, pp. 72–77, 1977.

[386] S. R. Gautam, F. F. Hopkins, R. Klinkseik, and I. L. Morgan, "Compton interaction tomography I. feasibility studies for applications in earth quake engineering," *IEEE Transactions on Nuclear Science*, vol. NS30, pp. 1680–1684, 1983.

[387] J. Maskowitz, "Nondestructive on-line density measurement of organic material on a steel conveyor belt using x-ray scattering," in *Neutrons, X Rays, and Gamma Rays: Imaging Detectors, Material Characterization Techniques, and Applications* (J. M. Carpenter, D. B. Cline, R. Lanza, and D. F. Mildner, eds.), vol. 1737, pp. 121–128, SPIE - International Society for Optical Engineering, 1993.

[388] W. L. Dunn and J. E. Hutchinson, "A nuclear technique for thin-lift gauging," *International Journal of Applied Radiation and Isotopes*, vol. 33, pp. 563–567, 1982.

[389] K. Yui, Y. Shirakawa, Y. Matsuo, T. Ozeki, H. Tominaga, T. Imahashi, and N. Tachikawa, "In-stream gamma-ray backscatter gauge for sintering plant control in the steel industry," *Transactions of the American Nuclear Society*, vol. 56, no. 3, pp. 21–22, 1988.

[390] C. G. Clayton and M. R. Wormald, "Coal analysis by nuclear methods," *Applied Radiation and Isotopes*, vol. 34, pp. 3 –22, 1983.

[391] A. J. Becker, "Bremsstrahlung source for density logging - lab study of influence of beam energy," *IEEE Transaction on Nuclear Science*, vol. 41, pp. 933–937, August 1994.

[392] A. J. Becker, J. R. Boyce, G. W. Corris, and G. I. King, "Detection of scattered x-rays from an electron linac in a borehole," *Nuclear Instruments and Methods in Physics Research B*, vol. 24-25, pp. 995–998, April 1986.

[393] G. I. King, A. J. Becker, G. W. Corris, J. R. Boyce, and R. Bramblett, "Density logging using an electron linear accelerator as the x-ray source," *Nuclear Instruments and Methods in Physics Research B*, vol. 24-25, pp. 990–994, April 1986.

[394] J. R. Boyce, W. King, George, A. J. Becker, J. Doucet, and R. L. Bramblett, "An electron linac as an x-ray source for measuring geological density," *Nuclear Instruments and Methods in Physics Research A*, vol. 242, pp. 507–510, January 1986.

[395] A. L. Huddleston and J. B. Weaver, "Dual-energy Compton-scatter densitometry," *Applied Radiation and Isotopes*, vol. 34, no. 7, pp. 997–1002, 1983.

[396] F. Hensel, "Investigation of a density measurement technique using a positron radiation," *Applied Radiation and Isotopes*, vol. 53, pp. 617–624, 2000.

[397] ASTM, "B568-98: Standard Test Method for Measurement of Coating Thickness by X-Ray Spectrometry." American Society for Testing and Materials, West Conshohocken, PA, 1998.

[398] ASTM, "B567-98 Standard Test Method for Measurement of Coating Thickness by the Beta Backscatter Method." American Society for Testing and Materials, West Conshohocken, PA.

[399] K. Muskalla, T. Pfeiffer, and W. Moller, "A comparison of optical and nuclear methods for the measurement of thin metal foils," *Thin Solid Films*, vol. 76, pp. 259–267, February 1981.

[400] A. Rytz, "Alpha particles as probes for the comparison of the circular diaphragms," *Applied Radiation and Isotopes*, vol. 35, pp. 311–311, 1984.

[401] R. Lijestrand, G. Blsnpied, and G. W. Hoffmann, "Target thickness uniformity gauge," *Nuclear Instruments and Methods*, vol. 138, pp. 471–477, 1976.

[402] V. I. Pankratov, I. T. Prilipko, M. L. Gomberg, V. V. Zhikharev, and A. A. Asamenko, "Radioisotope thickness gauge based on an stream of β particles deflected by a magnetic field," *Measurement Techniques*, vol. 19, pp. 824–826, 1976.

[403] T. D. T. Latter, "Measuring coating thicknesses by the beta-backscatter technique," *Electroplating & Metal Finishing*, vol. 28, pp. 7–14, October 1975.

[404] P. Taábor, L. Molnár, and D. Nagymihåalyi, "Microcomputer based thickness gauges," *Applied Radiation and Isotopes*, vol. 41, no. 10/11, pp. 1123–1129, 1990.

[405] Y. Shirakawa, "A build-up treatment for thickness gauging of steel plates based on gamma-ray transmission," *Applied Radiation and Isotopes*, vol. 53, pp. 581–586, 2000.

[406] C. Segebade, "The non-destructive investigation on medieval iron-based weapons," *Nuclear Instruments and Methods in Physics Research A*, vol. 353, pp. 654–657, 1994.

[407] H. B. Eldridge, H. L. Pray, and B. Yee, "An x-ray wood densitometer," *International Advances of Nondestructive Testing*, vol. 15, pp. 25–26, 1990.

[408] H. Mohammadi, "Thickness gaging with scattered β, γ, and x-rays," *International Journal of Applied Radiation and Isotopes*, vol. 32, pp. 524–526, 1981.

[409] M. Kato and H. Sato, "A study of the application of backscattered gamma-rays to gauging," in *Proc. ERDA X- and Gamma-Ray Symposium*, Conf. 760639, (Ann Arbor, MI), pp. 230–233, May 1976.

[410] R. B. J. Palmer, "Nucleonic instrumentation applied to the measurement of physical parameters by means of ionizing radiation," *Journal of Physics E*, vol. 15, pp. 873–883, 1982.

[411] G. B. Cook, Mellish, and C. E. Payne, "Measurement of tin metal layers fluorescent x-ray production by radioisotopic sources," *Analytical Chemistry*, vol. 32, pp. 590–593, 1960.

[412] T. D. T. Latter, "X-ray fluorescence for coating thickness measurement," *Circuit World*, vol. 10, no. 1, pp. 20–25, 1983.

[413] B. Rosner and Y. Segal, "Thickness gauging through the ratio of x-ray fluorescence lines," *Thin Solid Films*, vol. 27, pp. 141–147, May 1975.

[414] H. Kerkow, G. Kryesch, and R. Stolle, "Measurement of the thickness of thin layers by proton induced x-ray emission," *Thin Solid Films*, vol. 62, pp. 73–81, 1979.

[415] D. K. Kaush, S. P. Singh, C. Bhan, S. K. Chattopadhyaya, and N. Nath, "X-ray fluorescence to determine the thickness of single, double, and triple-layered films of Cu, Bi and Au on Mylar substrates," *Thin Solid Films*, vol. 67, pp. 353–356, 1980.

[416] G. Luzzi, A. Mazzel, A. Neri, M. Salmi, and G. Schirripa Spangnolo, "Measurement of film thickness by means of a simple non-destructive radioisotope technique," *Thin Solid Films*, vol. 68, pp. 347–351, 1980.

[417] S. Abdul-Majid, "Applications of neutron back-diffusion technique for wall thickness and scale measurements at desalination or chemical plants," *Desalination*, vol. 94, pp. 101–107, Sep 1993.

[418] P. Flormann and A. Shaw, "Minimization of composition sensitivity with a Strontium 90/Curium 244 (x-ray) thickness gauge in the measuring range 0 - 2 mm," *Applied Radiation and Isotopes*, vol. 44, no. 10/11, pp. 1255–1261, 1993.

[419] S. J. Outhwaite, "Dynamic alloy correction x-ray thickness gauge for measurement of aluminium sheet," *Applied Radiation and Isotopes*, vol. 44, no. 10/11, pp. 1263–1267, 1993.

[420] J. J. Allport, N. L. Browser, and R. A. Kramer, "Backscatter/transmission x-ray thickness gauge," *NDT International*, vol. 20, no. 4, pp. 217–223, 1987.

[421] D. V. Ellis, "Some insights on neutron measurements," *IEEE Transactions on Nuclear Science*, vol. 37, pp. 959–965, April 1990.

[422] M. Oraby, K. Verghese, and R. P. Gardner, "Investigation of an improved-sensitivity neutron-porosity oil-well logging tool," *Nuclear Instruments and Methods in Physics Research A*, vol. 299, pp. 674–681, 1990.

[423] J. D. Ferry, *Viscoeastic Properties of Polymers.* New York: John Wiley & Sons, Inc., 1961.

[424] K. Venkateswaran, K. L. Cheng, and Y. C. Jean, "Application of positron annihilation to study the surface properties of porous resins," *Journal of Physical Chemistry*, vol. 88, pp. 2465–2469, 1984.

[425] N. N. Barthakur, "The β-ray gauge as a leaf surface wetness detector," *International Journal of Applied Radiation and Isotopes*, vol. 34, pp. 1549–1552, 1983.

[426] T. Y. Bong, N. E. Wijeysundera, E. L. Saw, and K. O. Lau, "Comparison of beta and gamma-ray transmission methods for measurement of frost density distribution," *Experimental Heat Transfer*, vol. 4, pp. 567–576, 1991.

[427] K. Yoshioka, "Soil moisture gauge using terrestrial gamma-rays," *Nuclear Geophysics*, vol. 3, no. 4, pp. 397–401, 1989.

[428] D. R. Nielsen and D. K. Cassel, "Soil water management," in *Isotopes and Radiation in Agricultural Sciences* (M. F. L'Annunziata and J. O. Legg, eds.), vol. 1, Academic Press, 1984.

[429] E. Elias and T. Gozani, "Accurate hydrogen determination with a dual nuclear gauge," *Nuclear Instruments and Methods in Physics Research*, vol. 353, pp. 691–694, December 1994.

[430] H. W. Bonin and C. J. Thorp, "Design of a neutron gauge for the detection of water ingression in flat roofs," *Nuclear Technology*, vol. 95, pp. 337–348, September. 1991.

[431] P. J. Mathew, C. Ceravolo, P. Huppert, and J. G. Miles, "Continuous neutron level gauge," *Applied Radiation and Isotopes*, vol. 34, pp. 1377–1382, Sep 1983.

[432] D. Holdlin, F. Johansen, and V. Verbinski, "A moisture probe based upon neutron moderation for in situ use in a cone penetrometer system," *IEEE Transactions on Nuclear Science*, vol. 44, pp. 799–803, June 1997.

[433] R. B. Williams, "The applications of moisture measurement in materials used in iron making by neutron moderation," *Journal of Radioanalytical Chemistry*, vol. 48, pp. 49–63, 1979.

[434] J. S. Hewitt and V. M. Slobodian, "Measurement of consistency of pulpwood water slurry based on neutron slowing-down and diffusion," in *Practical Applications of Neutron Radiography and Gaging* (H. Berger, ed.), ASTM STP 586, pp. 292–302, American Society for Testing and Materials, 1976.

[435] M. Kodama and S. Kudo, "Response of atmospheric neutron fluxes to soil moisture content," *Scientific Papers of the Institute of Physical and Chemical Research (Japan)*, vol. 79, pp. 99–103, Sep 1985.

[436] L. West, R. J. Couvillion, R. Pedersen, B. Gartman, and J. Brantley, "Design of a gamma attenuation system to measure moisture content distribution neara heat source," *Nuclear Geophysics*, vol. 3, no. 4, pp. 387–396, 1989.

[437] C. G. Gurr, "Use of gamma rays in measuring water content and permeability in unsaturated columns of soil," *Soil Science*, vol. 94, pp. 224–229, 1962.

[438] B. D. Soane, "Dual energy gamma-ray transmission for coincident measurement of water content and dry bulk density of soil," *Nature*, vol. 214, pp. 1273–1274, June 1967.

[439] P. L. Reeder, D. C. Stromswold, R. L. Brodzinski, J. H. Reeves, and W. E. Wilson, "Moisture measurement for radioactive wastes using neutron activation of copper," *Applied Radiation and Isotopes*, vol. 48, pp. 247–256, February 1997.

[440] A. M. C. Chan and S. Banerjee, "Design aspects of gamma densitometers for void fraction measurements in small scale two-phase flows," *Nuclear Instruments and Methods in Physics Research*, vol. 190, pp. 135–148, November 1981.

[441] A. M. C. Chan, "Single-beam multi-detector gamma densitometer for void fraction and phase distribution measurements in transient two-phase flows," in *Measuring Techniques in Gas-Liquid Two-Phase Flows* (J. M. Delhaye and G. Cognet, eds.), pp. 281–304, Berlin: Springer-Verlag, 1984.

[442] T. J. Kennett, W. V. Prestwich, and A. Robertson, "Dynamic density measurement by high energy photon scattering," *International Journal of Applied Radiation and Isotopes*, pp. 529–532, 1976.

[443] A. Gayer, S. Bukshpan, and D. Kedem, "In situ density measurement in aqueous solutions by the gamma backscattering method," *Nuclear Instruments and Methods*, vol. 192, pp. 619–621, 1982.

[444] D. D. S. Liu, D. J. Patmore, T. S. Yuyitung, J. J. Lipsett, and K. Chapman, "Gamma-ray method for measuring liquid expansion coefficient at high temperatures and pressures," *The Canadian Journal of Chemical Engineering*, vol. 66, pp. 521–525, June 1988.

[445] P. Johnson, "Measurement of flow using radioactive tracers," in *Radioisotope Techniques for Problem-Solving in Industrial Process Plants* (J. S. Charlton, ed.), ch. 8, Glasgow: Leonard Hill, 1986.

[446] P. L. Airey, C. G. E., A. Davison, G. F. Easey, and A. W. Morley, "An evaluation of tracer dilution techniques for gauging of rivers in flood," *Journal of Hydrology*, vol. 74, pp. 105–118, 1984.

[447] P. L. Airey and J. S. Charlton, "Radiotracer applications in Australia," *Transactions of the American Nuclear Society*, vol. 70 (Suppl. 1), pp. 331–334, 1994.

[448] M. Perez-Griffo, R. C. Block, and R. T. Lahey, "Measurement of flow in large pipes by the pulsed neutron activation method," *Nuclear Science and Engineering*, vol. 82, pp. 19–33, 1982.

[449] H. A. Larson, C. C. Price, R. N. Curran, and J. I. Sackett, "Flow measurement in sodium and water using pulsed-neutron activation: Part 1, Theory," *Nuclear Technology*, vol. 57, pp. 264–271, May 1982.

[450] C. C. Price, H. A. Larson, R. N. Curran, and J. I. Sackett, "Flow measurement in sodium and water using pulsed-neutron activation: Part 2, Experiment," *Nuclear Technology*, vol. 57, pp. 272–284, May 1982.

[451] P. Kehler, "Pulsed neutron measurement of single- and two-phase liquid flow," *IEEE Transactions on Nuclear Science*, vol. NS-26, pp. 1627–1631, February 1979.

[452] P. Kehler, "Feasibility of using PNA techniques for in-situ calibration of UPTF two-phase flow instrumentation," Tech. Rep. ANL-CT-81-18; NUREG/CR-1971, Argonne National Laboratory, February 1981.

[453] G. E. Rochau, D. R. Hornsby, J. F. Mareda, and W. C. Riggan, "A pulsed neutron generator for mass flow measurement using the pulsed neutron activation technique," *IEEE Transactions on Nuclear Science*, vol. NS-28, pp. 165–1660, April 1981.

[454] D. C. McKeon, H. D. Scott, J.-R. Olesen, G. L. Patton, and R. Mitchell, "Improved oxygen-activation method for determining water flow behind casing," *SPE Formation Evaluation*, vol. 6, pp. 334–342, Sep 1991.

[455] T. F. Lin, R. C. Block, O. C. Jones, R. T. Lahey, and M. Murase, "Measurement of flow in a horizontal pipe using the pulsed photon activation technique," *Nuclear Science and Engineering*, vol. 91, pp. 235–247, 1985.

[456] H. Matsuda, M. Fukaya, and S. Mintao, "Feasibility study on a cosmic-ray level gauge," *Nuclear Geophysics*, vol. 3, no. 4, pp. 403–406, 1989.

[457] P. Dickstein and A. Notea, "Resolving power of radioactive solution level gauging," *Nuclear Instruments and Methods in Physics Research A*, vol. 252, pp. 95–100, November 1986.

[458] E. W. Okyere, A. J. Baratta, and W. A. Jester, "Response of ex-core neutron detectors to large- and small-break loss-of-coolant accidents in pressurized water reactors," *Nuclear Technology*, vol. 96, pp. 272–289, December 1991.

[459] J. P. Adams and V. T. Berta, "Monitoring reactor vessel liquid level with a vertical string of selfpowered neutron detectors," *Nuclear Science and Engineering*, vol. 88, pp. 367–375, November 1984.

[460] F. Reisch, "Meeting the need for unambiguous PWR coolant level measurement," *Nuclear Engineering International*, vol. 29, pp. 34–38, January 1984.

[461] C. H. Neuschaefer, "Reactor vessel level monitoring system, an aid to the operators in assessing an approach to inadequate core cooling," *IEEE Transactions on Nuclear Science*, vol. NS-29, pp. 669–673, February 1981.

[462] E. Elias, Y. Segal, and A. Notea, "Gamma gauge for the control of interzone layer in an extraction tower," *Nuclear Technology*, vol. 21, pp. 57–66, January 1974.

[463] R. Roper, "Leakage detection," in *Radioisotope Techniques for Problem-Solving in Industrial Process Plants* (J. S. Charlton, ed.), ch. 10, Glasgow: Leonard Hill, 1986.

[464] D. A. W. Bossus and R. Van Sluijs, "Leak detection in a nitric acid plant using Argon-41," *Applied Radiation and Isotopes*, vol. 41, no. 10-11, pp. 1103–1110, 1990.

[465] J. H. Beynon and M. A. Nicholson, "A radioactive ionization gauge and its application to the measurement of latent heat of vaporization," *Journal of Scientific Instruments*, vol. 33, pp. 376–380, 1956.

[466] R. P. Gardner, K. Verghese, and J. I. Cehn, "Mathematical model for the measurement of gas properties with alpha-particle gauges," *Nuclear Technology*, vol. 16, pp. 418–429, November 1972.

[467] S. Matsumoto and H. Kobayashi, "Humidity measurement using a double ionization chamber with α-rays," *Nuclear Instruments and Methods*, vol. 184, pp. 603–604, 1981.

[468] J. A. Holt and J. F. Easey, "Numbers and biomass of mound-building termites (isoptera) in a semi-arid tropical woodland near Charters Towers, North Queensland, Australia," *Social Biology*, vol. 2, no. 3, pp. 281–286, 1993.

[469] J. E. Martin, T. Stevens, G. E. Bellhen, L. W. King, and J. M. Hylko, "Carbon-14 tracer study of polyacrylate polymer in a waste water plant," *Applied Radiation and Isotopes*, vol. 41, no. 10-11, pp. 1165–1172, 1990.

[470] I. Villanueva, P. Heresi, P. Henriquez, and P. Vega, "Tracer techniques in the study of flow patterns in leaching tanks," *Applied Radiation and Isotopes*, vol. 41, no. 10/11, pp. 1159–1163, 1990.

[471] A. Das and N. Changdar, "Tracer diffusion of studies in the system phosphoric acid-disodium hydrogen phosphate-water by a radioactive method," *Applied Radiation and Isotopes*, vol. 45, no. 5, pp. 773–778, 1995.

[472] H. C. Perkins, M. Yusuf, and G. Leppert, "A void measurement technique for local boiling," *Nuclear Science and Engineering*, vol. 11, pp. 304–311, 1961.

[473] G. F. Hewitt, *Measurement of Two Phase Flow Parameters.* London: Academic Press, 1978.

[474] T. R. Heidrick, J. R. Saltvold, and S. Banerjee, "Application of a three-beam gamma densitometer to two-phase flow design and density measurements," in *AIChE Symposium Series*, vol. 73, pp. 248–255, American Institute of Chemical Engineers, 1975.

[475] Z. L. Miropolsky and R. I. Shneyerova, "Application of x-rays, excited by β-sources, to study hydrodynamics of two-phase media," *International Journal of Heat and Mass Transfer*, vol. 5, pp. 723–728, 1962.

[476] N. N. Kondic and G. D. Lassahn, "Nonintrusive density distribution measurement in dynamic high-temperature systems," *ISA Transactions*, vol. 19, no. 2, pp. 11–20, 1980.

[477] P. Han and U. Wesser, "Estimating bubble size, bubble velocity and void fraction of two-phase flow with x-ray scattering," in *Gas-Liquid Flows, Proc. Fluids Engineering Conference, June 1993, Washington,DC*, vol. FED-165, pp. 39–43, American Society of Mechanical Engineers, Fluids Engineering Division, New York, 1993.

[478] E. Åbbro and G. A. Johansen, "Improved void fraction determination by means of multibeam gamma-ray attenuation measurements," *Flow Measurement and Instrumentation*, vol. 10, pp. 99–108, 1999.

[479] S. Z. Rouhani, "Void measurement by the (γ,n) reaction," *Nuclear Science and Engineering*, vol. 14, pp. 414–419, 1962.

[480] G. Woiceshyn, P. Yuen, J. Manzano, W. Martin, and J. Montin, "Steam quality measurements by neutron transmission," in *Proc. Third Int. Conf. on Heavy Crude and Tar Sands, Long Beach, CA*, vol. 3, (New York, NY), pp. 1197–1207, UNITAR/UNDP Information Cent for Heavy Crude & Tar Sands, 1985.

[481] A. A. Harms, S. Lo, and W. T. Hancox, "Measurement of time-averaged voids by neutron diagnosis," *Journal of Applied Physics*, vol. 42, pp. 4080–4082, 1971.

[482] J. Oyeldele and A. A. Harms, "Measurement of void frequency distributions and void profiles in nucleate pool boiling using neutron beams," *Nuclear Instruments and Methods*, vol. 167, pp. 255–259, 1979.

[483] M. H. Younis, T. W. Hoffmann, and A. A. Harms, "Neutron diagnosis of two-phase flow," *Nuclear Instruments and Methods*, vol. 187, pp. 489–497, 1981.

[484] F. V. Frazzoli and A. Magrini, "Neutron gauge for measurement of high void fraction in water-steam mixture," *Nuclear Technology*, vol. 45, pp. 177–182, 1979.

[485] W. T. Sha and C. F. Bonilla, "Out-of-pile steam fraction determination by neutron-beam attenuation," *Nuclear Applications*, vol. 1, pp. 69–75, 1965.

[486] E. J. Waller and E. M. A. Hussein, "A portable device for void fraction measurement in a small-diameter pipe," *Nuclear Instruments and Methods in Physics Research A*, vol. 299, pp. 670–673, 1990.

[487] F. Frazzoli, A. Magrini, and C. Mancinic, "Void fraction measurement of water-steam mixture by means of a californium-252 source," *Applied Radiation and Isotopes*, vol. 29, pp. 311–314, 1978.

[488] E. M. A. Hussein and E. J. Waller, "A neutron steam quality meter for a fluidized bed plant," *Applied Radiation and Isotopes*, vol. 41, pp. 1049–1055, 1990.

[489] P. Han, E. M. A. Hussein, P. J. Ingham, and R. M. Henschell, "Nonintrusive measurement of transient flow boiling in rod-bundle channels using fast-neutron scattering," *Nuclear Instruments and Methods in Physics Research A*, vol. 353, pp. 695–698, 1994.

[490] J. R. Buell, D. P. Byskal, P. J. Ingham, R. S. Swartz, and E. M. Hussein, "A neutron scatterometer for void-fraction measurement in rod-bundle channels: From a concept to a practicable on-line system," *Transactions of American Nuclear Society*, vol. 86, pp. 380–381, June 2002.

[491] A. C. Butterworth and D. R. Weaver, "Void distribution estimation using a neutron scattering gauge," *Nuclear Instruments and Methods in Physics Research A*, vol. 336, pp. 278–284, 1993.

[492] A. Khachadour, "Interfacial area measurement using a radioisotopic technique," M. Eng., McMaster University, Hamilton, ON, Canada, 1979.

[493] D. I. Atkinson, M. Berard, and G. Segeral, "Qualification of a nonintrusive multiphase flow meter in viscous flows," in *Proceedings 2000 SPE Annual Technical Conference and Exhibition - Production Operations and Engineering General, Dallas, TX*, pp. 533–544, Society of Petroleum Engineers, 2000.

[494] G. J. Roach, J. S. Watt, H. W. Zastawny, P. E. Hartley, and W. K. Ellis, "Multiphase flowmeter for oil, water and gas in pipelines based on gamma-ray transmission techniques," *Nuclear Geophysics*, vol. 8, no. 3, pp. 225–242, 1994.

[495] P. E. Hartley, G. J. Roach, D. Stewart, J. S. Watt, H. W. Zastawny, and W. K. Ellis, "Trial of a gamma-ray multiphase flow meter on the West Kingfish oil platform," *Nuclear Geophysics*, vol. 9, pp. 533–552, December 1995.

[496] C. M. Bishop and G. D. James, "Analysis of multiphase flows using dual-energy gamma densitometry and neural networks," *Nuclear Instruments and Methods in Physics Research A*, vol. 327, pp. 580–593, April 1993.

[497] G. A. Johansen and P. Jackson, "Salinity independent measurement of gas volume fraction in oil/gas/water pipe flows," *Applied Radiation and Isotopes*, vol. 53, pp. 595–601, 2000.

[498] G. Chalier, V. Giry, K. Madaoui, S. Sakthikumar, and P. Maquignon, "Three phase oil relative permeability determination as a key point in the evaluation of a tertiary gas gravity drainage project," in *Proc. 1995 SPE Annual Technical Conference and Exhibition Dallas, TX*, pp. 607–615, Society of Petroleum Engineers, 1995.

[499] M. Oostrom and R. J. Lenhard, "Comparison of relative permeability-saturation-pressure parametric models for infiltration and redistribution of a light nonaqueous-phase liquid in sandy porous media," *Advances in Water Resources*, vol. 21, pp. 145–157, March 1998.

[500] M. Oostrom, C. Hofstee, and J. H. Dane, "Light nonaqueous-phase liquid movement in a variably saturated sand," *Soil Science Society of America Journal*, vol. 61, pp. 1547–1554, November/December 1997.

[501] C. Hofstee, R. C. Walker, and J. H. Dane, "Infiltration and redistribution of perchloroethylene in stratified water-saturated porous media," *Soil Science Society of America Journal*, pp. 13–22, January/February 1998.

[502] G. Barth, T. Illangasekare, H. Rajaram, and H. Ruan, "Model calibration and verification for entrapped NAPL using tracer tests in a large, two-dimensional tank with heterogeneous packing," in *Proc. of the ModelCARE'96 Conference, Golden, CO*, vol. 237, pp. 169–178, International Association of Hydrological Sciences, 1996.

[503] A. M. Scheers and W. F. J. Slijkerman, "Multiphase flow measurement using multiple energy gamma ray absorption (MEGRA) composition measurement," in *Proc. 1996 SPE Annual Technical Conference and Exhibition. Pi, Denver, CO*, vol. Pi, pp. 203–211, Society of Petroleum Engineers, 1996.

[504] R. D. Luggar, M. J. Key, E. J. Morton, and W. B. Gilboy, "Energy dispersive x-ray scatter of measurement of oil/water ratios," *Nuclear Instruments and Methods in Physics Research A*, vol. 422, pp. 938–941, 1999.

[505] E. M. A. Hussein and P. Han, "Phase volume-fraction measurement in oil-water-gas using fast neutrons," *Nuclear Geophysics*, vol. 9, no. 3, pp. 229–234, 1995.

[506] V. T. Dubinchuk, A. Plata-Bedmar, and K. Fröehlich, "Nuclear techniques for investigating migration of pollutants in groundwater," *International Atomic Energy Agency Bulletin*, vol. 32, no. 4, pp. 16–21, 1990.

[507] W. C. Sidle, "Environmental isotopes for resolution of hydrology problems," *Environmental Monitoring and Assessment*, vol. 52, pp. 389–410, Sep 1998.

[508] USGS, "Isotope interest group home page." U.S. Geological Survey. (http://wwwrcamnl.wr.usgs.gov/isoig/, accessed June 2002).

[509] P. I. Chalov and Tikhonov, "Application of the natural fractionation of ^{234}U and ^{238}U in studying underground-water dynamics under active water exchange conditions," *Nuclear Geophysics*, vol. 4, no. 1, pp. 1–15, 1990.

[510] C. M. Bethke, T. Torgersen, and J. Park, "Age of very old groundwater: Insights from reactive transport models," *Journal of Geochemical Exploration*, vol. 69, pp. 1–4, June 2000.

[511] I. Clark and P. Fritz, *Environmental Isotopes in Hydrogeology.* Boca Raton: CRC Press, 1997.

[512] P. Viitanen, "Tracer experiments in a research tunnel of a repository for low- and medium-active nuclear waste," *Nuclear Instruments and Methods in Physics Research A*, vol. 353, pp. 546–549, 1994.

[513] G. J. Nimz, "Lithogenic and cosmogenic tracers in catchment hydrology," in *Isotope Tracers in Catchment Hydrology* (C. Kendall and J. J. McDonnell, eds.), pp. 247–290, Elsevier, 1998.

[514] E. Caldwell and D. Snyder, "Periodic table - beryllium." U.S. Geological Survey. (http://wwwrcamnl.wr.usgs.gov/isoig/period/be_inig.html, accessed June 2002).

[515] "Radiocarbon web-info." (http://www.c14dating.com, accessed June 2002).

[516] D. E. Granger and A. L. Smith, "Dating buried sediments using radioactive decay and muogenic production of ^{26}Al and ^{10}Be," *Nuclear Instruments and Methods in Physics Research B*, vol. 172, pp. 822–826, October 2000.

[517] M. J. Aitken, *Thermoluminescence dating.* London: Academic Press, 1985.

[518] Quaternary, "A guide to TL date measurement." Quaternary TL Surveys. (http://www.users.globalnet.co.uk/~qtls/, accessed November 2001).

[519] T. Čechák, J. Gerndt, M. Kubelik, L. Musilek, and M. Pavlik, "Radiation methods in research of ancient monuments," *Applied Radiation and Isotope*, vol. 53, pp. 565–570, July 2000.

[520] M. S. Abdel-Wahab, S. A. El-Fiki, M. A. EL-Fiki, M. Gomma, S. Abdel-Karim, and N. El-Faramawy, "Annual dose measurement and TL-dating of ancient Egyptian pottery," *Radiation Physics and Chemistry*, vol. 47, no. 5, pp. 697–700, 1996.

[521] M. Fattahi and S. Stokes, "Extending the time range of luminescence dating using red TL (RTL) from volcanic quartz," *Radiation Measurements*, vol. 32, pp. 479–485, Sep 2000.

[522] V. Mejdahl, "Thermoluminescence dating of Late-Glacial sand sediments," *Nuclear Tracks and Radiation Measurements*, vol. 18, no. 1-2, pp. 71–75, 1991.

[523] V. Dynybyl, J. Gerndt, J. Kanaval, M. Kubelik, L. Musilek, M. Pavik, and J. Zyma, "Thermoluminescence reader for dating of brick structures," *Radiation Physics and Chemistry*, vol. 51, no. 4-6, pp. 715–716, 1998.

[524] K. Kanosue, H. Toda, M. Hirai, H. Kanamori, and M. Ikeya, "Thermoluminescence (TL) and ESR study of γ-irradiated SO_2 frost for future dating in outer planets," *Radiation Measurements*, vol. 27, pp. 399–403, April 1997.

[525] M. Sharifzadeh and M. Sohrabpour, "Investigation of the irradiation history of the Iranian dates and pistachio nuts using thermoluminescence technique," *Radiation Physics and Chemistry*, vol. 42, no. n 1-3 pt 1, pp. 407–411, 1993.

[526] M. Martini and E. Sibilia, "Radiation in archaeometry: archaeological dating," *Radiation Physics and Chemistry*, vol. 6, pp. 241–246, June 2001.

[527] H. Sato, T. Takatsuji, J. Takada, *et al.*, "Measuring the external exposure dose in the contaminated area near the Chernobyl Nuclear Power Station using the thermoluminescence of quartz in bricks," *Heatlth Physics*, vol. 83, no. 2, pp. 227–236, 2002.

[528] R. L. Paul and Lindstrom, "Prompt gamma-ray activation analysis: Fundamentals and applications," *Journal of Radioianalytical and Nuclear Chemistry*, vol. 243, no. 1, pp. 181–189, 2000.

[529] C. Yonezawa, "Prompt gamma neutron activation analysis with reactor neutrons," in *Non-destructive Elemental Analysis* (Z. B. Alfassi, ed.), ch. 2, Oxford: Blackwell Science, 2001.

[530] G. L. Molnár, Z. Révay, T. Belgya, and R. B. Firestone, "The new gamma-ray catalogue for PGAA," *Applied Radiation and Isotopes*, vol. 53, pp. 527–533, 2000.

[531] J. H. Marshall III and J. F. Zumberge, "On-line measurements of bulk coal using prompt gamma neutron activation analysis," *Nuclear Geophysics*, vol. 3, no. 4, pp. 455–459, 1989.

[532] M. R. Wormald, "A bulk materials analyser using pair Compton-suppressed gamma-ray spectrometry," *Nuclear Geophysics*, vol. 3, no. 4, pp. 461–466, 1989.

[533] W. Herzog, "Prompt gamma neutron activation analysis of hard coal, raw lignite and a raw glass mixture," *Nuclear Geophysics*, vol. 3, no. 4, pp. 467–473, 1989.

[534] M. Borsaru and Z. Jecny, "Application of PGNAA for bulk coal samples in a 4π geometry," *Applied Radiation and Isotope*, vol. 54, no. 3, pp. 519–526, 2001.

[535] M. Borsaru, M. Biggs, W. Nichols, and F. Bos, "The application of prompt-gamma neutron activation analysis to borehole logging for coal," *Applied radiation and isotopes*, vol. 54, no. 2, pp. 335–343, 2001.

[536] I. R. Supernaw and A. P. Richer, "A nuclear salt-in-crude monitor," *IEEE Transactions on Nuclear Science*, vol. 36, pp. 1193–1199, February 1989.

[537] J. Charbucinski, J. A. Aylmer, P. L. Eisler, and M. Borsaru, "Quantitative and qualitative applications of the neutron-gamma borehole logging," *Nuclear Geophysics*, vol. 3, no. 4, pp. 475–485, 1989.

[538] J. L. Mikesell, F. E. Senfle, R. N. Anderson, and M. Greenberg, "Elemental concentration in diabase determined by high-resolution borehole gamma-ray spectrometry," *Nuclear Geophysics*, vol. 3, no. 3, pp. 501–519, 1989.

[539] A. Eissa, N. B. Rofail, R. A. Ali, and A. M. Hassan, "Investigation of an Egyptian phosphate ore sample by neutron activation analysis technique," *Radiation Physics and Chemistry*, vol. 47, no. 5, pp. 705–708, 1996.

[540] J.-H. Chao and C. Chung, "In situ elemental measurements in an environmental water body by prompt gamma-ray spectrometry," *Nuclear Geophysics*, vol. 7, no. 3, pp. 469–478, 1993.

[541] J.-H. Chao and C. Chung, "Optimization of in situ prompt gamma-ray analysis of lake water using HPGe-^{252}Cf probe," *Nuclear Instruments and Methods in Physics Research A*, vol. A299, pp. 651–655, 1990.

[542] L. G. Evans, J. R. Lapides, J. I. Trombka, and D. H. Jensen, "In situ elemental analysis using neutron-capture gamma-ray spectroscopy," *Nuclear Instruments and Methods in Physics Research*, vol. 193, pp. 353–357, February/March 1981.

[543] B. W. Thomas, C. G. Clayton, V. V. C. Ranasinghe, and I. M. Blair, "Mineral exploration of the sea bed by towed sea bed spectrometers," *International Journal of Applied Radiation and Isotopes*, vol. 34, no. 1, pp. 437–449, 1983.

[544] T. Gozani, P. Ryge, P. Shea, C. Seher, and R. E. Morgado, "Explosive detection system based on thermal neutron activation," *IEEE Aerospace and Electronic Systems Magazine*, vol. 4, pp. 17–20, December. 1989.

[545] P. Shea, T. Gozani, and H. Bozorgmanesh, "A TNA explosive-detection system in airline baggage," *Nuclear Instruments and Methods in Physics Research A*, vol. 299, pp. 444–448, 1990.

[546] T. Cousins, T. A. Jones, J. R. Brisson, J. E. McFee, T. J. Jamieson, E. J. Waller, F. J. LeMay, H. Ing, E. T. H. Clifford, and E. B. Selkirk, "The development of a thermal neutron activation (TNA) system as a confirmatory non-metallic land mine detector," *Journal of Radioanalytical and Nuclear Chemistry*, vol. 235, no. 1-2, pp. 53–58, 1998.

[547] H.-H. Hsu and K. J. Kearfott, "Effects of neutron source selection on landmine detection efficiency," *Nuclear Instruments and Methods in Physics Research A*, vol. 422, pp. 914–917, 1999.

[548] J. Bendahan and T. Gozani, "Mobile TNA system to detect explosives and drugs concealed in cars and trucks," in *Proc. 1998 Enforcement and Security Technologies*, vol. 3575, (Boston, MA), pp. 363–367, SPIE - International Society for Optical Engineering, 1999.

[549] W. C. Lee, D. B. Mahood, P. Ryge, P. Shea, and T. Gozani, "Thermal neutron analysis (TNA) explosive detection based on electronic neutron generators," *Nuclear Instruments and Methods in Physics Research B*, vol. 99, pp. 739–742, 1995.

[550] R. A. M. Rizk, "A nuclear on-line sensor for continuous control of vanadium content in oil pipelines," *Nuclear Geophysics*, vol. 3, no. 4, pp. 309–312, 1989.

[551] J. S. Schweitzer and C. A. Peterson, "Elemental logging with a germanium spectrometer in the Continental Deep Drilling Project," *IEEE Transaction on Nuclear Science*, vol. 40, no. 4, pp. 920–923, 1992.

[552] M. Nomai, D. L. Hollis, and E. R. Van Artsdalen, "Neutron activation analysis with low neutron flux," *IEEE Transactions on Nuclear Science*, vol. NS-33, no. 5, pp. 1150–1153, 1986.

[553] R. P. Gardner, P. Guo, Y. Y. Wang, A. Sood, S. H. Lee, and C. L. Dobbs, "Feasibility of neutron activation methods for measurement of sodium and aluminum in green liquor," *Applied Radiation and Isotopes*, vol. 48, no. 10-12, pp. 355–1372, 1997.

[554] B. D. Sowerby, "Measurement of specific energy, ash and moisture in bulk coal samples by a combined neutron and gamma-ray method," *Nuclear Instruments and Methods*, vol. 160, pp. 173–182, 1979.

[555] C. Lim and B. Sowerby, "Neutron techniques of on-line analysis in the Australian mineral industry," in *Proc. International Conference on Neutrons and Their Applications* (G. Vourvopoulos and T. Paradellis, eds.), vol. 2339, pp. 419–423, 1994.

[556] W. D. James, "Activation analysis of coal and coal effluents," in *Activation Analysis* (Z. B. Alfassi, ed.), vol. II, ch. 8, Boca Raton: CRC Presss, 1990.

[557] L. Wielopolski, I. Oriorr, G. Hendrey, and H. Roger, "Soil carbon measurements using inelastic neutron scattering," *IEEE Transactions on Nuclear Science*, vol. 47, pp. 914–917, June 2000.

[558] B. D. Sowerby, "Elemental analysis by neutron inelastic scatter gamma rays with a radioisotope neutron source," *Nuclear Instruments and Methods*, vol. 166, pp. 571–579, 1979.

[559] R. A. Livingston, "Standarization of the neutron probe for the assessment of masonary deterioration," *Applied Radiation and Isotopes*, vol. 44, no. 10/11, pp. 1285–1300, 1993.

[560] C. S. Lim, J. R. Tickner, B. D. Sowerby, D. A. Abernethy, A. J. McEwan, Rainey, R. Stevens, C. Manias, and D. Retallack, "On-belt elemental analyser for the cement industry," *Applied Radiation and Isotopes.*, vol. 54, pp. 11–19, 2001.

[561] C. S. Lim, B. Sowerby, S. Rainey, and G. J. Hardie, "Plant trial of a neutron inelastic scattering gauge for the online analysis of hot reduced iron ores," in *Proc. International Conference on Neutrons in Research and Industry* (G. Vourvopoulos, ed.), vol. 2867, pp. 211–214, SPIE - International Society for Optical Engineering, 1977.

[562] Z. P. Sawa and T. Gozani, "PFNA technique for the detection of explosives," in *Proc. 1st Int. Conf. on Explosive Detection Technology* (S. Khan, ed.), DOT/FAA/CT-92/11, pp. 82–103, Federal Aviation Administration, Department of Transportation, May 1992.

[563] Z. Sawa, "PFN GASCA technique for detection of explosives and drugs," *Nuclear Instruments and Methods in Physics Research B*, vol. 79, pp. 593–596, 1993.

[564] C. M. Gordon and C. W. Peters, "Instrumental assay of kerogen in shale using fast neutrons," *Nuclear Geophysics*, vol. 2, no. 2, pp. 123–128, 1988.

[565] R. C. Smith, M. J. Hurwitz, and K. C. Tran, "System to detect contraband in cargo containers using fast and slow neutron irradiation and collimated gamma detectors," *Nuclear Instruments and Methods in Physics Research B*, vol. 99, pp. 733–735, 1995.

[566] E. M. A. Hussein, "Detection of explosive materials using nuclear radiation: A critical review," in *Aviation Security Problems and Related Technologies* (M. W. H., ed.), vol. CR42 of *Critical Reviews of Optical Science and Technology*, pp. 126–136, Bellingham: SPIE - International Society for Optical Engineering, 1992.

[567] T. Gozani, "Understanding the physics limitations of PFNA - the nanosecond pulsed fast neutron analysis," *Nuclear Instruments and Methods in Physics Research B*, vol. 99, pp. 743–747, 1995.

[568] D. R. Brown and T. Gozani, "Cargo inspection system based on pulsed fast neutron analysis," *Nuclear Instruments and Methods in Physics Research B*, vol. 99, pp. 753–756, 1995.

[569] D. R. Brown, T. Gozani, R. Loveman, J. Bendahan, P. Ryge, J. Stevenson, F. Liu, and M. Sivakumar, "Application of pulsed fast neutrons analysis to cargo inspection," *Nuclear Instruments and Methods in Physics Research A*, vol. 353, pp. 684–688, 1994.

[570] D. R. Brown, T. Gozani, J. Bendahan, F. Liu, R. Loveman, P. Ryge, P. Shea, M. Sivakumar, and J. Stevenson, "Pulsed fast neutron analysis for cargo inspection for drugs and terrorist threats," in *Proc. of the 1998 Enforcement and Security Technologies*, vol. 3575, (Boston, MA), pp. 342–347, 1999.

[571] P. Bach, J. L. Ma, D. Froment, and J. C. Jaureguy, "Chemical weapons detection by fast neutron activation analysis techniques," *Nuclear Instruments and Methods in Physics Research B*, vol. 79, pp. 605–610, 1993.

[572] S. Schneider, "Material characterization using nuclear sources," in *Neutrons, X Rays, and Gamma Rays: Imaging Detectors, Material Characterization*

Techniques, and Applications (J. M. Carpenter, D. B. Cline, R. Lanza, and D. F. Mildner, eds.), vol. 1737, pp. 152–159, SPIE - International Society for Optical Engineering, 1993.

[573] P. C. Womble, F. J. Schultz, and G. Vourvopoulos, "Non-destructive characterization using pulsed fast-thermal neutrons," *Nuclear Instruments and Methods in Physics Research B*, vol. 99, pp. 757–760, 1995.

[574] P. C. Womble, G. Vourvopoulos, J. Paschal, and P. Dokhale, "Multi-element analysis utilizing pulsed fast/thermal neutron analysis for contraband detection," in *Proc. 1999 Penetrating Radiation Systems and Applications*, vol. 3769, (Denver, CO), pp. 189–195, SPIE - International Society for Optical Engineering, 1999.

[575] G. Vourvopoulos and P. C. Womble, "Pulsed fast/thermal neutron analysis: A technique for explosives detection," *Talanta*, vol. 54, no. 3, pp. 459–468, 2001.

[576] P. C. Womble, F. J. Schultz, and G. Vourvopoulos, "Non-destructive characterization using pulsed fast-thermal neutrons," *Nuclear Instruments and Methods in Physics Research B*, vol. 99, pp. 757–760, 1995.

[577] P. C. Womble, G. Vourvopoulos, J. B. Howard, and J. Paschal, "Neutron-based portable drug probe," in *Application of Accelerators in Research and Industry*, vol. 2, p. 691, American Institute of Physics, 1999.

[578] G. Vourvopoulos, D. L. Humphrey, P. L. Setters, and K. Lamkin, "On-line elemental analysis of coal via high resolution PGNA," *Nuclear Instruments and Methods in Physics Research B*, vol. 40-41, pp. 853–856, 1989.

[579] G. Vourvopoulos and F. J. Schultz, "A pulsed fast-thermal neutron system for the detection of hidden explosives," *Nuclear Instruments and Methods in Physics Research B*, vol. 79, pp. 585–588, 1993.

[580] G. Vourvopoulos, L. Dep, S. Sudar, and P. C. Womble, "Neutron-generator based on-line coal analysis," *Coal science and technology*, no. 24, 1995.

[581] L. Dep and G. Vourvopoulos, "Pulsed fast and thermal neutron analysis for coal and cement industries," in *AIP conference proceedings*, vol. 392/PT2, p. 861, American Institute of Physics, 1997.

[582] A. Beyerle, J. P. Hurley, and L. Tunnell, "Design of an associated particle imaging system," *Nuclear Instruments and Methods in Physics Research A*, vol. 299, pp. 458–462, 1990.

[583] E. Rhodes, C. E. Dickerman, A. DeVolpi, and C. W. Peters, "APSTNG: radiation interrogation for verification of chemical and nuclear weapons," *IEEE Transactions on Nuclear Science*, vol. 39, pp. 1041–1045, 1992.

[584] C. M. Gordon, C. W. Peters, and T. K. Olson, "Fast-neutron diagnostic probe," *Journal of Research of the National Bureau of Standards*, vol. 93, pp. 484–486, May/June 1988.

[585] C. M. Gordon and C. W. Peters, "A fast-neutron probe for tomography and bulk analysis," *Applied radiation and isotopes*, vol. 41, no. 10-11, pp. 1111–1116, 1990.

[586] P. Rhodes and C. W. Peters, "APSTNG: neutron interrogation for detection of explosives and drugs and nuclear and CW materials," in *Neutrons, X Rays, and Gamma Rays: Imaging Detectors, Material Characterization Techniques, and Applications* (J. M. Carpenter, D. B. Cline, R. Lanza, and D. F. Mildner, eds.), vol. 1737, pp. 160–170, SPIE - International Society for Optical Engineering, 1993.

[587] A. J. Caffrey, J. D. Cole, R. J. Gehrke, and R. C. Greenwood, "Chemical warfare agent and high explosive identification by spectroscopy of neutron-induced gamma rays," *IEEE Transactions on Nuclear Science*, vol. 39, no. 5, pp. 1422 –1426, 1992.

[588] W. E. Parker, W. M. Buckley, S. A. Kreek, A. J. Caffrey, G. Mauger, A. D. Lavietes, and A. D. Dougan, "Portable system for nuclear, chemical agent and explosives identification," in *Proceedings of the 1999 Penetrating Radiation Systems and Applications*, vol. 3769, (Denver, CO), pp. 43–50, SPIE - International Society for Optical Engineering, 1999.

[589] J. L. Joness, "Detection of pulsed, bremsstrahlung-induced, prompt neutron-capture gamma-rays with a HPGe detector," in *Proc. Neutron in Research and Industry* (G. Vourvopoulos, ed.), vol. 2867, pp. 202–205, SPIE - International Society for Optical Engineering, 1997.

[590] R. W. Bild, "14 MeV neutron activation analysis of oxygen in Li-Si alloys for Li(Si)/FeS_2 thermally activated batteries," *IEEE Transactions on Nuclear Science*, vol. NS-28, pp. 1622–1625, April 1981.

[591] H. D. Scott, M. C. Pearson, S. M. Renke, D. C. McKeon, and J. P. Melsenhelder, "Applications of oxygen activation for injection and production profiling in the Kuparuk river field," *SPE Formation Evaluation*, vol. 8, pp. 103–111, June 1993.

[592] J. P. Biersack, D. Fink, R. Henkrlmann, and Múller, "The use of neutron induced reactions for light element proifiling and lattice localization," *Nuclear Instruments and Methods*, vol. 149, pp. 93–97, 1978.

[593] G. Tessler, B. R. Beaudion, W. J. Beggs, L. B. Freeman, A. Kahler, and S. W. C., "A gauge for nondestructive assay of irradiated fuel rods," *Nuclear Technology*, vol. 82, pp. 275–289, September 1988.

[594] H. O. Menlove, R. A. Forster, and D. L. Matthews, "A photoneutron antimony-124-beryllium system for fissile materials assay," *Nuclear Technology*, vol. 19, pp. 181–187, September 1973.

[595] J. A. Czubek, "Pulsed neutron method for uranium well logging," *Geophysics*, vol. 37, pp. 160–173, 1972.

[596] M. M. Givens, W. R. Mills, C. L. Dennis, and R. L. Caldwell, "Uranium assay logging using a pulsed 14 MeV neutron sources and detection of delayed fission neutrons," *Geophysics*, vol. 41, pp. 468–490, 1976.

[597] R. C. Smith, "Measured characteristic of pulsed-neutron assaying of uranium ore deposits," *IEEE Transactions on Nuclear Science*, vol. NS-26, no. 1, pp. 1579–1583, 1979.

[598] W. W. Givens and D. C. Stromswold, "Prompt fission neutron logging for uranium," *Nuclear Geophysics*, vol. 3, no. 4, pp. 299–307, 1989.

[599] F. C. Difilippo, "Applications of Monte Carlo simulations of thermalization processes to the nondestructive assay of graphite," *Nuclear Science and Engineering*, vol. 133, pp. 163–177, October 1999.

[600] B. H. Armitage, P. M. J. Chard, K. P. Croft, S. Lambert, and D. J. Lloyd, "Characterization of the Harwell model N93 versatile passive neutron monitor," *Nuclear Instruments and Methods in Physics Research A*, vol. 252, pp. 683–683, 1994.

[601] Y. D. Harker, "Performance evaluation of the passive active (PAN) radioassay system at the Idaho National Engineering Laboratory," in *Proc. Neutron in Research and Industry* (G. Vourvopoulos, ed.), vol. 2867, pp. 198–201, SPIE - International Society for Optical Engineering, 1997.

[602] P. N. Rinard, "Application guide to shufflers," Tech. Rep. LA-13819-MS, Los Alamos National Laboratory, 2001.

[603] P. M. Rinard, "Shuffler assays for uranium in mixed-oxide fuel with recycled weapons plutonium," Tech. Rep. LA-13856-MS, Los Alamos National Laboratory, 2001.

[604] P. M. Rinard, E. L. Adams, H. O. Menlove, and J. K. Sprinkle, "The nondestructive assay of 55-gallon drums containing uranium and transuranic waste using passive-active shufflers," Tech. Rep. LA-12446-MS, Los Alamos National Laboratory, 1992.

[605] P. M. Rinard, "Shuffler instruments for the nondestructive assay of fissile materials," Tech. Rep. LA-12105, Los Alamos National Laboratory, 1991.

[606] P. M. Horley, "Automated box/drum waste assay ^{252}Cf shuffler through the material access and accountability boundary," Tech. Rep. LA-UR-92-2360, Los Alamos National Laboratory, LA-UR-92-2360.

[607] V. K. Pare, "Portable subcriticality measurement system with calculational validation capability," in *Int. Conf. On Nuclear Criticality*, CONF-910993-8, 1991.

[608] J. T. Mihalczo, V. K. Pare, and T. E. Valentine, "Nuclear weapons identification system," in *Proc. 1996 IEEE Nuclear Science Symposium, November 1996, Anaheim, CA*, Part 1, pp. 44–45, IEEE, 1996.

[609] T. Uckan, M. S. Wyatt, J. T. Mihalczo, T. E. Valentine, J. Mullens, and T. F. Hannon, "^{252}Cf-source-correlated transmission measurements for uranyl fluoride deposit in a 24-in-OD process pipe," *Nuclear Instruments and Methods in Physics Research A*, vol. 422, no. 1, pp. 26–34, 1999.

[610] J. A. Williams, J. T. Mihalczo, C. W. Ricker, F. L. Glesius, and T. Kniss, "High-sensitivity position-sensitive fission chamber for subcriticality measurements of spent fuel," *Nuclear Instruments and Methods in Physics Research A*, vol. 299, pp. 187–190, December 1990.

[611] J. T. Mihalczo, E. D. Blakeman, G. E. Ragan, R. C. Kryter, H. Seino, and R. Robinson, "^{252}Cf-source-driven noise measurements of subcriticality for a slab tank of aqueous Pu-U nitrate," *Nuclear Science and Engineering*, vol. 107, pp. 35–57, January 1991.

[612] J. T. Mihalczo, E. D. Blakeman, G. E. Ragan, R. C. Kryter, H. Seino, and R. C. Robinson, "^{252}Cf-source-driven neutron noise measurements of subcriticality for an annular tank containing aqueous Pu-U nitrate," *Nuclear Technology*, vol. 94, pp. 336–360, June 1991.

[613] M. M. Chiles, J. T. Mihalczo, and C. E. Fowler, "Small, annular, double-contained ^{252}Cf fission chamber for source-driven subcriticality measurement," *IEEE Transactions on Nuclear Science*, vol. 40, pp. 816–818, August 1993.

[614] J. T. Mihalczo, E. D. Blakeman, V. K. Pare, T. E. Valentine, and D. J. Auslander, "Subcriticality of two uranyl nitrate flat cylindrical tanks spaced in air by ^{252}Cf-source-driven noise analysis," *Nuclear Technology*, vol. 103, no. 3, pp. 346–378, 1993.

[615] J. L. Munoz-Cobo, Y. Rugama, T. E. Valentine, J. T. Mihalczo, and R. B. Perez, "Subcritical reactivity monitoring in accelerator driven systems," *Annals of Nuclear Energy*, vol. 28, pp. 1519–1547, October 2001.

[616] M. Zucker, "Maximum information with minimum complexity from a coincidence assay system," *Applied Radiation and Isotopes*, vol. 41, no. 10/11, pp. 995–1001, 1990.

[617] K. Masumoto, T. Kato, and N. Suzuki, "Activation yield curves of photonuclear reactions for multielement photon activation analysis," *Nuclear Instruments and Methods*, vol. 157, pp. 567–577, 1978.

[618] L. Lakosi and Á. Veres, "Nondestructive assay of fission products in spent-fuel assemblies using gamma and photoneutron activation," *Nuclear Instruments and Methods in Physics Research A*, vol. 299, pp. 463–467, 1990.

[619] R. F. Overman and J. C. Corey, "Use of gamma-neutron reaction for determining D_2O in H_2O and in tracing soil moisture," *Transactions of American Society*, vol. 11, p. 74, 1968.

[620] J. C. Corey, R. H. Hawkins, and R. F. Overman, "Gamma-photoneutron method for laboratory studies of soil water," *Soil Science Society of America Proceedings*, vol. 34, no. 4, pp. 557–600, 1970.

[621] W. G. Winn and N. P. Baumann, "Tritiated water measurements with a $^2D(\gamma,n)^1H$ detector," *Nuclear Technology*, vol. 64, pp. 300–306, March 1984.

[622] D. L. Smith, F. Maekawa, and Y. Ikeda, "Observation of uranium photofission by ^{16}N decay gamma rays from water activated by D-T fusion neutrons," *Fusion Engineering and Design*, vol. 47, no. 4, pp. 403–424, 2000.

[623] J. R. Dherbey, A. Lyoussi, and A. Buissopn, "Measurement of fissile mass in large size containers with a transportable linear accelerator," in *Proc. Neutron in Research and Industry* (G. Vourvopoulos, ed.), vol. 2867, pp. 223–226, SPIE - International Society for Optical Engineering, 1997.

[624] T. Gozani, "Physics of nuclear materials safeguards techniques," *Nuclear Technology*, vol. 13, pp. 8–19, 1972.

[625] G. Engekmann, J. Filippi, F. Gosset, and F. Moreau, "New method to measure oxygen isotopic concentration ratios ^{18}O/^{16}O by gamma-activation analysis in biological media," *Journal of Radioanalytical Chemistry*, vol. 37, pp. 559–570, 1977.

[626] K. W. Habiger, J. R. Clifford, R. B. Miller, and W. F. McCullough, "Explosives detection with energetic photons," *Nuclear Instruments and Methods in Physics Research B*, vol. 46-57, pp. 834–838, 1991.

[627] K. Whitham, R. C. Miller, H. Anamkath, J. R. Clifford, R. B. Miller, and K. Habiger, "Linear accelerator for explosive detection," *Nuclear Instruments and Methods in Physics Research B*, vol. 56/57, pp. 825–828, 1991.

[628] T. Gozani, "Nuclear detection technology, form plutonium through coal to explosives," *Transactions of American Nuclear Society*, vol. 56, no. 3, pp. 38–40, 1988.

[629] J. P. F. Sellschop, R. J. Keddy, D. W. Mingay, M. J. Renan, and D. G. Schuster, "A non-destructive nuclear method for the determination of nitrogen in solids," *Applied Radiation and Isotopes*, vol. 26, pp. 640–647, 1975.

[630] D. J. Cherniak and W. A. Lanford, "Nuclear reaction analysis," in *Nondestructive Elemental Analysis* (Z. B. Alfassi, ed.), ch. 9, Oxford: Blackwell Science, 2001.

[631] W. B. Stannard, S. R. Walker, P. N. Johnston, and I. F. Bubb, "Carbon analysis in MOCVD grown HgCdMnTE by charged particle activation," *Nuclear Instruments and Methods in Physics Research*, vol. 353, pp. 593–596, 1994.

[632] NASA, "Mars pathfinder instrument descriptions, alpha proton x-ray spectrometer." NASA. (http://mars3.jpl.nasa.gov/mpf/sci_desc.html#APXS, accessed April 2001).

[633] R. Rieder, H. Wänke, T. Economou, and A. Turkevich, "Determination of the chemical composition of martian soil and rocks: The alpha-proton-x-ray spectrometer." (http://astro.uchicago.edu/home/web/papers/economou/pathfinder/apxs.html#chap9, accessed April 2001).

[634] T. Economou, "Chemical analyses of Martian soil and rocks obtained by the Pathfinder Alpha Proton X-ray spectrometer," *Radiation Physics and Chemistry*, vol. 61, pp. 191–197, June 2001.

[635] T. E. Economou and A. L. Turkevich, "An alpha particle instrument with alpha, proton, and x-ray modes for planetary chemical analysis," *Nuclear Instruments and Methods*, vol. 134, pp. 391–400, 1976.

[636] A. N. Singh and M. Rathnakaran, "An instrument for on-line monitoring of tritium-in-air in heavy water reactors," *Nuclear Instruments and Methods in Physics Research A*, vol. 258, pp. 250–254, 1987.

[637] A. N. Singh, M. Rathnakaran, and K. G. Vohra, "An on-line tritium in water monitor," *Nuclear Instruments and Methods in Physics Research A*, vol. 236, pp. 159–164, 1985.

[638] R. Sigg, J. E. McCarty, R. R. Livingston, and M. A. Sanders, "Real-time aqueous tritium monitor using liquid scintillation counting," *Nuclear Instruments and Methods in Science Research A*, vol. 353, pp. 494–498, 1994.

[639] M. Borsaru, C. Ceravolo, P. Waddington, and W. Gu, "A coal face ash analyser based on natural gamma-ray activity," *Nuclear Geophysics*, vol. 6, pp. 383–390, September 1992.

[640] D. B. Sowerby and J. S. Watt, "Development of nuclear techniques for on-line analysis in the coal industry," *Nuclear Instruments and Methods in Physics Research A*, vol. 229, pp. 642–647, 1990.

[641] P. Huppert, M. Borsaru, J. Charbucinski, C. Ceravolo, and P. L. Eisler, "Combined natural-gamma/gamma-gamma borehole lithology logging," *Nuclear Geophysics*, vol. 3, no. 4, pp. 381–386, 1989.

[642] M. Borsaru, R. Dixon, A. Rojc, R. Stehle, and Z. Jecny, "Coal face and stockpile ash analyser for the coal mining industry," *Applied Radiation and Isotopes*, vol. 55, no. 3, pp. 407–412, 2001.

[643] A. F. Oluwole, O. Ajayi, J. Ojo, F. A. Balogun, I. B. Obioh, J. A. Adejumo, O. J. Ogunsola, A. Adepetu, H. B. Olaniyi, and O. I. Asubjojo, "Characterization of pollutants around tin mining and smelting operations using EDXRF," *Nuclear Instruments and Methods in Physics Research A*, vol. 353, pp. 499–502, 1994.

[644] J. B. Olomo, M. K. Akinloye, and F. A. Balogun, "Distribution of gamma-emitting natural radionuclides in soils and water around nuclear-research establishments, Ile-Ife, Nigeria," *Nuclear Instruments and Methods in Physics Research A*, vol. 353, pp. 553–557, 1994.

[645] D. J. Lawrence, W. C. Feldman, B. L. Barraclough, A. B. Binder, R. C. Elphic, S. Maurice, M. C. Miller, and T. H. Prettyman, "High resolution measurements of absolute thorium abundances on the lunar surface," *Geophysical Research Letters*, vol. 26, no. 17, pp. 2681–2684, 1999.

[646] M. C. Gallardo, M. García-León, M. Mundi, and M. A. Respaldiaza, "Gamma-spectroscopic analysis of river sediment collected around phosphate fertilizer industries," *Nuclear Geophysics*, vol. 7, no. 3, pp. 461–468, 1993.

[647] J. A. Kirby and R. P. Lindquist, "Development of marijuana and tobacco detectors using potassium-40 gamma ray emissions," in *Proc. Cargo Inspection*

Technologies (A. H. Lawrence, ed.), vol. 2276, pp. 374–382, SPIE - International Society for Optical Engineering, 1994.

[648] N. V. Arendtsz, "Contraband detection with natural K-40 gamma-ray emissions: Numerical analysis of experimental gamma-ray signatures," in *Physics-Proc. Based Technologies for the Detection of Contraband* (L. O. Malotky and J. J. Pennella, eds.), pp. 63–74, 1997.

[649] B. D. Keele, R. G. Bauer, G. R. Blewett, and G. L. Troyer, "In situ characterization to support a test excavation at a liquid waste disposal site," *Nuclear Instruments and Methods in Physics Research A*, vol. 353, pp. 512–514, 1994.

[650] O. J. Yaroshevich, I. V. Zhuk, E. M. Lomonosova, N. N. Svetlakova, V. P. Mironov, V. P. Kudryashov, and A. V. Bushuev, "Determination of Pu and ^{241}Am in soils by instrumental methods," *Nuclear Geophysics*, vol. 9, no. 3, pp. 235–239, 1995.

[651] A. J. Schilk, M. A. Knopf, R. C. Thompson, C. W. Hubbard, K. Abel, D. Edwards, and J. R. Abraham, "Real-time in situ detection of ^{90}Sr and ^{238}U in soils via scintillating fiber-sensor technology," *Nuclear Instruments and Methods in Physics Research A*, vol. 353, pp. 477–481, 1994.

[652] R. Seymour, C. D. Hull, T. Crawford, B. Coyne, M. Bliss, and R. A. Craig, "Portal, freight and vehicle monitor performance using scintillating glass fiber detectors for the detection of plutonium in the Illicit Trafficking Radiation Assessment Program," *Journal of Radioanalytical and Nuclear Chemistry*, vol. 248, no. 3, pp. 699–705, 2001.

[653] H. O. Menlove, "Accurate plutonium waste measurements using the ^{252}Cf add-a-source technique for matrix corrections," Tech. Rep. LA-UR-92-2120, Los Alamos National Laboratory, 1992.

[654] B. H. Armitage, P. M. J. Chard, K. P. Croft, S. Lambert, and D. J. Lloyd, "A scan a source matrix correction technique for use with a reconfigurable passive plutonium assay system," *Nuclear Instruments and Methods in Physics Research*, vol. 354, pp. 522–526, 1995.

[655] R. A. August, G. W. Phillips, S. E. King, and J. H. Cutchin, "Treaty verification with "passive neutron signatures"," *Nuclear Instruments and Methods in Physics Research A*, vol. 353, pp. 712–715, 1994.

[656] W. C. Feldman, W. V. Boynton, B. M. Jakosky, and M. T. Mellon, "Redistribution of subsurface neutrons caused by ground ice on Mars," *Journal of Geophysical Research*, vol. 98, pp. 20855–20870, November 1993.

[657] W. C. Feldman, S. Maurice, A. B. Binder, B. L. Barraclough, R. Elpic, and D. J. Lawrence, "Fluxes of fast and epithermal neutrons from lunar prospector: Evidence of water ice at the lunar poles," *Science*, vol. 281, pp. 1496–1500, Dept 1998.

[658] W. C. Feldman, R. C. Reedy, and D. S. McKay, "Lunar neutron leakage fluxes as a function of composition and hydrogen content (paper 91GL02618)," *Geophysical Research Letters*, vol. 18, pp. 2157–2160, November 1991.

[659] W. C. Feldman, B. L. Barraclough, K. R. Fuller, *et al.*, "The lunar prospector gamma-ray and neutron spectrometers," *Nuclear Instruments and Methods in Physics Research A*, vol. 422, pp. 562–566, 337 1999.

[660] W. C. Feldman, W. V. Boynton, R. L. Tokar, *et al.*, "Global distribution of neutrons from Mars: results from Mars Odyssey," *Science*, vol. 297, pp. 75–78, May 2002. (www.sciencexpress.org, May 30, 2002; 10.1126/science.1073541).

[661] J. C. Overley, "Element-sensitive computed tomography with fast neutrons," *Nuclear Instruments and Methods in Physics Research B*, vol. 24-25, pp. 1058–1062, 197.

[662] T. G. Miller and W. S. H. Makky, "Application of fast neutron spectroscopy/radiography to airport security," in *Neutrons, X Rays, and Gamma Rays: Imaging Detectors, Material Characterization Techniques, and Applications* (J. M. Carpenter, D. B. Cline, R. C. Lanza, and D. F. Mildner, eds.), vol. 1737, pp. 184–196, SPIE - International Society for Optical Engineering, 1993.

[663] J. C. Overley, M. Chmelik, R. J. Rasmussen, R. M. S. Schofield, and H. W. Lefevre, "Explosives detection through fast-neutron time-of-night attenuation measurements," *Nuclear Instruments and Methods in Physics Research B*, vol. 99, pp. 728–732, 1995.

[664] C. L. Fink, T. J. Micklich, Y. T. J., H. P., L. Sagalovsky, and M. M. Martin, "Evaluation of neutron techniques for illicit substance detection," *Nuclear Instruments and Methods in Physics Research B*, vol. 99, pp. 748–752, 1995.

[665] B. J. Micklich, M. K. Harper, A. H. Novick, and D. L. Smith, "Illicit substance detection using fast-neutron transmission spectroscopy," *Nuclear Instruments and Method sin Physics Research A*, vol. 353, pp. 646–649, 1994.

[666] R. J. Rasmussen, W. S. Fanselow, H. W. Lefevre, M. Chmelik, J. C. Overley, A. P. Brown, G. E. Sieger, and R. M. S. Schofield, "Average atomic number of heterogeneous mixtures from the ratio of gamma to fast-neutron attenuation," *Nuclear Instruments and Methods in Physics Research B*, vol. 124, pp. 611–614, 1997.

[667] E. M. A. Hussein and E. J. Waller, "Direct use of ^{252}Cf for landmine detection," *Transactions of American Nuclear Society*, vol. 83, pp. 318–319, November 2000.

[668] E. M. A. Hussein, "What can an isotopic neutron source provide for landmine detection," in *First Research Co-ordination Meeting of the CRP on Application of Nuclear Techniques to Anti-Personnel Landmine Identification (F1.10.07)*, (Zagreb), International Atomic Energy Agency, November 1999. (http://www.iaea.org/programmes/ripc/physics/zagreb_crp_report.htm).

[669] B. Király, L. Olh́, and J. Csikai, "Neutron-based techniques for detection of explosives and drugs," *Radiation Physics and Chemistry*, vol. 61, pp. 781–784, June 2001.

[670] E. M. A. Hussein, P. M. Lord, and D. L. Bot, "An empirical fast-neutron technique for detection of explosive-like materials," *Nuclear Instruments and Methods in Physics Research*, vol. 299, pp. 453–457, 1990.

[671] J. W. Behrens, R. G. Johnson, and R. A. Schrack, "Neutron resonance transmission analysis of reactor fuel samples," *Nuclear Technology*, vol. 67, pp. 162–168, October 1984.

[672] H. G. Priesmeyer and U. Harz, "Isotopic content determination in irradiated fuel by neutron transmission analysis," *Atomkernenergie*, vol. 25, no. 2, pp. 109–113, 1975.

[673] A. A. Bergman, A. I. Isacoff, I. D. Murin, F. L. Shapiro, I. Shtranikh, and M. V. Cazarnovsky, "A neutron spectrometer based on measuring the slowing-down time of neutrons in lead," in *Proc. International Conf. on the Peaceful Uses of Atomic Energy, Geneva, 1955*, vol. 4, pp. 135–146, New York: United Nations, 1956.

[674] E. C. Vanterpool, R. E. Slovacek, D. R. Harris, and R. C. Block, "Fuel assembly assay by neutron interrogation in a lead slowing-down-time spectrometer," *Nuclear Science and Engineering*, vol. 110, February 1992.

[675] M. Sawan and C. R., "Neutron pulses slowing down in heavy media analysis with applications of the lead spectrometer," *Nuclear Science and Engineering*, vol. 54, pp. 127–142, 1974.

[676] D. Vartsky, G. Engler, and M. B. Goldberg, "A method for detection of explosives based on nuclear resonance absorption of gamma rays in ^{14}N," *Nuclear Instruments and Methods in Physics Research A*, vol. 348, pp. 688–691, 1994.

[677] B. D. Sowerby and W. K. Ellis, "Borehole analysis for copper and nickel using gamma-ray resonance scattering," *Nuclear Instruments and Methods*, vol. 115, pp. 511–523, 1974.

[678] A. W. Wylie, *Nuclear Assaying of Mining Borehole*, ch. 13. Amsterdam: Elsevier, 1984.

[679] B. Blanpain, P. Revesz, L. R. Doolittle, K. H. Purser, and J. W. Mayer, "The use of the 3.05 MeV oxygen resonance for ^{4}He backscattering near-surface analysis of oxygen-containing high Z compounds," *Nuclear Instruments and Methods in Physics Research B*, vol. 34, pp. 459–464, 1988.

[680] J. Räisäinen, "Charged particle-induced *gamma*-ray emission," in *Nondestructive Elemental Analysis* (Z. Alfassi, ed.), ch. 8, Oxford: Blackwell Science, 2001.

[681] A. Markwitz, J. L. Ruvallcaba, and G. Demortier, "Light element detection in heavy matrices by high energy backscattering spectroscopy," *Nuclear Instruments and Methods in Physics Research B*, vol. 122, pp. 685–688, 1997.

[682] H. J. Gomberg and B. G. Kushner, "Neutron elastic scatter (NES) for explosive detection systems (EDS)," in *Proc. 1st Int. Conf. on Explosive Detection Technology* (S. Khan, ed.), DOT/FAA/CT-92/11, pp. 123–139, Proc. 1st

Int. Conf. on Explosive Detection Technology, Department of Transportation, May 1992.

[683] F. D. Brooks, A. Buffler, M. S. Allie, K. Bharuth-Ram, M. Nchodu, and B. R. S. Simpson, "Determination of HCNO concentrations by fast neutron scattering analysis," *Nuclear Instruments and Methods in Physics Research A*, vol. 410, pp. 319–328, 1998.

[684] A. Buffler, F. D. Brooks, M. S. Allie, K. Bharuth-Ram, and M. Nchodu, "Material classification by fast neutron scattering," *Nuclear Instruments and Methods in Physics Research B*, vol. 173, pp. 483–502, 2001.

[685] J. Csikai and I. ElAgib, "Bulk media assay using backscattered Pu-Be neutrons," *Nuclear Instruments and Methods in Physics Research A*, vol. 432, pp. 410–414, 1999.

[686] I. El Agib and J. Csikai, "Validation of neutron data libraries by backscattered spectra of Pu-Be neutrons," *Nuclear Instruments and Methods in Physics Research A*, vol. 435, pp. 465–461, 1999.

[687] B. S. Eshardi and R. Venkateasan, "Transmission characteristics of ^{252}Cf neutrons passing through rare earth and boron loaded concrete slabs," *Nuclear Engineering and Design*, vol. 117, pp. 325–331, 1989.

[688] K. M. Glover, "Alpha-particle spectrometry and its applications," *Journal of Applied Radiation and Isotopes*, vol. 35, no. 4, pp. 239–250, 1984.

[689] K. A. Al-Saleh and N. S. Saleh, "Nuclear analysis of Jordanian tobacco," *Nuclear Instruments and Methods in Physics Research B*, vol. 18, pp. 77–79, 1986.

[690] H. J. Whitlow, J. E., P. A. Ingemarsson, and S. Hogmark, "Recoil spectrometry of oil additive associated compositional changes in sliding metal surfaces," *Nuclear Instruments and Methods in Physics Research B*, vol. 63, pp. 445–450, 1992.

[691] M. D. Strathman, "Rutherford backscattering in an industrial environment," *Nuclear Instruments and Methods in Physics Research B*, vol. 10-11, pp. 600–605, 1985.

[692] G. Demortier and G. Tewagne, "Backscattering and forward scattering for depth profiling of solids," in *Non-destructive Elemental Analysis* (Z. B. Alfassi, ed.), ch. 10, Oxford: Blackwell Science, 2001.

[693] A. L. Turkevich, "The chemical composition of the lunar surface," *Accounts of Chemical Research*, vol. 6, pp. 81–89, 1973.

[694] M. S. Soller, L. Cirignano, P. Lieberman, and M. R. Squillante, "System for precise determination of effective atomic number by beta backscatter," *IEEE Transactions in Nuclear Science*, vol. 37, pp. 230–233, 1990.

[695] S. R. Walker, P. N. Johnston, I. F. Bubb, W. B. Stannard, D. Cohen, N. Dytlewski, M. Hult, H. J. Whitlow, C. Zring, M. Östling, and M. Andersson, "Mass and energy dispersive recoil spectrometry of MOCVD grown

$Al_xGa_{1-x}As$," *Nuclear Instruments and Methods in Physics Research A*, vol. 353, pp. 563–567, 1994.

[696] R. D. Giauque, F. S. Goulding, J. M. Jaklevic, and P. R. H., "Trace element determination with semiconductor detector x-ray spectrometers," *Analytical Chemistry*, pp. 671–681, 1973.

[697] J. J. LaBrecque, P. A. Rosales, and G. Mejías, "Simultaneous determination of lanthanides by radioisotope x-ray fluorescence spectrometry based on characteristic K-radiation," *Analytica Chimica Acta*, vol. 188, pp. 9–14, 1986.

[698] J. Injuk and V. Grieken, "X-ray fluorescence," in *Non-destructive Elemental Analysis* (Z. B. Alfassi, ed.), ch. 4, Oxford: Blackwell Science, 2001.

[699] S. J. Outhwaite, "The determination of binary mixtures in metallic coatings," *Applied Radiation and Isotopes*, vol. 44, no. 10/11, pp. 1269–1277, 1993.

[700] F. A. Akeredolu, H. B. Olaniyi, J. A. Adejumo, I. B. Obioh, O. J. Ogunsola, O. I. Asubjojo, and A. F. Oluwole, "Determination of elemental composition of TSP from cement industries in Nigeria using EDXRF technique," *Nuclear Instruments and Methods in Physics Research A*, vol. 353, pp. 542–545, 1994.

[701] A. Almasoumi, M. Borsaru, and J. Charbucinski, "Determination of the lead concentration of Pb-Zn ores in laboratory boreholes using gamma-gamma techniques with very low activity sources," *Applied radiation and isotopes*, vol. 49, pp. 125–131, 1998.

[702] J. C. Lofts, H. P. K., M. A. Lovell, and J. Locke, "Application of induced gamma-ray spectroscopy measurements: Characterization of a North Sea reservoir," *IEEE Transaction on Nuclear Science*, vol. 44, pp. 360–363, February 1994.

[703] J. L. Szabo, A. C. Simon, and R. Junca, "Non-destructive analysis of uranium and/or plutonium using x-ray (K or L band) fluorescence excited by sealed sources or x-ray tubes," *Nuclear Instruments and Methods in Physics Research A*, vol. 353, pp. 668–671, December 1994.

[704] C. R. Appoloni, F. R. Espinoza Quiñones, P. H. A. Aragão, A. O. dos Santos, L. M. da Silva, P. F. Barbieri, V. F. do Nascimento Filho, and M. M. Coimbra, "EDXRF study of Tupi-Guarani archaeological ceramics," *Radiation Physics and Chemistry*, vol. 61, pp. 711–712, June 2001.

[705] R. D. Vis, "Charged particle-induced x-ray emission," in *Non-destructive Elemental Analysis* (Z. B. Alfassi, ed.), ch. 7, Oxford: Blackwell Science, 2001.

[706] T. A. Cahill, "Innovative aerosol sampling devices based upon PIXE capabilities," *Nuclear Instruments and Methods*, vol. 181, pp. 473–480, 1981.

[707] C. Swann, "Review of the recent application of the nuclear microprobe to art and archaeology," *Nuclear Instruments and Methods in Physics Research B*, vol. 104, pp. 576–583, Sep 1995.

[708] Y. Horino, Y. Mokuno, A. Kinomura, A. Chayahara, and K. Fujii, "Heavy ion microprobe for PIXE analysis of iron," *Nuclear Instruments and Methods in Physics Research*, vol. 353, pp. 619–622, 1994.

[709] N. P.-O. Homman, C. Yang, and K. Malmqvist, "A highly sensitive method for rare-earth element analysis using ionoluminescence combined with PIXE," *Nuclear Instruments and Methods in Physics Research A*, vol. 353, pp. 610–614, 1994.

[710] C. Yang, N. P.-O. Homman, L. Johansson, and K. G. Malmqvist, "Microcharacterizing zircon mineral grain by ionoluminescence combined with PIXE," *Nuclear Instruments and Methods in Physics Research B*, vol. 85, pp. 808–814, March 1994.

[711] A. A. Bettiol, D. N. Jamieson, S. Prawer, and M. G. Allen, "Ion beam induced luminescence from diamond and other crystals from a nuclear microbeam," *Nuclear Instruments and Methods in Physics Research B*, vol. 85, pp. 775–779, March 1994.

[712] C. G. Ryan, C. A. Heinrich, and T. P. Mernagh, "PIXE microanalysis of fluid inclusions and its application to study ore metal segregation between magmatic brine and vapor," *Nuclear Instruments and Methods in Physics Research B*, vol. 77, pp. 463–471, May 1993.

[713] L. Grodzins, "Critical absorption tomography of small samples," *Nuclear Instruments and Methods*, vol. 206, pp. 547–552, 1983.

[714] M. Zuick, T. J. Janik, M. Wilson, D. G. Torr, A. J. Fennelly, and E. L. Fry, "Chlorine, oxygen, nitrogen, and carbon (LONC-TOMOSCOP) detection system based on tomographic spectroscopy," in *Cargo Inspection Technology, San Diego, July 1994* (A. H. Lawrence, ed.), vol. 2276, pp. 392–403, SPIE - International Society for Optical Engineering, 1994.

[715] M. P. Maclenan, S. C. Oldham, and P. F. Nolan, "Dynamic elemental analysis during thermal decomposition using x-ray attenuation," *Nuclear Instruments and Methods in Physics Research A*, vol. 299, pp. 648–650, 1990.

[716] D. V. Ellis, "Gamma ray scattering measurement for density and lithology determination," *IEEE Transactions on Nuclear Science*, vol. 35, pp. 806–811, 1988.

[717] W. Bertozzi, D. V. Ellis, and J. S. Wahl, "The physical foundation for formation lithology logging with gamma rays," *Geophysics*, vol. 46, pp. 1439–1455, October 1981.

[718] E. M. A. Hussein, "Detection of explosive materials using gamma radiation," in *Proc. Cargo Inspection Technologies, San Diego, Ca.* (A. H. Lawrence, ed.), vol. 2276, pp. 321–325, SPIE - International Society for Optical Engineering, July 1994.

[719] E. M. A. Hussein and P. Gokhale, "Inspection of cargo containers using gamma radiation," in *Proc. Physics-Based Technologies for the Detection of*

Contraband (L. O. Malotky and J. J. Pennella, eds.), vol. 2936, pp. 210–218, SPIE - International Society for Optical Engineering, 1997.

[720] M. Borsaru, C. Ceravolo, G. Carson, and T. Tchen, "Low radioactivity portable coal face ash analyser," *Applied radiation and isotopes*, vol. 48, no. 6, 1997.

[721] M. Borsary, J. Charbucinski, P. Eisler, and S. F. Youl, "Determination of ash content in coal by borehole logging in dry boreholes using gamma-gamma methods," *Geoexploratoin*, vol. 23, pp. 503–518, 1985.

[722] M. Borsaru and C. Ceravolo, "Low activity spectrometric gamma-gamma borehole logging tool for the coal industry," *Nuclear Geophysics*, vol. 8, pp. 343–350, August 1994.

[723] M. Borsaru, C. Ceravolo, and T. Tchen, "The application of the low activity borehole logging tool to the iron ore mining industry," *Nuclear Geophysics*, vol. 9, no. 1, pp. 55–62, 1995.

[724] C. C. Watson and A. J. Becker, "The Compton and photoelectric spectroscopy of multiply scattered photons," *IEEE Transactions on Nuclear Science*, vol. 27, pp. 923–930, April 1990.

[725] S. M. McDaid, J. Hunt, M. J. Cooper, and G. Harding, "Fast x-ray scatter ratio method for security screening applications," in *Proc. Law Enforcement Technologies: Identification Technologies and Traffic Safety, Munich, Germany*, vol. 2511, pp. 79–85, SPIE - International Society for Optical Engineering, 1995.

[726] G. Harding, R. Armstrong, S. McDaid, and M. J. Cooper, "A K edge filter technique for optimization of the coherent-to-Compton scatter ratio method," *Medical Physics*, vol. 22, pp. 2007–2014, 1995.

[727] M. J. Cooper, A. J. Rollason, and R. W. Tuxworth, "Gamma-ray scattering studies of composition variations in alloys," *Journal of Physics E: Scientific Instruments*, vol. 15, pp. 568–572, 1982.

[728] L. Confalonieri, R. Crippa, and M. Milazzo, "An experimental approach to the analysis of binary alloys by the Rayleigh-Compton backscatter ratio method," *Applied Radiation and Isotopes*, vol. 38, pp. 139 –142, 1987.

[729] S. Manninen and S. Koikkalainen, "Determination of the effective atomic number using elastic and inelastic scattering of γ-rays," *Applied Radiation and Isotopes*, vol. 35, no. 10, pp. 965–968, 1984.

[730] H. P. Schatzle, "Basic aspects on the use of elastic and inelastic scattered gamma radiation for the determination of binary systems with effective atomic number of less than 10," *Applied Radiation and Isotopes*, vol. 30, pp. 115–121, 1979.

[731] A. G. Alvarez, M. Molina-Sabio, and F. Rodriguez-Reinoso, "An x-ray scattering investigation of the carbonization of olive stones," *Carbon*, vol. 36, no. 1-2, pp. 67–70, 1998.

[732] P. Arikan, A. Zararsiz, P. Kirmaz, and N. Efe, "Determination of high-Z materials in low-Z medium by x-ray scattering techniques," *Nuclear Instruments and Methods in Physics Research B*, vol. 95, pp. 82–86, January 1995.

[733] L. M. N. Távora, R. E. Morgado, R. J. Estep, M. Rawool-Sullivan, G. W. B., and M. E. J., "One-sided imaging of large, dense objects using 511-keV photons from induced-pair production," *IEEE Transactions on Nuclear Science*, vol. 45, no. 3 pt 1, pp. 970–975, 1998.

[734] W. B. Gilboy, G. Harding, L. M. N. Tavora, and B. Ulmer, "On the potential of photon induced annihilation photons for inspection of engineering structures," *Applied Radiation and Isotopes*, vol. 48, October/December 1997.

[735] B. D. Sowerby and V. N. Ngo, "Determination of the ash content of coal using annihilation radiation," *Nuclear Instruments and Methods in Physics Research*, vol. 188, pp. 429–437, Sep 1981.

[736] G. Harding, W. Gilboy, and B. Ulmer, "Photon-induced positron annihilation radiation (PIPAR) - A novel gamma-ray imaging technique for radiographically dense materials," *Nuclear Instruments and Methods in Physics Research A*, vol. 398, pp. 409–422, October 1997.

[737] R. A. Semmler, "Neutron-moderation moisture meters: Analysis and applications to coal and soil," Tech. Rep. COO-712-73, University of Chicago, September 1963. USAEC Report.

[738] InstroTek, Inc., "Understanding and operating nuclear gauges more effectively." InstroTek, Inc. , Raleigh, NC, January 1999.

[739] F. Brooks, A. Buffler, and S. Allie, "Using neutron backscattering to detect buried plastic anti-personnel landmines," in *First Research Co-ordination Meeting of the CRP on Application of Nuclear Techniques to Anti-Personnel Landmine Identification (F1.10.07)*, (Zagreb), International Atomic Energy Agency, November 1999. (http://www.iaea.org/programmes/ripc/physics/zagreb_crp_report.htm).

[740] C. P. Datema, V. R. Bom, and C. W. E. van Eijk, "Landmine detection with the neutron backscattering method," in *Nuclear Science Symposium Conference Record*, vol. 1, pp. 5(111) –5(114), IEEE, 2000.

[741] T. O. Tümer, R. M. Pierce, K. C. Dotson, J. R. Jadamec, and C. W. Su, "A portable narcotics detector with identification capability," in *Proc. Cargo Inspection Technologies* (A. H. Lawrence, ed.), vol. 2276, pp. 255–265, SPIE - International Society for Optical Engineering, 1994.

[742] T. O. Tümer, C. W. Su, and C. R. Kaplan, "A portable narcotics detector and the results obtained in field tests," in *Proc. Physics-Based Technologies for the Detection of Contraband* (L. O. Malotky and J. J. Pennella, eds.), vol. 2936, pp. 95–101, SPIE - International Society for Optical Engineering, 1997.

[743] T. O. Tümer, L. Doan, C. W. Su, B. Baritelle, and B. Rhoton, "A sensitive, selective, and portable detector for contraband: the compact integrated

narcotics instrument," *Transactions of American Nuclear Society*, vol. 82, pp. 96–97, 2000.

[744] H. Tominaga, M. Tsuchihiro, I. Tachikawa, N. Ishikawa, H. Saskai, and H. Yoshida, "Simultaneous measurement of oil and moisture in soil by a new technique of fast neutron moderation combined with resonant microwave attenuation," in *Proc. 4th Topical Meeting on Industrial Radiation and Radioisotope Measurement Applications, Raleigh, NC*, (LaGrange Park, Il), p. 61, American Nuclear Society, October 1999.

[745] R. A. Craig, A. J. Peurrung, and D. C. Stromswold, "Land-mine detection using timed neutron detection," *Transactions of American Nuclear Society*, vol. 83, pp. 267–268, November 2000.

[746] K. N. Black, "The nuclear asphalt content gauge for measuring asphalt content in mixes." Public Roads On-Line, Volume 58, No. 3 , US Federal Highway Administration, Construction and Maintenance Division, Office of Engineering, Washington, DC., Winter 1995.

[747] T. Ohnuki, T. Yamamoto, and Y. Wadachi, "Non-destructive method for determining the distribution of soil water content by measuring fast neutron transmission using an NE-213 organic scintillator," *Japanese Journal of Applied Physics*, vol. 23, pp. 1124–1128, August 1984.

[748] V. P. Smirnov and E. P. Klochkov, "Measurement of the hydrogen concentration in graphite and zirconium by the transmission of neutron beams with the aid of an isotope source," *Soviet Atomic Energy*, vol. 55, no. 5, pp. 792–794, 1983.

[749] S. Liu and Z. Lin, "Measurement of moisture in pottery materials by using thermal neutron transmission method," *Hejishu/Nuclear Techniques*, vol. 16, pp. 409–412, July 1993.

[750] S. Liu, Y. Yu, F. Zhang, and X. Bai, "Moisture measurement of glass materials using a thermal neutron transmission gauge," *HeJishu/Nuclear Techniques*, vol. 20, no. 9, pp. 546–550, 1997.

[751] H. Pleinert, H. Sadouki, and F. H. Wittmann, "Determination of moisture distributions in porous building materials by neutron transmission analysis," *Materials and Structures/Materiaux et Constructions*, vol. 31, pp. 218–224, May 1998.

[752] M. S. Gusarov, M. G. Zemlyanov, F. G. Reshetnikov, S. P. Mironov, and G. V. Afonin, "Determination of hydrogen density in matter and hydrogen interaction with metals from neutron transmission," *Soviet Atomic Energy*, vol. 39, pp. 724–726, August 1975.

[753] C. M. Bartle, "Simultaneous transmission of neutrons and gamma rays (NEUGAT) to measure fat in meat," *Applied Radiation and Isotopes*, vol. 42, no. 11, pp. 1115–1116, 1991.

[754] C. M. Bartle and C. R. Purcell, "Determination of the moisture content of wheat, using the NEUGAT technique," *Nuclear Instruments and Methods in Physics Research A*, vol. 291, pp. 655–661, 1990.

[755] C. M. Bartle, "Measurement of the composition of dairy products using the neutron/gammatransmission [NEUGAT) method," *Applied Radiation and Isotopes*, vol. 45, pp. 1039–1045, October 1994.

[756] H. D. Kosanke, "Hydrogen sensitive neutron radiography," *Transactions of the American Nuclear Society*, vol. 14, pp. 533–533, 1971.

[757] H. M. Miller, L. Li-Te, and B. R. M., "Detection of parts-per-billion levels of hydrogen in solids using a modified notched neutron spectrum technique," *Nuclear Technology*, vol. 99, pp. 252–257, 1992.

[758] W. H. Miller, M. Popovici, and L. Groza, "Non-destructive analysis of hydrogen in nuclear power plant materials." presented at 8th International Conference on Nuclear Engineering, ASME, Baltimore, MD (http://www.icone-conf.org/icone8/program/abstracts/8386.html, accessed February 2001), April 2000.

[759] W. Meyer and W. H. Miller, "Nondestructive analysis for hydrogen in steel using neutron scattering," *Transactions of the American Nuclear Society*, vol. 59, pp. 137–138, June 1989.

[760] V. N. Smith and J. W. Otvos, "Hydrogen determination and liquid analysis with a beta-particle absorption apparatus," *Analytical Chemistry*, vol. 26, pp. 359–366, 1954.

[761] R. A. Livingston, D. Neumann, S. FitzGerald, and J. Rush, "Quasi-elastic scattering study of the hydration of tricalcium silicate," in *Proc. 1999 1st Conference on Radiation Sources and Radiation Interactions, July 1999* (G. Vourvopoulos, ed.), vol. 2867, pp. 148–151, SPIE - International Society for Optical Engineers, 1997.

[762] J. L. Putman, "Development in thickness gauges and allied instruments," in *Int. Conf. Peaceful Uses of Atomic Energy*, vol. 15, (Geneva), pp. 119–123, United Nations, 1955.

[763] J. Fodor and C. Varga, "Determination of tungsten in steels by beta-gamma backscattering," in *Proc. Second Int. Conf. Peaceful Uses of Atomic Energy*, vol. 19, (Geneva), pp. 215–230, United Nations, 1958.

[764] A. Damkjaer, "Composition analysis of tin/lead plating on printed circuit boards by the beta-backscatter method," *Applied Radiation and Isotopes*, vol. 27, pp. 631–636, 1976.

[765] S. A. Husain, "Investigation of backscattered beta radiation and analysis of solutions and alloys," in *Proc. Second Int. Conf. Peaceful Uses of Atomic Energy*, vol. 19, (Geneva), pp. 213–214, United Nations, 1958.

[766] I. P. Alimarin, "Application of radioactive isotopes in chemical analysis," in *Proc. Int. Conf. Peaceful Uses of Atomic Energy*, vol. 15, (Geneva), pp. 60–71, United Nations, 1955.

[767] P. R. Gray, D. H. Clarey, and W. H. Bramer, "Quantitative determination of hydrogen, carbon, nitrogen, oxygen, and fluorine in materials," *Analytical Chemistry*, vol. 32, pp. 582–589, 1960.

[768] R. P. Gardner and J. W. Dunn, "Simultaneous use of beta-particle transmission and backscattering gauges for determining hydrogen, carbon, and oxygen percentages in liquids," *Analytical Chemistry*, vol. 37, pp. 528–536, 1965.

[769] M. Gehrke, S. Kapila, and V. Flanigan, "Enhanced selectivity of electron capture detector for nitroaromatic explosives through the application of electron attachment reactions," in *Proc. Detection and Remediation Technologies for Mines and Minelike Targets, April 2000, Orlando, FL*, pp. 539–546, SPIE - International Society for Optical Engineering, 2000.

[770] J. Charbucinski and P. J. Mathew, "A backscatter gamma ray spectrometric method for the determination of ash in coal," *Nuclear Instruments and Methods*, vol. 193, pp. 269–273, 1982.

[771] X. Qi and Z. Yongchuan, "Novel automated separator based on dual energy gamma-rays transmission," *Measurement Science and Technology*, vol. 11, pp. 1383–1388, Sep 2000.

[772] W. Gu, "Influence of coal thickness change on dual-energy gamma-ray ash gauge reading," *Applied Radiation and Isotopes*, vol. 49, pp. 1633–1639, December 1998.

[773] D. D. S. Liu and D. J. Patmore, "Application of gamma-ray densitometry in developing primary upgrading processes," *Applied Radiation and Isotopes*, vol. 41, no. 10/11, pp. 1003–1012, 1990.

[774] T. Suzuki, R. Kitsutaka, T. Muto, and S. Morisaki, "Determination of the purity of gold alloys using gamma-ray transmission techniques," *Japanese Journal of Applied Physics, Part 1*, vol. 37, pp. 6242–6247, November 1998.

[775] K. Drozdowicz, "Total cross section of plexiglass in the thermal neutron region," *Annals of Nuclear Energy*, vol. 16, no. 6, pp. 275–278, 1989.

[776] K. Kanda and O. Aizawa, "Total neutron cross section of lead," *Nuclear Science and Engineering*, vol. 60, pp. 230–234, July 1976.

[777] A. K. Freund, H. Friedrich, W. Nistler, and R. Scherm, "Neutron transmission properties of perfect silicon crystals," *Nuclear Instruments and Methods in Physics Research A*, vol. 234, pp. 116–121, January 1985.

[778] C. Mancini, "Neutron and gamma ray transmission: methods and accuracy for the analysis of silver and gold alloys," *Applied Radiation and Isotopes*, vol. 36, pp. 489–494, June 1985.

[779] R. Cesareo and C. Mancini, "Non-destructive analysis of silver alloys by means of low energy gamma -rays and neutron transmission measurements," *Applied Radiation and Isotopes*, vol. 30, pp. 589–594, October 1979.

[780] M. J. Greenhill-Hooper, "Role of borates as builders in laundry detergency," *Tenside, Surfactants, Detergents*, vol. 33, pp. 366–373, September/October 1996.

[781] E. M. A. Hussein, D. R. O'Connor, and M. E. Mosher, "A non-intrusive neutron method for gadolinium poison concentration monitoring in CANDU reactors," *Nuclear Journal of Canada*, vol. 1, pp. 46–52, 1987.

[782] D. R. O'Connor and E. M. A. Hussein, "A neutron scatterometer for gadolinium concentration measurement: Design and testing," *Applied Radiation and Isotopes*, vol. 41, pp. 1041–1047, 1990.

[783] D. C. Stromswold, W. R. Mills, R. D. Wilson, and T. K. Cook, "Formation porosity measurement using epithermal neutron lifetime," *IEEE Transactions on Nuclear Science*, vol. 36, pp. 1210–1214, February 1989.

[784] A. Stromswold and L. S. Allen, "Epithermal neutron lifetime logging in partially saturated formations and large-diameter holes," *IEEE Transactions in Nuclear Science*, vol. 39, no. 4, pp. 007–1001, 1992.

[785] E. A. Pentaleri and T. Gozani, "Complete characterization of containerized waste using a PFNA-based inspection system," *Nuclear Instruments and Methods in Science Research A*, vol. 353, pp. 489–493, 1994.

[786] T. Sumitra, N. Chankow, S. Punnachaiya, and S. Srisatit, "A compact XRF unit for determining total sulphur content in coals," *Transactions of the American Nuclear Society*, vol. 70 (Suppl. 1), pp. 387–390, 1994.

[787] D. Page and T. Piggins, "Development of an x-ray fluorescence method for on-line determination of sulphur in coal," *Applied Radiation and Isotopes*, vol. 36, no. 6, pp. 1–5, 1985.

[788] T. Čechák and L. Thinová, "Sulfur content measurement in coal by x-ray fluorescence method," *Radiation Physics and Chemistry*, vol. 61, pp. 759–761, 2001 2001.

[789] T. Sumitra, S. Punnachaiya, N. Chankow, and R. Laopaibul, "Determination of SO_2 in ambient air and in stacks of a tobacco curing plant using the XRF technique," *Nuclear Instruments and Methods in Physics Research A*, vol. 299, December 1990.

[790] W. D. Ruhter and R. Gunnink, "Application of cadmium-zinc-telluride detectors in ^{235}U enrichment measurements," *Nuclear Instruments and Methods in Physics Research A*, vol. 252, pp. 716–718, 1994.

[791] J. Lartigue, R. Padilla, T. Martínez, and M. Orta, "EDXRF determination of Pb in aerosol samples from Mexico Valley," *Radiation Physics and Chemistry*, vol. 61, pp. 681–682, June 2001.

[792] Y. J. Wang, J. S. Iwanczyk, and W. R. Graham, "Evaluation of HgI_2 detectors for lead detection in paint," *IEEE Transactions on Nuclear Science*, vol. 40, no. 4, pp. 846–850, 1993.

[793] A. G. Darnley, "The development of airborne gamma-ray spectrometry: case study in technological innovation and acceptance," *Nuclear Geophysics*, vol. 5, pp. 377–402, 1991.

[794] E. M. Elkattan, “Recognition of favorable zones for uranium and thorium accumulation, at UmAra-Um Shilman granitic pluton, South Eastern Desert, Egypt, using airborne spectrometric and magnetic data,” *Nuclear Geophysics*, vol. 9, no. 3, pp. 241–261, 1995.

[795] H. E. McGannon, ed., *The Making, Shaping and Treating of Steel*. Pittsburgh: United States Steel, ninth ed., 1971.

[796] R. D. Beshears and L. Hediger, “Radiographic analysis of the TSS-1R conductive tether,” *Materials Evaluation*, vol. 54, pp. 1348–1349, December 1996.

[797] I. G. Scott and C. M. Scala, “Review of non-destructive testing of composite materials,” *NDT International*, vol. 15, no. 2, pp. 75–86, 1982.

[798] T. L. Crane, “Measurement of composite ply orientation using a radiographic fringe technique,” *Materials Evaluation*, vol. 34, pp. 79–80, 1976.

[799] D. E. W. Stone, “Non-destructive inspection of composite materials for aircraft structure applications,” *British Journal of Nondestructive Testing*, vol. 20, pp. 43–49, 1978.

[800] A. McNulty, J. McNab, and K. Greenwood, “Small controlled area radiography (SCAR) system,” *Insight: Non-Destructive Testing and Condition Monitoring*, vol. 41, pp. 439–441, July 1999.

[801] ASTM, “E1000-98: Standard Guide for Radioscopy.” American Society for Testing and Materials, West Conshohocken, PA, 1998.

[802] ASTM, “E1255-96: Standard Practice for Radioscopy.” American Society for Testing and Materials, West Conshohocken, PA, West Conshohocken, PA, 1996.

[803] ASTM, “E1647-98a: Standard Practice for Determining Contrast Sensitivity in Radioscopy .” American Society for Testing and Materials, West Conshohocken, PA West Conshohocken, PA, 1998.

[804] S. I. Rokhlin and A. C. Guu, “Computerized radiographic sensing and control of an arc welding process,” *Welding Journal*, vol. 69, no. 3, pp. 83s–97s, 1990.

[805] S. I. Rokhlin and A. C. Guu, “Control of submerged arc weld penetration by radiographic means,” *NDT International*, vol. 22, pp. 74–80, April 1989.

[806] S. I. Rokhlin, K. Cho, and A. C. Guu, “Closed-loop process control of weld penetration using real-time radiography,” *Materials Evaluation*, vol. 47, pp. 363–369, 3 1989.

[807] S. I. Rokhlin, “In-process radiographic evaluation of arc welding,” *Materials Evaluation*, vol. 47, pp. 219–224, February 1989.

[808] X. E. Gros, “Real-time radiography of underwater pipelines,” *British Journal of Non-Destructive Testing*, vol. 35, pp. 492–495, Sep 1993.

[809] Anonymous, “Casting defect detection using real-time radioscopy,” *Foundry Management & Technology*, vol. 118, pp. 56–58, 60, May 1990.

[810] D. R. Korzekwa, P. S. Dunn, and P. K. Tubesing, "Casting of uranium and bismuth through filters: Experiments and simulations," in *Light Metals: Proc. 1997 127th TMS Annual Meeting, San Antonio, TX*, (Warrendale, PA), pp. 1371 –1372, Minerals, Metals and Materials Society (TMS), 1998.

[811] R. Derebail and J. N. Koster, "Visualization study of melting and solidification in convecting hypoeutectic Ga-In alloy," *Journal of Heat and Mass Transfer*, vol. 41, pp. 2537–2548, August 1998.

[812] R. Derebail and J. N. Koster, "Radioisotopic visualization of isothermal solidification of eutectic Ga-In alloy," *Metallurgical and Materials Transactions B: Process Metallurgy and Materials Processing Science*, vol. 27, August 1996.

[813] J. N. Koster, R. Derebail, and A. Grotzbach, "Visualization of convective solidification in a vertical layer of Eutectic Ga-In melt," *Applied Physics A: Materials Science & Processing*, vol. 64, pp. 45–54, January 1997.

[814] T. A. Campbell and J. N. Koster, "Interface dynamics during indium antimonide crystal growth," *Crystal Research and Technology*, vol. 33, no. 5, pp. 707–716, 1998.

[815] T. A. Campbell and J. N. Koster, "Compositional effects on solidification of congruently melting InSb," *Crystal Research and Technology*, vol. 33, no. 5, pp. 717–731, 1998.

[816] T. A. Campbell and J. N. Koster, "In-situ visualization of off-stoichiometric equilibrium crystal growth within indium antimonide: Antimony-rich composition," *Journal of Crystal Growth*, vol. 174, pp. 238–244, April 1997.

[817] T. A. Campbell and J. N. Koster, "Visualization of liquid-solid interface morphologies in gallium subject to natural convection," *Journal of Crystal Growth*, vol. 140, pp. 414–425, July 1994.

[818] J. N. Koster, "Visualization of Rayleigh-Benard convection in liquid metals," *European Journal of Mechanics, B/Fluids*, vol. 16, no. 3, pp. 447–454, 1997.

[819] J. Breard, A. Saouab, and G. Bouquet, "Permeability measurement of matrix polymer composite reinforcement," *EPJ Applied Physics*, vol. 1, no. 2, pp. 269–278, 1998.

[820] M. C. Maguire, M. D. Baldwin, and C. L. Atwood, "Fastcast: integration and application of rapid prototyping and computational simulation to investment casting," in *Proc. 1995 27th International SAMPE Technical Conference, Albuquerque, NM*, vol. 27, pp. 235–244, 1995.

[821] Y. Fu, A. W. Batchelor, and N. L. Loh, "Study on fretting wear behavior of laser treated coatings by x-ray imaging," *Wear*, vol. 218, pp. 250–260, July 1998.

[822] M. T. Kortschot and P. W. R. Beaumont, "Damage mechanics of composite materials. I. Measurements of damage and strength," *Composites Science and Technology*, vol. 39, no. 4, pp. 289–301, 1990.

[823] C. T. Liu, "Monitoring damage initiation and evolution in a filled polymeric material using nondestructive testing techniques," *Computers and Structures*, vol. 76, pp. 57–65, 1 2000.

[824] C. Bueno, M. D. Barker, R. C. Barry, and R. A. Betz, "Corrosion detection and characterization using high resolution real-time radiography," in *Proc. 41st International SAMPE Symposium and Exhibition. Part 1, Anaheim,CA*, vol. 41, pp. 230–242, 1996.

[825] C. Bueno, M. D. Barker, R. A. Betz, R. C. Barry, and R. A. Buchanan, "Nondestructive evaluation of aircraft structures using high-resolution real-time radiography," in *Proc. SPIE - International Society for Optical Engineering*, vol. 2455, pp. 114–124, 1995.

[826] J. M. Galbraith, W. T. Allen, K. L. Coleman, and V. J. Alreja, "Real-time radisotopic inspection of insulated piping systems," *Insight - Non-Destructive Testing and Condition Monitoring.*, vol. 37, pp. 417–420, June 1995.

[827] S. M. Walker, "New NDE developments support rapid, economical screening for flow-accelerated corrosion," in *Integrity of Structures and Components; Nondestructive Evaluations, Proc. 1998 ASME/JSME Joint Pressure Vessels and Piping Conference, San Diego, CA, USA*, vol. 375, pp. 165–168, American Society of Mechanical Engineers, 1998.

[828] M. Trezzi, F. Frangi, and S. Zannella, "Investigation of melt-textured YBCO by radioscopic technique," *Superconductor Science & Technology*, vol. 9, pp. 315–320, April 1996.

[829] W. A. Johnston, J. W. Murdock, S. Koshigoe, and P. T. Than, "Slag accumulation in the titan solid rocket motor upgrade," *Journal of Propulsion and Power*, vol. 11, pp. 1012–1020, September/October 1995.

[830] R. A. J. Frederick, B. M. Williams, and S. B. Farmer, "Predicting and analyzing x-rays to measure propellant crack propagation speed," *Journal of Propulsion and Power*, vol. 12, pp. 310–314, March/April 1996.

[831] M. Ong, T. Ferriera, R. Gilliam, J. Zentler, and P. Wargo, "FXR fast beam imaging diagnostics," in *Digest of Technical Papers - IEEE International Pulsed Power Conference*, vol. 2, (Piscataway, NJ), pp. 636–639, IEEE, 1999.

[832] D. Grady, "Impact failure and fragmentation properties of tungsten carbide," *International Journal of Impact Engineering*, vol. 23, pp. 307–317, 1999.

[833] D. E. Grady and M. E. Kipp, "Fragmentation properties of metals," *International Journal of Impact Engineering*, vol. 20, no. n 1-5 pt 1, pp. 293–308, 1997.

[834] D. E. Grady and M. E. Kipp, "Experimental measurement of dynamic failure and fragmentation properties of metals," *International Journal of Solids and Structures*, vol. 32, pp. 2779–279, 1995.

[835] D. E. Grady and M. E. Kipp, "Experimental and computational simulation of the high velocity impact of copper spheres on steel plates," *International Journal of Impact Engineering*, vol. 15, no. 5, pp. 645–660, 1994.

[836] Y. V. Bat'kov, N. P. Kovalev, A. D. Kovtun, V. G. Kuropatkin, A. I. Lebedev, Y. M. Makarov, S. F. Manachkin, S. A. Novikov, V. A. Raevsky, and Y. M. Styazhkin, "Explosive three-stage laughter to accelerate metal plates to velocities more than 10 km/s," *International Journal of Impact Engineering*, vol. 20, no. 1-5 pt 1, pp. 89–92, 1997.

[837] P. Lundberg, L. Westerling, and B. Lundberg, "Influence of scale on the penetration of tungsten rods into steel-backed alumina targets," *International Journal of Impact Engineering*, vol. 18, pp. 403–416, June 1996.

[838] J. D. Walker, D. J. Grosch, and S. A. Mullin, "Hypervelocity fragment launcher based on an inhibited shaped charge," *International Journal of Impact Engineering*, vol. 14, no. 1-4, pp. 763–774, 1993.

[839] T. G. Theofanous, S. Angelini, X. Chen, R. Luo, and W. W. Yuen, "Quantitative radiography for transient multidimensional, multiphase flows," *Nuclear Engineering and Design*, vol. 184, pp. 163–181, August 1998.

[840] M. Chick, R. B. Frey, and A. Bines, "Jet penetration in Plexiglas," *International Journal of Impact Engineering*, vol. 9, no. 4, pp. 433–439, 1990.

[841] E. Ferm and J. B. Ramsay, "Jet penetration of surrogate steel-explosive systems," *Propellants, Explosives, Pyrotechnics*, vol. 16, pp. 123–130, June 1991.

[842] R. L. Summers, W. P. Walters, and R. D. Dick, "Behavior of shaped charges with open-poled hemispherical liners," *Propellants, Explosives, Pyrotechnics*, vol. 16, pp. 140–144, June 1991.

[843] L. A. Parnell, R. S. Nelson, T. R. Ogden, J. P. Lucero, and D. Rogerson, "Flash and real-time radiographic diagnostics of closed liquid metal combustion for underwater propulsion," in *Proc. 25th Intersociety Energy Conversion Engineering Conference - IECEC'90, Reno, NV*, vol. 6, pp. 350–355, IEEE, 1990.

[844] K. A. Kodimer, L. A. Parnell, R. S. Nelson, and P. J. Papin, "Computer simulations of closed liquid metal combustor fluid flow," in *Energy Systems - New Technologies, Proc. 27th Intersociety Energy Conversion Engineering Conference, San Diego, CA*, vol. 4, pp. 259–263, IEEE, 1992.

[845] T. R. Neal, "Flash radiography as a quantitative tool," in *Proc. 16th International Congress on High Speed Photography and Photonics,Strasbourg, Fr*, vol. 491, pp. 123–129, SPIE - International Society for Optical Engineering, 1985.

[846] J. Geiswiller, E. Robert, L. Hure, C. Cachoncinlle, R. Viladrosa, and J. M. Pouvesle, "Flash x-ray radiography of argon jets in ambient air," *Measurement Science & Technology*, vol. 9, pp. 1537–1542, Sep 1998.

[847] N. G. Triantafilopoulos and T. E. J. Farrington, "Flash x-ray radiography techniques for visualizing coating flows," in *Proc. 1988 Coatings Conference, New Orleans, LA*, pp. 47–52, The Technical Association of the Pulp and Paper Industry, 1988.

[848] T. J. Heindel, "Bubble size measurements in a quiescent fibre suspension," *Journal of Pulp and Paper Science*, vol. 25, no. 3, pp. 104–110, 1999.

[849] T. J. Heindel, "Effect of fiber consistency on bubble size," *Nordic Pulp and Paper Research Journal*, vol. 14, no. 2, pp. 171–178, 1999.

[850] T. N. Claytor, H. M. Frost, T. H. Feiertag, G. A. Sheppard, and P. Shalek, "Nondestructive measurement of microstructure evolution in ceramics," *Materials Evaluation*, vol. 47, pp. 532–537, May 1989.

[851] D. J. Cotter, W. D. Koenigsberg, A. E. Pasto, and L. Bowen, "Predicting failure stress of silicon nitride ceramics using microfocus radiography," *Journal of the American Ceramic Society*, pp. 460–461, November 1988.

[852] T. J. Young, R. Lamp, Thomas, L. Ramalingam, Mysore, and B.-H. Tsao, "Failure analysis and materials performance of a heat coupled thermionic converter," in *Proc. ASME Winter Annual Meeting*, (Anaheim, CA), pp. 1–7, American Society of Mechanical Engineers, 1992.

[853] G. Y. Baaklini and P. B. Abel, "Radiographic and ultrasonic characterization of sintered silicon carbide," *Materials Evaluation*, vol. 46, pp. 477–1483, October 1988.

[854] B. Venkatraman, V. K. Sethi, T. Jayakumar, and B. Raj, "High-definition radiography of tube-to-tube sheet welds of steam generator of prototype fast breeder reactor," *Insight - Non-Destructive Testing and Condition Monitoring*, vol. 37, pp. 189–192, March 1995.

[855] B. Zoofan and S. I. Rokhlin, "Microradiographic detection of corrosion pitting," *Materials Evaluation*, vol. 56, pp. 191–194, February 1998.

[856] M. Wevers, P. De Meester, M. Lodewijckx, *et al.*, "Application of microfocus x-ray radiography in materials and medical research," *NDT & E International*, vol. 26, no. 3, pp. 135–140, 1993.

[857] D. Stegemann, W. Reimche, and J. Schmidbauer, "Investigation of light metal casting processes by realtime microfocus radioscopy," *Non-Destructive Testing*, vol. 1, pp. 107–117, January 1992.

[858] Y. Ikeda, Y. Mizuta, and K. Onda, "Rapid and sensitive radioscopy for fine ceramics using an image-subtraction method," *Journal of the American Ceramic Society*, vol. 82, pp. 882–888, 1999.

[859] P. Dick and I. Smith, "Real time radiography as an inspection tool on complex electronic assemblies," in *Through the Eyes of an Eagle: 11th World Conference on Nondestructive Testing.,Las Vegas, NV*, Int. Committee on Nondestructive Testing, 1985.

[860] A. S. Shiu, K. R. Hogstrom, N. A. Janjan, R. S. Fields, and L. J. Peters, "Technique for verifying treatment fields using portal images with diagnostic quality," *International Journal of Radiation Oncology Biology Physics*, vol. 13, no. 10, pp. 1589–1594, 1987.

[861] R. W. Volberding and S. M. Khan, "Cargo container inspection test program at ARPA's nonintrusive inspection technology testbed," in *Proc. Cargo Inspection Technologies* (A. H. Lawrence, ed.), vol. 2276, pp. 210–217, SPIE - International Society for Optical Engineering, 1994.

[862] R. W. Volberding, R. F. Reiter, and S. M. Khan, "Design and implementation of three-month intensive cargo inspection test at ARPA's nonintrusive inspection technology testbed," in *Proc. Public Safety/Law Enforcement Technology Orlando, FL*, pp. 2–11, SPIE - International Society for Optical Engineering, April 1995.

[863] R. F. Reiter and R. W. Volberding, "Results of three-month intensive cargo inspection test at ARPA's nonintrusive inspection technology testbed," in *Proc. Public Safety/Law Enforcement Technology, Orlando, FL*, pp. 12–23, SPIE - International Society for Optical Engineering, April 1995.

[864] V. V. Verbiniski and V. J. Orphan, "Gamma radiography cargo vehicle scanner," in *Proc. Physics-Based Technologies for the Detection of Contraband* (L. O. Malotky and J. J. Pennella, eds.), vol. 2936, pp. 112–123, SPIE - International Society for Optical Engineering, 1997.

[865] D. L. Smith, Y. Ikeda, and Y. Uno, "Investigation into the possibility of performing radiography with gamma rays emitted from water made radioactive by irradiation with 14 MeV D-T fusion neutrons," *Fusion Engineering and Design*, vol. 31, pp. 41–52, April 1996.

[866] R. D. Edwards, M. A. Sinclair, and T. J. Goldsack, "Laser radiography," in *Proc. 28th IEEE International Conference on Plasma Science/ 13th IEEE International Pulsed Power Conference, June 2001, Las Vegas, NV*, p. P3E01, 2001.

[867] D. H. Kalantar, M. H. Key, L. B. Da Silva, S. G. Glendinning, B. A. Remington, J. E. Rothenberg, F. Weber, S. V. Weber, E. Wolfrum, N. S. Kim, D. Neely, J. Zhang, J. Wark, A. Demir, J. Lin, *et al.*, "Measurements of direct drive laser imprint in thin foils by radiography using an x-ray laser backlighter," *Physics of Plasmas*, vol. 4, pp. 1985–1993, May 1997.

[868] D. H. Kalantar, T. W. J. Barbee, L. B. DaSilva, S. G. Glendinning, F. Weber, S. V. Weber, M. H. Key, and J. P. Knauer, "X-ray laser radiography of perturbations due to imprint of laser speckle in $0.35\mu m$ laser irradiation of a thin Si foil," *Review of Scientific Instruments*, vol. 67, pp. 781–785, March 1996.

[869] B. Zoofan and S. I. Rokhlin, "Microradiographic detection of corrosion pitting," *Materials Evaluation*, vol. 56, pp. 191–194, February 1998.

[870] A. Sehgal, G. S. Frankel, B. Zoofan, and S. Rokhlin, "Pit growth study in Al alloys by the foil penetration technique," *Journal of the Electrochemical Society*, vol. 147, no. 1, pp. 140–148, 2000.

[871] J. Benci and D. P. Pope, "Microradiography of creep damage in copper," *Materials Science and Engineering A*, vol. 117, pp. 51–58, Sep 1989.

[872] O. Ogbobe, "Study of the dispersion of low-density polyethylene additive master batches in polyolefins," *Journal of Applied Polymer Science*, vol. 49, pp. 381–389, July 1993.

[873] A. H. Wagner, D. M. Kalyon, R. Yazici, and T. J. Fiske, "Extensional flow of engineering plastics with glass fibers," in *Proc. Annual Technical Conference - ANTEC*, vol. 1, pp. 996–1000, Society of Plastics Engineers, 1997.

[874] I. E. Holloway, M. S. Choudhary, and W. B. Gilboy, "Reconstruction bremsstrahlung tomography as a method of imaging pure beta-ray emitters in matter," *Applied Radiation and Isotopes*, vol. 41, no. 10/11, pp. 967–974, 1990.

[875] M. Kallergi, W. Qian, L. P. Clarke, and A. R. Gondeck, "Bremsstrahlung imaging with the gamma camera," in *Proc. Medical Imaging VI: Instrumentation, February 1992, Newport Beach, CA*, pp. 212–221, SPIE - International Society for Optical Engineering, 1992.

[876] A. R. Kalukin and V. Sankaran, "Three-dimensional visualization of multilayered assemblies using x-ray laminography," *IEEE Transactions on Components, Packaging, and Manufacturing Technology Part A*, vol. 20, pp. 361–366, Sep 1997.

[877] S. Thomas and C. Thornton, "Process development and SPC implementation," *Circuits Assembly*, pp. 30–33, April 1996.

[878] S. M. Rooks, B. Benhabib, and K. C. Smith, "Development of an inspection process for ball-grid-array technology using scanned-beam x-ray laminography," *IEEE Transactions on Components, Packaging, and Manufacturing Technology Part A*, vol. 18, pp. 851–861, December 1995.

[879] P. O'Shea, "X-ray inspection leads the way to world-class results," *EE: Evaluation Engineering*, vol. 34, pp. 37–39, November 1995.

[880] C. Neubauer, "Intelligent x-ray inspection for quality control of solder joints," *IEEE Transactions on Components, Packaging, and Manufacturing Technology Part C: Manufacturing*, vol. 20, pp. 111–120, April 1997.

[881] B. Baker, "X-ray inspection of three-dimensional solder joints," *Electronic Manufacturing*, vol. 35, pp. 20–22, February 1989.

[882] B. Bolliger and M. Stewart, "X-ray laminography improves functional-test yield," *EE: Evaluation Engineering*, vol. 35, November 1996.

[883] G. Harding, H. Strecker, and R. Tischler, "X-ray imaging with Compton scatter radiation," *Philips Technical Review*, vol. 41, pp. 46–59, 1984.

[884] G. Harding and J. Kosanetzky, "Scattered x-ray beam nondestructive testing," *Nuclear Instruments and Methods in Physics Research A*, vol. 280, pp. 517–528, August 1989.

[885] W. Roye, "Compton-röeckstreuverfahren. Comscan - ein neues roentgenverfahren füer kunststoffe, Compton backscatter process. Comscan - a new x-ray

process for the testing of plastics," *Plastverarbeiter*, vol. 39, pp. 88–94, July 1988.

[886] H. Weiss, "Digital imaging for industrial applications," *Informationstechnik it: Computer, Systeme, Anwendungen*, vol. 33, pp. 15–18, January 1991.

[887] G. Harding, "On the sensitivity and application possibilities of a novel Compton scatter imaging system," *IEEE Transactions on Nuclear Science*, vol. NS-29, pp. 1260–1265, 1982.

[888] K. K. Yeung, P. Lambrineas, R. D. Finlayson, W. R. Sturrock, B. Suendermann, and J. R. Davis, "A COMSCAN investigation of glass-reinforced plastic/foam sandwich material," in *Second International Conference on Sandwich Construction*, (Gainsville, Florida), University of Florida, March 1992.

[889] P. Lambrineas, K. K. Yeung, and R. D. Finlayson, "Imaging of burned glass reinforced plastic (GRP) using Compton backscattering," in *Conf. on Characterization and NDE of Heat Damage in Graphite Epoxy Composites*, (Austin, Texas), Nondestructive Testing Information Center, April 1993.

[890] G. Harding, "Inelastic photon scattering: Effects and applications in biomedical science and industry," *Radiation Physics and Chemistry*, vol. 50, no. 1, pp. 91–111, 1997.

[891] L. Lawson, "Compton x-ray backscatter depth profilometry for aircraft corrosion inspection," *Materials Evaluation*, vol. 53, pp. 936–941, 1995.

[892] C. F. Poranski, Y. S. Ham, E. C. Greenawald, C. Draper, J. Chow, and L. Levenberry, "X-ray backscatter tomography for nondestructive evaluation at the Naval Research Lab.," in *Proc. Nondestructive Evaluation of Aging Maritime Applications, Oakland, CA*, vol. 2459, pp. 70–78, SPIE - International Society for Optical Engineering, 1995.

[893] C. F. Poranski, E. C. Greenawald, and Y. S. Ham, "X-ray backscatter tomography: NDT potential and limitations," *Materials Science Forum*, vol. 210-213, no. 11, pp. 211–218, 1996.

[894] E. C. Greenawald, J. B. Nagode, C. F. Poranski, and Y. S. Ham, "*In-Situ* NDE of navy sonar domes via x-ray backscatter," in *Review of Progress in Quantitative Nondestructive Evaluation* (D. O. Thompson and D. E. Chimenti, eds.), vol. 14, pp. 881–888, Plenum Press, 1995.

[895] E. C. Greenawald, C. F. Poranski, and D. Schafer, "X-ray backscatter tomography for NDE of steel reinforced rubber domes," in *Review of Progress in Quantitative Nondestructive Evaluation* (D. O. Thompson and D. E. Chimenti, eds.), vol. 10A, pp. 469–475, New York: Plenum Press, 1991.

[896] J. A. Stockes, K. R. Alvar, R. L. Corey, D. G. Costello, J. John, S. Kocimski, N. A. Lurie, A. P. Thayer, and J. C. Young, "Some new applications of collimated photon scattering for nondestructive examination," *Nuclear Instruments and Methods*, vol. 193, pp. 261–267, 1982.

[897] R. S. Holt, "Compton imaging," *Endeavour, New Series*, vol. 8, no. 2, pp. 97–105, 1985.

[898] V. E. Stepanov, O. P. Ivanov, V. N. Potapov, A. N. Sudarkin, and L. I. Urutskoev, "Application of gamma-ray imager for non-destructive testing," *Nuclear Instruments and Methods in Physics Research A*, vol. 422, pp. 724–728, 1999.

[899] S. W. Smith, "Digital image processing in the SECURE concealed-object detection system," in *Proc. Applications of Signal and Image Processing in Explosives Detection Systems, November 1992, Boston, MA*, pp. 194–198, SPIE - International Society for Optical Engineering, 1992.

[900] B. Bridge, F. Harirchian, D. C. Imrie, Y. Mehrabi, and A. R. Meragi, "Preliminary experiments with an automated three-dimensional Compton imaging system using a weak barium-133 source," *British Journal of Non-Destructive Testing*, vol. 31, pp. 134–139, March 1989.

[901] B. Bridge, F. Harirchian, D. C. Imrie, Y. Mehrabi, and A. R. Meragi, "Experiments in Compton-scatter imaging of materials with wide ranging densities using a low strength gamma source," *NDT International*, vol. 20, no. 6, pp. 339–346, 1987.

[902] D. R. Schafer and R. D. Swift, "AS&E cargo vehicle x-ray inspection system," in *Proc. Substance Identification Analytics, Innsbruck, Austria*, vol. 2093, pp. 472–482, SPIE - International Society for Optical Engineering, 1994.

[903] G. J. Smith, W. L. Adams, and S. Huang, "Mobile large vehicle inspection system design issues," in *Proc. 1998 Enforcement and Security Technologies, Boston, MA, USA*, vol. 3575, pp. 326–334, SPIE - International Society for Optical Engineering, 1999.

[904] R. D. Swift, "Mobile x-ray scatter imaging system for inspection of vehicles," in *Proc. Physics-Based Technologies for the Detection of Contraband* (L. O. Malotky and J. J. Pennella, eds.), vol. 2936, pp. 124–132, SPIE - International Society for Optical Engineering, 1997.

[905] R. D. Swift, "Conceptual development of a transportable/deployable x-ray inspection system for cars and vans," in *Proc. Physics-Based Technologies for the Detection of Contraband* (L. O. Malotky and J. J. Pennella, eds.), vol. 2936, pp. 133–139, SPIE - International Society for Optical Engineering, 1997.

[906] J. C. Campbell and A. M. Jacobs, "Detection of buried land mines by backscatter imaging," *Nuclear Science and Engineering*, vol. 110, pp. 417–424, 1992.

[907] C. Wells, Z. Su, J. Moore, E. Dugan, and A. Jacobs, "Lateral migration measured image signatures in the detection and identification of buried land mines," in *Proc. 1999 Detection and Remediation Technologies for Mines and Minelike Targets IV, Orlando, FL*, vol. 3710, pp. 906–916, SPIE - International Society for Optical Engineering, 1999.

[908] C. J. Wells, Z. Su, A. Allard, S. Salazar, E. Dugan, and A. Jacobs, "Suitability of simulated land mines for detection measurements using x-ray lateral migration radiography," in *Proc. Detection and Remediation Technologies for*

Mines and Minelike Targets V, Orlando, FL, vol. 4038, pp. 578–589, SPIE - International Society for Optical Engineering, 2000.

[909] Y. Watanabe, J. Monroe, S. Keshavmurthy, A. M. Jacobs, and E. Dugan, "Computational methods for shape restoration of buried objects in Compton backscatter imaging," *Nuclear Science and Engineering*, vol. 122, pp. 55–67, 1996.

[910] H. Lee and E. S. Kenney, "An improvement of Compton scatter imaging with wide aperture detectors - A Monte Carlo study," *IEEE Transactions on Nuclear Science*, vol. 38, pp. 812–827, April 1991.

[911] H. Lee and E. S. Kenney, "A new pipe wall thinning inspection system," *Nuclear Technology*, vol. 100, pp. 70–78, 1992.

[912] X. Xu, R. Gould, S. Khan, E. H. Klevans, and E. Kenney, "A high speed Compton scatter pipe wall imaging system," *Nuclear Instruments and Methods in Physics Research A*, vol. 353, pp. 334–337, December 1994.

[913] T. Zarabi and A. Kantzas, "Analysis of gamma-ray imaging data to describe hydrodynamic properties of gas-polyethylene fluidized beds," *Canadian Journal of Chemical Engineering*, vol. 78, pp. 849–857, October 2000.

[914] V. R. Bom, C. W. E. van Eijk, G. Jonkerst, P. T. Por, and J. G. G. van de Vorst, "Gamma-ray tomography in fluidized beds using a gamma camera," *Transactions of the Institute of Measurement and Control*, vol. 20, no. 4, pp. 178–185, 1998.

[915] V. Blet, P. Berne, S. Legoupil, and X. Vitart, "Radioactive tracing as aid for diagnosing chemical reactors," *Oil and Gas Science and Technology*, vol. 55, no. 2, pp. 171–183, 2000.

[916] G. Jonkers, E. A. van den Bergen, and P. A. Vermont, "Industrial applications of gamma-ray camera system - 1. Qualitative studies," *Applied Radiation and Isotopes*, vol. 41, no. 10/11, pp. 1023–1031, 1990.

[917] Y.-B. Huang and C. C. Gryte, "Gamma-camera imaging of oil displacement in thin slabs of porous media," *JPT, Journal of Petroleum Technology*, vol. 40, no. 10, pp. 1355–1360, 1988.

[918] R. Redus, M. Squillante, J. S. Gordon, P. Bennett, G. Entine, G. Knoll, D. Wehe, and S. Guru, "Imaging nuclear survey system," *IEEE Transactions on Nuclear Science*, vol. 43, pp. 1827–1831, June 1996.

[919] S. V. Guru, Z. He, D. K. Wehe, G. F. Knoll, R. H. Redus, and M. R. Squillante, "Portable high energy gamma ray imagers," *Nuclear Instruments and Methods in Physics Research A*, vol. 378, pp. 612–619, August 1996.

[920] O. Gal, F. Jean, F. Laine, and C. Leveque, "CARTOGAM portable gamma imaging system," *IEEE Transactions on Nuclear Science*, vol. 47, pp. 952–956, June 2000.

[921] O. Gal, C. Izac, F. Laine, and A. Nguyen, "CARTOGRAM: a portable gamma camera," *Nuclear Instruments and Methods in Physics Research A*, vol. 387, pp. 297–303, March 1997.

[922] G. W. Phillips, "Applications of Compton imaging in nuclear waste characterization and treaty verification," in *Proc. 1997 IEEE Nuclear Science Symposium, Albuquerque, NM*, vol. 1, pp. 362–364, IEEE, 1997.

[923] G. Phillips, "Applications of Compton imaging in nuclear waste characterization and treaty verification," in *Proc. 1997 IEEE Nuclear Science Symposium*, vol. 1, pp. 362–364, IEEE, 1997.

[924] G. J. Royle and R. D. Speller, "Compton scatter imaging of a nuclear industry site," in *Proceedings of the 1997 IEEE Nuclear Science Symposium, Albuquerque, NM*, vol. 1, pp. 365–368, IEEE, 1997.

[925] M. R. Hawkesworth, M. A. O'Dwyer, J. Walker, P. Fowles, J. Heritage, P. Stewart, R. Witcomb, J. E. Bateman, J. F. Connolly, and R. Stephenson, "A positron camera for industrial applications," *Nuclear Instruments and Methods in Physics Research A*, vol. 253, pp. 145–157, 1986.

[926] T. Wroblewski, "X-ray imaging using the radiation diffracted by polycrystalline materials," *Radiation Physics and Chemistry*, vol. 61, no. 3-6, pp. 329–332, 2001.

[927] T. Wroblewski, S. Geier, R. Hessmer, M. Schreck, and B. Rauschenbach, "X-ray imaging of polycrystalline materials," *Review of Scientific Instruments*, vol. 66, pp. 3560–3562, June 1995.

[928] B. Harold and F. Iddings, "Neutron radiography," Tech. Rep. NTIAC-SR-98-01, Nondestructive Testing Information Analysis Center, Austin, TX, 1998.

[929] R. C. Lanza, E. W. McFarland, and S. Shi, "Cooled CCD and amorphous silicon based neutron imaging system for low fluence neutron sources," in *Proc. Neutron in Research and Industry* (G. Vourvopoulos, ed.), vol. 2867, pp. 332–338, SPIE - International Society for Optical Engineering, 1996.

[930] E. McFarland and R. Lanza, "Recent advances in neutron radiography," in *Neutrons, X Rays, and Gamma Rays: Imaging Detectors, Material Characterization Techniques, and Applications* (J. M. Carpenter, D. B. Cline, R. Lanza, and D. F. Mildner, eds.), vol. 1737, pp. 101–111, SPIE - International Society for Optical Engineering, 1993.

[931] F. Groeschel, P. Schleuniger, A. Hermann, E. Lehmann, and L. Wiezel, "Neutron radiography of irradiated fuel rod segments at the SINQ: Loading, transfer and irradiation concept," *Nuclear Instruments and Methods in Physics Research A*, vol. 424, pp. 215–220, March 1999.

[932] J. F. W. Markgraf, S. McAllister, H. P. Leeflang, and K. H. van Otterdijk, "Non-destructive testing of light water reactor fuel rods at the HFR Petten," *Kerntechnik*, vol. 56, pp. 118–123, April 1991.

[933] A. Sinha, B. D. Bhawe, C. G. Panchal, A. Shyam, M. Srinivasan, and V. M. Joshi, "High sensitivity neutron imaging system for neutron radiography with

a small neutron source," *Nuclear Instruments and Methods in Physics Research B*, vol. 108, pp. 408–412, March 1996.

[934] W. E. Dance, J. R. Huriet, S. Cluzeau, H.-U. Mast, and F. Albisu, "DIANE: Advanced system for mobile neutron radiology," *Instruments and Methods in Physics Research B*, vol. 40-41, pp. 1316–1321, April 1989.

[935] W. E. Dance, S. Cluzeau, and H.-U. Mast, "Integration of an advanced sealed-tube neutron generator into a mobile neutron radiology system and resulting performance," *Nuclear Instruments and Methods in Physics Research B*, vol. 56/57, pp. 907–910, 1991.

[936] P. Bach, "Mobile neutron radiography systems and applications," *Insight: Non-Destructive Testing and Condition Monitoring*, vol. 42, no. 4, pp. 258–261, 2000.

[937] C. S., J. Huet, J. R. Huriet, and P. Le Tourneur, "Optimized thermal neutron source for neutron radiography with a static DIANE device," *Nuclear Instruments and Methods in Physics Research B*, vol. 79, pp. 851–854, 1993.

[938] S. Cluzeau, P. Le Tourneur, and W. E. Dance, "DIANE neutron radiography device performance comparison with reactor beam," *Nuclear Instruments and Methods in Physics Research B*, vol. 99, pp. 769–771, 1994.

[939] S. Cleazeau, J. Huet, and P. Le Tourneur, "DIANE stationary neutron radiography system image quality and industrial applications," *Nuclear Instruments and Methods in Physics Research B*, vol. 89, pp. 432–436, 1994.

[940] S. Cluzeau and P. Le Tourneur, "Stationary DIANE equipment description and performance of the thermal neutron source," *Nuclear Instruments and Methods in Physics Research B*, vol. 89, pp. 428–431, 1994.

[941] J. J. Antal and A. S. Marotta, "A neutron radiography system for field use," in *Nondestructive Testing and Evaluation for Manufacturing and Construction* (H. L. M. dos Reis, ed.), pp. 329–340, New York: Hemisphere, 1990.

[942] A. Yamaji, M. Koiso, Y. Kimoto, K. Mochiki, Y. Murata, and I. Ishikawa, "Cf-252 based neutron radiography using real-time image processing system," *Nondestructive Testing and Evaluation*, vol. 16, no. 2-6, pp. 435–444, 2001.

[943] ASTM, "E748-95: Standard Practices for Thermal Neutron Radiography of Materials." American Society for Testing and Materials, West Conshohocken, PA, 1995.

[944] ASTM, "E1496-97: Standard Test Method for Neutron Radiographic Dimensional Measurements." American Society for Testing and Materials, West Conshohocken, PA, 1997.

[945] ASTM, "E545-99: Standard Test Method for Determining Image Quality in Direct Thermal Neutron Radiographic Examination." American Society for Testing and Materials, West Conshohocken, PA, 1999.

[946] H. Sakaguchi, A. Kohzai, K. Hatakeyama, S. Fujine, K. Yoneda, K. Kanda, and T. Esaka, "Visualization of hydrogen in hydrogen storage alloys using

neutron radiography," *International Journal of Hydrogen Energy*, vol. 25, no. 12, pp. 1205–1208, 2000.

[947] B. F. Kammenzind, D. G. Franklin, H. R. Peters, and W. J. Duffin, "Hydrogen pickup and redistribution in alpha-annealed Zircaloy-4," in *Proc. 1995 11th International Symposium on Zirconium in the Nuclear Industry, Garmisch-Partenkirchen, Germany*, no. 1295 in ASTM Special Technical Publication, pp. 338–369, American Society for Testing Materials, 1996.

[948] M. Tamaki, S. Hara, I. Kondoh, K. Yoneda, A. Tsuruno, and H. Kobayashi, "Visualization and analysis of forced diffusion of hydrogen in palladium by electrotransport," *Nuclear Instruments and Methods in Physics Research A*, vol. 377, pp. 166–169, 1996.

[949] Y. Nir-El, V. Katz, J. Haddad, and D. Eliezer, "Determination of structure and composition in ceramics and aerospace materials by neutron radiography," *Nondestructive Testing and Evaluation*, vol. 11, no. 2-3, pp. 149–153, 1994.

[950] B. G. Martin, "An analysis of radiographic techniques for measuring resin content in graphite fiber reinforced epoxy resin composites," *Materials Evaluation*, vol. 35, pp. 65–68, 1977.

[951] W. E. Dance and J. B. Middlebrook, "Neutron radiographic non-destructive inspection for bonded composite structures," in *Non-destructive Evaluation and Flaw Criticality for Composite Materials* (R. B. Pipes, ed.), vol. ASTM of *STP 696*, pp. 57–71, American Society for Testing Materials, 1979.

[952] R. M. Tapphorn, P. Spencer, H. D. Beeson, and M. Kirsch, "Neutron radiography of composite materials - a comparison study of advanced NDE imaging techniques," *Nondestructive Testing and Evaluation*, vol. 11, no. 2-3, pp. 139–147, 1994.

[953] W. Lewis, B. Lewis, K. Jaansalu, D. Black, and C. Smith, "Suitability of neutron radiography for the inspection of advanced ceramic composites," *Nondestructive Testing and Evaluation*, vol. 11, no. 2-3, pp. 155–164, 1994.

[954] L. G. I. Bennett, M. L. Bickerton, and W. J. Lewis, "Development of techniques for the neutron radiography of CF188 flight control surfaces," *Nuclear Instruments and Methods in Physics Research A*, vol. 424, no. 1, pp. 104–112, 1999.

[955] H. P. Leeflang and J. F. W. Markgraf, "Detection of corrosion on aircraft components by neutron radiography," *Testing and Evaluation*, vol. 11, no. 2-3, pp. 115–128, 1994.

[956] K. Shields and W. J. Richards, "Aircraft inspection using neutron radioisotopic techniques," in *Proc. Nondestructive Evaluation of Aging Aircraft, Airports, Aerospace Hardware, and Materials, Oakland, CA*, vol. 2455, pp. 133–144, SPIE - International Society for Optical Engineering, 1995.

[957] V. R. Crispim and J. J. G. Da Silva, "Detection of corrosion in aircraft aluminum alloys," *Applied Radiation and Isotopes*, vol. 49, pp. 779–782, July 1998.

[958] L. Adams, J. Anderson, P. C. Ellen, H. P. Leeflang, and J. F. Markgraf, "Neutron radiography inspection of relays for satellite and space technology applications," in *Neutron Radiography (4)*, (Yverdon, Switzerland), pp. 115–122, Gordon and Breach, 1994.

[959] Y. Saito, K. Mishima, T. Hibiki, A. Yamamoto, J. Sugimoto, and K. Moriyama, "Application of high-frame-rate neutron radiography to steam explosion research," *Nuclear Instruments and Methods in Physics Research A*, vol. 424, pp. 142–147, 1999.

[960] H. Nakamura, Y. Sibamoto, Y. Anoda, K. Kukita, Yutakaand Mishima, and T. Hibiki, "Visualization of simulated molten-fuel behavior in a pressure vessel lower head using high-frame-rate neutron radiography," *Nuclear Technology*, vol. 125, pp. 213–224, February 1999.

[961] T. Nemec, J. Rant, V. Apih, and B. Glumac, "Study of building materials impregnation processes by quasi-real-time neutron radiography," *Nuclear Instruments and Methods in Physics Research A*, vol. 424, pp. 242–247, 1999.

[962] H. Justnes, K. Bryhn-Ingebrigtsen, and G. Rosvold, "Neutron radiography: An excellent method of measuring water penetration and moisture distribution in cementitious materials," *Advances in Cement Research*, vol. 6, pp. 67–72, April 1994.

[963] B. Tullis, J. T. Lindsay, and S. J. Wright, "Imaging of wetting front instabilities in porous media using neutron radioscopy," *Nondestructive Testing and Evaluation*, vol. 11, no. 2-3, pp. 97–106, 1994.

[964] J. Prazak, J. Tywoniak, F. Peterka, and T. Slonc, "Description of transport of liquid in porous media. a study based on neutron radiography data," *International Journal of Heat and Mass Transfer*, vol. 33, no. 6, pp. 1105–1120, 1990.

[965] J. K. Jasti and H. S. Fogler, "Application of neutron radiography to image flow phenomena in porous media," *AIChE Journal*, vol. 38, no. 4, pp. 481–488, 1992.

[966] L. Pel, A. A. J. Ketelaars, O. C. G. Adan, and A. A. Van Well, "Determination of moisture diffusivity in porous media using scanning neutron radiography," *International Journal of Heat and Mass Transfer*, vol. 36, pp. 1261–1267, March 1993.

[967] R. J. Bellows, M. Y. Lin, M. Arif, A. K. Thompson, and D. Jacobson, "Neutron imaging technique for in situ measurement of water transport gradients within Nafion in polymer electrolyte fuel cells," *Journal of the Electrochemical Society*, vol. 146, pp. 1099–1103, March 1999.

[968] T. Fujii, J. Ohta, N. Takenaka, and H. Asano, "Study of performance characteristics of nozzles for subcooled hot water. Effect of induced turbulence on

nozzle performance," in *Proc. 3rd ASME/JSME Thermal Engineering Joint Conference, Part 2 (of 5), Reno, NV*, pp. 175–181, 1991.

[969] T. Hibiki and K. Mishima, "Visualization and measurements of two-phase flows in metallic ducts using neutrons as microscopic probes (3rd report, quantitative measurement of neutron radiography image)," *Transactions of the Japan Society of Mechanical Engineers, Part B*, vol. 62, pp. 919–926, March 1996.

[970] T. Hibiki, K. Mishima, and H. Nishihara, "Measurement of radial void fraction distribution of two-phase flow in a metallic round tube using neutrons as microscopic probes," *Nuclear Instruments and Methods in Physics Research A*, vol. 399, pp. 432–438, November 1997.

[971] T. Hibiki, Y. Saito, K. Mishima, Y. Tobita, K. Konishi, and M. Matsubayashi, "Application of high-frame-rate neutron radiography with a steady thermal neutron beam to two-phase flow measurements in a metallic rectangular duct," *Nuclear Technology*, vol. 110, pp. 422–434, June 1995.

[972] T. Hibikit and K. Mishima, "Feasibility of high frame-rate neutron radiography by using a steady thermal neutron beam with 10^6 n/(cm^{2}s) flux," *Nuclear Instruments and Methods in Physics Research A*, vol. 369, pp. 186–194, 1996.

[973] K. Uchimura, G. D. Harvel, T. Matsumoto, M. Kanzaki, and J.-S. Chang, "Image processing approach for two-phase interfaces visualized by a real time neutron radiography technique," *Flow Measurement and Instrumentation*, vol. 9, pp. 203–210, December 1998.

[974] N. Takenaka, H. Asano, T. Fujii, and M. Matsubayashi, "Three-dimensional visualization of void fraction distribution in steady two-phase flow by thermal neutron radiography," *Nuclear Engineering and Design*, vol. 84, pp. 203–212, August 1998.

[975] N. Takenaka, H. Asano, T. Fujii, T. Wada, M. Matsubayashi, and A. Tsuruno, "Three-dimensional void fraction measurement of two-phase flow in a rod bundle by neutron radiography," *Nuclear Instruments and Methods in Physics Research A*, vol. 377, pp. 115–118, July 1996.

[976] N. Takenaka, H. Asano, T. Fujii, Y. Murata, K.-i. Mochiki, A. Taguchi, M. Matsubayashi, and A. Tsuruno, "Void fraction distribution measurement in two-phase flow by real-time neutron radiography and real-time image processing," *Nuclear Instruments and Methods in Physics Research A*, vol. 377, pp. 153–155, July 1996.

[977] K. Mishima and T. Hibiki, "Quantitative method to measure void fraction of two-phase flow using electronic imaging with neutrons," *Nuclear Science and Engineering*, vol. 124, pp. 327–338, October 1996.

[978] G. D. Harvel, K. Hori, K. Kawanishi, and J. S. Chang, "Real-time cross-sectional averaged void fraction measurements in vertical annulus gas-liquid two-phase flow by neutron radiography and x-ray tomography techniques," *Nuclear Instruments and Methods in Physics Research A*, vol. 371, pp. 544–552, March 1996.

[979] G. Harvel, J. Chang, and V. S. Krishnan, "Determination of time-dependent void fraction distribution in bubbly two-phase flow by a real-time neutron radiography technique," *Nuclear Technology*, vol. 109, pp. 132–141, December 1994.

[980] Y. Tsuji, S. Matsueda, M. Oda, M. Matsuda, T. Yagi, M. Tamaki, M. Matsubayashi, and S. Fujine, "Visualization and correlation analysis of counter-current two-phase flow in a thermosyphon by neuron radiography," *Nuclear Instruments and Methods in Physics Research A*, vol. 377, pp. 148–152, July 1996.

[981] T. Iwamura, H. Watanabe, and Y. Murao, "Critical heat flux experiments under steady-state and transient conditions and visualization of CHF phenomenon with neutron radiography," *Nuclear Engineering and Design*, vol. 149, pp. 195–206, September 1994.

[982] M. Ozawa, H. Umekawa, T. Matsuda, N. Takenaka, and M. Matsubayashi, "Flow pattern and heat transfer in tube banks of a simulated fluidized-bed heat exchanger," *JSME International Journal, Series B*, vol. 41, pp. 720–726, August 1998.

[983] M. Ozawa, H. Umekawa, T. Matsuda, N. Takenaka, A. Tsuruno, Matsubayashi, and Masahito, "Void fraction profile in tube-banks of a simulated fluidized-bed heat exchanger," *Nuclear Instruments and Methods in Physics Research A*, vol. 377, pp. 144–147, July 1996.

[984] T. Hibiki, Y. Saito, K. Mishima, Y. Tobita, K. Konishi, and M. Matsubayashi, "Study on flow characteristics in gas-molten metal mixture pool," *Nuclear Engineering and Design*, vol. 196, no. 2, pp. 233–245, 2000.

[985] K. Mishima, T. Hibiki, Y. Saito, H. Nishihara, Y. Tobita, K. Mishima, T. Hibiki, Y. Saito, H. Nishihara, and Y. Tobita, "Visualization and measurement of gas-liquid metal two-phase flow with large density difference using thermal neutrons as microscopic probes," *Nuclear Instruments and Methods in Physics Research A*, vol. 424, pp. 229–234, March 1999.

[986] N. Takenaka, T. Fujii, Y. Shibata, T. Ebisu, and M. Matsubayashi, "Visualization and measurement of refrigerant flow in compression-type refrigerator by neutron radiography," *Nuclear Instruments and Methods in Physics Research A*, vol. 424, pp. 98–103, March 1999.

[987] N. Takenaka, H. Asano, T. Fujii, T. Ushiro, J. Iwatani, Y. Murata, K.-I. Mochiki, A. Taguchi, M. Matsubayashi, and A. Tsuruno, "Application of neutron radiography to visualization of cryogenic fluid boiling two-phase flows," *Nuclear Instruments and Methods in Physics Research A*, vol. 377, pp. 174–176, July 1996.

[988] M. Balasko, E. Svab, A. Vida, and I. Szikra, "Neutron radiography visualization of internal processes in refrigerators," *Physica B: Condensed Matter*, vol. 234-236, pp. 1033–1034, June 1997.

[989] M. Balasko, E. Svab, and A. Nedelik, "Neutron radiography investigation of absorption and compression refrigerators," *Nondestructive Testing and Evaluation*, no. 2-3, pp. 69–76, 1994.

[990] H. Asano, N. Takenaka, T. Fujii, Y. Murata, K. Mochiki, A. Taguchi, M. Matsubayashi, and A. Tsuruno, "Application of neutron radiography to a diagnosis of refrigerator," *Nuclear Instruments and Methods in Physics Research A*, vol. 377, pp. 170–173, July 1996.

[991] J. Du Parquet and G. Bayon, "Contribution of neutron radiography to the conservation of the environment," *Nondestructive Testing and Evaluation*, vol. 11, no. 2-3, pp. 87–95, 1994.

[992] J. Furukawa, T. M. Nakanishi, and M. Matsubayashi, "Neutron radiography of a root growing in soil with vanadium," *Nuclear Instruments and Methods in Physics Research A*, vol. 424, pp. 116–121, March 1999.

[993] M. F. Middleton, "Petrophysical applications of neutron radiography," *Nondestructive Testing and Evaluation*, vol. 16, no. 2-6, pp. 321–333, 2001.

[994] F. Korosi, M. Balasko, and E. Svab, "Dynamic neutron radiography study of oil infiltration in sandstone," *Nondestructive Testing and Evaluation*, vol. 16, no. 2-6, pp. 309–319, 2001.

[995] G. Bayon, B. Winkler, A. Kahle, B. Hennion, and P. Boutrouille, "Application of dynamic neutron imaging in the earth sciences to determine viscosities and densities of silicate melts," *Nondestructive Testing and Evaluation*, vol. 6, no. 2-6, pp. 287–296, 2001.

[996] K. Kato, G. Matsumoto, Y. Haga, N. Niimura, and M. Matsubayashi, "Experiences of backward scattered neutron measurements using a neutron sensitive imaging plate," *Nondestructive Testing and Evaluation*, vol. 16, no. 2-6, pp. 413–422, 2001.

[997] T. Buchberger, H. Rauch, and E. Seidl, "Tritium and helium 3 in metals investigated by neutron radiography," *Kerntechnik*, vol. 53, no. 3, pp. 215–217, 1989.

[998] M. Kamata, T. Esaka, S. Fujine, K. Yoneda, and K. Kanda, "Lithium batteries: Application of neutron radiography," *Journal of Power Sources*, vol. 68, pp. 459–462, October 1997.

[999] T. Esaka, M. Kamata, K. Takami, S. Takai, S. Fujine, K. Yoneda, and K. Kanda, "Application of NR to study the lithium ion transfer in solid ionic conductors," *Key Engineering Materials*, vol. 132-136, no. Pt 2, pp. 1393–1396, 1997.

[1000] T. Esaka, "Ionic conduction in substituted scheelite-type oxides," *Solid State Ionics*, vol. 136-137, pp. 1–9, November 2000.

[1001] T. Esaka, M. Kamata, and H. Saito, "Alkali ion conduction in the substituted phase based on the scheelite-type oxide," *Solid State Ionics*, vol. 86-88, no. n pt 1, pp. 73–76, 1996.

[1002] M. Kamata, T. Esaka, K. Takami, S. Takai, S. Fujine, K. Yoneda, and K. Kanda, "Studies on the lithium ion conduction in $Ca_{0.95}Li_{0.10}WO_4$ using cold neutron radiography," *Solid State Ionics*, vol. 91, pp. 303–306, October 1996.

[1003] M. Kamata, T. Esaka, N. Kodama, S. Fujine, K. Yoneda, and K. Kanda, "Application of neutron radiography to visualize the motion of lithium ions in lithium-ionconducting materials," *Journal of the Electrochemical Society*, vol. 143, pp. 1866–1870, 1996.

[1004] M. Kamata, T. Esaka, S. Fujine, K. Yoneda, and K. Kanda, "Application of NR for research in electrochemical systems," *Nuclear Instruments and Methods in Physics Research A*, vol. 377, pp. 161–165, July 1996.

[1005] W.-M. Chia, C.-Y. Chuang, C.-S. Chen, T.-R. Yeh, and C.-H. Tseng, "Case studies of neutron radiography examination at INER," *Nuclear Instruments and Methods in Physics Research A*, vol. 377, pp. 16–19, July 1996.

[1006] R. R. Tsukimura, A. W. Meren, and V. P. Scott, "Detection of residual core in air-cooled turbine blades using neutron radiography," in *Proc. Nondestructive Evaluation of Aging Aircraft, Airports, Aerospace Hardware,and Materials, Oakland, CA*, vol. 2455, pp. 157–162, SPIE - International Society for Optical Engineering, 1995.

[1007] G. C. McClellan, "Neutron radiography techniques and applications," in *Nondestructive Testing and Evaluation for Manufacturing and Construction* (H. L. M. dos Reis, ed.), pp. 319–328, New York: Hemisphere, 1990.

[1008] R. Pugliesi and M. L. G. Andrade, "Study of cracking in concrete by neutron radiography," *Applied Radiation and Isotopes*, vol. 48, pp. 339–344, March 1997.

[1009] D. D. McRae, R. W. J. Jenkins, J. S. Brenizer, M. F. Sulcoski, and T. G. Williamson, "Real-time observation of aerosol deposition by neutron radiography," in *Industrial Applications of Airborne Particles, Proceedings of the First International Aerosol Conference, Minneapolis, MN*, pp. 241–244, Elsevier Conference, 1984.

[1010] J. S. Brenizer, K. W. J. Tobin, R. W. J. Jenkins, D. D. McRae, and M. F. Sulcoski, "Evaluation of filter behavior by neutron radiography," *Journal of Aerosol Science*, vol. 18, pp. 311–320, June 1987.

[1011] N. Takenaka, H. Asano, T. Fujii, Y. Motomura, A. Ono, M. Matsubayashi, and A. Tsuruno, "Liquid metal flow measurement by neutron radiography," *Nuclear Instruments and Methods in Physics Research A*, vol. 377, pp. 156–160, July 1996.

[1012] Y. Sibamoto, H. Nakamura, and Y. Anoda, "Neutron radiography flow visualization of liquid metal injected into an empty vessel and a vessel containing saturated water," *Nuclear Technology*, vol. 133, pp. 119–132, January 2001.

[1013] H. Umekawa, M. Ozawa, N. Takenaka, and M. Matsubayashi, "Visualization of bed material movement in a simulated fluidized bed heat exchanger by neutron radiography," *Nuclear Instruments and Methods in Physics Research A*, vol. 424, pp. 77–83, March 1999.

[1014] B. Kim and P. J. Prescott, "Neutron radiographic measurement of macrosegregation in an experimentally solidified binary metal alloy," in *Proc. 1996*

31st ASME National Heat 31st ASME National Heat Transfer Conference. Part 1 (of 8), Houston, TX, vol. 323 of *HTD*, pp. 9–16, American Society of Mechanical Engineers, Heat Transfer Division, 1996.

[1015] T. Guzel, S. Tretyakova, T. Akyuz, Y. T. Oganessian, C. Bolcal, P. Saritepe, N. Cagatay, and N. Kiratli, "Determination of concentrations of fissionable elements in the Black Sea sediment samples before and after Chernobyl using neutron radiography method," *Radiation Measurements*, vol. 28, no. 1-6, pp. 405–408, 1997.

[1016] J. Cassidy, "Pantex neutron radiography facility," in *Neutron Radiography (4)* (J. P. Bartyon, ed.), pp. 451–451, Gordon and Breach, Yverdon, Switzerland, 1994.

[1017] J. P. Barton, J. W. Bader, J. S. Brenizer, and B. Hosticka, "Feasibility of neutron radiography for space shuttle inspection," in *Neutron Radiography (4)* (J. P. Bartyon, ed.), pp. 123–131, Gordon and Breach, Yverdon, Switzerland, 1994.

[1018] M. Balasko, F. Korosi, E. Svab, and I. Eordogh, "Novel type epithermal neutron radiography detecting and imaging system," *Nuclear Instruments and Methods in Physics Research A*, vol. 424, no. 1, pp. 263–269, 1999.

[1019] K. Yoshii and K. Miya, "Fast neutron radiography at reactor YAYOI," *Progress in Nuclear Energy*, vol. 32, no. 1-2, pp. 123–139, 1998.

[1020] J. S. Brenizer, H. Berger, K. M. Gibbs, P. Mengers, C. T. Stebbings, D. Polansky, and D. J. Rogerson, "Development of a new electronic neutron imaging system," *Nuclear Instruments and Methods in Physics Research A*, vol. 424, pp. 9–14, November 1999.

[1021] R. Loveman, J. Bendahan, T. Gozani, and J. Stevenson, "Time of flight fast neutron radiography," *Nuclear Instruments and Methods in Physics Research B*, vol. 99, pp. 765–768, May 1995.

[1022] N. Takenaka, H. Asano, T. Fujii, M. Mizubata, and K. Yoshii, "Application of fast neutron radiography to three-dimensional visualization of steady two-phase flow in a rod bundle," *Nuclear Instruments and Methods in Physics Research A*, vol. 424, pp. 73–76, March 1999.

[1023] W. C. Sailor, R. C. Byrd, A. Gavron, and R. Hammock, "Neutron source imaging detector for nuclear arms treaty verification," *Nuclear Science and Engineering*, vol. 109, pp. 267–277, November 1991.

[1024] K. H. Kim, R. T. Klann, and B. B. Taju, "Fast neutron radiography for composite materials evaluation and testing," *Nuclear Instruments and Methods in Physics Research A*, vol. 422, pp. 929–932, 1999.

[1025] A. Gavron, K. Morley, C. Morris, S. Seestrom, J. Ullmann, G. Yates, and J. Zumbro, "High energy neutron radiography," in *Neutrons in Research and Industry* (G. Vourvopoulos, ed.), vol. 2867, pp. 326–331, SPIE - International Society for Optical Engineering, 1997.

[1026] C. Morris, V. Armijo, L. G. Atnecio, A. Bridge, A. Gavron, G. Hart, K. Morley, T. Mottershead, G. J. Yates, and J. Zumbro, "An integrating image detector for high energy neutrons," in *Proc. Neutron in Research and Industry* (G. Vourvopoulos, ed.), vol. 2867, pp. 351–357, SPIE - International Society for Optical Engineering, 1997.

[1027] Y. Ikeda, K. Ohkubo, G. Matsumoto, T. Nakamura, Y. Nozaki, S. Wakasa, Y. Toda, and T. Kato, "Fast neutron radiography testing for components of launch vehicles by a Baby-Cyclotron," in *Nondestructive Testing and Evaluation for Manufacturing and Construction* (H. L. M. dos Reis, ed.), pp. 349–359, New York: Hemisphere, 1990.

[1028] S. Yamamoto, K. Yoneda, S. A. Hayashi, I. Kobayashi, Katsuheiand Kimura, T. Suzuki, H. Nishihara, and S. Kanazawa, "Application of iron-filtered neutrons to radiography of a copper step within a large iron block and to computer tomography of metallic cylinders," *Nuclear Instruments and Methods in Physics Research A*, vol. 225, pp. 439–444, August 1984.

[1029] T. Wroblewski, E. Jansen, W. Schaefer, and R. Skowronek, "Neutron imaging of bulk polycrystalline materials," *Nuclear Instruments and Methods in Physics Research A*, vol. 423, pp. 428–434, March 1999.

[1030] W. K. Ng and C. T. J. Dodson, "Rapid formation performance test," in *Proc. 1995 International Paper Physics Conference, Niagara-on-the-Lake, ON*, pp. 49–53, 1995.

[1031] L. Wong, M. T. Kortschot, and C. T. J. Dodson, "Effect of formation on local strain fields and fracture of paper," in *Proc. 81st Annual Meeting on Technical Section Canadian Pulp and Paper Association, Montreal, QC*, pp. B215–B218, Canadian Pulp and Paper Association, 1995.

[1032] F. M. Kerr, "Using low energy β-sources for radiography," *Nuclear Instruments and Methods in Physics and Research A*, vol. 257, pp. 26–28, 1986.

[1033] J. Rant, G. Pregl, P. Leskovar, and F. Zitnik, "Radiography of surface layers with backdiffused β-rays and photon excited electrons," *Applied Radiation and Isotopes*, vol. 33, pp. 207–216, 1982.

[1034] G. K. Glass, R. Yang, T. Dickhaus, and N. R. Buenfeld, "Backscattered electron imaging of the steel-concrete interface," *Corrosion Science*, vol. 43, pp. 605–610, April 2001.

[1035] D. G. . A. Nelson, "Backscattered electron imaging of partially-demineralized enamel," *Scanning Microscopy*, vol. 4, pp. 31–42, March 1990.

[1036] N. S. P. King *et al.*, "800-MeV proton radiography facility for dynamic experiments," *Nuclear Instruments and Methods in Physics Research A*, vol. 424, pp. 84–91, March 1999.

[1037] G. E. Hogan *et al.*, "Proton radiography," in *Proc. 18th Biennial Particle Accelerator Conference, New York, NY*, vol. 1, pp. 579–583, IEEE, 1999.

[1038] O. J. Alford *et al.*, "Scrounge-atron: A phased approach to the advanced hydrotest facility utilizing proton radiography," in *Proc. 18th Biennial Particle Accelerator Conference, New York, NY*, vol. 4, pp. 2590–2592, IEEE, 1999.

[1039] N. S. P. King, E. Ables, K. Adams, K. R. Alrick, J. F. Amann, S. Balzar, P. D. J. Barnes, M. L. Crow, S. B. Cushing, J. C. Eddleman, T. T. Fife, P. Flores, D. Fujino, R. A. Gallegos, N. T. Gray, *et al.*, "800-MeV proton radiography facility for dynamic experiments," *Nuclear Instruments and Methods in Physics Research A*, vol. 424, no. 1, pp. 84–91, 1999.

[1040] J. Saudinos, G. Charpak, F. Sauli, D. Townsend, and J. Vinciarelli, "Nuclear scattering applied to radiography," *Physics in Medicine and Biology*, vol. 20, no. 6, pp. 890–905, 1975.

[1041] J. C. Duchazeaubeneix, J. C. Faivre, D. Garreta, B. Guillerminet, D. Legrand, M. Rouger, J. Saudinos, G. Charpak, G. Melchart, Y. Perrin, J. C. Santiard, and F. Sauli, "Proton nuclear scattering radiography," *IEEE Transactions on Nuclear Science*, vol. NS-30, pp. 601–604, February 1982.

[1042] J. Saudinos, "Imaging with high energy particles," *Journal of Optics*, vol. 18, pp. 193–198, July/August 1987.

[1043] C. Yang, N. P. O. Larrson, E. Swictlicki, K. G. Malmqvist, D. N. Jameison, and C. G. Ryan, "Imaging with ionoluminescence (IL) in a nuclear mircoprope," *Nuclear Instruments and Methods in Physics Research B*, vol. 77, pp. 188–194, 1993.

[1044] P. C. Ragland, "Autoradiographic investigations of naturally occurring materials," in *The Natural Radiation Environment* (J. A. S. Adams and W. M. Lowder, eds.), ch. 7, Chicago: The University of Chicago Press, 1964.

[1045] G. R. Chilton and K. J. Wilson, "Aspects of the certification of experimental mixed oxide fuel and fuel pins for use in the UK fast reactor programme," *Journal of Nuclear Materials*, vol. 153, pp. 127–132, April 1987.

[1046] J. K. Ghosh, J. P. Panakkal, K. N. Chandrasekharan, and P. R. Roy, "Combined x-radiography-gamma-autoradiography (XGAR) technique for characterisation of welded nuclear fuel pins," in *Proc. 11th World Conference on Nondestructive Testing, Las Vegas, NV*, (Columbus, OH), pp. 1174–1180, Int Committee on Nondestructive Testing, 1985.

[1047] B. B. Shriwastwa, B. Raghunath, and J. K. Ghosh, "Selective alpha autoradiography for monitoring thorium distribution in UO_2-ThO_2 fuel pellets," *Kerntechnik*, vol. 57, pp. 283–285, October 1992.

[1048] H. Furuya and M. Koizumi, "Surface diffusion of plutonium on uranium dioxide," in *Proc. Symp. Applied Physics, Thermodynamics of Nuclear Materials*, vol. STI/PUB/38, (Vienna, Austria), pp. 447–456, Int. Atomic Energy Agency, October 1974.

[1049] P. Vater, C. Laue, I. G. Bersina, H. H. Cui, S. Vonderau, R. Brandt, and J. Roesch, "Rapid determination of transuranium nuclides in the primary coolant of a nuclear power plant by using nuclear track micro filters and investigation of α-emitting particles on the loaded filters by α-autoradiography," *Radiation Measurements*, vol. 25, no. 1-4, pp. 377–380, 1995.

[1050] N. A. Keller, "Autoradiography of metallic matrices," tech. rep., Atomic Energy of Canada Limited, April 1982.

[1051] M. Au, "Visible technology for display of hydrogen distribution in metals," *Journal of Alloys and Compounds*, vol. 293, pp. 317–323, 1999.

[1052] S. Rosanvallon, G. Marbach, A. M. Brass, J. Chene, and J. P. Daclin, "Steel detritiation, optimization of a process," *Fusion Engineering and Design*, vol. 51-52, pp. 605–609, November 2000.

[1053] G. Katano, H. Saito, and M. Mori, "Observation of the hydrogen distribution in Ni_3Al intermetallic compound," *Journal of the Japan Institute of Metals*, vol. 64, no. 4, pp. 257–265, 2000.

[1054] H. Saitoh, H. Naganuma, T. Misawa, T. Tamura, Takuyaand Kuriiwa, A. Kamegawa, M. Okada, and T. Ohnishi, "Tritium radioluminography in $V_{3.5}$(Zr, Ti, Ni) protium absorption alloy," *Journal of the Japan Institute of Metals*, vol. 63, no. 9, pp. 1085–1088, 1999.

[1055] M. B. Jensen, H. C. B. Hansen, P. R. Jorgensen, and J. Magid, "Leaching of phosphate in structured soil - a two step process," *Nordic Hydrology*, vol. 30, no. 4-5, pp. 361–378, 1999.

[1056] I. E. Animitsa, L. I. Volosentseva, and V. M. Zhukovsky, "Diffusion behaviour of ^{22}Na in the poly (ethylene oxide)-poly (vinyl acetate) system," *Solid State Ionics*, vol. 119, no. 1, pp. 245–249, 1999.

[1057] M. Taniguchi, N. Gennai, and T. Makita, "Observation of distribution images of light elements in metals by the charged particle activation-autoradiography using imaging plate," *Journal of the Iron and Steel Institute of Japan*, vol. 85, pp. 174–179, February 1999.

[1058] J. Kloch, B. Billia, T. Okane, T. Umeda, and W. Wolczynski, "Experimental verification of the solute redistribution in cellular/dendritic solidification of the Al-3.5Li and Fe-4.34Ni alloys," *Materials Science Forum*, vol. 329, pp. 31–36, 2000.

[1059] D. Fink, S. Koul, and H. Kranz, "Spatial distributions of boron in silicon obtained by combination of two-dimensional bevelling and autoradiography," *Applied Physics A: Materials Science and Processing*, vol. 62, pp. 13–17, January 1996.

[1060] I. G. Berzina, A. V. Buryakin, E. B. Gusev, and G. N. Fedina, "Boron distribution in zones of laser hardening of certain alloys from (n, α) radiography data," *Soviet Physics Journal*, vol. 3, pp. 202–205, March 1984.

[1061] A. Lazaru, R. Ilic, J. Skvarc, E. Kristof, and T. Stafilov, "Neutron induced autoradiography of some minerals from the Allchar mine," *Radiation Measurements*, vol. 31, no. 1, pp. 677–682, 1999.

[1062] C. O. Fischer, M. Gallagher, C. Laurenze, C. Schmidt, and K. Slusallek, "Digital imaging of autoradiographs from paintings by Georges de La Tour (1593-1652)," *Nuclear Instruments and Methods in Physics Research A*, vol. 424, pp. 258–262, March 1999.

[1063] N. Eastaugh, "Examination of paintings by infra-red and other techniques," in *IEE Colloquium (Digest)*, pp. 6/1–6/3, IEE Science, Education and Technology Division, Colloquium in NDT in Archaeology and Art, London, UK, 1995.

[1064] A. Bernasik, J. Gilewicz-Wolter, and K. Kowalski, "Transport of oxygen in the scales growing on nickel and cobalt in SO_2 atmosphere," *Diffusion and Defect Data Pt.B: Solid State Phenomena*, vol. 72, pp. 41–46, 2000.

[1065] C. Rouleau, K. Borg-Neczak, J. Gottofrey, and H. Tjalve, "Accumulation of waterborne mercury (II) in specific areas of fish brain," *Environmental Science and Technology*, vol. 33, no. 19, pp. 3384–3389, 1999.

[1066] C. Rouleau, C. Gobeil, and H. Tjalva, "Pharmacokinetics and distribution of dietary tributyltin and methylmercury in the snowcrab (chionoecetes opilio)," *Environmental Science and Technology*, vol. 33, no. 19, pp. 3451–3457, 1999.

[1067] R. Pollanen, T. Ikaheimonen, S. Klemola, and J. Juhanoja, "Identification and analysis of a radioactive particle in a marine sediment sample," *Journal of Environmental Radioactivity*, vol. 45, no. 2, pp. 149–160, 1999.

[1068] W. R. Russ, W. Chung, and J. D. Valentine, "Determination of radiological contamination penetration in fernald transite panels," *Radioactive Waste Management and Environmental Restoration*, vol. 21, no. 3, pp. 155–174, 1999.

[1069] P. Klein, T. Aurisch, P. Buchholz, P. Fischer, M. Loecker, Neeser, W., L. Strueder, M. Trimpl, J. Ulrici, J. Vocht, and N. Wermes, "A DEPFET pixel Bioscope for the use in autoradiography," *Nuclear Instruments and Methods in Physics Research A*, vol. 454, pp. 152–157, November 2000.

[1070] W. Neeser, M. Boecker, P. Buchholz, P. Fischer, P. Holl, J. Kemmer, P. Klein, H. Koch, M. Loecker, G. Lutz, H. Matthaey, L. Strueder, M. Trimpl, J. Ulrici, and N. Wermes, "DEPFET pixel bioscope," *IEEE Transactions on Nuclear Science*, vol. NS 47, pp. 1246–1250, June 2000.

[1071] H. E. Martz, S. G. Azevedo, J. Braske, K. Waltjen, and D. Schneberk, "Computed tomography systems and their industrial applications," *Applied Radiation and Isotopes*, vol. 41, pp. 943–961, 10-11 1990.

[1072] E. A. Sivers, "Use of multiple CT scans to accommodate large objects and stretch dynamic range of detectablity," *Nuclear Instruments and Methods in Physics Research B*, vol. 99, pp. 761–764, 1995.

[1073] J. E. Youngberg and P. Burstein, "Three-dimensional computed tomography for rocket motor inspection," in *Proc. Nondestructive Evaluation of Aging Aircraft, Airports, Aerospace Hardware,and Materials, Oakland, CA*, vol. 2455, pp. 291–298, SPIE - International Society for Optical Engineering, 1995.

[1074] J. T. De Assis, R. T. Lopes, and J. L. Rodrigues, "Microtomography using microfocus radiography system," in *Proc. 1993 IEEE Nuclear Science Symposium and Medical Imaging Conference, San Francisco, CA*, pp. 1731–1733, IEEE, 1994.

[1075] S. F. Burch and P. F. Lawrence, "Recent advances in computerised x-ray tomography using real-time radiography equipment," *British Journal of Non-Destructive Testing*, vol. 34, pp. 129–133, March 1992.

[1076] G. L. Hoeft, S. H. C. Hughes, and R. E. Slocum, "Low cost computed tomography system using an image intensifier," in *Proc. Applications of Optical Engineering: Proceedings of OE/Midwest '90, Rosemont, IL*, vol. 1396, pp. 638–645, SPIE - International Society for Optical Engineering, 1991.

[1077] U. Kumar, G. S. Ramakrishna, S. S. Datta, and V. R. Ravindran, "Prototype gamma-ray computed tomographic imaging system for industrial applications," *Insight: Non-Destructive Testing and Condition Monitoring*, vol. 42, pp. 662–666, October 2000.

[1078] I. L. Morgan, "Real-time digital gauging for process control," *Applied Radiation and Isotopes*, vol. 41, no. 10/11, pp. 935–942, 1990.

[1079] ASTM, "E1441-00 Standard Guide for Computed Tomography (CT) Imaging." American Society for Testing and Materials, West Conshohocken, PA, 2000.

[1080] ASTM, "E1570-00 Standard Practice for Computed Tomographic (CT) Examination." American Society for Testing and Materials, West Conshohocken, PA, 2000.

[1081] ASTM, "E1672-95 (2001) Standard Guide for Computed Tomography (CT) System Selection." American Society for Testing and Materials, West Conshohocken, PA, 2001.

[1082] ASTM, "E1695-95 (2001) Standard Test Method for Measurement of Computed Tomography (CT) System Performance." American Society for Testing and Materials, West Conshohocken, PA, 2001.

[1083] ASTM, "E1935-97 Standard Test Method for Calibrating and Measuring CT Density." American Society for Testing and Materials, West Conshohocken, PA, 1997.

[1084] ASTM, "E1814-96 Standard Practice for Computed Tomographic (CT) Examination of Castings." American Society for Testing and Materials, West Conshohocken, PA, 1996.

[1085] N. MacCuaigm, "The industrial applications of low cost gamma-ray tomography," *British Journal of NDT*, vol. 33, pp. 111–114, March 1991.

[1086] R. Polichar and D. Kedem, "Application of x-ray tomography to crack imaging in steel," in *Proc. Through the Eyes of an Eagle: 11th World Conference on Nondestructive Testing., Las Vegas, NV*, Int. Committee on Nondestructive Testing, 1985.

[1087] B. D. Sawicka and R. L. Tapping, "CAT scanning of hydrogen-induced cracks in steel," *Nuclear and Methods in Nuclear Methods in Physics Research A*, vol. 256, pp. 103–111, 1987.

[1088] T. L. Kohutek and K. H. Frank, "Application of photon tomography to the inspection of bridge weldments," in *Presented at the Winter Annual Meeting, Miami Beach*, 85-WA/NDE-2, pp. 1–6, American Society of Mechanical Engineers, November 1985.

[1089] K. R. Squire, C. L. Vorres, D. E. Yuhas, and C. S. Lee, "Experimental verification of fiber orientation predicted by computer simulation," in *ANTEC 89 - 47th Annual Technical Conference of SPE, New York, NY*, pp. 253–256, Society of Plastics Engineers., 1989.

[1090] T. Rabe, J. Goebbels, and A. Kunzmann, "Control of density distribution in green components," *Industrial Ceramics*, vol. 20, pp. 193–195, December 2000.

[1091] B. D. Sawicka, J. G. Murphy, F. Taheri, and L. E. Kanary, "X-ray CT examination of density gradients in slip cast ceramics. correlation with stress distribution," *Nuclear Instruments and Methods in Physics Research B*, vol. 69, pp. 365–369, June 1992.

[1092] M. C. Stinson, D. A. Holloway, and W. A. Ellingson, "Quantification of volume flaw sizes in ceramics by images processing of 3D x-ray tomographic data sets," in *Proc. 17th Annual Conference on Composites and Advanced Ceramic Materials,Cocoa Beach, FL*, vol. 14, pp. 485–491, July/August 1993.

[1093] G. Y. Baaklini, R. T. Bhatt, A. J. Eckel, P. Engler, M. G. Castelli, and R. W. Rauser, "X-ray microtomography of ceramic and metal matrix composites," *Materials Evaluation*, vol. 53, pp. 1040–1044, Sep 1995.

[1094] V. I. Barakhov, V. A. Chernyaeva, V. S. Kiselev, and A. P. Stepanov, "Aspects of the study of moisture transfer in composite materials with the help of x-ray computational tomography," *Journal of Engineering Physics*, vol. 55, pp. 1359–1363, June 1989.

[1095] P. Lambrineas, J. R. Davis, B. Suendermann, P. Wells, K. Thomson, R. L. Woodward, G. T. Egglestone, and K. Challis, "X-ray computed tomography examination of inshore mine-hunter hull composite material," *NDT & E International*, vol. 24, no. 4, pp. 207–213, 1991.

[1096] C. Bathias and A. Cagnasso, "Application of x-ray tomography to the nondestructive testing of high-performance polymer composites," in *Damage Detection in Composite Materials* (J. E. Masters, ed.), STP1128, p. 3554, Philadelphia: ASTM, 1992.

[1097] S. R. Stock, T. M. Breunig, A. Guvenilir, J. H. Kinney, and M. Nichols, "Nondestructive x-ray tomographic microscopy of damage in various continuous-fiber metal matrix composites," in *Damage Detection in Composite Materials* (J. E. Masters, ed.), STP1128, pp. 25–34, Philadelphia: ASTM, 1992.

[1098] E. Scheinman and F. L. Roder, "Real-time CT of composites during destructive testing," in *Neutrons, X Rays, and Gamma Rays: Imaging Detectors, Material Characterization Techniques, and Applications* (J. M. Carpenter, D. B. Cline, R. Lanza, and D. F. Mildner, eds.), vol. 1737, pp. 171–183, SPIE - International Society for Optical Engineering, 1993.

[1099] J. H. Kinney, T. M. Breunig, T. L. Starr, D. Haupt, M. C. Nichols, S. R. Stock, M. D. Butts, and R. A. Saroyan, "X-ray tomographic study of chemical vapor infiltration processing of ceramic composites," *Science*, vol. 260, pp. 789–792, May 1993.

[1100] T. M. Breuning, M. C. Nichols, J. H. Kinney, and D. L. Haupt, "Quantitative microstructural characterization of a composite material using destructive and non-invasive techniques," in *Proc. 17th Annual Conference on Composites and Advanced Ceramic Materials, Cocoa Beach, FL*, vol. 14, pp. 500–509, July/August 1993.

[1101] B. Mattson, B. Stenberg, K. T. Gillen, R. L. Clough, and E. Ostman, "Novel techniques used to assess the ageing of carbon-black-filled materials," *Polymer Degradation and Stability*, vol. 41, no. 2, pp. 211–221, 1993.

[1102] E. Oestman, S. Persson, and H. Lavebratt, "Application of computed x-ray tomography scanning for studying the aging behaviour of thick-walled rubber," *NDT International*, vol. 22, pp. 359–365, December 1989.

[1103] B. Mattson and B. Stenberg, "Thermo-oxidative degradation and stabilization of rubber materials," *Progress in Rubber and Plastics Technology*, vol. 9, pp. 1–53, 1993.

[1104] H. Lavebratt, E. Oestman, S. Persson, and B. Stenberg, "Application of computed x-ray tomography scanning in the study of thermo-oxidative degradation of thick-walled filled natural rubber vulcanizates," *Journal of Applied Polymer Science*, vol. 44, pp. 83–94, January 1992.

[1105] S. Persson, "How computed tomography can be used to study cross link density in non-filled peroxide cured polyisoprene rubber," *Polymer*, vol. 29, no. 5, pp. 802–807, 1988.

[1106] J. H. Dunsmuir, A. J. Dias, D. G. Peiffer, R. Kolb, and G. Jones, "Microtomography of elastomers for tire manufacture," in *Proc. 1999 Developments in X-Ray Tomography II, Denver, CO*, vol. 3772, pp. 87–96, SPIE - International Society for Optical Engineering, 1999.

[1107] C. V. Kropas-Hughes and S. T. Neel, "Basics of computed tomography," *Materials Evaluation*, vol. 58, pp. 630–633, May 2000.

[1108] D. Toye, P. Marchot, M. Crine, and G. L'Homme, "Modelling of multiphase flow in packed beds by computer-assisted x-ray tomography," *Measurement Science and Technology*, vol. 7, pp. 436–443, March 1996.

[1109] P. Marchot, D. Toye, M. Crine, A.-M. Pelsser, and G. L'Homme, "Investigation of liquid maldistribution in packed columns by x-ray tomography,"

Chemical Engineering Research and Design A, vol. 77, no. 6, pp. 511–518, 1999.

[1110] D. Toye, P. Marchot, M. Crine, and G. L'Homme, "Analysis of liquid flow distribution in trickling flow reactor using computer assisted x-ray tomography," *Chemical Engineering Research and Design A*, vol. 73, pp. 258–262, April 1995.

[1111] Y. Akiyama, K. Hori, K. Miyazaki, K. Mishima, and S. Sugiyama, "Pressurized water reactor fuel assembly subchannel void fraction measurement," *Nuclear Technology*, vol. 112, pp. 412–421, 1995.

[1112] H. V. Kok, T. H. J. J. van der Hagen, and R. F. Mudde, "Subchannel void-fraction measurements in a 6×6 rod bundle using a simple gamma-transmission method," *International Journal of Multiphase Flow*, vol. 27, pp. 147–170, 2001.

[1113] K. A. Shollenberger, J. R. Torczynski, D. R. Adkins, and T. J. O'Hern, "Bubble column measurements using gamma tomography," in *Proc. 1995 ASME/JSME Fluids Engineering and Laser Anemometry Conference and Exhibition, Hilton Head, SC*, vol. FED-211, pp. 25–30, American Society of Mechanical Engineers, Fluids Engineering Division, 1997.

[1114] M. M. Hytros, I. M. Jureidini, J.-H. Chun, R. C. Lanza, and N. Saka, "High-energy x-ray computed tomography of the progression of the solidification front in pure aluminum," *Metallurgical and Materials Transactions A*, vol. 30, pp. 1403–1410, 1999.

[1115] M. M. Hytros, J.-H. Chun, R. C. Lanza, and N. Saka, "Solidification front monitor for metal casting," in *Advanced Materials: Development, Characterization Processing, and Mechanical Behavior, Proc. 1996 ASME International Mechanical Engineering Congress and Exposition, Atlanta, GA*, vol. 74, pp. 145–146, American Society of Mechanical Engineers, Materials Division, 1996.

[1116] J.-H. Chun, R. C. Lanza, N. Saka, and M. M. Hytros, "Online monitoring of the solidification front in metal casting," *CIRP Annals - Manufacturing Technology*, vol. 44, pp. 181–184, 1995.

[1117] J.-H. Chun, M. M. Hytros, I. M. Jureidini, N. Saka, and R. C. Lanza, "Noninvasive solidification sensor for metal casting," *CIRP Annals - Manufacturing Technology*, vol. 48, no. 1, pp. 147–150, 1999.

[1118] C. F. Chen, "Experimental study of convection in a mushy layer during directional solidification," *Journal of Fluid Mechanics*, vol. 293, pp. 81–98, June 1995.

[1119] T. Sumitra, S. Srisatit, A. Pattarasumunt, S. Punnachaiya, N. Chankow, and M. Wannaprapa, "A mobile computed tomography unit for inspecting reinforced concrete columns," *Transactions of the American Nuclear Society*, vol. 70 (Suppl. 1), pp. 405–414, 1994.

[1120] F. H. Hopkins, I. L. Morgan, H. D. Ellinger, R. V. Klinksiek, G. A. Meyer, and J. N. Thompson, "Industrial tomography applications," *IEEE Transactions on Nuclear Science*, vol. NS-28, no. 2, pp. 1717–1720, 1981.

[1121] E. N. Landis and D. T. Keane, "X-ray microtomography for fracture studies in cement-based materials," in *Proc. 1999 Developments in X-Ray Tomography II, Denver, CO*, vol. 3772, pp. 105–113, SPIE - International Society for Optical Engineering, 1999.

[1122] N. Shashidhar, "X-ray tomography of asphalt concrete," *Transportation Research Record*, vol. 1681, pp. 186–192, 1999.

[1123] D. Braz, R. T. Lopes, and L. M. G. da Motta, "Computed tomography: Evaluation of stability tests and indirect tensile strength of field asphaltic mixtures," *NDT & E International*, vol. 33, pp. 517–522, December 2000.

[1124] D. Braz, R. T. Lopes, and L. M. G. de Motta, "Analysis of the percentage voids of test and field specimens using computed tomography," *Nuclear Instruments and Methods in Physics Research A*, vol. 422, pp. 942–948, 1999.

[1125] E. Masad, B. Muhunthan, N. Shashidhar, and T. Harman, "Internal structure characterization of asphalt concrete using image analysis," *Journal of Computing in Civil Engineering*, vol. 13, pp. 88–95, April 1999.

[1126] D. Braz, R. T. Lopes, and L. M. G. da Motta, "Computed tomography: an evaluation of the effect of adding polymer SBS to asphaltic mixtures used in paving," *Applied Radiation and Isotopes*, vol. 53, pp. 725–729, 2000.

[1127] A. Onaisi, A. Audibert, M. T. Bieber, L. Bailey, J. Denis, and P. S. Hammond, "X-ray tomography visualization and mechanical modelling of swelling shale around the wellbore," *Journal of Petroleum Science and Engineering*, vol. 9, pp. 313–329, July 1993.

[1128] J. Cook, G. Goldsmith, and F. Auzerais, "X-ray tomography studies of wellbore processes under *in situ* stress conditions," *International Journal of Rock Mechanics and Mining Sciences and Geomechanics Abstracts*, vol. 30, pp. 1091–1094, December 1993.

[1129] R. T. Bernardi and H. J. Martz, "Nuclear waste drum characterization with 2 MeV x-ray and gamma-ray tomography," *Materials Evaluation*, vol. 53, pp. 1121–1126, October 1995.

[1130] B. D. Sawicka, R. V. Murphy, G. Tosello, P. W. Reynolds, and T. Romaniszyn, "Computed tomography of radioactive objects and materials," *Nuclear Instruments and Methods in Physics Research A*, vol. 299, pp. 468 –479, December 1990.

[1131] P. D. Tonner and J. H. Stanley, "Supervoltage computed tomography for large aerospace structures," *Materials Evaluation*, vol. 50, no. 12, pp. 1434–1438, 1992.

[1132] C. F. Buynak and R. H. Bossi, "Applied x-ray computed tomography," *Nuclear Instruments and Methods in Physics Research B*, vol. 99, pp. 772–774, 1995.

[1133] Invision, "EDS facts sheet." Invision Technologies, Inc., Newark, CA, May 2001. (http://ftp.invision-tech.com/products/pdf/EDS%20FS%20V12.pdf, accessed July 2001).

[1134] H. Sipila, "Moving object computer-tomography for luggage inspection," in *Proc. Applications of Signal and Image Processing in Explosives Detection Systems, Boston, MA*, vol. 1824, pp. 39–40, SPIE - International Society for Optical Engineering, 1993.

[1135] R. J. Kruse and R. H. Bossi, "X-ray tomographic inspection techniques on electrical components and PWA's," in *Proc. Technical Program of the National Electronic Packaging and Production Conference - NEPCON West '91, Anaheim, CA*, vol. 3, pp. 1770–1787, Proceedings of the Technical Program of the National Electronic Packaging and Production Conference - NEPCON West '91, Anaheim, CA, 1991.

[1136] J. Davis, R. Davies, P. Wells, and N. Benci, "A field transportable computerized tomography scanner for the nondestructive testing of wooden power poles," *Materials Evaluation*, vol. 51, no. 3, pp. 332–337, 1993.

[1137] K. Ueta, K. Tani, and T. Kato, "Computerized x-ray tomography analysis of three-dimensional fault geometries in basement-induced wrench faulting," *Engineering Geology*, vol. 56, no. 1, pp. 197–210, 2000.

[1138] E. J. Morton, R. D. Luggar, M. J. Key, A. Kundu, L. M. Távora, and W. B. Gilboy, "Development of a high speed x-ray tomography system for multiphase flow imaging," *IEEE Transaction on Nuclear Science*, vol. 46, pp. 380–384, June 1999.

[1139] K. Hori, T. Fujimoto, and K. Kawanishi, "Development of ultra-fast x-ray computed tomography scanner system," *IEEE Transactions on Nuclear Science*, vol. 45, pp. 2089–2094, August 1998.

[1140] E. J. Morton, M. J. Luggar, R.D. Key, A. Kundu, and W. Tavora, L. M. N.Gilboy, "Development of a high speed x-ray tomography system for multiphase flow imaging," *IEEE Transactions on Nuclear Science*, vol. 46, no. 3, 1999.

[1141] P. D. Tonner, B. D. Sawica, G. Tosello, and T. Romaniszyn, "Region-of-interest tomography imaging for product and material characterization," in *Proc. Industrial Computerized Tomography, Seattle, WA*, pp. 160–165, American Society of Nondestructive Testing, July 1989.

[1142] V. V. Nagarkar, J. S. Gordon, K. Daley, G. Entine, and M. R. Squillante, "Automated imaging of semiconductor crystal/melt interface zone," *Applied Radiation and Isotopes*, vol. 44, pp. 1301–1311, October/November 1993.

[1143] J. Zhou, M. Maisl, H. Reiter, and W. Arnold, "Computed laminography for materials testing," *Applied Physics Letters*, vol. 68, no. 24, pp. 3500–3502, 1996.

[1144] S. Gondrom, J. Zhou, M. Maisl, H. Reiter, K. M., and W. Arnold, "X-ray computed laminography: an approach of computed tomography for applications with limited access," *Nuclear Engineering and Design*, vol. 190, pp. 141–147, 1999.

[1145] Z. Tabatabaian and N. M. Spyrou, "Determination of minimum detectable mass of impurities in industrial objects using neutrons in transmission," *Nuclear Instruments and Methods in Physics Research A*, vol. 424, pp. 252–257, March 1999.

[1146] R. T. Lopes, A. P. Bessa, D. Braz, and E. F. O. de Jesus, "Neutron computerized tomography in compacted soil," *Applied Radiation and Isotopes*, vol. 50, pp. 451–458, February 1999.

[1147] W. J. Richards, J. C. I. Crump, and K. Shields, "Neutron tomography investigation of aircraft fan blades," in *Proc. 41st International SAMPE Symposium and Exhibition. Part 1 (of 2), Anaheim, CA*, vol. 41, pp. 243–250, SAMPE, 1996.

[1148] M. R. Gibbons, W. J. Richards, and K. Shields, "Optimization of neutron tomography for rapid hydrogen concentration inspection of metal castings," *Nuclear Instruments and Methods in Physics Research A*, vol. 424, pp. 53–57, March 1999.

[1149] R. C. Lanza, "Use of neutron tomographic techniques for the detection of corrosion damage in aircraft structures," in *Proc. 1995 ASME International Mechanical Engineering Congress and Exposition,San Francisco, CA*, vol. 47, pp. 253–259, American Society of Mechanical Engineers, Aerospace Division, 1995.

[1150] E. W. McFarland, J. Leigh, and R. C. Lanza, "Detection and characterization of the heterogeneous distribution of hydrogen in titanium compressor blades by neutron computed tomography," *Journal of Advanced Materials*, vol. 26, pp. 3–10, April 1995.

[1151] M. Zanarini, P. Chirco, M. Rossi, G. Baldazzi, G. Guidi, E. Querzola, M. Scannavini, F. Casali, A. Garagnani, and A. Festinesi, "Evaluation of hydrogen content in metallic samples by neutron computed tomography," *IEEE Transactions on Nuclear Science*, vol. 42, pp. 580–584, August 1995.

[1152] G. Pfister, A. K. Schatz, C. Siegel, E. Steichele, W. Waschkowski, and Bücherl, T., "Nondestructive testing of materials and components by computerized tomography with fast and thermal reactor neutrons," *Nuclear Science and Engineering*, vol. 110, pp. 303–315, 1992.

[1153] T. M. Nakanishi, J. Furukawa, and M. A. Matsubayashi, "Preliminary study of CT imaging of water in a carnation flower," *Nuclear Instruments and Methods in Physics Research A*, vol. 424, pp. 136–141, March 1999.

[1154] M. Rossi, F. Casali, P. Chirco, M. P. Morigi, E. Nava, E. Querzola, and M. Zanarini, "X-ray 3D computed tomography of bronze archaeological samples," *IEEE Transactions on Nuclear Science*, vol. 46, no. 4 I, pp. 897–903, 1999.

[1155] P. Chirco, M. Zanarini, C. Valeriani, G. Zambelli, M. Rossi, E. Querzola, and F. Casali, "Neutron computed tomography: a tool for the testing of advanced materials," in *Proc. Neutron in Research and Industry* (G. Vourvopoulos, ed.), vol. 2867, pp. 339–342, SPIE - International Society for Optical Engineering, 1997.

[1156] P. Zhu, P. Duvauchelle, G. Peix, and D. Babot, "X-ray Compton backscattering techniques for process tomography: imaging and characterization of materials," *Measurement Science & Technology*, vol. 7, pp. 281–286, March 1996.

[1157] ASTM, "E1931-97 Standard Guide for X-Ray Compton Scatter Tomography." American Society for Testing and Materials, West Conshohocken, PA, 1997.

[1158] B. L. Evans, J. B. Martin, L. W. Burggraf, and M. C. Roggemann, "Nondestructive inspection using Compton scatter tomography," *IEEE Transactions on Nuclear Science*, vol. 45, pp. 950–956, June 1998.

[1159] B. L. Evans, J. B. Martin, and L. W. Burggraf, "Nondestructive inspection of thin, low-Z samples using multiplexed Compton scatter tomography," in *Proc. 1997 MRS Fall Meeting, Boston, MA*, vol. 503, pp. 297–308, Materials Research Society, 1998.

[1160] LLNL, "Sensor technology at the Lawrence Livermore National Laboratory: Compton backscatter tomography." (www.llnl.gov/sensor_technology/STR21.html, accessed July 2001), August 1995.

[1161] T. O. Tümer, "A three-dimensional container and cargo inspection system," in *Proc. Physics-Based Technologies for the Detection of Contraband* (L. O. Malotky and J. J. Pennella, eds.), vol. 2936, pp. 48–57, SPIE - International Society for Optical Engineering, 1997.

[1162] E. A. van den Bergen, G. Jonkers, K. Strijckmans, and P. Goethals, "Industrial applications of positron emission computed tomography," *Nuclear Geophysics*, vol. 3, no. 4, pp. 407–418, 1989.

[1163] C. Bemrose, P. Fowles, M. R. Hawkesworth, and M. A. O'Dwyer, "Application of positron emission tomography to particulate flow measurement in chemical engineering processes," *Nuclear Instruments and Methods in Physics Research A*, vol. 273, pp. 874–880, December 1988.

[1164] R. I. Ewing and K. W. Marlo, "A fast-neutron detector used in verification of the INF Treaty," *Nuclear Instruments and Methods in Physics Research A*, vol. 299, pp. 559–561, 1990.

[1165] S. Legoupil, G. Pascal, D. Chambellan, and D. Bloyetm, "Determination of the detection process in an experimental tomography for industrial flow visulaization using radioactive tracers," *IEEE Transaction on Nuclear Science*, vol. 43, pp. 751–760, April 1996.

[1166] K. M. Hanson, "Proton computed tomography," *IEEE Transactions on Nuclear Science*, vol. NS-26, pp. 1635–1640, February 1978.

[1167] A. J. Antolak and G. S. Bench, "PIXE tomography of samples with inhomogeneous elemental composition," *Nuclear Instruments and Methods in Physics Research B*, vol. 88, pp. 297–307, May 1994.

[1168] R. M. S. Schofield and H. W. Lefevre, "PIXE-STIM microtomography. Zinc and manganese concentrations in a scorpionstinger," *Nuclear Instruments and Methods in Physics Research B*, vol. 72, pp. 104–110, October 1992.

[1169] K. D. Krug and J. A. Stein, "Advanced dual energy x-ray for explosive detection," in *Proc. 1st Int. Conf. on Explosive Detection Technology* (S. Khan, ed.), DOT/FAA/CT-92/11, pp. 282–284, Federal Aviation Administration, Department of Transportation, May 1992.

[1170] R. F. Eilbert and K. D. Krug, "Aspects of image recognition in Vivid technologies' dual-energy x-ray system for explosives detection," in *Proc. Applications of Signal and Image Processing in Explosives Detection Systems, Boston, MA*, vol. 1824, pp. 127–143, The International Society for Optical Engineering, 1993.

[1171] R. F. Eilbert and K. D. Krug, "Dual energy x-ray scanner for detecting contraband," in *Proc. Contraband and Cargo Inspection Technology International Symposium*, (Washington, D.C.), pp. 209–217, Office of National Drug Control Policy and National Institute of Justice, October 1992.

[1172] R. Paranjape, M. Sluser, and K. Runtz, "Segmentation of handguns in dual energy x-ray imagery of passenger carry-on baggage," in *Proc. 1998 11th Canadian Conference on Electrical and Computer Engineering, CCECE. Part 1 (of 2), Toronto, ON*, pp. 377–380, 1998.

[1173] P. J. Bjorkholm and T. R. Wang, "Explosive detection using three-dimensional computer-assisted image analysis," in *Proc. Applications of Signal and Image Processing in Explosives Detection Systems, Boston, MA*, vol. 1824, pp. 122–126, SPIE - International Society for Optical Engineering, 1993.

[1174] N. K. Gupta and B. D. Howard, "Non-intrusive container inspection concept," in *Detection and Analysis of Subsurface Objects and Phenomena, Monterey, CA* (I. A. S. Institute, ed.), October 1998.

[1175] C. Rizescu, C. Besliu, and A. Jipa, "Determination of local density and effective atomic number by the dual-energy computerized tomography method with the ^{192}Ir radioisotope," *Nuclear Instruments and Methods in Physics Research A*, vol. 465, pp. 584–599, June 2001.

[1176] E. Rhodes, D. S. Kupperman, and R. L. Hitterman, "Dual-energy neutron tomography of water in rock using the Argonne IPNS," in *Neutron Radiography (4)* (J. P. Barton, ed.), pp. 621–625, Gordon and Breach, Yverdon, Switzerland, 1994.

[1177] H. Rogasik, J. W. Crawford, O. Wendroth, I. M. Young, M. Joschko, and K. Ritz, "Discrimination of soil phases by dual energy x-ray tomography," *Soil Science Society of America Journal*, vol. 63, no. 4, pp. 741–751, 1999.

[1178] A. R. Kalukin, M. Van Geet, and R. Swennen, "Principal components analysis of multienergy x-ray computed tomography of mineral samples," *IEE Transactions on Nuclear Science*, vol. 47, pp. 1729–1736, October 2000.

[1179] M. Van Geet, R. Swennen, and M. Wevers, "Quantitative analysis of reservoir rocks by microfocus x-ray computerized tomography," *Sedimentary Geology*, vol. 132, pp. 25–36, 2000.

[1180] M. Bertschy, J. Jolie, and W. Mondelaers, "Heavy-element tomography using tunable gamma-ray beams," *Applied Physics A*, vol. 62, pp. 437–443, May 1996.

[1181] M. Bertschy, M. Crittin, J. Jolie, W. Mondelaers, and N. Warr, "Tunable monochromatic gamma-ray source. Part 3. Feasibility study of heavy element tomography," *Nuclear Instruments and Methods in Physics Research B*, vol. 103, pp. 3330–338, November 1995.

[1182] M. Bertschy, J. Jolie, and W. Mondelaers, "Improvement of heavy element tomography results using tunable gamma-ray beams," *Nuclear Instruments and Methods in Physics Research B*, vol. 113, pp. 146–149, June 1996.

[1183] C. Lopes, E. B. Costa, and E. F. O. de Jesus, "Computed tomography with monochromatic bremsstrahlung radiation," *Applied Radiation and Isotopes*, vol. 53, pp. 665–671, 2000.

[1184] T. H. Prettyman, R. P. Gardner, J. C. Russ, and K. Verghese, "Combined transmission and scattering tomographic approach to composition and density imaging," *Applied Radiation and Isotopes*, vol. 44, pp. 1327–1341, October/November 1993.

[1185] D. Schafer, M. Annis, and M. Hacker, "New x-ray technology for the detection of explosives," in *Proc. 1st Int. Conf. on Explosive Detection Technology* (S. Khan, ed.), DOT/FAA/CT-92/11, pp. 269–281, Federal Aviation Administration, Department of Transportation, May 1992.

[1186] T. Miller, P. K. Van Staagen, B. C. Gibson, and J. L. Orthel, "Contraband detection using neutron transmission," in *Proc. Physics-Based Technologies for the Detection of Contraband* (L. O. Malotky and J. J. Pennella, eds.), vol. 2936, pp. 102–109, SPIE - International Society for Optical Engineering, 1997.

[1187] J. J. Sredniawski, T. Debiak, E. Kamykowski, J. Rathke, P. Schmor, A. Altman, J. Rogers, and J. Boyd, "A new proof-of-principle contraband detection system," in *Proc. International Technology Symposium, Nashua, N. H.*, Office of National Drug Control Policy, October 1995.

[1188] S. Melnychuk, E. Kamykowski, J. Sredniawski, T. Debiak, R. Ruegg, and B. Milton, "Operating characteristics of a high current electrostatic accelerator for a contraband detection system," in *Proc. IEEE 18th Biennial Particle Accelerator Conference, New York, NY*, pp. 587–589, IEEE Nuclear and Plasma Sciences, March/April 1999.

[1189] S. Yamamoto, K. Yoneda, S. A. Hayashi, I. Kobayashi, Katsuheiand Kimura, T. Suzuki, H. Nishihara, and S. Kanazawa, "Application of iron-filtered neutrons to radiography of a copper step within a large iron block and to computer tomography of metallic cylinders," *Nuclear Instruments and Methods in Physics Research A*, vol. 225, pp. 439–444, August 1984.

[1190] T. O. McDonald, T. E. Jr Brun, T. N. Claytor, E. H. Farnum, G. Greene, and C. Morris, "Time-gated energy-selected cold neutron radiography," *Nuclear Instruments and Methods in Physics Research A*, vol. 424, pp. 235–241, March 1999.

[1191] G. Harding, "Optimization criteria for CXRS baggage inspection," in *Proc. Law Enforcement Technologies: Identification Technologies and Traffic Safety, Munich, Germany*, vol. 2511, pp. 64–70, SPIE - International Society for Optical Engineering, a, 1995.

[1192] H. Strecker, G. Harding, H. Bomsdorf, R. Kanzenbach, Jurgenand Linde, and G. Martens, "Detection of explosives in airport baggage using coherent x-ray scatter," in *Proc. Substance Detection Systems, Innsbruck, Austria*, vol. 2092, pp. 399–410, SPIE - International Society for Optical Engineering, 1994.

[1193] H. Strecker, "Simulation-based training and testing of classification schemes for CXRS explosives detection," in *Proc. Law Enforcement Technologies: Identification Technologies and Traffic Safety, Munich, Germany*, vol. 2511, pp. 88–98, SPIE - International Society for Optical Engineering, 1995.

[1194] S. Hnatnicky, "CXRS explosive-detection airport prototype," in *Proc. Physics-Based Technologies for the Detection of Contraband* (L. O. Malotky and J. J. Pennella, eds.), vol. 2936, pp. 180–190, SPIE - International Society for Optical Engineering, 1997.

[1195] T. Blaffert, "Theory of stochastic signal processing for the optimization of CXRS explosives detection," in *Proc. Law Enforcement Technologies: Identification Technologies and Traffic Safety, Munich, Germany*, vol. 2511, pp. 108–119, SPIE - International Society for Optical Engineering, 1995.

[1196] H. Ries, "Two-level x-ray-based explosives detection system," in *Proc. X-Ray and UV Detectors, San Diego, CA*, vol. 2278, pp. 58–63, SPIE - International Society for Optical Engineering, 1994.

[1197] R. D. Luggar, J. A. Horrocks, R. D. Speller, and R. J. Lacey, "Low angle x-ray scatter for explosives detection: A geometry optimization," *Applied Radiation and Isotopes*, vol. 48, pp. 215–224, February 1997.

[1198] G. Martens, H. Bomsdorf, G. Harding, J. Kanzenbach, and R. Linde, "Coherent x-ray scatter imaging for foodstuff contamination detection," in *Proc. Substance Detection Systems, Innsbruck, Austria*, vol. 2092, pp. 387–398, SPIE - International Society for Optical Engineering, 1994.

[1199] R. D. Luggar and W. B. Gilboy, "Applications of Rayleigh scattered photons to substance identification," *Nuclear Instruments and Methods in Physics Research A*, vol. 353, pp. 650–653, 1994.

[1200] G. Harding, J. Kosanetzky, and U. Neitze, "X-ray diffraction computed tomography," *Medical Physics*, vol. 14, pp. 515–525, July/August 1987.

[1201] G. H. J. Kosanetzky, B. Knoerr and U. Neitzel, "X-ray diffraction measurement of some plastics materials and body tissues," *Medical Physics*, vol. 14, pp. 526–533, July/August 1987.

[1202] R. D. Luggar and W. B. Gilboy, "Recent developments in industrial applications of elastic scatter x-ray inspection," *Radiation Physics and Chemistry*, vol. 56, no. 1, pp. 213–227, 1999.

[1203] T. Gozani, "Elemental imaging for contraband detection," in *Proc. Substance Identification Analytics, Innsbruck, Austria*, vol. 2093, pp. 288–300, SPIE - International Society for Optical Engineering, 1994.

[1204] W. P. Trower, "The nitrogen camera and the detection of concealed explosives," *Nuclear Instruments and Methods in Physics Research B*, vol. 79, pp. 589–592, 1993.

[1205] E. A. Knapp, R. B. Moler, A. W. Saunders, and W. P. Trower, "Direct imaging of explosives," *Applied Radiation and Isotopes*, vol. 53, pp. 711–716, 2000.

[1206] S. Kawasaki, M. Kondo, S. Izumi, and M. Kikuchi, "Radioactivity measurement of drum package waste by a computed-tomography technique," *Applied Radiation and Isotopes*, vol. 41, no. 10/11, pp. 983–987, 1990.

[1207] S. Charlton, "The application of nucleonic instruments in measurement and control," *Nucleoincs*, pp. 49–51, August/Sep 1984.

[1208] IAEA, "Bulk hydrogen analysis, using neutrons: Final report of the first research co-ordination meeting, remarks made by C. M. Bartle," Tech. Rep. IAEA/PS/RCM97-1, International Atomic Energy Agency, Vienna, July 1997.

[1209] A. Notea and Y. Segal, "General approach to the design of radiation gauges," *Nuclear Technology*, vol. 24, pp. 73–80, October 1974.

[1210] P. Tabor and L. Lakosi, "High-accuracy fluid density gauge," in *Radioisotope Instruments in Industry and Geophysics*, vol. 1, pp. 487–503, Vienna: International Atomic Energy Agency, 1966.

[1211] L. Grodzins, "Optimum energies for x-ray transmission tomography of small-samples," *Nuclear Instruments and Methods*, vol. 206, pp. 541–545, 1983.

[1212] J. van Santen, Z. I. Kolar, and A. M. Scheers, "Photon energy selection for dual energy γ- and x-ray absorption composition measurements in oil-water-gas mixtures," *Nuclear Geophysics*, vol. 9, no. 3, pp. 193–202, 1995.

[1213] P. Duvauchelle, P. Girier, and G. Peix, "Development of high resolution focusing collimators intended for nondestructive testing by the Compton scattering tomography technique," *Applied Radiation and Isotopes*, vol. 41, no. 2, pp. 199–205, 1990.

[1214] M. C. Wrobel, C.-Y. Ng, N. H. Clinthorne, T. A. Webster, Y. Zhang, and W. L. Rogers, "Design, construction and performance of one-dimensional lead foil parallel plate and fan-beam collimators for gamma ray imaging," *Nuclear Instruments and Methods in Physics Research*, vol. 376, pp. 477–483, July 1996.

[1215] G. Hehn and R. Becker, "Optimization of collimator for 14 MeV neutrons," in *First Proc. Symposium on Neutron Dosimetry in Biology and Medicine* (G. Burger, H. Schuraube, and H. G. Ebert, eds.), (Munchen), pp. 529–539, Commission of the European Communities, Luxembourg, 1972.

[1216] W. Soller, "A new precision x-ray spectrometer," *Physics Review*, vol. 24, pp. 158–167, 1924.

[1217] L. D. Cussen, "Design for improved neutron collimators," *Nuclear Instruments and Methods in Physics Research, A*, vol. 414, pp. 365–371, Sep 1998.

[1218] K. Al-Hiti, K. V. Eitinger, and H. A. Khan, "The use of nuclear track detectors in the design of neutron collimators and scatterers," *Physics in Medicine and Biology*, vol. 19, pp. 879–883, 1974.

[1219] ASTM, "E803-91(2002) Standard Test Method for Determining the L/D Ratio of Neutron Radiography Beams." American Society for Testing and Materials, West Conshohocken, PA, 2002.

[1220] J. S. Brenizer Jr., D. A. Raine, J. Gao, and J. Chen, "Design of a divergence and alignment indicator," *Nuclear Instruments and Methods in Physics Research A*, vol. 377, pp. 58–62, July 1996.

[1221] T. J. Jordan and P. C. Williams, "The design and performance characteristics of a multileaf collimator," *Physics in Medicine and Biology*, vol. 39, pp. 231–251, February 1994.

[1222] C. X. Yu, "Design considerations for the sides of multileaf collimator leaves," *Physics in Medicine and Biology*, vol. 43, no. 5, pp. 1335–1342, 1998.

[1223] P. B. Greer and T. van Doorn, "A design for a dual assembly multileaf collimator," *Medical Physics*, vol. 27, no. 10, pp. 2242–2255, 2000.

[1224] J. M. Galvin, A. R. Smith, and R. D. Moeller, "Evaluation of multileaf collimator design for a photon beam.," *International Journal of Radiation Oncology*, vol. 23, no. 4, pp. 789 –801, 1992.

[1225] G. Loi, E. Pignoli, M. Scorsetti, *et al.*, "Design and characterization of a dynamic multileaf collimator," *Physics in Medicine and Biology*, vol. 43, no. 10, pp. 3149–3155, 1998.

[1226] J. L. Cason, "Design and fabrication of a portable device for neutron radiography," Tech. Rep. BNWL-1329, Battelle-Memorial Institute, Pacific Northwest Laboratory, Washington, 1971. USAEC.

[1227] "Californium-252 progress." No. 7: 37-41, No. 8: 27-28, US Atomic Energy Commission, 1971.

[1228] D. Greene and D. Major, "Collimation of 14 MeV neutron beams," *European Journal of Cancer*, vol. 7, pp. 121–127, 1971.

[1229] D. Greene and D. Major, "Development in achieving optimum beam profiles for 15 MeV neutron beams," *European Journal of Cancer*, vol. 10, pp. 339–342, 1974.

[1230] F. Grunauer, "Calculation of neutron collimator for therapeutical purposes," in *Proc. First Symposium on Neutron Dosimetry in Biology and Medicine, Munich, Germany* (G. Burger, H. Schuraube, and H. G. Ebert, eds.), pp. 511–525, Commission of the European Communities, Luxembourg, 1972.

[1231] A. Knight, D. A. Lundeberg, and T. C. Marshall, "The development of 15 MeV neutron beam collimator for radiotherapy," in *Proc. First Symposium on Neutron Dosimetry in Biology and Medicine* (G. Burger, H. Schuraube, and H. G. Ebert, eds.), (Munchen), pp. 542–557, Commission of the European Communities, Luxembourg, 1972.

[1232] K. M. Jones and C. A. Kelsey, "Experimental determination of the need for multiple collimator for neutron radiotherapy," *Medical Physics*, vol. 1, pp. 215–218, 1974.

[1233] J. L. Beach, G. Shani, and G. A. Dissamaiki, "Evaluation of a continuously variable collimator for 14 MeV neutrons," *Physics in Medicine and Biology*, vol. 20, pp. 47–53, 1975.

[1234] R. L. Maughan, M. Yudelev, A. Aref, *et al.*, "Design considerations for a computer controlled multileaf collimator for the Harper Hospital fast neutron therapy facility," *Medical Physics*, vol. 29, no. 4, pp. 499–508, 2002.

[1235] M. Cooper, R. S. Holt, and G. Harding, "Elastic and Compton scattering with W Kα x-radiation," *Journal of Physics E: Scientific Instruments*, vol. 18, pp. 354–357, 1985.

[1236] T. Matsumoto, "Neutron beam design for neutron radiography at the Musashi reactor," *Nuclear Instruments and Methods in Physics Research A*, vol. 377, pp. 48–51, July 1996.

[1237] D. G. Lewis and S. S. A. Natto, "Monte Carlo design study of a moderated ^{252}Cf source for *in vivo* neutron activation analysis of aluminum," *Physics in Medicine and Biology*, vol. 42, pp. 625–636, April 1997.

[1238] S.-O. Lindahl and J. K. Shultis, "Effect of the 2. 37 MeV oxygen antiresonance in a fast neutron filter," *Nuclear Instruments and Method*, vol. 144, pp. 317–321, July 1977.

[1239] H. S. and M. C. Scott, "BNCT neutron beam characterization using a resonance absorption filter method," *Nuclear Instruments and Methods in Physics Research A*, vol. 476, pp. 99–105, January 2002.

[1240] G. Skarnemark, M. Skalberg, R. K. Tokay, P. Holtta, and R. Rosenberg, "Influence of an additional filter in epithermal neutron activation analysis," *Applied Radiation and Isotopes*, vol. 40, no. 1, pp. 37–40, 1989.

[1241] R. A. Forster and H. O. Menlove, "Neutron source moderator shield assemblies," in *Nuclear Safeguards Research and Development Program Status Report*, no. LA-4525-MS in USAEC Report, pp. 11–13, Los Alamos, New Mexico: Los Alamos National Laboratory, 1970.

[1242] J. R. Powell, M. Ludewig, H. Todosow, and M. Reich, "Target and filter concepts for accelerator-driven boron neutron capture therapy applications," *Nuclear Technology*, vol. 125, pp. 104–115, January 1999.

[1243] H. B. Liu, "PbF_2 compared to Al_2O_3 and AlF_3 to produce an epithermal neutron beam for radiotherapy," *Medical Physics*, vol. 23, no. 2, pp. 279 – 280, 1996.

[1244] C. K. Wang, J. F. Zino, and G. Kessler, "Enhancement of a ^{252}Cf-based neutron beam via subcritical multiplication for neutron capture therapy," *Applied Radiation and Isotopes*, vol. 53, pp. 811–814, November 2000.

[1245] R. C. Greenwood and R. E. Chrien, "Filtered reactor beams for fast neutron capture gamma-ray experiments," *Nuclear Instruments and Methods*, vol. 138, pp. 125–143, October 1976.

[1246] H. Aminfar and R. M. Brugger, "2 keV scandium filtered neutron beam using separated isotopes as a secondary filter," *Nuclear Instruments and Methods in Physics Research*, vol. 188, pp. 597–601, October 1981.

[1247] Y. Sakurai and T. Kobayashi, "Characteristics of the KUR heavy water neutron irradiation facility as a neutron irradiation field with variable energy spectra," *Nuclear Instruments and Methods in Physics Research A*, vol. 453, pp. 569–596, October 2000.

[1248] M. Balasko, E. Svab, and J. Palfalvi, "Neutron radiography with optimized beam formation and filtering unit," *Nondestructive Testing and Evaluation*, vol. 16, no. 2-6, pp. 297–307, 2001.

[1249] H. Kobayashi, T. Matsumoto, M. Matsubayashi, J. S. J. Brenizer, and J. T. Lindsay, "Experimental study of modification of neutron spectrum using filters," *Nuclear Instruments and Methods in Physics Research A*, vol. 377, pp. 37–40, July 1996.

[1250] J. P. Barton, "Filters for thermal neutron radiography," *Nondestructive Testing and Evaluation*, vol. 16, no. 2-6, pp. 95–100, 2001.

[1251] J. P. Barton, "Long Be filter compared with Be-Bi filter for dedicated cold neutron radiography beam," *Nondestructive Testing and Evaluation*, vol. 16, no. 2-6, pp. 85–94, 1999.

[1252] M. T. Rekveldt, P. Verkerk, and A. A. Van Wel, "Stacked neutron mirrors as efficient filter for use in thermal and subthermal neutron beams," *Nuclear Instruments and Methods in Physics Research B*, vol. 34, pp. 285–289, August 1988.

[1253] P. Verkerk, A. A. Van Well, and M. T. Rekveldt, "Transmission of thermal and fast neutrons and gamma rays in a novel filter," *Physica B: Condensed Matter*, vol. 156-57, pp. 544–546, January/February 1989.

[1254] P. Mikula, J. Kulda, J. Cervena, B. Chalupa, B. Haskova, and J. Vavra, "Statically bent silicon single crystal as a thermal neutron filter," *Nuclear Instruments and Methods in Physics Research*, vol. 211, pp. 519–523, June 1983.

[1255] C. R. Tipton, "Materials, review of moderator materials," in *Reactor Handbook*, vol. 1, ch. 14, New York: Interscience Publishers, 1960-1964.

[1256] A. X. da Silva and V. R. Crispim, "Moderator-collimator-shielding design for neutron radiography systems using ^{252}Cf," *Applied Radiation and Isotopes*, vol. 54, pp. 217–225, February 2001.

[1257] T. Matsumoto, "Neutron beam design for both medical neutron capture therapy and industrial neutron radiography for TRIGA reactor," *Progress in Nuclear Energy*, vol. 37, no. 1-4, pp. 327–332, 2000.

[1258] E. P. Shabalin, *Fast Pulsed and Burst Reactors, Boosters*, ch. 9. Oxford: Pergamon Press, 1979.

[1259] H. B. Liu, R. M. Brugger, D. C. Rorer, and P. R. Tichler, "Design of a high-flux epithermal neutron beam using ^{235}U fission plates at the Brookhaven Medical Research Reactor," *Medical Physics*, vol. 21, no. 10, pp. 1627–1631, 1994.

[1260] H. B. Liu and F. J. Patti, "Epithermal neutron beam upgrade with a fission plate converter at the Brookhaven Medical Research Reactor," *Nuclear Technology*, vol. 116, no. 3, pp. 373–377, 1996.

[1261] D. C. McKeon and H. D. Scott, "SNUPAR - a nuclear parameter code for nuclear geophysics applications," *IEEE Transactions in Nuclear Science*, vol. 36, pp. 1215–1219, February 1989.

[1262] S. A. Dupree and S. K. Fraley, *A Monte Carlo Primer: A Practical Approach to Radiation Transport.* Dordrecht: London: Kluwer Academic/Plenum, 2001.

[1263] I. Lux and L. Koblinge, *Monte Carlo Particle Transport Methods: Neutron and Photon Calculations.* Boca Raton: CRC Press, 1991.

[1264] R. L. Morin, ed., *Monte Carlo Simulation in the in Radiological Sciences.* Boca Rato: CRC Press, 1988.

[1265] RSICC, *Monte Carlo N-Particle Transport Code System.* Radiation Safety Information Computational Center, Oak Ridge National Laboratory, Oak Ride, TN. Computer Code Number: C00701.

[1266] T. Wilcox and E. Lent, *COG - A Particle Transport Code Designed to Solve the Boltzmann Equation for Deep-Penetration (Shielding) Problems.* Lawrence Livermore National Laboratory, Livermore, CA, 1989. (http://www-phys.llnl.gov/N_Div/COG/, accessed January 2002).

[1267] J. M. Hall, J. F. Morgan, and K. E. Sale, "Numerical modeling of nonintrusive inspection systems," in *Substance Detection Systems*, vol. 2092, pp. 342–352, SPIE - International Society for Optical Engineering, 1994.

[1268] T. H. Prettyman, R. Gardner, and K. Verghese, "The specific purpose Monte Carlo code McENL for simulating the response of epithermal neutron lifetime well logging tools," *IEEE Transactions on Nuclear Science*, vol. 40, no. 4 pt 1, pp. 933–938, 1993.

[1269] E. M. A. Hussein, "Approximate estimators for fluence at a point in center-of-mass Monte Carlo neutron transport," *Nuclear Science and Engineering*, vol. 109, pp. 416–422, 1991.

[1270] J. Baro, M. Roteta, J. M. Fernandez-Varea, and F. Salvat, "Analytical cross sections for Monte Carlo simulation of photon transport," *Radiation Physics and Chemistry*, vol. 44, no. 5, pp. 531–552, 1994.

[1271] G. R. Imel, J. R. C. Singleterry, G. Gillespie, and G. E. McMichael, "Moderator design for accelerator driven neutron radiography sources," *Instruments and Methods in Physics Research A*, vol. 377, pp. 20–22, July 1996.

[1272] R. C. J. Singleterry, G. R. Imel, and G. E. McMichael, "Moderator/collimator for a proton/deuteron linac to produce a high-intensity, high-quality thermal neutron beam for neutron radiography," in *Proc. 1995 16th IEEE Particle Accelerator Conference*, vol. 1, pp. 149–151, 1995.

[1273] D. P. Lewis, B. M. W. Tsui, C. Tocharoenchai, and E. C. Frey, "Characterization of medium and high energy collimators using ray-tracing and Monte Carlo methods," in *Proc. 1998 IEEE Nuclear Science Symposium*, vol. 3, pp. 2026–2030, 1999.

[1274] J. C. Yanch and A. B. Dobrzeniecki, "Monte Carlo simulation in SPECT: Complete 3D modeling of source, collimator and tomographic data acquisition," *IEEE Transactions on Nuclear Science*, vol. 40, pp. 198–203, April 1993.

[1275] E. C. Selcow, A. B. Dobrzeniecki, J. C. Yanch, A. Lu, and M.-J. Belanger, "Evaluation of the Monte Carlo simulation of SPECT projection data using MCNP and SimSPECT," in *Proc. 1996 4th ASME/JSME International Conference on Nuclear Engineering, ICONE-4*, vol. 1B, pp. 1015–1018, 1996.

[1276] R. D. Lewis, S. J. S. Ryde, D. A. Hancock, and C. J. Evans, "MCNP-based model of a linear accelerator x-ray beam," *Physics in Medicine and Biology*, vol. 44, pp. 1219–1230, May 1999.

[1277] A. W. Davis, C. R. Hills, M. J. Sheats, and T. Claytor, "High-energy x-ray and neutron modeling and digital imaging for nondestructive testing applications," in *Proc. Penetrating Radiation Systems and Application, August 2000, San Diego, CA*, vol. 4142, pp. 108–118, SPIE - International Society for Optical Engineering, 2000.

[1278] R. M. Buck and J. M. Hall, "Applications of the COG multiparticle Monte Carlo transport code to simulated imaging of complex objects," in *Proc. 1999 1st Conference on Radiation Sources and Radiation Interactions*, vol. 3771, pp. 127–134, SPIE - International Society for Optical Engineering, July 1999.

[1279] G. W. Griffith and H. O. Menlove, "Design of active-neutron fuel rod scanner," *Nuclear Instruments and Methods in Physics Research A*, vol. 378, pp. 552–560, August 1996.

[1280] Z. Su, A. Jacobs, E. T. Dugan, J. Howley, and J. Jacobs, "Lateral migration radiography application to land mine detection, confirmation and classification," *Optical Engineering*, vol. 39, pp. 2472–2479, Sep 2000.

[1281] J. E. McKisson, P. S. Haskins, D. P. J. Henderson, K. C. Neelands, C. K. Wang, and T. Evans, "3-D imaging of complex source fields with a Compton camera imager," in *Proc. 1996 IEEE Nuclear Science Symposium*, vol. 2, pp. 825–829, 1996.

[1282] M. A. Borsaru and J. L. B. Pinault, "Simulation of the neutron - gamma and gamma - gamma Sirolog tools response for coal logging using the MOCA Monte Carlo code," *Nuclear Geophysics*, vol. 9, pp. 387–399, October 1995.

[1283] J. M. Hall and B. A. Jacoby, "Analysis of a proposed Compton backscatter imaging technique," in *Substance Detection Systems*, vol. 2092, pp. 448–459, SPIE - International Society for Optical Engineering, 1994.

[1284] H. T. Watson, H. D., F. Johansen, T. Stokes, G. Vargo, and V. Verbinski, "Results of modelling and experimental measurements for the design of a neutron surface moisture measurement sensor," *IEEE Transactions on Nuclear Science*, vol. 44, pp. 764–773, June 1997.

[1285] Q. Ao, R. P. Gardner, and K. Verghese, "Combined weight window and biasing approach based on a subspace importance map in the Monte Carlo simulation of gamma-gamma lithodensity logging tool responses with the revised McLDL code," *Transactions on Nuclear Science*, vol. 42, pp. 626–633, August 1995.

[1286] K. E. Sale, "Signal predictions for a proposed fast neutron interrogation method," in *Substance Detection Systems*, vol. 2092, pp. 263–270, SPIE - International Society for Optical Engineering, 1994.

[1287] B. J. Micklich, C. L. Fink, and L. Sagalovsky, "Transport simulation and image reconstruction for fast-neutron detection of explosives and narcotics," in *Proc. Law Enforcement Technologies: Identification Technologies and Traffic Safety, Munich*, vol. 2511, pp. 33–44, SPIE - International Society for Optical Engineering, 1995.

[1288] B. J. Micklich, M. K. Harper, A. H. Novick, and D. L. Smith, "Illicit substance detection using fast-neutron transmission spectroscopy," *Nuclear Instruments and Methods in Physics Research A*, vol. 353, pp. 646–649, December 1994.

[1289] S. Locke, "A case study in modelling a nuclear formation evaluation sub," *IEEE Transactions on Nuclear Science*, vol. 39, no. 4, pp. 1019–1024, 1992.

[1290] M. Shahriari and M. Sohrabpour, "Borehole parametric study for neutron induced capture γ-ray spectrometry using the MCNP code," *Applied Radiation and Isotopes*, vol. 52, no. 1, pp. 127–135, 2000.

[1291] C. Oliveira, J. Salgado, I. F. Goncalves, F. G. Carvalho, and F. Leitao, "Monte Carlo study of the influence of the geometry arrangements and structural materials on a PGNAA system performance for cement raw material analysis," *Applied Radiation and Isotopes*, vol. 48, pp. 1349–1354, October/December 1997.

[1292] C. Oliveira, J. Salgado, and F. Leitao, "Density and water content corrections in the gamma count rate of a PGNAA system for cement raw material analysis using the MCNP code," *Applied Radiation and Isotopes*, vol. 49, pp. 923–930, August 1998.

[1293] C. Oliveira, J. Salgado, and F. Carvalho, "Optimization of PGNAA instrument design for bulk coal analysis," *Nuclear Geophysics*, vol. 7, pp. 285–303, June 1993.

[1294] J. Oliveira, C. Salgado and F. Carvalho, "Optimized geometry for bulk coal PGNAA with external moderation of the source neutrons," *Nuclear Geophysics*, vol. 9, no. 5, pp. 401–412, 1995.

[1295] C. Evans, S. J. S. Ryde, D. A. Hancock, and F. Al-Agel, "Monte Carlo simulation of prompt gamma neutron activation analysis using MCNP code," *Applied Radiation and Isotopes*, vol. 49, pp. 541–543, May/June 1998.

[1296] R. C. Odom, S. M. Bailey, and R. D. Wilson, "Benchmarking computer simulations of neutron-induced, gamma-ray spectroscopy for well logging," *Applied Radiation and Isotopes*, vol. 48, pp. 1321–1328, October/December 1997.

[1297] M. L. Gartner and L. A. Jacobson, "Dependence of elemental yield variance on detector type through mathematical modeling," *IEEE Transactions on Nuclear Science*, vol. 37, pp. 931–935, April 1990.

[1298] F. Inanc, "Analysis of x-ray and gamma ray scattering through computational experiments," *Journal of Nondestructive Evaluation*, vol. 18, no. 2, pp. 73–82, 1999.

[1299] R. M. Ambrosi, J. I. W. Watterson, and B. R. K. Kala, "Monte Carlo study of the effect of neutron scattering in a fast neutron radiography facility," *Nuclear Instruments and Methods in Physics Research B*, vol. 139, pp. 286–292, April 1998.

[1300] C. J. Leliveld, J. G. Maas, C. W. E. van Eikk, and V. R. Bom, "On the significance of scattered radiation in industrial x-ray computed tomography imaging," *IEEE Transaction on Nuclear Science*, vol. 41, pp. 290–294, February 1994.

[1301] D. J. Picton, R. G. Harris, K. Randle, and D. R. Weaver, "Effect of basic design parameters on the characteristics of a dual-detector density tool," *Nuclear Geophysics*, vol. 9, pp. 283–289, August 1995.

[1302] J. S. Petler, "Modelling the spatial response of a compensated density tool," *IEEE Transactions on Nuclear Science*, vol. 37, pp. 954–958, April 1990.

[1303] B. Couët and R. Hertzog, "Modeling analysis of environmental effects on accelerator porosity logging measurements," *IEEE Transaction on Nuclear Science*, vol. 39, no. 4, pp. 1022–1006, 1992.

[1304] J. Butler, A. K. McCracken, P. B. Kemshell, and A. K. Ziver, "A new approach to the characterization of rock formation for the use in logging tool calibration facilities," *IEEE Transactions in Nuclear Science*, vol. 37, pp. 912–922, April 1990.

[1305] I. F. Gonçalves, J. Salgado, J. Neves, and F. G. Carvalho, "Calibration and design optimization of a nuclear surface moisture-density gauge by Monte Carlo simulation," *Nuclear Geophysics*, vol. 8, pp. 527–537, December 1994.

[1306] C. Oliveira and J. Salgado, "Calibration curves of a PGNAA system for cement raw material analysis using the MCNP code," *Applied Radiation and Isotopes*, vol. 49, pp. 1685–1689, December 1998.

[1307] M. Sohrabpour, M. Shahriari, V. Zarifian, and K. K. Moghadam, "Borehole prompt gamma neutron activation and comparison with Monte Carlo simulation using MCNP code. Borehole PGNAA experiment comparison with MCNP," *Applied Radiation and Isotopes*, vol. 50, pp. 805–810, April 1999.

[1308] A. Nordlund, S. Avdic, B. Dahl, and N. Kaiser, "Transmission measurement of porosity with a neutron generator in geophysics applications," *Nuclear Instruments and Methods in Physics Research A*, vol. 462, pp. 451–456, April 2001.

[1309] A. Nordlund, P. Linden, G. Por, M. Solymar, and B. Dahl, "Measurements of water content in geological samples using transmission of fast neutrons," *Nuclear Instruments and Methods in Physics Research A*, vol. 462, pp. 457–462, April 2001.

[1310] B. Couët and C. Watson, "Applications of Monte Carlo differential neutron sensitivity calculations to geophysical measurements," *Nuclear Geophysics*, vol. 7, pp. 215–229, June 1993.

[1311] B. Couet and C. Watson, "Monte Carlo differential neutron sensitivity calculations for nuclear well-logging," *IEEE Transaction on Nuclear Science*, vol. 40, no. 4, pp. 928–932, 1993.

[1312] D. J. Picton, R. G. Harris, K. Randle, and D. R. Weaver, "Depth of investigation of density tools," *IEEE Transactions on Nuclear Science*, vol. 39, pp. 1014–1018, August 1992.

[1313] J. A. Kluson, "Environmental monitoring and in situ gamma spectrometry," *Radiation Physics and Chemistry*, vol. 61, pp. 209–216, June 2001.

[1314] T. Hamid and D. J. Thomas, "Calibration and Monte Carlo modelling of neutron long counters," *Nuclear Instruments and Methods in Physics Research, A*, vol. 452, pp. 470–483, October 2000.

[1315] E. A. Abdelbaky and E. M. A. Hussein, "A decoupled photon-electron transport method for calculation of energy deposition in gamma-ray detectors,"

Nuclear Instruments and Methods in Physics Research A, vol. 372, pp. 425–430, 1996.

[1316] I. V. Serov, T. M. John, and J. E. Hoogenboom, "Midway forward-adjoint coupling method for neutron and photon Monte Carlo transport," *Nuclear Science and Engineering*, vol. 133, pp. 55–72, Sep 1999.

[1317] I. V. Serov, T. M. John, and J. E. Hoogenboom, "New effective Monte Carlo midway coupling method in MCNP applied to a well logging problem," *Applied Radiation and Isotopes*, vol. 49, pp. 1737–1744, December 1998.

[1318] R. Jeraj, B. Glumac, and M. Maucec, "Determination of the response function of the proton recoil detector using MCNP-4A," in *Proc. 1995 2nd Regional Meeting on Nuclear Energy in Central Europe, Portoroz, Slovenia*, pp. 131–137, 1995.

[1319] R. P. Gardner, E. El Sayyed, Y. Zheng, S. Hayden, and C. W. Mayo, "NaI detector neutron activation spectra for PGNAA applications," *Applied Radiation and Isotopes*, vol. 53, pp. 483–497, 2000.

[1320] L. C.-A. Bourva and S. Croft, "Monte Carlo calculations of the neutron coincidence gate utilisation factor for passive neutron coincidence counting," *Nuclear Instruments and Methods in Physics Research A*, vol. 431, pp. 485–508, July 1999.

[1321] L. C.-A. Bourva, S. Croft, H. Ottmar, and D. R. Weaver, "MCNP modelling of a combined neutron/gamma counter," *Nuclear Instruments and Methods in Physics Research A*, vol. 426, no. 2, 1999.

[1322] S. Croft and L. C.-A. Bourva, "Measurement of passive neutron multiplicity counter gate utilization factors and comparisons with theory," *Nuclear Instruments and Methods in Physics Research A*, vol. 453, October 2000.

[1323] A. J. Peurrung, P. L. Reeder, E. A. Lepel, and D. C. Stromswold, "Location of neutron sources using moderated-free directional thermal neutron detectors," *IEEE Transactions on Nuclear Science*, vol. 44, no. 3, pp. 543–550, 1997.

[1324] O. F. Smidts, "Analysis of the transport of radionuclide chains in a stochastic geological medium by a biased Monte Carlo simulation," *Nuclear Science and Engineering*, vol. 129, pp. 224–245, July 1998.

[1325] R. P. Gardner and L. Liu, "Monte Carlo simulation of neutron porosity oil well logging tools: combining the geometry-independent fine-mesh importance map and one-dimensional diffusion model approaches," *Nuclear Science and Engineering*, vol. 133, pp. 80–91, Sep 1999.

[1326] J. K. Shultis and R. E. Faw, *Radiation Shielding.* La Grange Park, IL: American Nuclear Society, 2000.

[1327] J. I. Wood, *Computational Methods in Reactor Shielding.* Oxford: Pergamon Press, 1982.

[1328] J. Shapiro, *Radiation Protection: A Guide for Scientists and Physicians.* Cambridge, Mass: Harvard University Press, 2nd ed., 1981.

[1329] N. M. Schaeffer, *Reactor Shielding for Nuclear Engineers.* U.S. Dept. of Commerce, Springfield, Va: National Technical Information Service, 1973.

[1330] E. P. Blizard and L. S. Abbott, eds., *Reactor Handbook: Shielding*, vol. 3 of *pt. 2.* New York: Interscience Publishers, 2nd ed., 1962.

[1331] Kaman Sciences Corporation, Colorado Springs, *Model A-125 Neutron Generator Instruction Manual*, Sep 1974.

[1332] NCRP, "Structural shielding design and evaluation for medical x rays and gamma rays of energies up to 10 MeV: Recommendation of the National Council on Radiation Protection and Measurements." National Council on Radiation Protection and Measurement, Washington, DC, NCRP Report No. 49, September 1975.

[1333] "Safety Code 27. Requirements for industrial x-ray equipment use and installation." Environmental Health Directorate, Health Canada, Ottawa, 87-EHD-130, 1987.

[1334] D. H. Stoddard and H. E. Hootman, "^{252}Cf shielding guide," Tech. Rep. DP-12 46, Aiken, S. C. : E. I. Du Pont de Nemours & Co., Savannah River Laboratory, March 1971.

[1335] IAEA, "Manual for troubleshooting and upgrading neutron generators." IAEA-TECDOC-913, International Atomic Energy Agency, November 1996.

[1336] ICRP, "Data for protection against ionizing radiation from external sources - supplement to ICRP publication 15, ICRP publication 21." Pergamon for the I.C.R.P., Oxford, 1971.

[1337] RSICC, *QAD-QC: Three-Dimensional Point Kernel Gamma-Ray Shielding Code.* Radiation Safety Information Computational Center, Oak Ridge National Laboratory, Oak Ride, TN. RSICC Code Package CCC-401.

[1338] RSICC, *DOORS3.2: One, Two- and Three-Dimensional Discrete Ordinates Neutron/PhotonTransport Code System.* Radiation Safety Information Computational Center, Oak Ridge National Laboratory, Oak Ridge, TN. (RSICC Code Package CCC-650).

[1339] ANS, "Working Group ANS-6.1.1: American national standard neutron and gamma-ray flux-to-dose-rate factors." American Nuclear Society, La Grange Park, Ill, 1977.

[1340] NDDH, "North Dakota Department of Health: Radioactive material." Licensing and Inspection Program. (http://www.health.state.nd.us/ndhd/environ/ee/rad/materials.htm, accessed August 2002).

[1341] R. Brodzinski, D. P. Brown, J. C. Evans Jr., W. K. Hensley, J. H. Reeves, N. A. Wogman, F. T. I. Avignone, and H. S. Miley, "Ultralow background germanium gamma-ray spectrometer," *Nuclear Instruments and Methods in Physics Research A*, vol. 239, pp. 207–213, Sep 1985.

[1342] M. Tamaki, M. Oda, K. Takahashi, W. Tanimoto, and T. Funahashi, "Study of neutron imaging techniques and processing for developing quantitative neutron radiography," *Nuclear Instruments and Methods in Physics Research A*, vol. 377, pp. 102–106, 1996.

[1343] G. Barnea, C. E. Dick, A. Ginzburg, E. Navon, and S. M. Seltzer, "A study of multiple scattering background in Compton scatter imaging," *NDT & E International*, vol. 28, no. 3, pp. 155–162, 1995.

[1344] A. L. Huddleston, J. P. Sackler, and S. E. Dunn, "Multiple scattering in dual-energy Compton-scatter densitometry," *Nuclear Instruments and Methods in Physics Research A*, vol. 255, no. 351-354, 1987.

[1345] M. H. Younis, A. A. Harms, and T. W. Hoffman, "Source fluctuating effects in radiation diagnosis of voided fluidic systems," *Nuclear Engineering and Design*, vol. 24, pp. 145–151, 1973.

[1346] D. C. Champeney, *Fourier Transforms and Their Physical Applications.* London: Academic Press, 1973.

[1347] R. S. Figliola and D. E. Beasley, *Theory and Design for Mechanical Measurements.* New York: John Wiley & Sons, 3rd ed., 2000.

[1348] W. A. Curby, "Crossing the barrier between the laboratory working model and the practicable production model," in *Aviation Security Problem and Related Technologies* (W. H. Makky, ed.), vol. CR42, pp. 167–181, SPIE - International Society for Optical Engineering, December 1992.

[1349] FAA, "Test & evaluation process guidelines: FAST: The federal aviation administration acquisition toolset," tech. rep., Federal Aviation Administration, 1999. (http://fast.faa.gov/archive/v299/test_evaluation/pg2.html, accessed December 2001).

[1350] American Intellectual Property Law Association, "An overview of intellectual property: What is a patent, a trademark, and a copyright?." (http://www.aipla.org/html/whatis.html, accessed December 2001), 1995.

[1351] K. H. Jackson, *Patent Strategy for Researchers and Research Managers.* New York: Wiley and Sons, 2nd ed., 2001.

[1352] F. E. Grissom and D. Pressman, *The Inventor's Notebook.* Berkeley, CA: Nolo Press, 3rd ed., 2000.

[1353] D. G. Rosenbaum, *Patents, Trademarks, and Copyrights.* Philadelphia: Chelsea House Publishers, 1997.

[1354] AIPLA, *How to Protect and Benefit from your Ideas: a book intended to assist the independent or novice inventor in protecting, evaluating, and commercializing new ideas and inventions.* Arlington, Va: American Intellectual Property Law Association, 1995.

[1355] R. D. Foltz and T. Penn, *Protecting Scientific Ideas & Inventions.* Boca Raton: CRC Press, 2nd ed., 1990.

[1356] R. D. Foltz and T. Penn, *Handbook for Protecting Ideas & Inventions.* Cleveland, Ohio: Penn Institute, 1992.

[1357] J. C. Paul and M. W. Lauroesch, "Protecting your inventions with patents: The engineer as inventor and expert witness." presentation at 1995 International Mechanical Engineering Congress and Exposition, San Francisco California, American Society of Mechanical Engineers, November 1995.

[1358] R. Stim, P. Gima, and P. Pakroo, *License your Invention.* Berkeley, CA: Nolo Press, 1st ed., 1998.

[1359] R. Stim and S. Fishman, *Nondisclosure Agreements.* Berkeley, CA: Nolo Press, 2001.

[1360] BIPM, "The international system of units (SI)." Bureau International des Poids et Mesures (BIPM). (http://www.bipm.fr/enus/3_SI/, accessed August 2002).

[1361] NIST, "International system of units (SI)." National Institute of Standard and Technology. (http://physics.nist.gov/cuu/Units/index.html, accessed August 2002).

[1362] NIST, "Links to on-line unit conversions." National Institute of Standard and Technology. (http://physics.nist.gov/cuu/Reference/unitconversions.html, accessed December 2002).

[1363] NIST, "Fundamental physical constants." National Institute of Standard and Technology. (http://physics.nist.gov/cuu/Constants/index.html, accessed August 2002).

[1364] J. K. Tuli, "Nuclear wallet card (sixth edition, January 2000)." (http://www.nndc.bnl.gov/wallet/, accessed August 2002).

[1365] R. B. Leighton, *Principles of Modern Physics.* New York: McGraw-Hill, 1959.

[1366] N. C. Yang, P. K. Leichner, and W. G. Hawkins, "Effective atomic numbers for low-energy total photon interactions in human tissues," *Medical Physics*, vol. 14, pp. 759–766, September/October 1987.

[1367] V. R. K. Murty, D. P. Winkoun, and K. R. S. Devan, "Effective atomic numbers for W/Cu alloy using transmission experiments," *Applied Radiation and Isotopes*, vol. 53, pp. 945–948, 2000.

[1368] L. Lyons, *Statistics for Nuclear and Particle Physicists.* Cambridge (Cambridgeshire): Cambridge University Press, 1989.

[1369] R. Barlow, *Statistics: A Guide to the Use of Statistical Methods in the Physical Sciences.* Chichester, England: Wiley, 1989.

[1370] W. T. Eadie, *Statistical Methods in Experimental Physics.* Amsterdam: North-Holland, 1982.

[1371] J. R. Taylor, *An introduction to Error Analysis : the Study of Uncertainties in Physical Measurements.* Sausalito, CA: University Science Books, 1997.

[1372] A. J. Wheeler and A. R. Ganji, *Introduction to Engineering Experimentation.* Englewood Cliffs, N. J.: Prentice Hall, 1996.

[1373] H. D. Young, *Statistical Treatment of Experimental Data.* New York: McGraw-Hill, 1962.

About the Author

Esam M. A. Hussein, Ph.D., P.Eng.

is presently Professor and Chair of the Department of Mechanical Engineering at the University of New Brunswick, Fredericton, Canada. After completing his undergraduate studies and a master's degree in nuclear engineering at Alexandria University, Egypt, he earned a PhD in nuclear engineering from McMaster University, Canada. Prior to joining the University of New Brunswick in November 1984, he was employed for four years as a Nuclear Design Engineer at Ontario Hydro (now Ontario Power Generation). Currently, he is leading a research program that focuses on the industrial and medical uses of nuclear and atomic radiation, and the detection of threat-materials. His research work has been funded by a number of government agencies, national laboratories, and industrial firms in Canada and abroad. He has published numerous scientific papers and industrial reports, and has two patents. Dr. Hussein is a registered professional engineer in the provinces of New Brunswick and Ontario, and a member of ANS, ASME, ASNT, CNS and IEEE-NPSS. He can be reached by e-mail at hussein@unb.ca.

About the Editor

Application Index

Index

Batch number: 10725820

Printed by Printforce, the Netherlands